# Producing Safe Eggs

# Producing Safe Eggs
## Microbial Ecology of *Salmonella*


*Edited by*

**Steven C. Ricke**

**Richard K. Gast**



ELSEVIER

AMSTERDAM • BOSTON • HEIDELBERG • LONDON
NEW YORK • OXFORD • PARIS • SAN DIEGO
SAN FRANCISCO • SINGAPORE • SYDNEY • TOKYO

Academic Press is an imprint of Elsevier

Academic Press is an imprint of Elsevier
125 London Wall, London EC2Y 5AS, United Kingdom
525 B Street, Suite 1800, San Diego, CA 92101-4495, United States
50 Hampshire Street, 5th Floor, Cambridge, MA 02139, United States
The Boulevard, Langford Lane, Kidlington, Oxford OX5 1GB, United Kingdom



**Library of Congress Cataloging-in-Publication Data**
A catalog record for this book is available from the Library of Congress

**British Library Cataloguing-in-Publication Data**
A catalogue record for this book is available from the British Library

ISBN: 978-0-12-802582-6

For information on all Academic Press publications visit our website at https://www.elsevier.com/



*Publisher:* Nikki Levy
*Acquisition Editor:* Patricia Osborn
*Editorial Project Manager:* Jaclyn Truesdell
*Production Project Manager:* Caroline Johnson
*Designer:* Matthew Limbert

Typeset by TNQ Books and Journals

# Contents

## SECTION 2 *SALMONELLA* CONTAMINATION IN LAYER FLOCKS: PATHOGENESIS, DISSEMINATION, AND CURRENT CONTROL STRATEGIES

# List of Contributors

**Sandra Aehle**
Elanco Animal Health, Inc., Pacific, MO, United States

**Paul Barrow**
University of Nottingham, Loughborough, Leicestershire, United Kingdom

**Dante Javier Bueno**
National Agricultural Technology Institute EAA Conception of Uruguay, Conception of Uruguay, Entre Rios, Argentina

**James A. Byrd**
USDA-ARS, College Station, TX, United States

**Todd R. Callaway**
USDA/ARS, College Station, TX, United States

**Kim L. Chiok**
Washington State University, Pullman, WA, United States

**Kapil K. Chousalkar**
The University of Adelaide, Roseworthy, SA, Australia

**Philip G. Crandall**
University of Arkansas, Fayetteville, AR, United States

**Roy Curtiss III**
University of Florida, Gainesville, FL, United States

**Rob Davies**
Animal and Plant Health Agency (APHA), Addlestone, United Kingdom

**Turki M. Dawoud**
University of Arkansas, Fayetteville, AR, United States; King Saud University, Riyadh, Saudi Arabia

**Wolfgang De Cort**
Ghent University, Gent, Belgium

**Richard Ducatelle**
Ghent University, Gent, Belgium

**Tom S. Edrington**
USDA/ARS, College Station, TX, United States

**Jacob R. Elder**
Washington State University, Pullman, WA, United States

**Steven L. Foley**
U.S. Food and Drug Administration, Jefferson, AR, United States

**Richard K. Gast**
US National Poultry Research Center, Athens, GA, United States

**Eduardo Ignacio Godano**
Tecnovo S.A., Industrial Park, Crespo, Entre Ríos, Argentina

**Anup Kollanoor Johny**
University of Minnesota, Saint Paul, MN, United States

**Pravin R. Kaldhone**
University of Arkansas, Fayetteville, AR, United States

**Michael H. Kogut**
USDA-ARS, College Station, TX, United States

**Doug Korver**
University of Alberta, Edmonton, AB, Canada

**Young Min Kwon**
University of Arkansas, Fayetteville, AR, United States

**Soo-Kyoung Lee**
Konkuk University, Seoul, South Korea; Animal and Plant Quarantine Agency, Gimcheon, South Korea

**Francesca Martelli**
Animal and Plant Health Agency (APHA), Addlestone, United Kingdom

**Lynn McMullen**
University of Alberta, Edmonton, AB, Canada

**Andrea McWhorter**
The University of Adelaide, Roseworthy, SA, Australia

**Meera Surendran Nair**
University of Connecticut, Storrs, CT, United States

**David J. Nisbet**
USDA/ARS, College Station, TX, United States

**Corliss A. O'Bryan**
University of Arkansas, Fayetteville, AR, United States

**Narayan C. Paul**
Washington State University, Pullman, WA, United States

**Francisco Procura**
National Agricultural Technology Institute EAA Conception of Uruguay, Conception of Uruguay, Entre Rios, Argentina; National Council of Scientific and Technical Research (CONICET), Argentina

**Steven C. Ricke**
University of Arkansas, Fayetteville, AR, United States

**Francisco Isabelino Rodriguez**
National Agricultural Technology Institute EAA Conception of Uruguay, Conception of Uruguay, Entre Rios, Argentina; National Council of Scientific and Technical Research (CONICET), Argentina

**Kun-Ho Seo**
Konkuk University, Seoul, South Korea

**Devendra H. Shah**
Washington State University, Pullman, WA, United States

**Zhaohao Shi**
University of Arkansas, Fayetteville, AR, United States

**María Cecilia Soria**
INIQUI-CONICET, National University of Salta, Salta, Argentina

**Mario Alberto Soria**
National Agricultural Technology Institute EAA Conception of Uruguay, Conception of Uruguay, Entre Rios, Argentina

**Indu Upadhyaya**
University of Arkansas, Fayetteville, AR, United States

**Filip Van Immerseel**
Ghent University, Gent, Belgium

**Kumar Venkitanarayanan**
University of Connecticut, Storrs, CT, United States

**Andrew Wales**
Animal and Plant Health Agency (APHA), Addlestone, United Kingdom

**Doug Waltman**
Georgia Poultry Laboratory Network, Gainesville, GA, United States

**Paul Wigley**
University of Liverpool, Neston, Cheshire, United Kingdom

**Hsin-Bai Yin**
University of Connecticut, Storrs, CT, United States

# About the Editors

**Dr. Steven C. Ricke** received his BS degree in Animal Science (1979) and MS degree in Ruminant Nutrition (1982) from the University of Illinois and his PhD degree (1989) from the University of Wisconsin with a co-major in Animal Science and Bacteriology. From 1989 to 1992 Dr. Ricke was a US Department of Agriculture (USDA)-Agricultural Research Service (ARS) postdoctorate in the Microbiology Department at North Carolina State University. He was at Texas A&M University for 13 years and was a professor in the Poultry Science Department. In 2005 he joined the Department of Food Science at the University of Arkansas and became the first holder of the new Donald "Buddy" Wray Endowed Chair in Food Safety and Director of the Center for Food Safety at the University of Arkansas. He received the Poultry Science Association Research Award in 1999 and the American Egg Board Award in 2006. He was also honored in 2002 by Texas A&M University as a Texas Agricultural Experiment Station Faculty Fellow and in 2012 with the University of Arkansas Division of Agriculture–John White Outstanding Research Award. Dr. Ricke's research program is primarily focused on virulence and pathogenic characteristics of food-borne salmonellae. Dr. Ricke's *Salmonella* research projects have emphasized studies on the growth, survival, and pathogenesis of the organism under conditions encountered during food animal production and processing.

**Dr. Richard K. Gast** is a Microbiologist and Research Leader for the Egg Safety and Quality Research Unit at the US National Poultry Research Center in Athens, Georgia, United States. The mission of this group is to develop improved technologies for egg production and processing that reduce or eliminate microorganisms that can transmit disease to humans or cause spoilage. Richard's personal research program focuses on detecting and controlling *Salmonella* infections in poultry and *Salmonella* contamination of eggs. He received MS and PhD degrees in Poultry Science from The Ohio State University.

# Preface

In his preface to the proceedings of an international conference in 1984, Dr. Glenn Snoeyenbos noted that "raw foods of animal origin are frequently contaminated by salmonellae and serve as a major source of human infection" (Snoeyenbos, 1984). However, textbooks on poultry science or food microbiology that were published before the 1980s typically contained only very brief considerations of eggs as potential sources of food-borne illness, as "the contents of normal fresh eggs are, as a rule, sterile" (Nesheim et al., 1972). Although eggs were long known to be subject to external bacterial contamination of shells, occurring as they pass through the same exit portal of the chicken as voided feces, the edible contents were likely to become contaminated only in cracked or dirty eggs. Elimination of these from table egg market channels, along with the institution of consistently dependable pasteurization standards for liquid egg products, reduced the association of eggs with disease to negligible levels for many years.

The international emergence of *Salmonella* Enteritidis as a leading cause of human illness in the mid-1980s represented a major turning point in the history of egg safety. Within a few years, this serovar became a preeminent food-borne disease agent in many countries, and eggs were consistently identified as the principal source of these infections. Considerable investments of public and private resources were directed toward this problem, in terms of government regulatory programs, risk reduction practices for egg producers, and research to develop improved tools for disease control. All of these efforts have achieved tangible positive results, but the continuing occurrence of egg-transmitted illness around the world calls for an ongoing commitment to finding better solutions to this problem.

This volume explores several of the most important categories of issues that still confront us regarding *Salmonella* contamination of eggs. The first section reviews the nature of this problem as it is seen in a variety of individual nations or regions around the world. This geographical presentation of the problem represents an effort to define its scope and ongoing significance, and identifies important international similarities and distinctions in both epidemiology and control practices. The second section seeks to explore the mechanisms by which egg contamination occurs as a consequence of infections of laying hens with salmonellae, and assesses current efforts to harness this information for implementing effective control programs. The third section presents a wide range of ideas and concepts being pursued by researchers with the goal of improving our ability to prevent *Salmonella* infection in egg-laying chickens, to prevent the production of contaminated eggs, or to eliminate contaminants after they are deposited on or inside eggs. Although it is not yet possible to identify a single treatment or practice that is likely to serve as a unilateral or complete solution to disease transmission via contaminated eggs, the sustained application of a coordinated and comprehensive control program with multiple risk reduction components offers much promise. We are grateful to the numerous authors from around the world who have contributed their expertise to this book. We hope that it will be the basis for a continuing dialogue and many further explorations about this still evolving topic.

**Steven C. Ricke**
**Richard K. Gast**
**May 2016**

## REFERENCES

Nesheim, M.C., Austic, R.E., Card, L.E., 1972. Poultry Production, twelfth ed. Lea and Febiger, Philadelphia, PA.

Snoeyenbos, G.H., 1984. Proceedings of the International Symposium on Salmonella. American Association of Avian Pathologists, Kennett Square, PA.

SECTION 1

# *Salmonella* in Egg Production Systems: International Prevalence, Issues, and Challenges

CHAPTER

# Of Mice and Hens—Tackling *Salmonella* in Table Egg Production in the United Kingdom and Europe

# 1

**Francesca Martelli, Andrew Wales, Rob Davies**

*Animal and Plant Health Agency (APHA), Addlestone, United Kingdom*

## 1. *SALMONELLA* CONTROL IN TABLE EGG PRODUCTION IN EUROPE

In the European Union in 2014, 44% of the human cases of salmonellosis associated with a known food source were linked with eggs and egg products (EFSA, 2015). In Europe, the two *Salmonella* serovars that are specifically regulated in laying flocks in all Member States are Enteritidis and Typhimurium (EC, 2003), but in the Nordic Countries all serovars in poultry are subject to a slaughter policy, and *Salmonella* Kentucky, which has been spreading rapidly in poultry in some countries and is highly resistant to fluoroquinolones, has been added to National Control Programmes (NCPs) in France. The European Union level of prevalence of adult laying hen flocks that tested positive for *Salmonella* in 2014 was 2.54%. *Salmonella* Enteritidis was isolated from 0.7% of laying hen flocks and *Salmonella* Typhimurium from 0.2% (EFSA, 2015). *S.* Enteritidis has a special ability to colonize the ovary/oviduct of laying hens for long periods and therefore to internally contaminate eggs, and has been the most frequent serovar associated with egg-related foodborne outbreaks in Europe since the mid-1980s (Thorns, 2000). *S.* Typhimurium is a common cause of foodborne outbreaks in humans worldwide originating predominantly from cattle, pigs, and poultry (EFSA, 2010). Foodborne outbreaks related to the consumption of *S.* Enteritidis-contaminated eggs have been widely reported in Europe (Gormley et al., 2012; Harker et al., 2014; Liu et al., 2012; Zenner et al., 2013; Zielicka-Hardy et al., 2012), but have been decreasing since the introduction of harmonized monitoring and control programs in laying hens across Europe in 2008.

The role of *S.* Typhimurium in egg-related foodborne infections appears to be less significant than the one of *S.* Enteritidis (Martelli and Davies, 2011; Wales and Davies, 2011), except in Australia, where *S.* Enteritidis never became established in laying hens, since no infected primary breeding or parent stock was imported from either Europe or the United States.

**Producing Safe Eggs. http://dx.doi.org/10.1016/B978-0-12-802582-6.00001-X**

Other *Salmonella* serovars can be linked to poultry products, egg contamination, and foodborne outbreaks (Okamura et al., 2001). For example, in the United States, *Salmonella* Heidelberg and *S.* Kentucky are frequently associated with eggs, broilers, turkeys, and poultry products and have resulted in large product recalls and outbreaks of food poisoning in the human population (Foley et al., 2011; Reddy et al., 2016).

*S.* Enteritidis has been reported to have a greater ability to colonize avian reproductive organs and contaminate eggs than other zoonotic serovars (e.g., *S.* Typhimurium, *Salmonella* Infantis, *Salmonella* Hadar, *S.* Heidelberg, *Salmonella* Montevideo) in intravenously infected laying hens (Okamura et al., 2001). This ability to vertically transmit into the contents of forming eggs is shared with both biovars of *Salmonella* Gallinarum, which are derived from a common *S.* Enteritidis ancestor (Yao et al., 2016).

*S.* Enteritidis is often persistent in cage layer holdings where cleaning and disinfection procedures are not carried out appropriately and where there is a resident infected rodent population (Snow et al., 2010). This can also happen in large non-cage units, but is less common. Flocks infected with *S.* Enteritidis produce a variable proportion of contaminated eggs, depending on the strain, level of contamination of the flock, and time of the production period in which the eggs are laid (Braden, 2006). In a study conducted in France, *S.* Enteritidis was detected on up to 8.6% of the shells of eggs produced by an infected flock (Chemaly et al., 2009). In a study conducted in the United States, the estimated overall prevalence of egg contents from flocks infected with *S.* Enteritidis was 2.64/10,000 eggs (varying from 0 to 62.5/10,000) (Henzler et al., 1998). In both studies, a positive correlation between the level of environmental contamination and the proportion of positive eggs was noted (Chemaly et al., 2009; Henzler et al., 1998).

*Salmonella* serovars differ in their ability to cause contamination on the eggshell or the egg contents. In one study, a significant difference was found in the rate of egg contamination between serovars, with *S.* Enteritidis causing a higher rate of contamination of egg contents and a lower rate of contamination of eggshells per infected hen (0.32% and 0.34%, respectively) compared with *S.* Typhimurium (0.23% and 0.94%, respectively) and non–*S.* Enteritidis non–*S.* Typhimurium serovars (0.23% and 2.5%, respectively) (Arnold et al., 2013).

NCPs for the control of *Salmonella* in commercial scale flocks have been implemented in the European Union since 2008 (as detailed in regulations EC 2160/2003 and EC 517/2011). Flocks of layers on holdings with more than 1000 birds (National Regulations specify smaller numbers in some countries, e.g., 350 birds in the United Kingdom) are sampled at day old and 2 weeks before they are moved to the laying phase or laying unit and every 15 weeks during the laying phase. NCPs include operator sampling by means of boot swabs (two pairs per flock) or naturally pooled feces (300 g) during the period of egg production of each holding, supplemented by official sampling of one flock per holding of more than 1000 birds per year, carried out by agents of the competent authority, using the same method plus one additional dust or fecal/boot swab sample. Official samples are collected by employees of the competent authority or independent control bodies (in the United Kingdom only), who also check movement records, medical records, and sampling records at the time of their visit (Gosling et al., 2014). The introduction of yearly official sampling of one random flock of adult laying hens per holding and operator sampling within the NCP testing scheme provides a more detailed picture of the *Salmonella* prevalence in laying hen flocks as the pre-NCP voluntary surveillance data had very low sensitivity (Arnold et al., 2010). After detection of *S.* Enteritidis and *S.* Typhimurium, the eggs produced by the positive flock can be sold only after heat treatment, dramatically reducing their commercial value. This commercial incentive leads many producers (all producers in Great Britain) to slaughter confirmed *S.* Enteritidis- or Typhimurium-positive flocks, therefore reducing the possibility of perpetuating resident environmental contamination or infecting other flocks on site. Early removal of infected flocks also gives plenty of downtime to facilitate effective pest control and cleaning and disinfection. Official sampling of new flocks placed in a previously positive house is also carried out to identify potential carryover of infection at an early stage.

The efficiency of sampling programs has a large effect on the detection of *Salmonella* and therefore estimations of prevalence (Fletcher, 2006) and it is difficult to design an optimal sample size when the within-flock prevalence and number of organisms is unknown (Altekruse et al., 2003). It is recognized that thorough environmental sampling is usually the most effective way to detect *Salmonella* in a poultry flock (Aho, 1992; Johansson et al., 1996; Musgrove and Jones, 2005) and normally the occurrence of *Salmonella* in the occupied part a house reflects infection in the flock (Arnold et al., 2009). Dust is a useful sample for identifying previous excretion of *Salmonella* by a poultry flock (Riemann et al., 1998). It is normally best to take both fresh fecal and dust samples (Davies and Wray, 1996) to help compensate for variable detection in either sample.

Since rodents within a poultry house can carry *Salmonella* at a higher prevalence than birds, the monitoring on trapped and poisoned rodents and their droppings can also help to assess risk (Lapuz et al., 2012). Egg collection trays and equipment can also be a useful indicator of likely contamination (Davies and Breslin, 2001).

## 2. INFECTION DYNAMICS AND RISK FACTORS FOR ACQUISITION/PERSISTENCE OF *SALMONELLA* IN LAYING HEN FARMS

Several factors can affect *Salmonella* colonization in poultry, such as age and genetic susceptibility of the birds, stress, level of exposure to the pathogen, and *Salmonella* serovar and strain (Foley et al., 2011). Several risk factors have been associated with the introduction of *Salmonella* in a flock of laying hens. Breeding flocks, feed mills (*S.* Enteritidis and *S.* Typhimurium do occasionally contaminate feed and mash for layer breeders or laying hens that are not normally heat treated, but other serovars are much more commonly found) and hatcheries (*S.* Enteritidis and *S.* Typhimurium only rarely contaminate hatcheries in the absence of an infected breeding flock) act as focal points of general *Salmonella* contamination. *Salmonella* normally enters the hatchery through contaminated fertilized eggs, usually on the shells, and may subsequently become a resident contaminant in the ventilation systems, where cleaning and disinfection is less easy to carry out (Davies and Breslin, 2004b). Some *Salmonella* serovars (e.g., *Salmonella* Senftenberg) are able to persist in hatcheries longer than others, probably as a result of their ability to form biofilms (Mueller-Doblies et al., 2013). Hatchery-acquired *Salmonella* substantially reduces the effectiveness of interventions on farm aimed at preventing *Salmonella* from colonizing young chicks (Chao et al., 2007). It has been speculated that the *S.* Enteritidis pandemic in the 1980s was caused by vertical transfer of the bacterium from infected primary breeding flocks via international trade in hatching eggs and chicks (Cogan and Humphrey, 2003; Thorns, 2000). In most of Europe, since the introduction of the NCPs, the isolation of *S.* Enteritidis or Typhimurium from breeding flocks is a rare event, often relating to monophasic *S.* Typhimurium contamination of feed resulting from environmental occurrence of the organism that is disseminated from the pig reservoir into the wider environment via manure, run-off, and infection of wildlife (Wright et al., 2016). Such infections are normally stamped out effectively in European Union Member States, but there may be issues in other Eastern European countries (EFSA, 2015).

Previous *Salmonella* infection in the farm has been strongly associated with the contamination of the laying hen environment and infection of follow-on flocks (Chemaly et al., 2009; Schulz et al., 2011). Fecal shedding of *Salmonella* by infected hens is an important source of environmental contamination and can result in contamination of floors, manure belts, egg belts, feeders, and drinkers (Dewaele et al., 2012). Dust is an important medium for the survival and persistence of *Salmonella* between flocks of laying hens, and *Salmonella* can be isolated from dust samples for long periods, often years (Gole et al., 2014). Infection of flocks can therefore occur months or years after a previous infection if contaminated dust is dislodged from wall or roof spaces during building projects, such as installation of solar panels. The persistence of *Salmonella* infection after cleaning and disinfection in a poultry house previously occupied by an infected flock can also be facilitated by the presence of rodents, red mites, and flies that are able to carry *Salmonella* (Carrique-Mas et al., 2009a; Holt et al., 2007; Moro et al., 2007). The presence of rodents is strongly associated with the *Salmonella* contamination of laying hen farms (Carrique-Mas et al., 2009a; Garber et al., 2003; Snow et al., 2010). Flock size has also been related to the level of *Salmonella* contamination (Huneau-Salaun et al., 2009; Snow et al., 2010). The size of the farm can be linked with hygiene practices as large farms are more likely to only be dry cleaned, rather than washed and disinfected, between flocks (Aimey et al., 2013). High stocking densities are also associated with the suppression of both cellular and humoral immunity, which can lead to increased systemic invasion by *S.* Enteritidis (Gomes et al., 2014). The effect of housing types on *Salmonella* infection in laying hens has been evaluated in several studies (Denagamage et al., 2015). Some studies have observed an increased risk of *Salmonella* infection in flocks housed in cage systems, particularly in the presence of large populations of rodents

(EFSA, 2007; Namata et al., 2008; Snow et al., 2010; Van Hoorebeke et al., 2010b). Other studies have reported a higher frequency of transmission in noncage systems (De Vylder et al., 2011; Hannah et al., 2011), whereas no difference between the two systems was observed by others (Jones et al., 2012; Siemon et al., 2007). A ban on conventional cages has been in place in the European Union since January 2012, and this has led to the conversion of old cage systems to enriched colony cages. This required removal of all battery-style cages and elimination of deep pits, provided the opportunity to deep clean houses and eliminate rodents, and has been a major factor in the successful control of resident *S.* Enteritidis in the European egg industry. In the enriched cage system, groups of 30–80 birds are housed in a larger cage that provides more space (laying hens have at least 750 $cm^2$ of cage area per hen), perches, and a "nest-box" area (Council directive 1999/74/EC). Studies comparing traditional cage systems with conventional cage systems did not find differences between the two with regards to environmental contamination and frequency of transmission (De Vylder et al., 2009; Gast et al., 2014a,b; Van Hoorebeke et al., 2011). One study found that the frequency of positive fecal cultures after experimental infection of laying hens with *S.* Enteritidis was significantly greater for conventional cages than for enriched colony cages (Gast et al., 2015). Multiage management has also been identified as a risk factor for *Salmonella* contamination in the laying hen environment (Huneau-Salaun et al., 2009; Snow et al., 2010). Susceptibility to *Salmonella* infection and likelihood of establishment of a persistent colonization decrease with age, but sometimes fecal shedding in older birds continues for an extended interval of time (Li et al., 2007). If young chicks become infected with *S.* Enteritidis, the colonization of their intestinal tract can persist until maturity (Gast and Holt, 1998). The age of the poultry house is also a significant risk factor since most *S.* Enteritidis infections have persisted in laying farms for decades after the original introduction with birds from infected breeding flocks (Van Hoorebeke et al., 2010a).

Once *Salmonella* is introduced into a flock of laying hens, further transmission occurs via contact with infected individuals, ingestion of fecally contaminated materials (Holt et al., 1998), feed and water (Holt, 1995; Nakamura et al., 1997), and through aerosols and dust (Baskerville et al., 1992; Gast and Holt, 1998; Holt, 1995; Nakamura et al., 1997). The potential for contact transmission of *Salmonella* may be greater when birds are subjected to stress, especially induced molting (Gomes et al., 2014; Holt et al., 1998). In particular, young pullets that are moved to the laying houses just before the onset of lay are more susceptible. Stress related to transport and handling and relocation/remixing stress at a time of hormonal changes associated with the onset of lay can increase susceptibility to infection (Line et al., 1997). The introduction of susceptible young pullets in contaminated laying houses can result in the infection of the flock, followed by an early peak of infection within 3 weeks of housing (Gradel et al., 2002). Most hens stop shedding *Salmonella* after approximately 3 weeks, but a proportion of the flock may remain active or latent carriers of *Salmonella* for their whole life (Gast et al., 2005; Shivaprasad et al., 1990). However, if the hens are under stress, they may resume shedding, either as a result of reactivation of shedding in latent carriers or because of increased susceptibility to reinfection from the environment (Barrow, 1992; Skov et al., 2002). In addition to the early peak soon after introduction of birds into the laying house, the level of infection is commonly higher at the end of the laying period (Garber et al., 2003; van de Giessen et al., 2006; Wales et al., 2007). However, some studies have not found an association between prevalence of *Salmonella* and flock age (Schulz et al., 2011) or have reported a decline in *Salmonella* prevalence as the birds aged (Gole et al., 2014). This variability can be influenced by flock size and management, the type of housing, and the efficiency of vaccination.

## 3. FOCAL POINTS OF CONTROL FOR *SALMONELLA* IN LAYING HEN FARMS

To tackle *Salmonella* infection in laying flocks, effective rodent control and decontamination between crops is essential. Furthermore, vaccines against *S.* Enteritidis and *S.* Typhimurium are available and the implementation of more effective vaccination programs has been key in the reduction of the burden of infection. In the following paragraphs, a detailed description of these intervention measures is presented.

### 3.1 RODENT CONTROL

Laying houses and free-range units are prone to intractable problems with rodent control by virtue of several factors. These include numerous inaccessible niches within and around cages, banks of autonests, mini-pits, etc., and within and beneath the fabric of buildings and sheds, especially when there has been damage allowing access by rodents to voids and insulation. Open feeders, poorly secured feed hoppers, and spillages arising from feeding and from feed conveyers can provide readily available food sources for rodents and other wildlife, whereas the long production cycles associated with egg laying flocks prevent frequent and thorough cleaning, and can hinder repairs to building structures, allowing rodents to access wall and roof spaces, damaging insulation and threatening the integrity of electrical wiring in the process.

There is a strong association between *Salmonella* in the stock and environment of laying units and *Salmonella* in associated mice (Davies and Wray, 1995a; Henzler and Opitz, 1992; Kinde et al., 1996; Nahms, 2000; Wales et al., 2006b). This is particularly so for *S.* Enteritidis; in poultry units with infected stock, a high proportion of mice have also been found to carry the same serovar as the infected chickens (Davies and Wray, 1995a; Guard-Petter et al., 1997). Findings in smaller noncage laying units in the United States concur (Wallner-Pendleton et al., 2014). It has been shown that indistinguishable subtypes of *Salmonella* serovars Enteritidis and Infantis are shared between the laying farm environment and resident rodents (Lapuz et al., 2008, 2012).

Wild mice can become systemically infected and shed *S.* Enteritidis for weeks to several months after a single oral exposure, or following cross-contamination between colonies of mice (Davies and Wray, 1995a). This same study also demonstrated there was little resistance of mice to reinfection, and that growing chickens were readily infected by *S.* Enteritidis shed in mouse droppings. Similarly, Japanese roof rats from a laying farm and naturally infected with *Salmonella* serovars Infantis and Enteritidis shed highly similar strains of *S.* Infantis intermittently for up to 43 weeks after trapping, with between $10^2$ and $10^8$ colony-forming units per dropping (Umali et al., 2012). In the same study *S.* Enteritidis was shed less frequently, but still for up to 33 weeks.

European Union survey data indicate that poultry house mouse feces occasionally contains a high density of *Salmonella* cells (Davies and Wray, 1995a; Wales et al., 2006b), and supports the experimental findings that rodents can become colonized and systemically infected for many months (Rabie et al., 2010). Moreover, the *Salmonella* density in viscera from laying house mice can be similar to, or sometimes well above, that of other sample types (Davies and Wray, 1995b; Wales et al., 2006b), and mouse passage may select isolates of *S.* Enteritidis that show more virulence and invasiveness for chickens and a greater ability to contaminate forming eggs (Guard-Petter, 2001; Guard-Petter et al., 1997).

Thus, rodent infestations appear to act to sustain and intensify environmental contamination of laying houses by *Salmonella*, with perhaps a particularly significant interaction for *S.* Enteritidis. This is a serovar that was historically associated with mice (Edwards and Bruner, 1943) before its epidemic rise within egg production, and it may be that *S.* Enteritidis is particularly well adapted for maintenance, multiplication, and spread within a rodent reservoir. Similarly, on Japanese poultry farms there may be a particular epidemiological link between the frequent problems of *S.* Infantis contamination and roof rat infestation (Umali et al., 2012). In this context, the tendency for rodents to seek out feed destined for the laying hens, in hoppers, troughs, etc., and to defecate in the feed (Daniels et al., 2003), provides a ready route for multiplying and distributing *Salmonella* within, or sometimes between, laying houses.

Field studies indicate that the biological interactions between *Salmonella* and rodents on egg laying premises, as previously outlined, are indeed significant in respect to persistent flock infections and environmental contamination by *S.* Enteritidis (Denagamage et al., 2015). A European Union-wide baseline study of environmental *Salmonella* contamination among laying flocks of at least 1000 hens (EFSA, 2007) provided data for an epidemiological investigation in the United Kingdom (Snow et al., 2010). This indicated that a lower frequency of rodent sightings was partially protective for *S.* Enteritidis flock infection, with significant independent findings for either rats or mice. Another study in the United Kingdom concluded that excellent or improved rodent control was significantly associated with the elimination of *S.* Enteritidis during the life of a flock from laying houses holding *S.* Enteritidis-vaccinated laying hens (Carrique-Mas et al., 2009a). This was not the case with most other serovars. This finding, in 2007, was highly instrumental in convincing the egg industry that very intensive rodent control, including crating baiting ports with wall and roof spaces, was worthwhile, since elimination of *S.* Enteritidis infection (verified by intensive sampling) before the end of lay reduced the need for costly and complex intensive cleaning and disinfection between flocks (ACMSF, 2016).

A 1999 national survey of table egg laying birds in the United States, using environmental swabs, identified risk factors for *S.* Enteritidis-positive flock status that included wildlife vector access to feed and a relatively higher intensity of rodent infestation (Garber et al., 2003). Rodents also proved to be a risk factor for *S.* Enteritidis-positive premises status in a large Californian study of 133 laying flocks, using feces drag swabs (Castellan et al., 2004; Kinde et al., 2004), and similarly in the most recent US national laying hen survey, using varied environmental samples (Nahms, 2014).

Therefore, excellent rodent control is considered to be an important element in protecting laying flocks from *Salmonella* infection, or in moving toward *Salmonella*-free status. This is particularly the case for *S.* Enteritidis in regions where this serovar is endemic. Furthermore, given that total and permanent elimination of all rodents is not usually feasible on the premises, the protection of feed supplies from rodent and other wildlife incursions that might reestablish breeding populations on site should also be a high priority. Periods of depopulation offer additional opportunities for intensive baiting and for repairs to buildings that deny rodents access to the inside and the fabric of structures, but modern colony cage houses include a network of manure belts that must be protected against rodents by frequent placement of palatable fresh bait at their entry points into houses. Similarly, bait should be present at all potential entry points to the houses and must be palatable for the specific rodent populations present. In practice, it is often most successful to place a variety of baits, such as cut wheat, canary seed, pasta/peanut oil, or wax block formulations simultaneously if a significant infestation is present. Whole wheat should not be used for mice if more palatable poultry feed is present and care must be taken to bait very intensively and effectively, as suboptimal baiting over a long period selects for rodents that become behaviorally less likely to take bait. Baits that kill with a single feed are preferable if this is a severe infestation but should not be continued indefinitely unless resistance to other second-generation anticoagulants is present. Alpha-chloralose (an anesthetic compound also used as a rodenticide) baits can be useful in cold empty houses and trapping can be used to supplement baiting and to monitor progress in reducing rodent populations. Contact rodenticides can also be useful but cats cannot deal with established rodent populations within their protective harborage, although they may act as a deterrent to initial entry of rodents (and wild birds) to a pest-free site. Cats can also become infected with *Salmonella* and represent a reservoir of infection between flocks on the farm (Snow et al., 2010).

## 3.2 CLEANING AND DISINFECTION

In many laying flocks, *Salmonella* infections with non-Enteritidis serovars will resolve within the lifetime of a flock, especially if the source is transient, such as a contaminated batch of feed. However, *S.* Enteritidis typically is more persistent, and prevention of carryover to the next flock usually requires removal of environmental contamination during depopulation (Carrique-Mas et al., 2009a). Therefore, effective cleaning and disinfection is of major importance, as illustrated by a large-scale study in the United States indicating that failing to clean and disinfect between flocks was associated with *S.* Enteritidis-positive flock status (Garber et al., 2003). Another large US study (Nahms, 2014) showed a similar effect for units having a depopulated downtime of 10 days or less, which may indicate poorer decontamination regimens among other factors such as rodent control. When cleaning and disinfection of laying houses was verified by bacteriological sampling, effective removal of *Salmonella* proved to be positively associated with clearance of *Salmonella* infection between successive laying flocks in a study of 60 houses (Carrique-Mas et al., 2009b). Similar findings were reported for turkey facilities (Mueller-Doblies et al., 2010).

*S.* Enteritidis can survive for at least 1 year in poultry accommodations after depopulation, having been found in feed troughs, nest boxes, unbedded floor areas, and dust from sheds and free-range houses (Davies and Breslin, 2003b; Davies and Wray, 1996). This environmental resilience of *Salmonella*, coupled with the long production cycles of laying hens, leads to the accumulation of the organism in protected microenvironments including dust, organic soil, and spilled feed.

The intermittent detection of the same *Salmonella* serovar on poultry premises suggests that it may be present continuously, despite cleaning and disinfection. This view is supported by longitudinal studies involving detailed subtyping of *S.* Senftenberg isolates, which showed that the same subtypes persisted on the farm (Pedersen et al., 2008). When cleaning and disinfection of poultry houses, including those of laying flocks, has been objectively assessed by detailed sampling, there have often been only modest reductions, or occasionally even increases, observed in environmental contamination by *Salmonella* and coliforms (Davies and Breslin, 2003b; Davies and Wray, 1995b; Mueller-Doblies et al., 2010; Wales et al., 2007). Particular problem areas in laying houses include feeders, drinkers, nest boxes, and cage furniture (Wales et al., 2006a), all of which are in close contact with new flocks, thus potentially exposing newly placed pullets to early challenge by *Salmonella*. Ventilation ducting may also remain contaminated unless suitable cleaning towers and angled lances are used.

There are a number of practical and technical issues that together pose a severe challenge for effective cleaning and disinfection of laying flock accommodation, not the least of which are the multiple (often inaccessible) surfaces and niches. Difficulties with sealing ventilation systems for thermal fogging can be another obstacle to comprehensive coverage of microbicidal treatment. Large-scale field survey data from 2004 to 2005 indicated that caged flocks in the United Kingdom and Belgium were at higher risk of being *S.* Enteritidis- and *Salmonella*-positive, respectively, compared with noncaged flocks, independent of flock size (Namata et al., 2008; Snow et al., 2010). This may in part be because of challenges in effective cleaning and disinfection of cages.

Wet cleaning may cause increased mobilization and activation of *Salmonella* (Davies and Wray, 1995b), and bacteria in wet matter may benefit from increased protection against penetration by disinfectants, or indeed may multiply if in a favorable matrix such as feed residue. Field studies show that starting with wet cleaning can be associated with a poorer overall *Salmonella* reduction in laying houses than an initial dry clean if subsequent disinfection is insufficiently robust (Wales et al., 2006a), but that omitting dry cleaning before washing can be a risk factor for shedding of *S.* Enteritidis or *S.* Typhimurium by laying hens (Van Hoorebeke et al., 2010a). It is therefore necessary to carry out each stage: dry cleaning, washing, and disinfection effectively or the whole process may be undermined.

The protective effect of organic residues, plus survival features of *Salmonella* including oxidative resistance and biofilm formation commonly render disinfectants less effective than laboratory assessments may predict (McLaren et al., 2011; Stringfellow et al., 2009). This may be compounded by the use of inappropriate disinfectant dilutions, loss of disinfectant potency during storage, interfering substances in diluent water, and inadequate application technique (Davies and Wray, 1995b; Huneau-Salaün et al., 2010; Stringfellow et al., 2009). Different disinfectant preparations are affected to differing degrees by the conditions of application such as deviations from optimal dilution, residual soil, surface materials, etc. (McLaren et al., 2011; Russell, 2003). Thus, field experience and assessment using simulated "in use" conditions are important elements in predicting efficacy. Some field and laboratory investigations in the United Kingdom and Denmark have indicated that aldehyde- and phenolic compound–based preparations appear to be more consistently efficacious (Berchieri and Barrow, 1996; Davies and Wray, 1995b; Gradel et al., 2004; McLaren et al., 2011). However, availability and licensing of disinfectant components varies between territories.

Success in cleaning and disinfection is greatly aided by a rigorous and thorough approach, plus careful attention to the details of technique and to the selection and application of agents. Power washing of houses with 10% formalin solution at high pressure has proved to be the most effective program for persistently infected cage laying houses, and this can be effective even when thorough cleaning before disinfection is difficult, but good safety precautions, including breathing apparatus, are needed for this. In the European Union, formaldehyde has been classified as a class 1 carcinogen and its agricultural use is under threat. Glutaraldehyde is the next most effective agent, but this does depend on a good standard of cleaning before use. The available evidence strongly indicates that for persistent serovars such as *S.* Enteritidis, effective cleaning and disinfection is an essential element for timely elimination of flock infections, as well as having allied benefits in the control of other contaminant organisms and pests.

## 3.3 VACCINATION OF FLOCKS

Vaccination of chickens, together with other control measures as part of a comprehensive *Salmonella* control program, is an important component of strategies to prevent infection and reduce the prevalence of *Salmonella* in poultry flocks (Desin et al., 2013). Vaccination of laying hens against *S.* Enteritidis and *S.* Typhimurium has been shown to confer protection against *Salmonella* infection and to decrease the level of on-farm contamination (Van Immerseel et al., 2005). Vaccination of flocks of laying chickens started in some countries during the 1990s in response to the *S.* Enteritidis pandemic (Thorns, 2000), and was linked to a decrease in human cases of salmonellosis related to *S.* Enteritidis infections in the United Kingdom (O'Brien, 2012).

As *S.* Enteritidis and *S.* Typhimurium are considered to be the most important serovars for public health in Europe, existing commercially available live and inactivated *Salmonella* vaccines for poultry are intended for use against one or both of these serovars. There is little evidence of cross-protection between *Salmonella* serogroups in chickens (Curtiss and Hassan, 1996) but a partial cross-immunity effect between serogroups B (*S.* Typhimurium) and D (*S.* Enteritidis) has been suggested (Parker et al., 2001; Springer et al., 2011). After the introduction of large-scale vaccination using live *S.* Enteritidis vaccine in poultry breeding farms (layers and broilers) the detection of both *S.* Enteritidis and *S.* Typhimurium was reduced considerably (EFSA, 2004).

In some European countries (Austria, Belgium, the Czech Republic, Germany, and Hungary) vaccination of laying flocks is compulsory, in others vaccination is permitted and recommended (Bulgaria, Belgium, Cyprus, Estonia, France, Greece, Italy, Latvia, Lithuania, The Netherlands, Poland, Portugal, Romania, Slovakia, Slovenia, Spain, and the United Kingdom), whereas in others it is banned (Denmark, Finland, Sweden, and Ireland) (Galis et al., 2013). In the United Kingdom, most commercial scale egg producers subscribe to the British Egg Industry Council (BEIC) Quality Assurance Scheme that provides a code of practice (Lion Code) on farm hygiene and welfare standards. Vaccination against *Salmonella* began in broiler breeding flocks in 1994 (it was never routinely used for layer breeder flocks) and in laying flocks in 1998 for farms that subscribe to the BEIC Lion Code Scheme (Cogan and Humphrey, 2003; O'Brien, 2012; Ward et al., 2000). In the United Kingdom, five live vaccines and two killed vaccines are currently available (Clifton-Hadley et al., 2002; Gantois et al., 2006; Springer et al., 2011). Table 1.1 summarizes the vaccines available for poultry in Europe. These vaccines are used singly or combined and a three-dose course of one of the live *S.* Enteritidis vaccines has also been licensed for protection against *S.* Typhimurium. To maximize protection, vaccination programs that combine live and killed vaccines are often used. Within these programs, oral vaccines are administered as three doses during the rearing period of the pullets and this is complemented by one or, rarely, two injections of killed vaccine (normally close to point of lay) (EFSA, 2004).

Because of the intracellular location of part of the *Salmonella* population within infected animals, it is believed that *Salmonella* vaccines must have the potential to induce a cell-mediated response, which is more achievable with potent live vaccines (Gantois et al., 2006) and most live vaccines have to be extremely attenuated because of concerns about safety and environmental persistence. Killed vaccines have been shown to be partially protective against *Salmonella* challenge in poultry (Clifton-Hadley et al., 2002; Feberwee et al., 2000). An inactivated bivalent *S.* Enteritidis and *S.* Typhimurium vaccine was able to reduce the number of internal organs colonized, the number of ceca colonized, and the level of shedding in an experimental seeder

**Table 1.1** Vaccines Currently Available for Chicken Layers in Europe

| Name | Type of Vaccine | Administration Route | Timing | Notes |
|---|---|---|---|---|
| Salenvac T[a] | Killed SE/ST | IM injection | 12 weeks (w) & 16 w | Alhydrogel adjuvant, high circulating/maternal antibody |
| Avipro Salmonella Vac E[b] | Live SE | Water | Day old (D.O.); 7 w; 3 w prelay | Metabolic drift mutant |
| Avipro Salmonella Vac T[b] | Live ST | Water | D.O.; 7 w; 3 w prelay | Metabolic drift mutant |
| Avipro Duo[b] | Live SE/ST mix | Water/(spray) | D.O.; 7 w; 3 w prelay | Metabolic drift mutants |
| Gallivac SE (2-dose course versus normal challenge)[c] | Live SE | Water/(spray) | D.O.; +2 w | Auxotrophic mutant |
| Gallivac SE (3-dose course versus high challenge/ST)[c] | Live SE | Water/(spray) | D.O.; 2 w; 3 w prelay | Auxotrophic mutant |
| Gallimune SE+ST[c] | Killed SE/ST | IM injection | From 6 w, & 16 w | Mineral oil adjuvant |

IM, *Intramuscular;* SE, Salmonella *Enteritidis;* ST, Salmonella *Typhimurium.*
[a]*Intervet UK Ltd. Walton Manor, Walton, Milton Keynes MK7 7AJ.*
[b]*Lohmann Animal Health GmbH, Heinz-Lohmann-Straße 4, 27,472 Cuxhaven, Germany.*
[c]*Merial GmbH, Am Söldnermoos 6, 85,399 Hallbergmoos, Germany.*

challenge model (Clifton-Hadley et al., 2002). Killed vaccines have been shown to be effective in preventing the exacerbation of *S.* Enteritidis infection within infected flocks, if administered premolting (Nakamura et al., 2004).

Vaccination is used to prevent systemic infection (and localization in the reproductive tract) and to reduce fecal shedding (and consequently carcass and/or egg contamination). Vaccination is regarded only as an additional measure to increase resistance of chicks against *Salmonella*, especially if the flock prevalence is already high. Vaccination reduces the risk of inter- and intra-flock *Salmonella* contamination but it must be combined with other measures (such as rodent control and effective cleaning and disinfection) to effectively prevent perpetuation of infection on the farm (Davies and Breslin, 2003a; Woodward et al., 2002). In the presence of high levels of environmental contamination, hens may ingest high numbers of salmonellae and vaccination can be insufficient to provide protection (Atterbury et al., 2009; De Buck et al., 2005; Woodward et al., 2002). Vaccinated hens can become infected with *S.* Enteritidis and the bacteria can be found in their ceca up to the end of the laying period (Dewaele et al., 2012). In the field, poor injection technique in the case of live vaccines and failure to ensure sufficient uptake of the vaccine by the birds, proper distribution of the full dose of live vaccines throughout drinker lines and its uptake by birds within a 2-h window before deterioration of viability of the vaccine is a further impediment to efficacy.

Although vaccination is not fully protective, especially in the case of laying hens placed in a previously contaminated laying house that contains infected rodents, it is likely to reduce fecal shedding, ovarian transmission, and the within-flock prevalence, thereby reducing contamination of table eggs and the environment. Most importantly, the use of vaccination against *S.* Enteritidis and *S.* Typhimurium results in lower internal-egg contamination levels thereby directly contributing to public health even when flocks remain infected, as with most large cage flocks in Europe before the introduction of NCPs (Davies and Breslin, 2004a; Gantois et al., 2006). Vaccination was effective in reducing the number of internally contaminated eggs and eggshells produced by hens experimentally challenged with *S.* Enteritidis, *S.* Typhimurium, and monophasic *S.* Typhimurium variants (Arnold et al., 2014). In the European Union-wide baseline study conducted in 2004–5, vaccination of laying flocks was found to decrease the risk of *S.* Enteritidis infection when compared with unvaccinated flocks. Vaccination was demonstrated to be particularly effective in Member States with high holding prevalence (>15%) of *Salmonella* infection (EFSA, 2007).

There is some indication in the literature that eggs laid by chickens vaccinated with killed vaccines may be more resistant to *Salmonella* contamination and further multiplication, as maternal anti-*Salmonella* antibodies can be present in the egg (Hassan and Curtiss, 1996). In a study conducted in broilers, the samples collected from flocks that were progeny of vaccinated broiler breeders had a 62% lower chance of being *Salmonella* positive than chicks collected from equivalent flocks that were progeny of unvaccinated breeders (Berghaus et al., 2011).

## 4. CONCLUDING REMARKS

Tackling *Salmonella* infection in laying hen flocks can prove challenging, especially in the presence of rodents. It is essential to control rodent populations on farm, as mice and rats can amplify the infection and favor carryover between flocks. If breeding populations of rodents are eliminated and flocks are vaccinated, *S.* Enteritidis or *S.* Typhimurium infections can disappear (or at least become undetectable by repeated intensive sampling) over a period of weeks or months. This was the most important finding of research on *Salmonella* control in the United Kingdom and led to the successful efforts toward the current status of virtual eradication of *S.* Enteritidis in poultry and the conclusion that infection was more a problem of rodents than birds all along. There are still issues in people, due to foreign travel and imported foods, and occasionally in imported broiler hatching eggs from those countries that have diverted table eggs from infected flocks for heat treatment rather than slaughtering the flocks as in United Kingdom and Nordic countries. Imported infection is therefore a constant threat that necessitates continuation of vaccination and high farm standards despite the current absence of *S.* Enteritidis infection in commercial scale laying flocks. Vaccination and effective cleaning and disinfection are important for infection control, but their effect can be undermined by the presence of rodents and poor vaccination technique. Effective sampling strategies are important to detect infection in flocks. In the European Union, a combination of these approaches including legislation penalizing the sale of eggs from infected flocks has resulted in a progressive decrease of prevalence of *S.* Enteritidis and *S.* Typhimurium in laying hen flocks, accompanied by a concomitant reduction of human cases related to the consumption of nonimported chicken eggs. The European experience demonstrates that the control of *Salmonella* in laying hens flocks is possible, in particular in the context of national control programs that introduce effective sampling schemes and financial penalties related to noncompliance with the requirements of the programs.

## REFERENCES


ACMSF, 2016. Available online: https://www.food.gov.uk/sites/default/files/consultation/pack/acmsf-egg-report.pdf.

Aho, M., 1992. Problems of Salmonella sampling. Int. J. Food Microbiol 15, 225–235.

Aimey, V., Hunte, K., Whitehall, P., Sanford, B., Trotman, M., Delgado, A., Lefrancois, T., Shaw, J., Hernandez, J., 2013. Prevalence of and examination of exposure factors for *Salmonella* on commercial egg-laying farms in Barbados. Prev. Vet. Med. 110, 489–496.

Altekruse, S.F., Elvinger, F., Wang, Y., Ye, K., 2003. A model to estimate the optimal sample size for microbiological surveys. Appl. Environ. Microbiol. 69, 6174–6178.

Arnold, M.E., Gosling, R.J., La Ragione, R.M., Davies, R.H., Martelli, F., 2014. Estimation of the impact of vaccination on faecal shedding and organ and egg contamination for *Salmonella* Enteritidis, *Salmonella* Typhiumurium and monophasic *Salmonella* Typhimurium. Avian Pathol. 43, 155–163.

Arnold, M.E., Martelli, F., McLaren, I., Davies, R.H., 2013. Estimation of the rate of egg contamination from *Salmonella*-infected chickens. Zoonoses Public Health 61 (1), 18–27.

Arnold, M.E., Mueller-Doblies, D., Carrique-Mas, J.J., Davies, R.H., 2009. The estimation of pooled-sample sensitivity for detection of *Salmonella* in turkey flocks. J. Appl. Microbiol. 107, 936–943.

Arnold, M.E., Papadopoulou, C., Davies, R.H., Carrique-Mas, J.J., Evans, S.J., Hoinville, L.J., 2010. Estimation of *Salmonella* prevalence in UK egg-laying holdings. Prev. Vet. Med. 94, 306–309.

Atterbury, R.J., Carrique-Mas, J.J., Davies, R.H., Allen, V.M., 2009. *Salmonella* colonisation of laying hens following vaccination with killed and live attenuated commercial *Salmonella* vaccines. Vet. Rec. 165, 493–496.

Barrow, P.A., 1992. ELISAs and the serological analysis of *Salmonella* infections in poultry: a review. Epidemiol. Infect. 109, 361–369.

Baskerville, A., Humphrey, T.J., Fitzgeorge, R.B., Cook, R.W., Chart, H., Rowe, B., Whitehead, A., 1992. Airborne infection of laying hens with *Salmonella enteritidis* phage type 4. Vet. Rec. 130, 395–398.

Berchieri Jr., A., Barrow, P.A., 1996. The antibacterial effects for *Salmonella* Enteritidis phage type 4 of different chemical disinfectants and cleaning agents tested under different conditions. Avian Pathol. 25, 663–673.

Berghaus, R.D., Thayer, S.G., Maurer, J.J., Hofacre, C.L., 2011. Effect of vaccinating breeder chickens with a killed *Salmonella* vaccine on *Salmonella* prevalences and loads in breeder and broiler chicken flocks. J. Food Prot. 74, 727–734.

Braden, C.R., 2006. *Salmonella enterica* serotype Enteritidis and eggs: a national epidemic in the United States. Clin. Infect. Dis. 43, 512–517.

Carrique-Mas, J.J., Breslin, M., Snow, L., McLaren, I., Sayers, A.R., Davies, R.H., 2009a. Persistence and clearance of different *Salmonella* serovars in buildings housing laying hens. Epidemiol. Infect. 137, 837–846.

Carrique-Mas, J.J., Marin, C., Breslin, M., McLaren, I., Davies, R., 2009b. A comparison of the efficacy of cleaning and disinfection methods in eliminating *Salmonella* spp. from commercial egg laying houses. Avian Pathol. 38, 419–424.

Castellan, D.M., Kinde, H., Kass, P.H., Cutler, G., Breitmeyer, R.E., Bell, D.D., Ernst, R.A., Kerr, D.C., Little, H.E., Willoughby, D., Riemann, H.P., Ardans, A., Snowdon, J.A., Kuney, D.R., 2004. Descriptive study of California egg layer premises and analysis of risk factors for *Salmonella enterica* serotype Enteritidis as characterized by manure drag swabs. Avian Dis. 48, 550–561.

Chao, M.R., Hsien, C.H., Yeh, C.M., Chou, S.J., Chu, C., Su, Y.C., Yu, C.Y., 2007. Assessing the prevalence of *Salmonella enterica* in poultry hatcheries by using hatched eggshell membranes. Poult. Sci. 86, 1651–1655.

Chemaly, M., Huneau-Salaun, A., Labbe, A., Houdayer, C., Petetin, I., Fravalo, P., 2009. Isolation of *Salmonella enterica* in laying-hen flocks and assessment of eggshell contamination in France. J. Food Prot. 72, 2071–2077.

Clifton-Hadley, F.A., Breslin, M., Venables, L.M., Sprigings, K.A., Cooles, S.W., Houghton, S., Woodward, M.J., 2002. A laboratory study of an inactivated bivalent iron restricted *Salmonella enterica* serovars Enteritidis and Typhimurium dual vaccine against Typhimurium challenge in chickens. Vet. Microbiol. 89, 167–179.

Cogan, T.A., Humphrey, T.J., 2003. The rise and fall of *Salmonella* Enteritidis in the UK. J. Appl. Microbiol. 94 (Suppl), 114S–119S.

Curtiss 3rd, R., Hassan, J.O., 1996. Nonrecombinant and recombinant avirulent *Salmonella* vaccines for poultry. Vet. Immunol. Immunopathol. 54, 365–372.

Daniels, M.J., Hutchings, M.R., Greig, A., 2003. The risk of disease transmission to livestock posed by contamination of farm stored feed by wildlife excreta. Epidemiol. Infect. 130, 561–568.

Davies, R., Breslin, M., 2001. Environmental contamination and detection of *Salmonella enterica* serovar Enteritidis in laying flocks. Vet. Rec. 149, 699–704.

Davies, R., Breslin, M., 2003a. Effects of vaccination and other preventive methods for *Salmonella* Enteritidis on commercial laying chicken farms. Vet. Rec. 153, 673–677.

Davies, R.H., Breslin, M., 2003b. Persistence of *Salmonella* Enteritidis phage type 4 in the environment and arthropod vectors on an empty free-range chicken farm. Environ. Microbiol. 5, 79–84.

Davies, R., Breslin, M., 2004a. Observations on *Salmonella* contamination of eggs from infected commercial laying flocks where vaccination for *Salmonella enterica* serovar Enteritidis had been used. Avian Pathol. 33, 133–144.

Davies, R.H., Breslin, M.F., 2004b. Observations on the distribution and control of *Salmonella* contamination in poultry hatcheries. Br. Poult. Sci. 45 (Suppl. 1), S12–S14.

Davies, R.H., Wray, C., 1995a. Mice as carriers of *Salmonella enteritidis* on persistently infected poultry units. Vet. Rec. 137, 337–341.

Davies, R.H., Wray, C., 1995b. Observations on disinfection regimens used on *Salmonella enteritidis* infected poultry units. Poult. Sci. 74, 638–647.

Davies, R.H., Wray, C., 1996. Persistence of *Salmonella enteritidis* in poultry units and poultry food. Br. Poult. Sci. 37, 589–596.

De Buck, J., Van Immerseel, F., Haesebrouck, F., Ducatelle, R., 2005. Protection of laying hens against *Salmonella* Enteritidis by immunization with type 1 fimbriae. Vet. Microbiol. 105, 93–101.

De Vylder, J., Dewulf, J., Van Hoorebeke, S., Pasmans, F., Haesebrouck, F., Ducatelle, R., Van Immerseel, F., 2011. Horizontal transmission of *Salmonella* Enteritidis in groups of experimentally infected laying hens housed in different housing systems. Poult. Sci. 90, 1391–1396.

De Vylder, J., Van Hoorebeke, S., Ducatelle, R., Pasmans, F., Haesebrouck, F., Dewulf, J., Van Immerseel, F., 2009. Effect of the housing system on shedding and colonization of gut and internal organs of laying hens with *Salmonella* Enteritidis. Poult. Sci. 88, 2491–2495.

Denagamage, T., Jayarao, B., Patterson, P., Wallner-Pendleton, E., Kariyawasam, S., 2015. Risk factors associated with *Salmonella* in laying hen farms: systematic review of observational studies. Avian Dis. 59, 291–302.

Desin, T.S., Koster, W., Potter, A.A., 2013. *Salmonella* vaccines in poultry: past, present and future. Expert Rev. Vaccines 12, 87–96.

Dewaele, I., Van Meirhaeghe, H., Rasschaert, G., Vanrobaeys, M., De Graef, E., Herman, L., Ducatelle, R., Heyndrickx, M., De Reu, K., 2012. Persistent *Salmonella* Enteritidis environmental contamination on layer farms in the context of an implemented national control program with obligatory vaccination. Poult. Sci. 91, 282–291.

EC, 2003. Regulation (EC) 2160/2003 of the European Parliament and of the Council of 17 November 2003 on the control of *Salmonella* and other specified food-borne zoonotic agents. Official Journal of the European Union 325 (1).

Edwards, P.R., Bruner, D.W., 1943. The occurrence and distribution of *Salmonella* types in the United States. J. Infect. Dis. 72, 58–67.

EFSA, 2004. The use of vaccines for the control of *Salmonella* in poultry. EFSA J. 114, 1–74.

EFSA, 2007. Report of the Task Force on Zoonoses Data Collection on the Analysis of the baseline study on the prevalence of *Salmonella* in holdings of laying hen flocks of *Gallus gallus*. EFSA J. 97, 1–84.

EFSA, 2010. Scientific opinion on a quantitative estimation of the public health impact of setting a new target for the reduction of *Salmonella* in laying hens. EFSA J. 8, 1546.

EFSA, 2015. The European Union summary Report on trends and sources of Zoonoses, Zoonotic agents and food-borne outbreaks in 2014. EFSA J. 13, 4329.

Feberwee, A., de Vries, T.S., Elbers, A.R., de Jong, W.A., 2000. Results of a *Salmonella enteritidis* vaccination field trial in broiler-breeder flocks in The Netherlands. Avian Dis. 44, 249–255.

Fletcher, D.L., 2006. Influence of sampling methodology on reported incidence of *Salmonella* in poultry. J. AOAC Int. 89, 512–516.

Foley, S.L., Nayak, R., Hanning, I.B., Johnson, T.J., Han, J., Ricke, S.C., 2011. Population dynamics of *Salmonella enterica* serotypes in commercial egg and poultry production. App. Environ. Microbiol. 77, 4273–4279.

Galis, A., Marcq, C., Marlier, D., Portetelle, D., Van, I., Beckers, Y., Thewis, A., 2013. Control of *Salmonella* contamination of shell eggs – preharvest and postharvest methods: a review. Compr. Rev. Food Sci. Food Saf. 12, 155–182.

Gantois, I., Ducatelle, R., Timbermont, L., Boyen, F., Bohez, L., Haesebrouck, F., Pasmans, F., van Immerseel, F., 2006. Oral immunisation of laying hens with the live vaccine strains of TAD *Salmonella* vac E and TAD *Salmonella* vac T reduces internal egg contamination with *Salmonella* Enteritidis. Vaccine 24, 6250–6255.

Garber, L., Smeltzer, M., Fedorka-Cray, P., Ladely, S., Ferris, K., 2003. *Salmonella enterica* serotype enteritidis in table egg layer house environments and in mice in U.S. layer houses and associated risk factors. Avian Dis. 47, 134–142.

Gast, R.K., Guard-Bouldin, J., Holt, P.S., 2005. The relationship between the duration of fecal shedding and the production of contaminated eggs by laying hens infected with strains of *Salmonella* Enteritidis and *Salmonella* Heidelberg. Avian Dis. 49, 382–386.

Gast, R.K., Guraya, R., Jones, D.R., Anderson, K.E., 2014a. Contamination of eggs by *Salmonella* Enteritidis in experimentally infected laying hens housed in conventional or enriched cages. Poult. Sci. 93, 728–733.

Gast, R.K., Guraya, R., Jones, D.R., Anderson, K.E., 2014b. Horizontal transmission of *Salmonella* Enteritidis in experimentally infected laying hens housed in conventional or enriched cages. Poult. Sci. 93, 3145–3151.

Gast, R.K., Guraya, R., Jones, D.R., Anderson, K.E., 2015. Persistence of fecal shedding of *Salmonella* Enteritidis by experimentally infected laying hens housed in conventional or enriched cages. Poult. Sci. 94, 1650–1656.

Gast, R.K., Holt, P.S., 1998. Persistence of *Salmonella enteritidis* from one day of age until maturity in experimentally infected layer chickens. Poult. Sci. 77, 1759–1762.

Gole, V.C., Caraguel, C.G., Sexton, M., Fowler, C., Chousalkar, K.K., 2014. Shedding of *Salmonella* in single age caged commercial layer flock at an early stage of lay. Int. J. Food Microbiol. 189, 61–66.

Gomes, A.V., Quinteiro-Filho, W.M., Ribeiro, A., Ferraz-de-Paula, V., Pinheiro, M.L., Baskeville, E., Akamine, A.T., Astolfi-Ferreira, C.S., Ferreira, A.J., Palermo-Neto, J., 2014. Overcrowding stress decreases macrophage activity and increases *Salmonella* Enteritidis invasion in broiler chickens. Avian Pathol. 43, 82–90.

Gormley, F.J., Rawal, N., Little, C.L., 2012. Choose your menu wisely: cuisine-associated food-poisoning risks in restaurants in England and Wales. Epidemiol. Infect. 140, 997–1007.

Gosling, R.J., Martelli, F., Sayers, R., Larkin, L., Davies, R.H., 2014. A review of the official sampling of flocks of laying hens in the *Salmonella* National Control Programme in Great Britain. Br. Poult. Sci. 55, 569–575.

Gradel, K.O., Andersen, J., Madsen, M., 2002. Comparisons of sampling procedures and time of sampling for the detection of *Salmonella* in Danish infected chicken flocks raised in floor systems. Acta Vet. Scand. 43, 21–30.

Gradel, K.O., Sayers, A.R., Davies, R.H., 2004. Surface disinfection tests with *Salmonella* and a putative indicator bacterium, mimicking worst-case scenarios in poultry houses. Poult. Sci. 83, 1636–1643.

Guard-Petter, J., 2001. The chicken, the egg and *Salmonella enteritidis*. Environ. Microbiol. 3, 421–430.

Guard-Petter, J., Henzler, D.J., Rahman, M.M., Carlson, R.W., 1997. On-farm monitoring of mouse-invasive *Salmonella enterica* serovar Enteritidis and a model for its association with the production of contaminated eggs. Appl. Environ. Microbiol. 63, 1588–1593.

Hannah, J.F., Wilson, J.L., Cox, N.A., Richardson, L.J., Cason, J.A., Bourassa, D.V., Buhr, R.J., 2011. Horizontal transmission of *Salmonella* and *Campylobacter* among caged and cage-free laying hens. Avian Dis. 55, 580–587.

Harker, K.S., Lane, C., Gormley, F.J., Adak, G.K., 2014. National outbreaks of *Salmonella* infection in the UK, 2000–2011. Epidemiol. Infect. 142, 601–607.

Hassan, J.O., Curtiss 3rd, R., 1996. Effect of vaccination of hens with an avirulent strain of *Salmonella typhimurium* on immunity of progeny challenged with wild-Type *Salmonella* strains. Infect. Immun. 64, 938–944.

Henzler, D.J., Kradel, D.C., Sischo, W.M., 1998. Management and environmental risk factors for *Salmonella enteritidis* contamination of eggs. Am. J. Vet. Res. 59, 824–829.

Henzler, D.J., Opitz, H.M., 1992. The role of mice in the epizootiology of Salmonella enteritidis infection on chicken layer farms. Avian Dis. 36, 625–631.

Holt, P.S., 1995. Horizontal transmission of Salmonella enteritidis in molted and unmolted laying chickens. Avian Dis. 39, 239–249.

Holt, P.S., Geden, C.J., Moore, R.W., Gast, R.K., 2007. Isolation of *Salmonella enterica* serovar Enteritidis from houseflies (*Musca domestica*) found in rooms containing *Salmonella* serovar Enteritidis-challenged hens. Appl. Environ. Microbiol. 73, 6030–6035.

Holt, P.S., Mitchell, B.W., Gast, R.K., 1998. Airborne horizontal transmission of Salmonella enteritidis in molted laying chickens. Avian Dis. 42, 45–52.

Huneau-Salaun, A., Marianne, C., Sophie le, B., Francoise, L., Isabelle, P., Sandra, R., Virginie, M., Philippe, F., Nicolas, R., 2009. Risk factors for *Salmonella enterica* subsp. *enterica* contamination in 519 French laying hen flocks at the end of the laying period. Prev. Vet. Med. 89, 51–58.

Huneau-Salaün, A., Michel, V., Balaine, L., Petetin, I., Eono, F., Ecobichon, F., Bouquin, S.L., 2010. Evaluation of common cleaning and disinfection programmes in battery cage and on-floor layer houses in France. Br. Poult. Sci. 51, 204–212.

Johansson, T.M., Schildt, R., Ali-Yrkko, S., Siitonen, A., Maijala, R.L., 1996. The first Salmonella enteritidis phage type 1 infection of a commercial layer flock in Finland. Acta Vet. Scand 37, 471–479.

Jones, D.R., Anderson, K.E., Guard, J.Y., 2012. Prevalence of coliforms, *Salmonella*, *Listeria*, and *Campylobacter* associated with eggs and the environment of conventional cage and free-range egg production. Poult. Sci. 91, 1195–1202.

Kinde, H., Castellan, D.M., Kass, P.H., Ardans, A., Cutler, G., Breitmeyer, R.E., Bell, D.D., Ernst, R.A., Kerr, D.C., Little, H.E., Willoughby, D., Riemann, H.P., Snowdon, J.A., Kuney, D.R., 2004. The occurrence and distribution of *Salmonella enteritidis* and other serovars on California egg laying premises: a comparison of two sampling methods and two culturing techniques. Avian Dis. 48, 590–594.

Kinde, H., Read, D.H., Chin, R.P., Bickford, A.A., Walker, R.L., Ardans, A., Breitmeyer, R.E., Willoughby, D., Little, H.E., Kerr, D., Gardner, I.A., 1996. *Salmonella enteritidis*, phase type 4 infection in a commercial layer flock in southern California: bacteriologic and epidemiologic findings. Avian Dis. 40, 665–671.

Lapuz, R., Tani, H., Sasai, K., Shirota, K., Katoh, H., Baba, E., 2008. The role of roof rats (*Rattus rattus*) in the spread of *Salmonella* Enteritidis and *S.* Infantis contamination in layer farms in eastern Japan. Epidemiol. Infect. 136, 1235–1243.

Lapuz, R.R., Umali, D.V., Suzuki, T., Shirota, K., Katoh, H., 2012. Comparison of the prevalence of *Salmonella* infection in layer hens from commercial layer farms with high and low rodent densities. Avian Dis. 56, 29–34.

Li, X., Payne, J.B., Santos, F.B., Levine, J.F., Anderson, K.E., Sheldon, B.W., 2007. *Salmonella* populations and prevalence in layer feces from commercial high-rise houses and characterization of the *Salmonella* isolates by serotyping, antibiotic resistance analysis, and pulsed field gel electrophoresis. Poult. Sci. 86, 591–597.

Line, J.E., Bailey, J.S., Cox, N.A., Stern, N.J., 1997. Yeast treatment to reduce *Salmonella* and *Campylobacter* populations associated with broiler chickens subjected to transport stress. Poult. Sci. 76, 1227–1231.

Liu, Y.L., Schmid, D., Voss, A.S., Kasper, S., Lassnig, H., Ableitner, O., Kornschober, C., Karnthaler, U., Allerberger, F., 2012. A 2010 Austrian *Salmonella* Enteritidis PT4 outbreak associated with a laying hen holding previously involved in an *S.* Enteritidis PT4 cluster: pitfalls of regulatory responses in risk management. J. Infect. Public Health 5, 332–339.

Martelli, F., Davies, R.H., 2011. *Salmonella* serovars isolated from table eggs: an overview. Food Res. Int. 45, 745–754.

McLaren, I., Wales, A., Breslin, M., Davies, R., 2011. Evaluation of commonly-used farm disinfectants in wet and dry models of *Salmonella* farm contamination. Avian Pathol. 40, 33–42.

Moro, C.V., Fravalo, P., Amelot, M., Chauve, C., Zenner, L., Salvat, G., 2007. Colonization and organ invasion in chicks experimentally infected with *Dermanyssus gallinae* contaminated by *Salmonella* Enteritidis. Avian Pathol. 36, 307–311.

Mueller-Doblies, D., Carrique-Mas, J.J., Sayers, A.R., Davies, R.H., 2010. A comparison of the efficacy of different disinfection methods in eliminating *Salmonella* contamination from turkey houses. J. Appl. Microbiol. 109, 471–479.

Mueller-Doblies, D., Clouting, C., Davies, R.H., 2013. Investigations of the distribution and persistence of *Salmonella* and ciprofloxacin-resistant *Escherichia coli* in turkey hatcheries in the UK. Zoonoses Public Hhealth 60, 296–303.

Musgrove, M.T., Jones, D.R., 2005. Recovery of Salmonella from nest run egg cart shelves. Poult. Sci 84, 77.

Nahms, 2000. Layers '99: *Salmonella enterica* Serotype Enteritidis in Table Egg Layers in the U.S. United States Department of Agriculture, National Animal Health Monitoring Service, Fort Collins, Colorado, USA.

Nahms, 2014. Layers 2013 Part II: Control and Prevention of *Salmonella* Enteritidis on Table-Egg Farms in the United States, 2013. United States Department of Agriculture/National Animal Health Monitoring System, Fort Collins, Colorado, USA.

Nakamura, M., Nagata, T., Okamura, S., Takehara, K., Holt, P.S., 2004. The effect of killed *Salmonella enteritidis* vaccine prior to induced molting on the shedding of *S.* enteritidis in laying hens. Avian Dis. 48, 183–188.

Nakamura, M., Takagi, M., Takahashi, T., Suzuki, S., Sato, S., Takehara, K., 1997. The effect of the flow of air on horizontal transmission of Salmonella enteritidis in chickens. Avian Dis. 41, 354–360.

Namata, H., Meroc, E., Aerts, M., Faes, C., Abrahantes, J.C., Imberechts, H., Mintiens, K., 2008. *Salmonella* in Belgian laying hens: an identification of risk factors. Prev. Vet. Med. 83, 323–336.

O'Brien, S.J., 2012. The "decline and fall" of nontyphoidal *Salmonella* in the United Kingdom. Clin. Infect. Dis. 56 (5), 705–710.

Okamura, M., Kamijima, Y., Miyamoto, T., Tani, H., Sasai, K., Baba, E., 2001. Differences among six *Salmonella* serovars in abilities to colonize reproductive organs and to contaminate eggs in laying hens. Avian Dis. 45, 61–69.

Parker, C., Asokan, K., Guard-Petter, J., 2001. Egg contamination by *Salmonella* serovar Enteritidis following vaccination with Delta-aroA *Salmonella* serovar Typhimurium. FEMS Microbiol. Lett. 195, 73–78.

Pedersen, T.B., Olsen, J.E., Bisgaard, M., 2008. Persistence of *Salmonella* Senftenberg in poultry production environments and investigation of its resistance to desiccation. Avian Pathol. 37, 421–427.

Rabie, A., Carrique-Mas, J., Davies, R., 2010. Persistence of *Salmonella* in rodent populations from infected farms in Great Britain. In: Proceedings of the International Conference of Salmonella and Salmonellosis, 2010, pp. 219–222.

Reddy, S.P., Wang, H., Adams, J.K., Feng, P.C., 2016. Prevalence and characteristics of *Salmonella* serotypes isolated from fresh produce marketed in the United States. J. Food Prot. 79, 6–16.

Riemann, H., Himathongkham, S., Willoughby, D., Tarbell, R., Breitmeyer, R., 1998. A survey for Salmonella by drag swabbing manure piles in California egg ranches. Avian Dis 42, 67–71.

Russell, A.D., 2003. Biocide use and antibiotic resistance: the relevance of laboratory findings to clinical and environmental situations. Lancet Infect. Dis. 3, 794–803.

Schulz, J., Van Hoorebeke, S., Hald, B., Hartung, J., Van Immerseel, F., Radtke, I., Kabell, S., Dewulf, J., 2011. The dynamics of *Salmonella* occurrence in commercial laying hen flocks throughout a laying period. Avian Pathol. 40, 243–248.

Shivaprasad, H.L., Timoney, J.F., Morales, S., Lucio, B., Baker, R.C., 1990. Pathogenesis of Salmonella enteritidis infection in laying chickens. Studies on egg transmission, clinical signs, fecal shedding, and serologic responses. Avian Dis. 34, 548–557.

Siemon, C.E., Bahnson, P.B., Gebreyes, W.A., 2007. Comparative investigation of prevalence and antimicrobial resistance of *Salmonella* between pasture and conventionally reared poultry. Avian Dis. 51, 112–117.

Skov, M.N., Feld, N.C., Carstensen, B., Madsen, M., 2002. The serologic response to Salmonella enteritidis and Salmonella typhimurium in experimentally infected chickens, followed by an indirect lipopolysaccharide enzyme-linked immunosorbent assay and bacteriologic examinations through a one-year period. Avian Dis. 46, 265–273.

Snow, L.C., Davies, R.H., Christiansen, K.H., Carrique-Mas, J.J., Cook, A.J., Evans, S.J., 2010. Investigation of risk factors for *Salmonella* on commercial egg-laying farms in Great Britain, 2004–2005. Vet. Rec. 166, 579–586.

Springer, S., Lindner, T., Ahrens, M., Woitow, G., Prandini, F., Selbitz, H.J., 2011. Duration of immunity induced in chickens by an attenuated live *Salmonella* Enteritidis vaccine and an inactivated *Salmonella* Enteritidis/Typhimurium vaccine. Berl Munch. Tierarztl. Wochenschr. 124, 89–93.

Stringfellow, K., Anderson, P., Caldwell, D., Lee, J., Byrd, J., McReynolds, J., Carey, J., Nisbet, D., Farnell, M., 2009. Evaluation of disinfectants commonly used by the commercial poultry industry under simulated field conditions. Poult. Sci. 88, 1151–1155.

Thorns, C.J., 2000. Bacterial food-borne zoonoses. Rev. Sci. Tech. 19, 226–239.

Umali, D.V., Lapuz, R.R., Suzuki, T., Shirota, K., Katoh, H., 2012. Transmission and shedding patterns of *Salmonella* in naturally infected captive wild roof rats (*Rattus rattus*) from a *Salmonella*-contaminated layer farm. Avian Dis. 56, 288–294.

van de Giessen, A.W., Bouwknegt, M., Dam-Deisz, W.D., van Pelt, W., Wannet, W.J., Visser, G., 2006. Surveillance of *Salmonella* spp. and *Campylobacter* spp. in poultry production flocks in The Netherlands. Epidemiol. Infect. 134, 1266–1275.

Van Hoorebeke, S., Van Immerseel, F., De Vylder, J., Ducatelle, R., Haesebrouck, F., Pasmans, F., de Kruif, A., Dewulf, J., 2010a. The age of production system and previous *Salmonella* infections on-farm are risk factors for low-level *Salmonella* infections in laying hen flocks. Poult. Sci. 89, 1315–1319.

Van Hoorebeke, S., Van Immerseel, F., Schulz, J., Hartung, J., Harisberger, M., Barco, L., Ricci, A., Theodoropoulos, G., Xylouri, E., De Vylder, J., Ducatelle, R., Haesebrouck, F., Pasmans, F., de Kruif, A., Dewulf, J., 2010b. Determination of the within and between flock prevalence and identification of risk factors for *Salmonella* infections in laying hen flocks housed in conventional and alternative systems. Prev. Vet. Med. 94, 94–100.

Van Hoorebeke, S., Van Immerseel, F., Haesebrouck, F., Ducatelle, R., Dewulf, J., 2011. The influence of the housing system on *Salmonella* infections in laying hens: a review. Zoonoses Public Health 58, 304–311.

Van Immerseel, F., Methner, U., Rychlik, I., Nagy, B., Velge, P., Martin, G., Foster, N., Ducatelle, R., Barrow, P.A., 2005. Vaccination and early protection against non-host-specific *Salmonella* serotypes in poultry: exploitation of innate immunity and microbial activity. Epidemiol. Infect. 133, 959–978.

Wales, A., Breslin, M., Carter, B., Sayers, A.R., Davies, R., 2007. A longitudinal study of environmental *Salmonella* contamination in caged and free-range layer flocks. Avian Pathol. 36, 187–197.

Wales, A., Breslin, M., Davies, R., 2006a. Assessment of cleaning and disinfection in *Salmonella*-contaminated poultry layer houses using qualitative and semi-quantitative culture techniques. Vet. Microbiol. 116, 283–293.

Wales, A., Breslin, M., Davies, R., 2006b. Semiquantitative assessment of the distribution of *Salmonella* in the environment of caged layer flocks. J. Appl. Microbiol. 101, 309–318.

Wales, A.D., Davies, R.H., 2011. A critical review of *Salmonella* Typhimurium infection in laying hens. Avian Pathol. 40, 429–436.

Wallner-Pendleton, E.A., Patterson, P.H., Kariyawasam, S., Trampel, D.W., Denagamage, T., 2014. On-farm risk factors for *Salmonella* Enteritidis contamination. J. Appl. Poult. Res. 23, 345–352.

Ward, L.R., Threlfall, J., Smith, H.R., O'Brien, S.J., 2000. *Salmonella enteritidis* epidemic. Science 287, 1753–1754 author reply 1755–1756.

Woodward, M.J., Gettinby, G., Breslin, M.F., Corkish, J.D., Houghton, S., 2002. The efficacy of Salenvac, a *Salmonella enterica* subsp. *enterica* serotype Enteritidis iron-restricted bacterin vaccine, in laying chickens. Avian Pathol. 31, 383–392.

Wright, A., Ginn, A., Luo, Z., 2016. Molecular tools for monitoring and source-tracking *Salmonella* in wildlife and the environment. In: Food Safety Risks from Wildlife. Springer International Publishing, pp. 131–150.

Yao, K., Muruvanda, T., Roberts, R.J., Payne, J., Allard, M.W., Hoffmann, M., 2016. Complete genome and methylome sequences of two *Salmonella enterica* spp. Genome Announc. 4.

Zenner, D., Janmohamed, K., Lane, C., Little, C., Charlett, A., Adak, G.K., Morgan, D., 2013. The serotype case-case design: a direct comparison of a novel methodology with a case-control study in a national *Salmonella* Enteritidis PT14b outbreak in England and Wales. Epidemiol. Infect. 141, 2346–2353.

Zielicka-Hardy, A., Zarowna, D., Szych, J., Madajczak, G., Sadkowska-Todys, M., 2012. Ensuring safety of home-produced eggs to control salmonellosis in Poland: lessons from an outbreak in September 2011. Eurosurveillance 17.

CHAPTER

# Microbiology of Shell Egg Production in the United States

# 2

**Richard K. Gast**

*US National Poultry Research Center, Athens, GA, United States*

## 1. INTRODUCTION: *SALMONELLA* AND EGGS: A PUBLIC HEALTH AND ECONOMIC PROBLEM

Despite ambitious public health goals for the first decade of the 21st century, the incidence of human *Salmonella* infections remained nearly constant and a 44% increase was reported for infections with *Salmonella enterica* subspecies *enterica* serotype Enteritidis (Centers for Disease Control and Prevention, 2011; Chai et al., 2012). Since the 1980s, a high international prevalence of *Salmonella* Enteritidis infections has been principally linked to the consumption of contaminated eggs (Braden, 2006; Jackson et al., 2013; Pires et al., 2014). On the basis of both retrospective epidemiological analysis and active disease surveillance, the incidence of human salmonellosis has been strongly associated with the prevalence of *S.* Enteritidis in commercial flocks of egg-laying chickens (Havelaar et al., 2013; Arnold et al., 2014a). For example, a study involving 24 European nations (De Knegt et al., 2015) identified laying hens as the leading reservoir of human salmonellosis, responsible for 42% of all cases (and 96% of these were caused by *S.* Enteritidis). Human salmonellosis in the United States has been estimated to have costs to society, industry, and government as high as $11 billion per year for medical treatment, lost wages, legal actions, market restriction and reduced product value, risk reduction efforts, and regulatory programs (Scharff, 2012). Federal egg safety regulations (U.S. Food and Drug Administration, 2009) suggested that provisions to control *S.* Enteritidis would cost the egg industry approximately $81 million annually, but could reduce human health costs (due to 79,000 egg-related illnesses) by $1.4 billion.

## 2. *SALMONELLA* ENTERITIDIS AND LAYING HENS: INFECTION AND EGG CONTAMINATION

### 2.1 INTESTINAL COLONIZATION AND FECAL SHEDDING

The exposure of highly susceptible young chicks to *S.* Enteritidis can result in the establishment of intestinal tract colonization, which continues until maturity

Producing Safe Eggs. http://dx.doi.org/10.1016/B978-0-12-802582-6.00002-1

(Gast and Holt, 1998; Van Immerseel et al., 2004). Intestinal colonization by *Salmonella* usually declines steadily following the exposure of mature chickens (Gast et al., 2005, 2011b), but infection can sometimes persist for an extended period of time (Li et al., 2007; Gast et al., 2009). Bacterial persistence in even a small percentage of hens could prolong opportunities for horizontal transmission within flocks. The horizontal dissemination of *Salmonella* in laying flocks depends directly on both the frequency and the duration of infection in individual hens.

Fecal shedding of salmonellae is an important consequence of their ability to adhere to cells of the avian intestinal tract and can be a prominent source of bacterial contamination of the poultry housing environment (Trampel et al., 2014). Both the frequency and duration of fecal shedding are directly correlated with the oral exposure dose (Gast and Holt, 2000a; Gast et al., 2011b). Mature hens infected with large doses can sometimes shed *S.* Enteritidis in their feces for several months (Gast et al., 2013b). The prevalence of fecal shedding of *S.* Enteritidis in commercial laying flocks may fluctuate over time (Wales et al., 2007; Schulz et al., 2011), but appears to peak just before egg laying commences and gradually declines thereafter (Li et al., 2007; Gole et al., 2014). Persistent fecal shedding of *S.* Enteritidis does not necessarily predict the likelihood of systemic infection or egg contamination (Gast and Holt, 2000a; Gast et al., 2005).

## 2.2 INTERNAL ORGAN COLONIZATION

Within only a few hours after laying hens are orally inoculated, *S.* Enteritidis can invade beyond the intestinal tract to colonize the liver and spleen (He et al., 2010). The frequency of *S.* Enteritidis isolation from the internal organs of mature birds declines steadily over the first few weeks following oral exposure (Gast et al., 2007b, 2011a), but infection is sometimes persistent in individual birds (Gast et al., 2009). The deposition of *S.* Enteritidis inside the edible interior contents of eggs is a consequence of reproductive organ colonization in systemically infected hens (Gantois et al., 2009; Gast et al., 2011a). However, the invasion of reproductive tissues at high frequencies or involving large loads of *Salmonella* cells does not always lead to a similarly high incidence of egg contamination (Gast et al., 2004, 2007b, 2011c). Both the ovary (where yolks mature before release) and oviduct (where albumen is secreted around descending yolks) can harbor *S.* Enteritidis, thereby leading to bacterial deposition at corresponding locations—yolk or albumen—within eggs (Humphrey et al., 1989; Gast and Holt, 2000a; De Buck et al., 2004; Gast et al., 2004, 2007b).

## 2.3 EGG CONTAMINATION

The dose of *S.* Enteritidis cells administered experimentally to hens has been shown to affect the frequency of systemic infection, pattern of internal organ involvement, and likelihood of deposition in eggs (Gast et al., 2011a, 2013a). Nevertheless, even very large oral doses typically cause rather infrequent egg contamination, at generally low initial bacterial population levels (Humphrey et al., 1991; Gast and Beard, 1992; Gast and Holt, 2000a). Naturally occurring *S.* Enteritidis infections in commercial flocks, often acquired from environmental sources at low exposure doses and spread by horizontal contact, are generally associated with a very low prevalence of egg contamination (Gast and Holt, 1999; Ebel and Schlosser, 2000; DeWinter et al., 2011; Esaki et al., 2013). Similarly, intestinal colonization, organ invasion, and egg contamination have all occurred less frequently among experimentally contact-exposed birds than among birds inoculated with large oral doses (Gast and Beard, 1990; Gast and Holt, 1999).

The initial site of *Salmonella* deposition inside eggs influences the potential for the pathogen to multiply during storage. Egg yolk provides abundant nutrients to support rapid and prolific microbial growth at warm temperatures (Gurtler and Conner, 2009), but albumen contains proteins that limit iron availability and disrupt bacterial membranes (Baron et al., 2016). Although *S.* Enteritidis is most commonly deposited either in the albumen or on the outside surface of the vitelline (yolk) membrane of contaminated eggs, it can rapidly migrate across this membrane to reach the nutrient-dense interior contents of the yolk at warm temperatures (Gast and Holt, 2001a; Gast et al., 2003, 2010). However, refrigeration temperatures can reduce or prevent both *Salmonella* multiplication in egg yolks (Gast and Holt, 2000a; Gurtler and Conner, 2009) and penetration of vitelline membranes (Gast et al., 2006, 2007a).

## 3. INFECTION OF LAYING HENS AND EGG CONTAMINATION BY DIFFERENT *SALMONELLA* STRAINS AND SEROTYPES

### 3.1 PATHOBIOLOGY OF *SALMONELLA* ENTERITIDIS STRAINS AND PHAGE TYPES

The public health consequences of infections in laying hens often vary significantly between *Salmonella* strains or serotypes. In experimental infection studies, *S.* Enteritidis strains have sometimes differed significantly in both invasion to reproductive organs and contamination of eggs (Gast and Holt, 2000a, 2001a; Guard et al., 2011). *S.* Enteritidis strains that are efficient at causing egg contamination have been distinguished from other environmental salmonellae by their capabilities for adherence to reproductive tract mucosa and for survival in forming eggs. These traits are associated with the expression of genes found in the major pathogenicity islands, involved in cell wall or lipopolysaccharide structure, or related to stress responses (Guard-Bouldin et al., 2004; Wales and Davies, 2011; Coward et al., 2013; Raspoet et al., 2014). The deposition of salmonellae within developing eggs may require the expression of a sequence of phenotypic properties as they become necessary at successive stages of infection (Gast et al., 2002a; Guard et al., 2010). Differential accumulation of small genetic changes may result in divergent abilities to invade internal organs and eggs among isolates of the same *S.* Enteritidis phage type (Guard et al., 2010, 2011). No affinities for specific reproductive tract sites or consistent patterns of deposition in either egg yolk or albumen have been attributed to individual *Salmonella* isolates (Gast and Holt, 2000a, 2001a; Gast et al., 2007b). Phage typing of *S.* Enteritidis has been valuable for identifying epidemiological relationships between isolates from different sources, but it has not consistently correlated with the egg contamination potential of individual strains (Gast and Holt, 2000a; Gantois et al., 2009; Guard et al., 2011).

Environmental conditions (particularly pH and temperature) have been demonstrated to influence expression of potential *Salmonella* virulence factors such as flagella, fimbria, outer membrane proteins, and iron uptake systems (McDermid et al., 1996; Walker et al., 1999). A greater diversity of *S.* Enteritidis phage types has been found in poultry houses than in contaminated eggs, suggesting that the environment may serve as a reservoir from which strains with a heightened ability to cause systemic infection and associated egg contamination may periodically emerge (Henzler et al., 1994, 1998; Dewaele et al., 2012a,b). Conditions in the tissues of infected hens may also exert selective pressures that can influence the expression of bacterial virulence factors. For example, genes that were highly expressed in *S.* Enteritidis oviduct isolates from infected hens were also highly expressed in egg isolates (Gantois et al., 2008). The ability of *S.* Enteritidis strains to cause egg contamination was experimentally increased by repeated passage through infected hens (Gast et al., 2003, 2005), although virulence was unaffected by a single passage through reproductive tissues (Gast et al., 2009).

Individual *Salmonella* strains can sometimes differ in their growth properties in eggs (Gast and Holt, 2001b; Cogan et al., 2004; Gast and Guraya, 2013) or in their ability to penetrate through yolk membranes (Gast et al., 2007a; Gantois et al., 2008), but these differences do not coincide with phage typing distinctions (Gast and Holt, 2001b; Gast and Guraya, 2013). Isolates of *S.* Enteritidis have been observed to survive better in egg albumen than strains of other serotypes (De Vylder et al., 2012). *S.* Enteritidis strains sensitive to both acidic and oxidative stress exhibited reduced survival and growth in egg albumen (Shah et al., 2012). The frequency at which *S.* Enteritidis is able to migrate into yolks and multiply can vary in eggs from different genetic lines of commercial laying hens (Gast et al., 2010).

## 3.2 *SALMONELLA* ENTERITIDIS AND OTHER *SALMONELLA* SEROTYPES

The epidemiological relationship between *S.* Enteritidis and eggs is uniquely important, but strains of some other *Salmonella* serotypes can colonize reproductive tissues in laying hens and contaminate the contents of developing eggs (Keller et al., 1997; Okamura et al., 2001; Gast et al., 2004). In North America, eggs contaminated by *Salmonella* Heidelberg (a serotype that is common in the housing environment of laying flocks) have been implicated as vehicles of human illness, although far less often than *S.* Enteritidis (Chittick et al., 2006). *Salmonella* Typhimurium has been identified as a source of sporadic egg-transmitted disease in Australia (Gole et al., 2014). Most other serotypes commonly found in commercial flocks, such as Kentucky in the United States, have not been associated with meaningful levels of egg contamination. Experimental infection studies involving strains of serotypes Heidelberg and Typhimurium have sometimes produced high frequencies of systemic infection and reproductive organ colonization (Gantois et al., 2008), but these serotypes caused egg contamination far less often than *S.* Enteritidis (Gast et al., 2005, 2007b, 2011b). Possible explanations for these trends include the stronger adherence of *S.* Enteritidis to reproductive tract mucosa and the elicitation of more intense immune responses by other serotypes (Wales and Davies, 2011). In most instances around the world, laying flock testing programs that focus on *S.* Enteritidis as being epidemiologically preeminent constitute a cost-effective use of limited resources for protecting public health. Nevertheless, monitoring for the emergence of previously infrequent or inconsequential serotypes can also have proactive public health value.

## 4. ENVIRONMENTAL INFLUENCES ON *SALMONELLA* INFECTIONS IN EGG-LAYING FLOCKS

Opportunities for the introduction and dissemination of pathogens in laying flocks largely depend on environmental conditions in egg production facilities (Trampel et al., 2014). The poultry house environment often serves as a reservoir for *Salmonella* strains that can infect egg-producing chickens. Persistent environmental contamination is sometimes responsible for the transmission of infection into successive laying flocks over extended periods of time (Davies and Breslin, 2003; Dewaele et al., 2012a,b; Lapuz et al., 2012). *S.* Enteritidis isolates with impaired resistance to environmental stressors also had reduced pathogenicity for chickens (Shah, 2014). *Salmonella* contamination is often widely distributed throughout laying houses in association with dust and feces (Garber et al., 2003; Kinde et al., 2005; Im et al., 2015), and can be perpetuated and amplified by severe rodent or insect infestations to levels capable of surviving standard cleaning and disinfection methods (Carrique-Mas et al., 2009b; Snow et al., 2010; Lapuz et al., 2012; Wallner-Pendleton et al., 2014).

After salmonellae are introduced into laying houses, rapid and extensive horizontal dissemination throughout flocks is mediated by direct contact between hens, ingestion of contaminated feed or feces, movement of personnel and equipment, and airborne circulation of contaminated dust and aerosols (Gast et al., 1998, 2014b; Thomas et al., 2009, 2011). Environmental stressors, including feed deprivation, water deprivation, or excessive heat, can increase the susceptibility of hens to horizontally transmitted infections (Asakura et al., 2001; Humphrey, 2006; Okamura et al., 2010). Commercial egg production facilities represent highly complex environments, so the potential influences on the presence of food-borne pathogens in egg-producing flocks are correspondingly diverse. The most commonly identified risk factors linked to increased *Salmonella* prevalence in egg-laying chickens are larger flock size, greater flock age, housing in older facilities, and multiple-age stocking (Mollenhorst et al., 2005; Namata et al., 2008; Huneau-Salaün et al., 2009; Snow et al., 2010; Van Hoorebeke et al., 2010a; Pitesky et al., 2013; Denagamage et al., 2015). However, within-flock variations in the prevalence of infection over time can confound definitive risk assessment (Wales et al., 2007). The presently incomplete understanding of environmental risk factors constitutes an important data gap for the development of effective and sustainable long-term measures for controlling egg-transmitted disease.

## 5. EGG PRODUCTION HOUSING SYSTEMS AND *SALMONELLA*

### 5.1 HOUSING SYSTEMS FOR EGG-LAYING HENS

The animal welfare, economic viability, and public health implications of different production housing systems for commercial laying hens have been widely discussed. The various housing options have numerous complex facility characteristics and management practices that can influence the persistence and transmission of infections with pathogens such as *S.* Enteritidis (Carrique-Mas et al., 2009a; Jones et al., 2015). Conventional laying cages, predominant in the commercial egg production for many years, house small groups of laying hens at relatively high stocking densities (expressed as the amount of available floor space per bird). Newer alternative caging systems include enriched colony cages (substantially larger units providing lower stocking densities plus environmental enhancements such as perches, nesting areas, and scratching pads) and aviaries in which birds can move freely among multiple open levels of enriched cage and floor areas within houses. Cage-free or free-range housing, offering greater opportunities for freedom of movement via varying degrees of access to outdoor forage or pasture areas, are also becoming increasingly common in the egg industry.

### 5.2 HOUSING INFLUENCES ON *SALMONELLA* IN LAYING FLOCKS

Reviews of the published scientific literature regarding the food safety consequences of different types of housing for egg-laying poultry have not identified any definitive overall consensus viewpoint (Holt et al., 2011; Whiley and Ross, 2015). Highly diverse results have been obtained in studies comparing the prevalence of *Salmonella* infection or environmental contamination in flocks housed in various types of cage-based or cage-free systems, and no consistent advantage has been documented for any one class of housing. In some studies, cage-based housing systems have been linked to a higher frequency of *Salmonella* infection in poultry flocks, especially when large populations of rodents are present (Huneau-Salaün et al., 2009; Snow et al., 2010; Van Hoorebeke et al., 2010b; Denagamage et al., 2015). In other studies, cage-free housing systems have been associated with greater numbers of Enterobacteriaceae on egg shells, a higher incidence of environmental contamination with *Salmonella*, and more frequent horizontal transmission of *Salmonella* infection within flocks (De Vylder et al., 2011; Hannah et al., 2011; Watanabe et al., 2012;

Jones and Anderson, 2013; Parisi et al., 2015). Access to outdoor areas, which are potentially vulnerable to the introduction of pathogens from external sources, may be an important source of risk in free-range housing (Mollenhorst et al., 2005). In a third group of studies, no significant differences in either *Salmonella* fecal shedding or environmental contamination have been reported between cage-based and cage-free flocks (Siemon et al., 2007; Jones et al., 2012) or between conventional cage and enriched colony cage systems (De Vylder et al., 2009; Nordentoft et al., 2011; Van Hoorebeke et al., 2011). In a large field study conducted under commercial egg production conditions, no broad trends in either total bacterial levels or the presence of pathogens distinguished environmental samples and eggs from conventional cages, enriched colony cages, and aviaries, although specific management concerns (such as contaminated scratch pads in enriched cages and contaminated floor eggs in aviaries) were identified as challenges pertinent to each system (Jones et al., 2015).

## 5.3 POTENTIAL MECHANISMS FOR HOUSING EFFECTS

In a series of experimental infection studies, *S.* Enteritidis was isolated from internal organs and voided feces at significantly higher overall frequencies from hens in conventional cages than in enriched colony cages, but no differences between the two housing systems were found for the persistence of fecal shedding, the frequency of horizontal transmission of infection, or the production of internally contaminated eggs (Gast et al., 2013b, 2014a,b, 2015). One possible explanation for these results is that some intrinsic characteristic of conventional cage housing (such as bird density or behavioral restriction) was a stressor that compromised immunity and thereby increased susceptibility to the systemic dissemination of *S.* Enteritidis infection. The stocking density of laying hens in production systems has been identified as a parameter capable of influencing flock infections with *Salmonella*, perhaps by diminishing immune responses or increasing opportunities for horizontal contact transmission of the pathogen. Crowded and unsanitary conditions are known to decrease the resistance of chickens to infection (Asakura et al., 2001). Housing chickens at high stocking densities has been associated with the suppression of both humoral and cellular immunity and to increased *S.* Enteritidis invasion of internal organs (Gomes et al., 2014). The impairment of lymphocyte function in intestinal lymphoid tissues such as Peyer patches and cecal tonsils could compromise effective clearance of infection from the gut (Holt et al., 2010). Significant housing system effects on colonization of the spleen (an important secondary lymphatic organ) in experimental infection studies are consistent with a stress-related explanation (Gast et al., 2013b). Nevertheless, other secondary manifestations of *S.* Enteritidis infections (including the most important food safety parameter, egg contamination) may not always be directly affected by these same influences. In the course of *S.* Enteritidis infection in laying hens, systemic dissemination is a necessary component in the process that leads to bacterial deposition inside eggs, but the frequency or magnitude of reproductive organ colonization does not consistently predict the likelihood of subsequent egg contamination (Gast et al., 2004, 2007b, 2011b).

## 6. CONTROLLING *SALMONELLA* INFECTION AND EGG CONTAMINATION IN LAYING FLOCKS

### 6.1 *SALMONELLA* RISK REDUCTION IN LAYING FLOCKS

Reducing the prevalence of *S.* Enteritidis infection in flocks of commercial laying hens is critical for reducing egg-transmitted human illness. The application of multiple interventions distributed throughout the egg production cycle is the most promising strategy for *S.* Enteritidis control (Trampel et al., 2014). Many recommended *S.* Enteritidis risk reduction practices are also applicable against other salmonellae as well as a broad spectrum of other poultry pathogens. Preventing the introduction of *Salmonella* into poultry facilities requires measures such as the implementation of stringent biosecurity measures, stocking exclusively with replacement pullets that are demonstrably uninfected, controlling population levels of rodent and insect pests, and securing farms against access by wildlife (Davies and Breslin, 2003). The possibility of indirect horizontal transmission of infections to subsequent flocks via contaminated environmental sources can be minimized by thorough cleaning and disinfection of laying houses between flocks (Carrique-Mas et al., 2009b). In addition to the various serotype-independent strategies, the three most prominent risk reduction practices that are specifically applicable to *S.* Enteritidis in laying hens are vaccination, flock testing, and egg refrigeration.

### 6.2 VACCINATION

The goals of *S.* Enteritidis vaccination of laying hens are to reduce the susceptibility of individual birds to infection, vertical and horizontal transmission between birds, the pathogen load in poultry house environments, and the frequency of egg contamination (Gast, 2007). Both inactivated (killed) and attenuated (live) *Salmonella* vaccine products are commercially available, eliciting varying degrees of mucosal and systemic immunity. Administering either type of *S.* Enteritidis vaccine preparation to pullets or hens has characteristically reduced, but seldom prevented entirely, fecal shedding, organ invasion, and egg contamination after experimental challenge (Gast et al., 1992, 1993). Moreover, protection by *Salmonella* vaccines can be overcome by high challenge doses (Barrow, 2007) and may be compromised by stressors such as feed deprivation, water deprivation, or environmental heat (Nakamura et al., 1994). Nevertheless, the inclusion of vaccination protocols in risk reduction programs for egg-laying hens has been associated with significantly lower frequencies of egg contamination and human *S.* Enteritidis infections (Cogan and Humphrey, 2003; Toyota-Hanatani et al., 2009).

### 6.3 GENERAL TESTING CONSIDERATIONS

Testing has played a crucial but often controversial role in programs for controlling *S.* Enteritidis in poultry and eggs. Preliminary efforts (in the late 1980s and early 1990s) to identify and eradicate infected laying flocks responsible for human disease outbreaks were not highly effective because they could not keep pace with a rapidly spreading problem associated with continuous diverse environmental sources of infection. Moreover, testing to decide the fate of flocks or eggs involves considerable uncertainty because of both assay sensitivity limitations and fluctuations over time in the prevalence of *S.* Enteritidis in the environment, hens, and eggs. Nevertheless, a testing component remains essential for effective *S.* Enteritidis control, both to identify flocks that pose a threat to public health and to verify the cost-effectiveness of resource investments in risk reduction (Gast, 2007).

## 6.4 ENVIRONMENTAL TESTING

The detection of *S.* Enteritidis in laying house environments is epidemiologically correlated with the production of contaminated eggs. Because of this relationship, and the infrequency at which contaminated eggs are typically produced by infected flocks, testing environmental samples for *S.* Enteritidis is often the primary screening method used to identify potentially infected flocks for further scrutiny (Gast, 2007; Trampel et al., 2014). Fecal shedding by infected hens is a principal contributor to *Salmonella* contamination levels in poultry houses, although the magnitude of fecal shedding alone does not always predict environmental sampling results (Wales et al., 2006). Fecal shedding appears to peak just before the commencement of egg laying in commercial flocks and then declines steadily thereafter (Li et al., 2007; Gole et al., 2014). Cloacal swabs from individual birds have the unique potential to indicate the precise prevalence of gastrointestinal infection within flocks, but they are relatively insensitive in comparison with other options for detecting salmonellae (García et al., 2011; Schulz et al., 2011). Among environmental testing options, dust samples have been shown to yield both more frequent isolation of *S.* Enteritidis and a longer duration of positive results than fecal samples (Gole et al., 2014; Arnold et al., 2014b). However, the sensitivity limit for *Salmonella* detection in chicken feces is lower than for poultry house dust (Martelli et al., 2014). Testing both dust and feces appears to be particularly effective for detecting environmental contamination (Carrique-Mas and Davies, 2008; Arnold et al., 2010). Common methods for collecting environmental samples in poultry facilities include dragging gauze swab assemblies across floor surfaces, walking through houses while wearing absorbent fabric shoe covers, and collecting litter material or dust from locations such as egg belts, fan blades, or nest boxes (Lungu et al., 2010; Davies and Breslin, 2001). *Salmonella* isolation and identification from environmental samples is usually accomplished by traditional selective enrichment culturing methods, followed by biochemical and serological confirmation, although rapid assays (based on the recognition of specific genetic sequences or antigenic molecules) are becoming increasingly popular (Waltman and Gast, 2008).

## 6.5 TESTING FOR SPECIFIC ANTIBODIES

Another option for detecting *S.* Enteritidis in laying flocks is based on the production of specific antibodies by infected chickens. Antibodies are present in the blood and deposited in egg yolks, sometimes at high titers, for extended intervals following the exposure of hens to *S.* Enteritidis (Gast et al., 2002b; Gast, 2007). Although the high sensitivity of serological methods makes them attractive as screening methods for flock infection, they have not been incorporated into most contemporary *S.* Enteritidis testing programs in the United States due to concerns about antigenic cross-reactivity with other serotypes and antibody persistence long after the clearance of active infection. Nevertheless, egg yolk antibody assays have detected infected flocks with a sensitivity that is similar to culturing environmental samples (Klinkenberg et al., 2011).

## 6.6 EGG TESTING

The presence of *S.* Enteritidis in the edible internal contents of eggs is the most unequivocally relevant measure of the public health risk posed by laying flocks, so egg culturing is a central component step in the testing protocols of most control programs. However, even in flocks known to be infected with *S.* Enteritidis, the infrequent, sporadic, and transient occurrence of egg contamination limits the diagnostic sensitivity of egg testing (Gantois et al., 2009). *Salmonella* contaminants are typically found inside eggs at both a low frequency and in very small population numbers, so the most common testing approach is to pool together the entire liquid contents (yolk plus albumen) of up to 20 eggs to generate a manageable number of samples. These egg contents pools are sometimes preincubated or supplemented with a source of iron before further culturing to support the expansion of small *S.* Enteritidis populations to more consistently detectable levels (Gast and Holt, 2003). Eggshells can also be examined for external surface contamination (García et al., 2011).

## 6.7 EGG REFRIGERATION

Risk assessment studies have identified egg refrigeration as one of the most consistently effective strategies for mitigating the egg-borne transmission of *S.* Enteritidis to consumers (Schroeder et al., 2006; Latimer et al., 2008), and most risk reduction plans accordingly incorporate egg refrigeration requirements or recommendations. The prompt refrigeration of shell eggs after laying can restrict multiplication by small initial contaminant levels to higher levels more likely to cause illness in humans (Gast and Holt, 2000b). Improper storage practices are not common in modern commercial egg production, but Canadian investigators suggested that the relatively rare (0.6%) occurrence of poor temperature management of eggs was a factor in nearly half of egg-associated illnesses (DeWinter et al., 2011). The efficacy of refrigeration for preventing *S.* Enteritidis growth in eggs depends on the initial level and location of contamination, the potential for migration of bacteria or nutrients within eggs during storage, and the rate at which growth-limiting temperatures are achieved (Gast and Guraya, 2013).

## 6.8 US FEDERAL EGG SAFETY PROGRAM

Both government and industry have invested substantial resources in controlling *S.* Enteritidis in egg-laying poultry in the United States. Sustained commitment to comprehensive risk reduction programs has already led to reported declines in both egg contamination and human illnesses in several other countries (Esaki et al., 2013; O'Brien, 2013). Regulatory authority over shell production in the United States was established by the rule for Prevention of *S.* Enteritidis in Shell Eggs during Production, Storage, and Transportation (U.S. Food and Drug Administration, 2009). This program mandates specific risk reduction practices plus a testing program in all flocks. As the first step toward regulatory compliance, all commercial egg producers must institute a written *S.* Enteritidis prevention plan. They must purchase all chicks from breeder flocks certified as uninfected under provisions of the National Poultry Improvement Plan (U.S. Department of Agriculture, 2014). Comprehensive facility biosecurity must be enforced, and stringent efforts made to control rodents and insect pests. All eggs must be stored and transported under refrigeration at 7.2°C (45°F) beginning 36h after laying (although eggs may be equilibrated to room temperature for an additional 36h before processing to prevent sweating). Testing for *S.* Enteritidis is carried out during both pullet rearing and egg laying. Environmental samples are collected and tested from pullet flocks at 14–16 weeks of age. Likewise, environmental samples are again collected from laying houses and tested at 40–45 weeks of age and also at 4–6 weeks after an induced molt. If positive results for *S.* Enteritidis are obtained from environmental testing of a laying flock, several lots of 1000 eggs each are also evaluated. The positive isolation of *S.* Enteritidis from any sampled eggs requires all eggs to be diverted for pasteurization until repeated negative testing results are achieved. Any affected poultry houses must be thoroughly cleaned and disinfected between flocks. A significant aspect of this US federal rule is that it does not require egg producers to vaccinate laying flocks against *S.* Enteritidis infection. This contrasts with the Lion Code of Practice for Shell Egg Producers in the United Kingdom (O'Brien, 2013), which incorporates risk reduction practices and testing protocols similar to those used in the United States, but also mandates *Salmonella* vaccination for all egg-producing flocks.

# 7. CONCLUSIONS

The international emergence of pandemic, egg-transmitted salmonellosis in the 1980s posed new challenges to public health authorities and the poultry industry. Caused by the propensity of some salmonellae (particularly strains of *S.* Enteritidis) to exhibit invasive behavior in infected laying hens, the deposition of this pathogen inside eggs is also directly related to its overall prevalence in poultry flocks and their environment. Because *Salmonella* can be highly persistent in both infected birds and diverse environmental reservoirs, effective disease control requires intervention at more than one step in the complex egg production cycle. Microbial quality assurance programs of demonstrated efficacy apply a broad range of risk reduction strategies and practices, including biosecurity, sanitation, pest control, and egg refrigeration. In many instances, vaccination of breeding and laying hens is also included to elicit protective immunity against infection. Bacteriological monitoring of poultry houses and eggs is used to identify problems requiring intervention and to evaluate the cost-effectiveness of investments in risk reduction. Although this approach does not constitute an impenetrable barrier against the production of contaminated eggs, it provides the best available protection for both the health of consumers and the economic viability of the egg industry.

## REFERENCES

Arnold, M.E., Carrique-Mas, J.J., Davies, R.H., 2010. Sensitivity of environmental sampling methods for detecting *Salmonella* Enteritidis in commercial laying flocks relative to the within-flock prevalence. Epidemiol. Infect. 138, 330–339.

Arnold, M.E., Martelli, F., McLaren, I., Davies, R.H., 2014a. Estimation of the rate of egg contamination from *Salmonella*-infected chickens. Zoonoses Public Health 61, 18–27.

Arnold, M.E., Martelli, F., McLaren, I., Davies, R.H., 2014b. Estimation of the sensitivity of environmental sampling for detection of *Salmonella* in commercial layer flocks post-introduction of national control programmes. Epidemiol. Infect. 142, 1061–1069.

Asakura, H., Tajima, O., Watarai, M., Shirahata, T., Kurazono, H., Makino, S., 2001. Effects of rearing conditions on the colonization of *Salmonella enteritidis* in the cecum of chicks. J. Vet. Med. Sci. 63, 1221–1224.

Baron, F., Nau, F., Guérin-Dubiard, C., Bonnassie, S., Gautier, M., Andrews, S.C., Jan, S., 2016. Egg white versus *Salmonella* Enteritidis! A harsh medium meets a resilient pathogen. Food Microbiol. 53, 82–93.

Barrow, P.A., 2007. Salmonella infections: immune and non-immune protection with vaccines. Avian Pathol. 36, 1–13.

Braden, C.R., 2006. *Salmonella enterica* serotype Enteritidis and eggs: a national epidemic in the United States. Clin. Infect. Dis. 43, 512–517.

Carrique-Mas, J.J., Breslin, M., Snow, L., McLaren, I., Sayers, A.R., Davies, R.H., 2009a. Persistence and clearance of different *Salmonella* serovars in buildings housing laying hens. Epidemiol. Infect. 137, 837–846.

Carrique-Mas, J.J., Davies, R.H., 2008. Sampling and bacteriological detection of *Salmonella* in poultry and poultry premises: a review. Rev. Sci. Tech. 27, 665–677.

Carrique-Mas, J.J., Marín, C., Breslin, M., McLaren, I., Davies, R.H., 2009b. A comparison of the efficacy of cleaning and disinfection methods in eliminating *Salmonella* spp. from commercial egg laying houses. Avian Pathol. 38, 419–424.

Centers for Disease Control and Prevention, 2011. Vital signs: incidence and trends of infection with pathogens transmitted commonly through food—foodborne diseases active surveillance network, 10 U.S. sites, 1996–2010. MMWR. Morb. Mortal. Wkly. Rep. 60, 749–755.

Chai, S.J., White, P.L., Lathrop, S.L., Solghan, S.M., Medus, C., McGlinchey, B.M., Tobin-D'Angelo, M., Marcus, R., Mahon, B.E., 2012. *Salmonella enterica* serotype Enteritidis: increasing incidence of domestically acquired infections. Clin. Infect. Dis. 54, S488–S497.

Chittick, P., Sulka, A., Tauxe, R.V., Fry, A.M., 2006. A summary of national reports of foodborne outbreaks of *Salmonella* Heidelberg infections in the United States: clues for disease prevention. J. Food Prot. 69, 1150–1153.

Cogan, T.A., Humphrey, T.J., 2003. The rise and fall of *Salmonella* Enteritidis in the UK. J. Appl. Microbiol. 94, 114S–119S.

Cogan, T.A., Jørgensen, F., Lappin-Scott, H.M., Benson, C.E., Woodward, M.J., Humphrey, T.J., 2004. Flagella and curli fimbriae are important for the growth of *Salmonella enterica* serovars in hen eggs. Microbiology 150, 1063–1071.

Coward, C., Sait, L., Cogan, T., Humphrey, T.J., Maskell, D.J., 2013. O-antigen repeat number in *Salmonella enterica* serovar Enteritidis is important for egg contamination, colonisation of the chicken reproductive tract and survival in egg albumen. FEMS Microbiol. Lett. 343, 169–176.

Davies, R., Breslin, M., 2001. Environmental contamination and detection of *Salmonella enterica* serovar enteritidis in laying flocks. Vet. Rec. 149, 699–704.

Davies, R.H., Breslin, M., 2003. Persistence of *Salmonella* Enteritidis phage type 4 in the environment and arthropod vectors on an empty free-range chicken farm. Environ. Microbiol. 5, 79–84.

De Buck, J., Pasmans, F., Van Immerseel, F., Haesebrouck, F., Ducatelle, R., 2004. Tubular glands of the isthmus are the predominant colonization site of *Salmonella* Enteritidis in the upper oviduct of laying hens. Poult. Sci. 83, 352–358.

De Knegt, L.V., Pires, S.M., Hald, Y., 2015. Attributing foodborne salmonellosis in humans to animal reservoirs in the European Union using a multi-country stochastic model. Epidemiol. Infect. 143, 1175–1186.

Denagamage, T., Jayarao, B., Patterson, P., Wallner-Pendleton, E., Kariyawasam, S., 2015. Risk factors associated with *Salmonella* in laying hen farms: systematic review of observational studies. Avian Dis. 59, 291–302.

De Vylder, J., Dewulf, J., Van Hoorebeke, S., Pasmans, F., Haesebrouck, F., Ducatelle, R., Van Immerseel, F., 2011. Horizontal transmission of *Salmonella* Enteritidis in groups of experimentally infected hens housed in different housing systems. Poult. Sci. 90, 1391–1396.

De Vylder, J., Raspoet, R., Dewulf, J., Haesebrouck, F., Ducatelle, R., Van Immerseel, F., 2012. *Salmonella* Enteritidis is superior in egg white survival compared with other *Salmonella* serotypes. Poult. Sci. 92, 842–845.

De Vylder, J., Van Hoorebeke, S., Ducatelle, R., Pasmans, F., Haesebrouck, F., Dewulf, J., Van Immerseel, F., 2009. Effect of the housing system on shedding and colonization of gut and internal organs of laying hens with *Salmonella* Enteritidis. Poult. Sci. 88, 2491–2495.

Dewaele, I., Rasschaert, G., Wildemauwe, C., Van Meirhaeghe, H., Vanrobaeys, M., De Graef, E., Herman, L., Ducatelle, R., Heyndrickx, M., De Reu, K., 2012a. Polyphasic characterization of *Salmonella* Enteritidis isolates on persistently contaminated layer farms during the implementation of a national control program with obligatory vaccination: a longitudinal study. Poult. Sci. 91, 2727–2735.

Dewaele, I., Van Meirhaeghe, H., Rasschaert, G., Vanrobaeys, M., De Graef, E., Herman, L., Ducatelle, R., Heyndrickx, M., De Reu, K., 2012b. Persistent *Salmonella* Enteritidis environmental contamination on layer farms in the context of an implemented national control program with obligatory vaccination. Poult. Sci. 91, 282–291.

DeWinter, L.M., Ross, W.H., Couture, H., Farber, J.F., 2011. Risk assessment of shell eggs internally contaminated with *Salmonella* Enteritidis. Int. Food Risk Anal. J. 1, 40–81.

Ebel, E., Schlosser, W., 2000. Estimating the annual fraction of eggs contaminated with *Salmonella enteritidis* in the United States. Int. J. Food Microbiol. 61, 51–62.

Esaki, H., Shimura, K., Yamazaki, Y., Eguchi, M., Nakamura, M., 2013. National surveillance of *Salmonella* Enteritidis in commercial eggs in Japan. Epidemiol. Infect. 141, 941–943.

Gantois, I., Ducatelle, R., Pasmans, F., Haesebrouck, F., Gast, R., Humphrey, T.J., Van Immerseel, F., 2009. Mechanisms of egg contamination by *Salmonella* Enteritidis. FEMS Microbiol. Rev. 33, 718–738.

Gantois, I., Eeckhaut, V., Pasmans, F., Haesebrouck, F., Ducatelle, R., Van Immerseel, F., 2008. A comparative study on the pathogenesis of egg contamination by different serotypes of *Salmonella*. Avian Pathol. 37, 399–406.

Garber, L., Smeltzer, M., Fedorka-Cray, P., Ladely, S., Ferris, K., 2003. *Salmonella enterica* serotype enteritidis in table egg layer house environments and in mice in U.S. layer houses and associated risk factors. Avian Dis. 47, 134–142.

García, J., Soriano, M., Benítez, V., Catalá-Gregori, P., 2011. Assessment of *Salmonella* spp. in feces, cloacal swabs, and eggs (eggshell and content separately) from a laying hen farm. Poult. Sci. 90, 1581–1585.

Gast, R.K., 2007. Serotype-specific and serotype-independent strategies for preharvest control of food-borne *Salmonella* in poultry. Avian Dis. 51, 817–828.

Gast, R.K., Beard, C.W., 1990. Isolation of *Salmonella enteritidis* from internal organs of experimentally infected hens. Avian Dis. 34, 991–993.

Gast, R.K., Beard, C.W., 1992. Detection and enumeration of *Salmonella enteritidis* in fresh and stored eggs laid by experimentally infected hens. J. Food Prot. 55, 152–156.

Gast, R.K., Guard-Bouldin, J., Guraya, R., Holt, P.S., 2009. Effect of prior passage through laying hens on invasion of reproductive organs by *Salmonella enteritidis*. Int. J. Poult. Sci. 8, 116–212.

Gast, R.K., Guard-Bouldin, J., Holt, P.S., 2004. Colonization of reproductive organs and internal contamination of eggs after experimental infection of laying hens with *Salmonella heidelberg* and *Salmonella enteritidis*. Avian Dis. 48, 863–869.

Gast, R.K., Guard-Bouldin, J., Holt, P.S., 2005. The relationship between the duration of fecal shedding and the production of contaminated eggs by laying hens infected with strains of *Salmonella* Enteritidis and *Salmonella* Heidelberg. Avian Dis. 49, 382–386.

Gast, R.K., Guard-Petter, J., Holt, P.S., 2002a. Characteristics of *Salmonella enteritidis* contamination in eggs after oral, aerosol, and intravenous inoculation of laying hens. Avian Dis. 46, 629–635.

Gast, R.K., Guard-Petter, J., Holt, P.S., 2003. Effects of prior serial in vivo passage on the frequency of *Salmonella enteritidis* contamination in eggs from experimentally infected laying hens. Avian Dis. 47, 633–639.

Gast, R.K., Guraya, R., 2013. Continuing multiplication of *Salmonella enteritidis* in egg yolk during refrigeration at 7.2°C. Int. J. Poult. Sci. 12, 622–627.

Gast, R.K., Guraya, R., Guard, J., 2013a. *Salmonella* Enteritidis deposition in eggs after experimental infection of laying hens with different oral doses. J. Food Prot. 76, 108–113.

Gast, R.K., Guraya, R., Guard, J., Holt, P.S., 2011a. Frequency and magnitude of internal organ colonization following exposure of laying hens to different oral doses of *Salmonella enteritidis*. Int. J. Poult. Sci. 10, 325–331.

Gast, R.K., Guraya, R., Guard, J., Holt, P.S., 2011b. The relationship between the numbers of *Salmonella* Enteritidis, *Salmonella* Heidelberg, or *Salmonella* Hadar colonizing reproductive tissues of experimentally infected laying hens and deposition inside eggs. Avian Dis. 55, 243–247.

Gast, R.K., Guraya, R., Guard-Bouldin, J., Holt, P.S., 2007a. In vitro penetration of egg yolks by *Salmonella* Enteritidis and *Salmonella* Heidelberg strains during thirty-six-hour ambient temperature storage. Poult. Sci. 86, 1431–1435.

Gast, R.K., Guraya, R., Holt, P.S., 2011c. Frequency and persistence of fecal shedding following exposure of laying hens to different oral doses of *Salmonella enteritidis*. Int. J. Poult. Sci. 10, 750–756.

Gast, R.K., Guraya, R., Guard-Bouldin, J., Holt, P.S., Moore, R.W., 2007b. Colonization of specific regions of the reproductive tract and deposition at different locations inside eggs laid by hens infected with *Salmonella enteritidis* or *Salmonella heidelberg*. Avian Dis. 51, 40–44.

Gast, R.K., Guraya, R., Jones, D.R., Anderson, K.E., 2013b. Colonization of internal organs by *Salmonella* Enteritidis in experimentally infected laying hens housed in conventional or enriched cages. Poult. Sci. 92, 468–473.

Gast, R.K., Guraya, R., Jones, D.R., Anderson, K.E., 2014a. Contamination of eggs by *Salmonella* Enteritidis in experimentally infected laying hens housed in conventional or enriched cages. Poult. Sci. 93, 728–733.

Gast, R.K., Guraya, R., Jones, D.R., Anderson, K.E., 2014b. Horizontal transmission of *Salmonella* Enteritidis in experimentally infected laying hens housed in conventional or enriched cages. Poult. Sci. 93, 3145–3151.

Gast, R.K., Guraya, R., Jones, D.R., Anderson, K.E., 2015. Persistence of fecal shedding of *Salmonella* Enteritidis by experimentally infected laying hens housed in conventional or enriched cages. Poult. Sci. 94, 1650–1656.

Gast, R.K., Holt, P.S., 1998. Persistence of *Salmonella enteritidis* from one day of age until maturity in experimentally infected layer chickens. Poult. Sci. 77, 1759–1762.

Gast, R.K., Holt, P.S., 1999. Experimental horizontal transmission of *Salmonella enteritidis* strains (phage types 4, 8, and 13a) in chicks. Avian Dis. 43, 774–778.

Gast, R.K., Holt, P.S., 2000a. Deposition of phage type 4 and 13a *Salmonella enteritidis* strains in the yolk and albumen of eggs laid by experimentally infected hens. Avian Dis. 44, 706–710.

Gast, R.K., Holt, P.S., 2000b. Influence of the level and location of contamination on the multiplication of *Salmonella enteritidis* at different storage temperatures in experimentally inoculated eggs. Poult. Sci. 79, 559–563.

Gast, R.K., Holt, P.S., 2001a. Assessing the frequency and consequences of *Salmonella enteritidis* deposition on the egg yolk membrane. Poult. Sci. 80, 997–1002.

Gast, R.K., Holt, P.S., 2001b. Multiplication in egg yolk and survival in egg albumen of *Salmonella enterica* serotype Enteritidis strains of phage types 4, 8, 13a, and 14b. J. Food Prot. 64, 865–868.

Gast, R.K., Holt, P.S., 2003. Incubation of supplemented egg contents pools to support rapid detection of *Salmonella enterica* serovar Enteritidis. J. Food Prot. 66, 656–659.

Gast, R.K., Holt, P.S., Guraya, R., 2006. Effect of refrigeration on in vitro penetration of *Salmonella* Enteritidis through the egg yolk membrane. J. Food Prot. 69, 1426–1429.

Gast, R.K., Jones, D.R., Anderson, K.E., Guraya, R., Guard, J., Holt, P.S., 2010. In vitro penetration of *Salmonella* Enteritidis through yolk membranes of eggs from 6 genetically distinct commercial lines of laying hens. Poult. Sci. 89, 1732–1736.

Gast, R.K., Mitchell, B.W., Holt, P.S., 1998. Airborne transmission of *Salmonella enteritidis* infection between groups of chicks in controlled-environment isolation cabinets. Avian Dis. 4, 315–320.

Gast, R.K., Nasir, M.S., Jolley, M.E., Holt, P.S., Stone, H.D., 2002b. Serologic detection of experimental *Salmonella enteritidis* infections in laying hens by fluorescence polarization and enzyme immunoassay. Avian Dis. 46, 137–142.

Gast, R.K., Stone, H.D., Holt, P.S., Beard, C.W., 1992. Evaluation of the efficacy of an oil-emulsion bacterin for protecting chickens against *Salmonella enteritidis*. Avian Dis. 36, 992–999.

Gast, R.K., Stone, H.D., Holt, P.S., 1993. Evaluation of the efficacy of oil-emulsion bacterins for reducing fecal shedding of Salmonella enteritidis by laying hens. Avian Dis. 37, 1085–1091.

Gole, V.C., Caraguel, C.G.B., Sexton, M., Fowler, C., Chousalkar, K.K., 2014. Shedding of *Salmonella* in single age caged commercial layer flock at an early stage of lay. Int. J. Food Microbiol. 189, 61–66.

Gomes, A.V.S., Quinteiro-Filho, W.M., Ribeiro, A., Ferraz-de-Paula, V., Pinheiro, M.L., Baskeville, E., Akamine, A.T., Astolfi-Ferreira, C.S., Perreira, A.J.P., Palermo-Neto, J., 2014. Overcrowding stress decreases macrophage activity and increases *Salmonella* Enteritidis invasion in broiler chickens. Avian Pathol. 43, 82–90.

Guard, J., Gast, R.K., Guraya, R., 2010. Colonization of avian reproductive-tract tissues by variant subpopulations of *Salmonella* Enteritidis. Avian Dis. 54, 857–861.

Guard, J., Morales, C.A., Fedorka-Cray, P., Gast, R.K., 2011. Single nucleotide polymorphisms that differentiate two subpopulations of *Salmonella enteritidis* within phage type. BMC Res. Notes 369.

Guard-Bouldin, J., Gast, R.K., Humphrey, T.J., Henzler, D.J., Morales, C., Coles, K., 2004. Subpopulation characteristics of egg-contaminating *Salmonella enterica* serovar Enteritidis as defined by the lipopolysaccharide O chain. Appl. Environ. Microbiol. 70, 2756–2763.

Gurtler, J.B., Conner, D.E., 2009. Survival and growth of *Salmonella* Enteritidis in liquid egg products varying by temperature, product composition, and carbon dioxide concentration. Foodborne Pathog. Dis. 6, 561–567.

Hannah, J.F., Wilson, J.L., Cox, N.A., Richardson, L.J., Cason, J.A., Bourassa, D.V., Buhr, R.J., 2011. Horizontal transmission of *Salmonella* and *Campylobacter* among caged and cage-free laying hens. Avian Dis. 55, 580–587.

Havelaar, A.H., Ivarsson, S., Löfdahl, M., Nauta, M.J., 2013. Estimating the true incidence of campylobacteriosis and salmonellosis in the European Union, 2009. Epidemiol. Infect. 141, 293–302.

He, G.Z., Tian, W.Y., Qian, N., Cheng, A.C., Deng, S.X., 2010. Quantitative studies of the distribution pattern for *Salmonella* Enteritidis in the internal organs of chicken after oral challenge by a real-time PCR. Vet. Res. Commun. 34, 669–676.

Henzler, D.J., Ebel, E., Sanders, J., Kradel, D., Mason, J., 1994. *Salmonella enteritidis* in eggs from commercial chicken layer flocks implicated in human outbreaks. Avian Dis. 38, 37–43.

Henzler, D.J., Kradel, D.C., Sischo, W.M., 1998. Management and environmental risk factors for *Salmonella enteritidis* contamination of eggs. Am. J. Vet. Res. 59, 824–829.

Holt, P.S., Davies, R.H., Dewulf, J., Gast, R.K., Huwe, J.K., Jones, D.R., Waltman, D., Willian, K.R., 2011. The impact of different housing systems on egg safety and quality. Poult. Sci. 90, 251–262.

Holt, P.S., Vaughn, L.E., Gast, R.K., 2010. Flow cytometric characterization of Peyer's patch and cecal tonsil T lymphocytes in laying hens following challenge with *Salmonella enterica* serovar Enteritidis. Vet. Immunol. Immunopathol. 133, 276–281.

Humphrey, T.J., 2006. Are happy chickens safer chickens? Poultry welfare and disease susceptibility. Brit. Poult. Sci. 47, 379–391.

Humphrey, T.J., Baskerville, A., Mawer, S., Rowe, B., Hopper, S., 1989. *Salmonella enteritidis* phage type 4 from the contents of intact eggs: a study involving naturally infected hens. Epidemiol. Infect. 103, 415–423.

Humphrey, T.J., Whitehead, A., Gawler, A.H.L., Henley, A., Rowe, B., 1991. Numbers of *Salmonella enteritidis* in the contents of naturally contaminated hens' eggs. Epidemiol. Infect. 106, 489–496.

Huneau-Salaün, A., Marianne, C., Le Bouquin, S., Lalande, F., Petetin, I., Rouxel, S., Virginie, M., Fravalo, P., Rose, N., 2009. Risk factors for *Salmonella enterica* subsp. *enterica* contamination in 519 French laying hen flocks at the end of the laying period. Prev. Vet. Med. 89, 51–58.

Im, M.C., Jeong, S.J., Kwon, Y.-K., Jeong, O.-M., Kang, M.-S., Lee, Y.J., 2015. Prevalence and characteristics of *Salmonella* spp. isolated from commercial layer farms in Korea. Poult. Sci. 94, 1691–1698.

Jackson, B.R., Griffin, P.M., Cole, D., Walsh, K.A., Chai, S.J., 2013. Outbreak-associated *Salmonella enterica* serotypes and food commodities, United States, 1998–2008. Emerg. Infect. Dis. 19, 1239–1244.

Jones, D.R., Anderson, K.E., 2013. Housing system and laying hen strain impacts on egg microbiology. Poult. Sci. 92, 2221–2225.

Jones, D.R., Anderson, K.E., Guard, J.Y., 2012. Prevalence of coliforms, *Salmonella*, *Listeria*, and *Campylobacter* associated with eggs and the environment of conventional cage and free-range egg production. Poult. Sci. 91, 1195–1202.

Jones, D.R., Cox, N.A., Guard, J., Fedorka-Cray, P.J., Buhr, R.J., Gast, R.K., Abdo, Z., Rigsby, L.L., Plumblee, J.R., Karcher, D.M., Robison, C.I., Blatchford, R.A., Makagon, M.M., 2015. Microbiological impact of three commercial laying hen housing systems. Poult. Sci. 94, 544–551.

Keller, L.H., Schifferli, D.M., Benson, C.E., Aslam, S., Eckroade, R.J., 1997. Invasion of chicken reproductive tissues and forming eggs is not unique to *Salmonella enteritidis*. Avian Dis. 41, 535–539.

Kinde, H., Castellan, D.M., Kerr, D., Campbell, J., Breitmeyer, R., Ardans, A., 2005. Longitudinal monitoring of two commercial layer flocks and their environments for *Salmonella enterica* serovar Enteritidis and other Salmonellae. Avian Dis. 49, 189–194.

Klinkenberg, D., Thomas, E., Artavia, F.F.C., Bouma, A., 2011. *Salmonella* Enteritidis surveillance by egg immunology: impact of the sampling scheme on the release of contaminated table eggs. Risk Anal. 31, 1260–1270.

Lapuz, R.R., Umali, D.V., Suzuki, T., Shirota, K., Katoh, H., 2012. Comparison of the prevalence of *Salmonella* infection in layer hens from commercial layer farms with high and low rodent densities. Avian Dis. 56, 29–34.

Latimer, H.K., Marks, H.M., Coleman, M.E., Schlosser, W.D., Golden, N.J., Ebel, E.D., Kause, J., Schroeder, M., 2008. Evaluating the effectiveness of pasteurization for reducing human illnesses from *Salmonella* spp. in egg products: results of a quantitative risk assessment. Foodborne Pathog. Dis. 5, 59–68.

Li, X., Payne, J.B., Santos, F.B., Levine, J.F., Anderson, K.E., Sheldon, B.W., 2007. *Salmonella* populations and prevalence in layer feces from commercial high-rise houses and characterization of the *Salmonella* isolates by serotyping, antibiotic resistance analysis, and pulsed field gel electrophoresis. Poult. Sci. 86, 591–597.

Lungu, B., Waltman, W.D., Berghaus, R.D., Hofacre, C.L., 2010. Comparison of a real-time PCR method with a culture method for the detection of *Salmonella enterica* serotype Enteritidis in naturally contaminated environmental samples from integrated poultry houses. J. Food Prot. 75, 743–747.

Martelli, F., Gosling, R., McLaren, I., Wales, A., Davies, R., 2014. Development and testing of external quality assessment samples for *Salmonella* detection in poultry samples. Lett. Appl. Microbiol. 59, 443–448.

McDermid, A.S., McKee, A.S., Dowsett, A.B., Marsh, P.D., 1996. The effect of environmental pH on the physiology and surface structures of *Salmonella* serotype Enteritidis phage type 4. J. Med. Microbiol. 45, 452–458.

Mollenhorst, H., van Woudenbergh, C.J., Bokkers, E.G.M., de Boer, I.J.M., 2005. Risk factors for *Salmonella enteritidis* infections in laying hens. Poult. Sci. 84, 1308–1313.

Nakamura, M., Nagamine, N., Takahashi, T., Suzuki, S., Sato, S., 1994. Evaluation of the efficacy of a bacterin against *Salmonella enteritidis* infection and the effect of stress after vaccination. Avian Dis. 38, 717–724.

Namata, H., Méroc, E., Aerts, M., Faes, C., Cortiñas, J., Imberechts, H., Mintiens, K., 2008. *Salmonella* in Belgian laying hens: an identification of risk factors. Prevent. Vet. Med. 83, 323–336.

Nordentoft, S., Mølbak, L., Bjerrum, L., De Vylder, J., Van Immerseel, F., Pedersen, K., 2011. The influence of the cage system and colonisation of *Salmonella* Enteritidis on the microbial gut flora of laying hens studied by T-RFLP and 454 pyrosequencing. BMC Microbiol. 11, 187.

O'Brien, S.J., 2013. The "decline and fall" of nontyphoidal *Salmonella* in the United Kingdom. Clin. Infect. Dis. 56, 705–710.

Okamura, M., Kamijima, Y., Miyamoto, T., Tani, H., Sasai, K., Baba, E., 2001. Differences among six *Salmonella* serovars in abilities to colonize reproductive organs and to contaminate eggs in laying hens. Avian Dis. 4, 61–69.

Okamura, M., Sonobe, M., Obara, S., Kubo, T., Nagai, T., Noguchi, M., Takehara, K., Nakamura, M., 2010. Potential egg contamination by *Salmonella enterica* serovar Typhimurium definitive type 104 following experimental infection of pullets at the onset of lay. Poult. Sci. 89, 1629–1634.

Parisi, M.A., Northcutt, J.K., Smith, D.P., Steinberg, E.L., Dawson, P.L., 2015. Microbiological contamination of shell eggs produced in conventional and free-range housing systems. Food Control 47, 161–165.

Pires, S.M., Vieira, A.R., Hald, T., Col, D., 2014. Source attribution of human salmonellosis: an overview of methods and estimates. Foodborne Pathog. Dis. 11, 667–676.

Pitesky, M., Charlton, B., Bland, M., Rolfe, D., 2013. Surveillance of *Salmonella* Enteritidis in layer houses: a retrospective comparison of the Food and Drug Administration's egg safety rule (2010–2011) and the California Egg Quality Assurance Program (2007–2011). Avian Dis. 57, 51–56.

Raspoet, R., Appia-Ayme, C., Shearer, N., Martel, A., Pasmans, F., Haesebrouck, F., Ducatelle, R., Thompson, A., Van Immerseel, F., 2014. Microarray-based detection of *Salmonella enterica* serovar Enteritidis genes involved in chicken reproductive tract colonization. Appl. Environ. Microbiol. 80, 7710–7716.

Scharff, R.L., 2012. Economic burden from health losses due to foodborne illness in the United States. J. Food Prot. 75, 123–131.

Schroeder, C.M., Latimer, H.K., Schlosser, W.D., Golden, N.J., Marks, H.M., Coleman, M.E., Hogue, A.T., Ebel, E.D., Quiring, N.M., Kadry, A.-R.M., Kause, J., 2006. Overview and summary of the Food Safety and Inspection Service risk assessment for *Salmonella* Enteritidis in shell eggs, October 2005. Foodborne Pathog. Dis. 3, 403–412.

Schulz, J., Van Hoorebeke, S., Hald, B., Hartung, J., Van Immerseel, F., Radtke, I., Kabell, S., Dewulf, J., 2011. The dynamics of *Salmonella* occurrence in commercial laying hen flocks throughout a laying period. Avian Pathol. 40, 243–248.

Shah, D.H., 2014. RNA sequencing reveals differences between the global transcriptomes of *Salmonella enterica* serovar Enteritidis strains with high and low pathogenicities. Appl. Environ. Microbiol. 80, 896–906.

Shah, D.H., Casavant, C., Hawley, Q., Addwebi, T., Call, D.R., Guard, J., 2012. *Salmonella* Enteritidis strains from poultry exhibit differential responses to acid stress, oxidative stress, and survival in the egg albumen. Foodborne Pathog. Dis. 9, 258–265.

Siemon, C.E., Bahnson, P.B., Gebreyes, W.A., 2007. Comparative investigation of prevalence and antimicrobial resistance of *Salmonella* between pasture and conventionally reared poultry. Avian Dis. 51, 112–117.

Snow, L.C., Davies, R.H., Christiansen, K.H., Carrique-Mas, J.J., Cook, A.J.C., Evans, S.J., 2010. Investigation of risk factors for *Salmonella* on commercial egg-laying farms in Great Britain, 2004–2005. Vet. Rec. 166, 579–586.

Thomas, E., Bouma, A., Klinkenberg, D., 2011. A comparison of transmission characteristics of *Salmonella enterica* serovar Enteritidis between pair-housed and group-housed laying hens. Vet. Res. 42, 40.

Thomas, M.E., Klinkenberg, D., Ejeta, G., Van Knapen, F., Bergwerff, A.A., Stegeman, J.A., Bouma, A., 2009. Quantification of horizontal transmission of *Salmonella enterica* serovar Enteritidis bacteria in pair-housed groups of laying hens. Appl. Environ. Microbiol. 75, 6361–6366.

Toyota-Hanatani, Y., Ekawa, T., Ohta, H., Igimi, S., Hara-Kudo, Y., Sasai, K., Baba, E., 2009. Public health assessment of *Salmonella enterica* serovar Enteritidis inactivated-vaccine treatment in layer flocks. Appl. Environ. Microbiol. 75, 1005–1010.

Trampel, D.W., Holder, T.G., Gast, R.K., 2014. Integrated farm management to prevent *Salmonella* Enteritidis contamination of eggs. J. Appl. Poult. Res. 23, 353–365.

U.S. Department of Agriculture, 2014. National Poultry Improvement Plan Program Standards. https://www.poultryimprovement.org/documents/ProgramStandardsAugust2014.pdf.

U.S. Food and Drug Administration, 2009. Prevention of *Salmonella* Enteritidis in shell eggs during production, storage, and transportation; final rule. Fed. Regist. 74, 33039–33101.

Van Hoorebeke, S., Van Immerseel, F., Haesebrouck, F., Ducatelle, R., Dewulf, J., 2011. The influence of the housing system on *Salmonella* infections in laying hens. Zoonoses Public Health 58, 304–311.

Van Hoorebeke, S., Van Immerseel, F., De Vylder, J., Ducatelle, R., Haesebrouck, F., Pasmans, F., de Kruif, A., Dewulf, J., 2010a. The age of production system and previous *Salmonella* infections in laying hen flocks. Poult. Sci. 89, 1315–1319.

Van Hoorebeke, S., Van Immerseel, F., Schulz, J., Hartung, J., Harisberger, M., Barco, L., Ricci, A., Theodoropoulos, G., Xylouri, E., De Vylder, J., Ducatelle, R., Haesebrouck, F., Pasmans, F., de Kruif, A., Dewulf, J., 2010b. Determination of the within and between flock prevalence and identification of risk factors for *Salmonella* infections in laying hen flocks housed in conventional and alternative systems. Prev. Vet. Med. 94, 94–100.

Van Immerseel, F., De Buck, J., Pasmans, I.F., Bohez, L., Boyen, F., Haesbrouck, F., Ducatelle, R., 2004. Intermittent long-term shedding and induction of carrier birds after infection of chickens early posthatch with a low or high dose of *Salmonella* Enteritidis. Poult. Sci. 83, 1911–1916.

Wales, A., Breslin, M., Carter, B., Sayers, R., Davies, R., 2007. A longitudinal study of environmental *Salmonella* contamination in caged and free-range layer flocks. Avian Pathol. 36, 187–197.

Wales, A., Breslin, M., Davies, R., 2006. Semiquantitative assessment of the distribution of *Salmonella* in the environment of caged layer flocks. J. Appl. Microbiol. 101, 309–318.

Wales, A.D., Davies, R.H., 2011. A critical review of *Salmonella* Typhimurium infection in laying hens. Avian Pathol. 40, 429–436.

Walker, S.L., Sojka, M., Dibb-Fuller, M., Woodward, M.J., 1999. Effect of pH, temperature and surface contact on the elaboration of fimbriae and flagella by *Salmonella* serotype Enteritidis. J. Med. Microbiol. 48, 253–261.

Wallner-Pendleton, E.A., Patterson, P.H., Kariyawasam, S., Trampel, D.W., Denagamage, T., 2014. On-farm risk factors for *Salmonella* Enteritidis contamination. J. Appl. Poult. Res. 23, 345–352.

Waltman, W.D., Gast, R.K., 2008. Salmonellosis. In: Dufour-Zavala, L., Swayne, D.E., Glisson, J.R., Pearson, J.E., Reed, W.M., Jackwood, M.W., Woolcock, P.R. (Eds.), A Laboratory Manual for the Isolation and Identification of Avian Pathogens, fifth ed. American Association of Avian Pathologists, Athens, Georgia, pp. 3–9.

Watanabe, O., Ogino, M., Iwamoto, A., Akiyama, T., Miki, T., 2012. Henhouse feeding style and *Salmonella* Enteritidis contamination in unvaccinated flocks of egg farms, April 1994–March 2001. J. Vet. Med. Sci. 74, 575–582.

Whiley, H., Ross, K., 2015. *Salmonella* and eggs: from production to plate. Int. J. Environ. Res. Public Health 12, 2543–2556.

CHAPTER

# Egg Production Systems and *Salmonella* in Korea

# 3

**Soo-Kyoung Lee[1,2], Kun-Ho Seo[1]**

*[1]Konkuk University, Seoul, South Korea; [2]Animal and Plant Quarantine Agency, Gimcheon, South Korea*

## 1. EGG PRODUCTION INDUSTRY: OVERVIEW

The poultry egg industry in South Korea has developed rapidly over the past few decades. Until 1960, laying chickens were reared in open barns located in the corners of farmyards. Production in closed barns began in the early 1960s, when poultry farming began to become a full-time operation. The battery cage system was introduced into Korea in the mid-1960s, and barns without open side curtain walls were first built in 1964 by several poultry farmers. Since then, commercial egg production systems have moved toward bigger farms and a large part of the egg production costs can be reduced through the automation of facilities. Since 1980, there has been a change from full-time poultry farms to poultry factories. With the modernization of the egg production system, specialization in hatching, rearing, egg grading, and packing has taken place and vertically and horizontally integrated systems have been established.

## 2. SIZE AND DISTRIBUTION OF EGG-LAYING FLOCKS

The Korean layer industry grew significantly over the last two decades, producing over 70 million laying hens by 2015, in response to an increase in the per capita consumption of eggs (Fig. 3.1). Although the total number of laying birds can vary significantly from year to year, the general trend over the last two decades has been toward an increase in laying bird numbers. The total number of laying hens in Korea increased from 45 million in 1995 to 70 million in 2015. There has also been significant growth in the size of poultry farms.

As shown in Fig. 3.2, the proportion of poultry farms raising more than 50,000 layers has increased from 10% to 34% between 2005 and 2015. Meanwhile, the number of holdings with egg layer flocks under 10,000 birds has decreased from 968 (43% of all holdings) in 2005 to 178 (15% of all holdings) in 2015. Although the total number of laying birds increased, the number of layer holdings has declined from 2005 to 2015 due to the increase in average farm size (Fig. 3.3).

**Producing Safe Eggs. http://dx.doi.org/10.1016/B978-0-12-802582-6.00003-3**

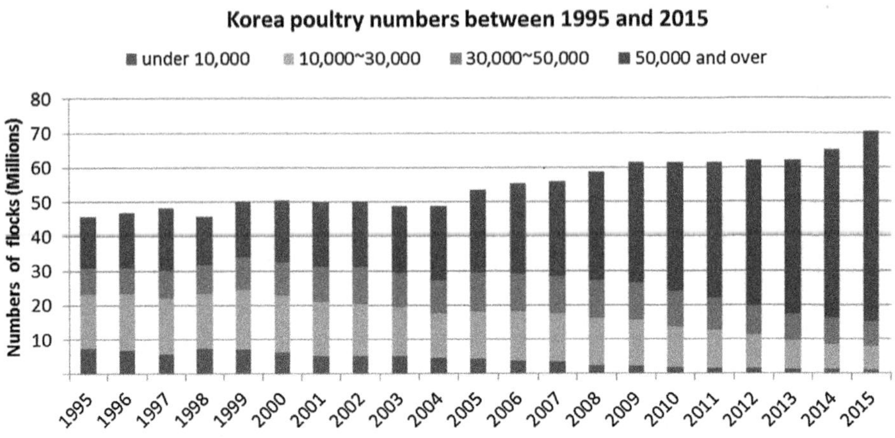


**FIGURE 3.1**

Numbers of Laying Flock by Group Size Between 1995 and 2015 in Korea.

*Data from KOSIS (KOSTAT's Korean Statistical Information Service), 2015.*

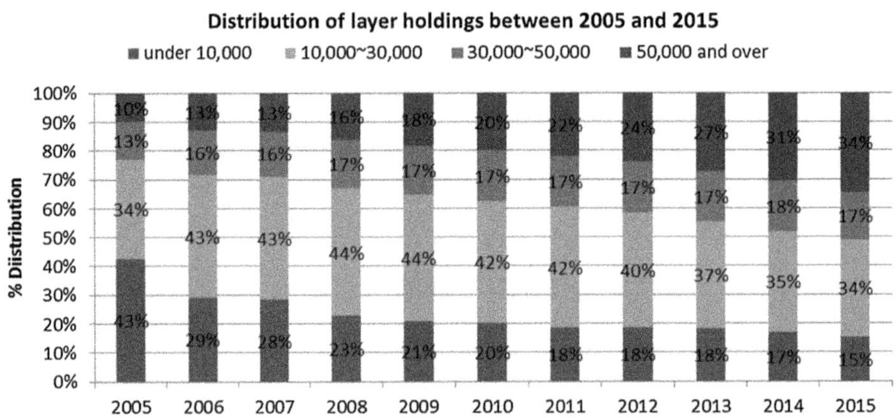


**FIGURE 3.2**

Distribution of Egg-Laying Hen Holdings (%) by Flock Size Between 2005 and 2015.

*Data from KOSIS (KOSTAT's Korean Statistical Information Service), 2015.*

According to the Korean Statistical Information Service (KOSIS), the production of laying fowls was distributed among 1155 laying fowl holdings in 2015. Table 3.1 shows how these units were distributed over flock size groups, where flock size is defined by the number of laying birds. In terms of production volumes, the laying fowl sector is heavily dominated by the largest producers, with 78.4% of all production concentrated on the largest 399 holdings, representing 34.6% of all holdings.

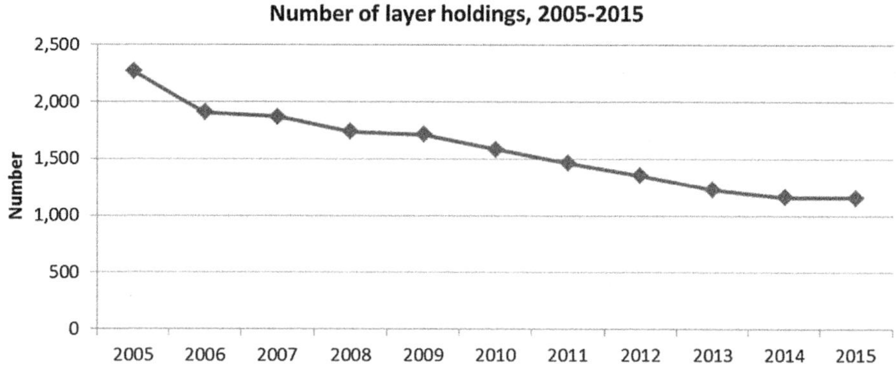


**FIGURE 3.3**

Numbers of Holdings with Egg Layer Flocks Between 2005 and 2015 in Korea.

*Data from KOSIS (KOSTAT's Korean Statistical Information Service), 2015.*

**Table 3.1** Distribution of Layer Holdings and Production Volumes by Flock Size, 2015

| Flock Size | Number of Holdings (%) | Total Number of Laying Birds (%) |
|---|---|---|
| Under 10,000 layers | 178 (15.4) | 1,087,561 (1.5) |
| 10,000–29,999 | 382 (33.0) | 6,846,860 (9.8) |
| 30,000–49,999 | 196 (17.0) | 7,259,086 (10.3) |
| 50,000 and over | 399 (34.6) | 54,994,185 (78.4) |
| Total | 1155 (100) | 70,187,692 (100) |

*Data from KOSIS (KOSTAT's Korean Statistical Information Service), 2015.*

The demand for poultry eggs increased over the last few decades; the consumption of eggs per capita was at 4.2 kg in 1970 and increased to 12.2 kg in 2015. On a per capita basis, the consumption of eggs accounted for 30.7% of all livestock product consumption in 2010. Around one-third of all eggs produced annually are sold via retail channels as shell eggs for home use, with the remainder being used for the manufacturing or processing of egg products. These egg products are used in the commercial food industry as ingredients in a myriad of products. Egg prices have shown an upward trend since 1970, and reached a peak in 2014. Egg production has also been increasing steadily and reached 638,000 tons in 2014 (Table 3.2).

**Table 3.2** Production, Consumption, and Value of Eggs Produced in Korea, 1970–2014

| Flock Size | 1970 | 1975 | 1980 | 1985 | 1990 | 1995 | 2000 | 2005 | 2010 | 2014 |
|---|---|---|---|---|---|---|---|---|---|---|
| Production (thousands of tons) | 123 | 148 | 236 | 286 | 386 | 454 | 486 | 591 | 578 | 638 |
| Consumption per capita (ea) | 77 | 83 | 119 | 131 | 167 | 184 | 184 | 220 | 236 | 254 |
| Consumption per capita (kg) | 4.2 | 4.6 | 6.5 | 7.2 | 9.2 | 10.1 | 10.1 | 12.1 | 11.9 | 12.2 |
| Price (won/10 ea) | 117 | 218 | – | 475 | 576 | 681 | 698 | 1054 | 1134 | 1374 |

*Data from the MAFRA's Agriculture, Food and Rural Affairs Statistics Yearbook, 1970–2014.*

## 3. EGG PRODUCTION SYSTEMS: HACCP AND EGG GRADING SYSTEM IN KOREA

In comparison with Western countries, more intensive livestock operation systems are currently being utilized in Korea to improve productivity within a limited area of land (Kim et al., 2005). Commercial egg production systems in Korea include caged, barn, free-range, and organic types. Conventional laying cages (traditional cages) have predominantly been used in Korea. However, to a very little extent, these have been replaced by alternative systems such as barn and free-range types, which may appear attractive to consumers because of their perceptions of such systems as being "animal welfare friendly" or providing better "well-being" of animals. The trend toward an increased use of barn and free-range systems is projected to continue and is mainly driven by a subset of consumers who are willing to pay a premium for free-range eggs.

Temperature/humidity condition under conventional laying cages is a major factor influencing growth rates, hatchability, egg size, egg shell quality, and egg production. The climate from late June to July in Korea has high temperatures and humidity averaging above 28°C and 90% relative humidity, respectively, which leads to lowering the production or quality of eggs. The optimal laying temperature is between 11°C and 26°C. Failure of humidity and temperature management in the traditional cage systems can have a severe impact on poultry performance. A humidity level above 75% will cause a reduction in egg laying, and seasonal temperature increases can reduce egg production by about 10% (Kekeocha, 1985).

The safety of livestock products has become another major issue of concern, which means that the livestock market has moved toward demanding higher safety food products "from farm to table." According to the recommendation of Codex with regard to hygiene and safety of products (Codex Alimentarius Commission, 2001), Korea has introduced Hazard analysis critical control point (HACCP) systems to ensure food safety and to provide a specific and systematic approach to food quality control. In Korea, the HACCP system was implemented on poultry farms (broilers and laying hens) in 2008 and on edible egg gathering centers in 2011 (MAFRA, 2008, 2011).

The Shell Egg Grading System (EGS) has been regulated through the Korea Institute for Animal Products Quality Evaluation by Ministry of Agriculture and Forestry (KIAPQE) since 2001. The EGS is a voluntary organization directed and controlled by the members of the KIAPQE to promote the availability of high-quality egg supply. Some manufacturers have expressed complaints about this system for not meeting its original official function of selecting various graded shell eggs in the market and providing enough profit to producers that produce top-quality graded shell eggs. The portion of graded shell eggs has increased every year, but due to low participation by manufacturers, graded shell eggs only represented approximately 6.8% of the egg supply in Korea in 2014 and most of them were graded 1+ among grades 1+, 1, 2, and 3 (Table 3.3; Fig. 3.4). During the manufacturing of eggs graded by EGS, eggs contaminated with feces were removed from the human food supply under the direction of the KIAPQE under the Ministry of Agriculture, Food, and Rural Affairs (MAFRA) at the point of production and may be subject to the standards recommended by the MAFRA.

It is important to recognize that the samples and tests reported do not necessarily represent 100% of the egg supply from every market. As industry participation in the EGS has increased, reporting of the number of samples and tests has similarly increased. Continuing efforts are being made to ensure that there is uniform reporting among all the industry members and that the test methods reported are correct.

The distribution of eggs without quality maintenance because of the complicated system and the absence of cold storage distribution are regarded as problems of the distribution system. It is important to supply eggs having the same quality throughout the year in Korea despite the seasonal changes in temperature. By employing cleaning and coating as basic sanitation methods and selling shell eggs at outlets with a cold storage system, the egg production system would be better set up for the regulation of supply.

## 4. EPIDEMIOLOGY OF *SALMONELLA* IN KOREA

The number of reported *Salmonella* clinical isolates in Korea fluctuated between 300 and 1500 from 2000 to 2014. After reaching a peak in 1999 (2252 isolates), the prevalence of *Salmonella* clinical isolates gradually declined until 2006 (Fig. 3.5).

Among all *Salmonella* serovars, *Salmonella* Enteritidis and *Salmonella* Typhimurium have been the most commonly reported serovars associated with human illnesses in South Korea (GFN, 2015; KCDC, 2011–2015). However, an increasing number of outbreaks caused by unusual serovars are being experienced in

**Table 3.3** Grading Ratio Among Eggs Produced in Korea, 2007–2014

| Grade | 2007 | 2008 | 2009 | 2010 | 2011 | 2012 | 2013 | 2014 |
|---|---|---|---|---|---|---|---|---|
| Production [a] | 11,927,241 | 11,995,918 | 13,346,581 | 13,413,745 | 13,644,600 | 13,576,616 | 14,113,246 | 13,644,515 |
| Graded eggs [a] | 284,865 | 344,511 | 418,604 | 521,005 | 588,119 | 710,136 | 858,048 | 927,874 |
| Ratio (%) | 2.4 | 2.9 | 3.1 | 3.9 | 4.3 | 5.2 | 6.3 | 6.8 |

[a] *Unit: thousand.*

*Data from KOSIS (KOSTAT's Korean Statistical Information Service), 2015.*

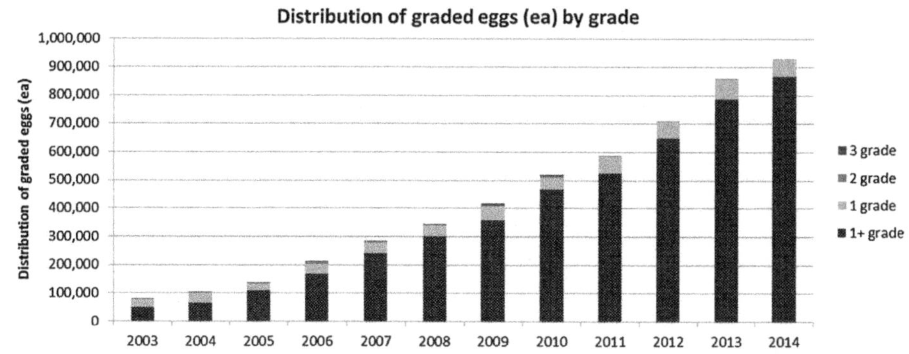


**FIGURE 3.4**

Distribution of Eggs (ea) by Grade, 2003–2014.

*Data from KAPE (Korea Institute for Animal Products Quality Evaluation), 2015.*

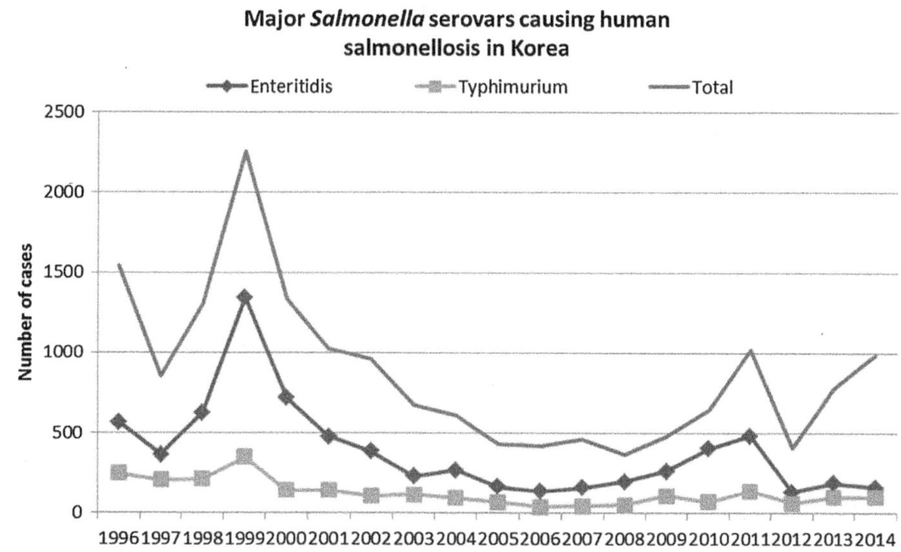


**FIGURE 3.5**

*Salmonella* Isolates from Human Sources by Serovar and Year, 1996–2014.

*Data from GFN (Global Foodborne Infections Network), 2015 and KCDC (Korea Centers for Disease Control and Prevention), 2011–2015.*

humans in Korea. In 2014, other serovars such as *Salmonella* Thompson, *Salmonella* Livingstone, *Salmonella* Bareilly, and *Salmonella* Montevideo were reported to be isolated at increasing frequencies and were implicated in several food-borne disease outbreaks (KCDC, 2015).

**Table 3.4** Food-Borne Disease Outbreaks Caused by *Salmonella*, 2007–2015 (Total Number of Patients per Outbreak)

| Year | 2007 | 2008 | 2009 | 2010 | 2011 | 2012 | 2013 | 2014 | 2015 |
|---|---|---|---|---|---|---|---|---|---|
| Number of *Salmonella* outbreaks (total number of patients) | 42 (1497) | 22 (387) | 17 (477) | 27 (677) | 24 (1065) | 9 (147) | 13 (690) | 24 (1416) | 11 (197) |

*Data from Food Safety Korea, 2015.*

Most human infections with *Salmonella* are food borne. *Salmonella* are among the leading causative agents reported most often in bacterial food-borne disease outbreaks (Foley et al., 2007). In Korea, approximately 200–300 food-borne disease outbreaks are reported annually. In previous studies, *Salmonella* spp. were the causative agents reported most often in outbreaks between 1993 and 1996 in Korea and were responsible for 55.1% of the recorded food-borne disease outbreaks in which the etiologic agent was determined (Bajk and Roh, 1998). Recent trends of food-borne disease outbreaks in Korea showed a decline in the frequency of food-borne outbreaks by *Salmonella* spp. (Table 3.4). In 2015, 0.03% of the reported food-borne outbreaks caused by an identified agent in Korea were confirmed to be caused by *Salmonella* spp., which ranked fourth among the identified agents in terms of outbreak frequency, below norovirus, pathogenic *E. coli*, and *Campylobacter* spp. (Food Safety Korea, 2015).

## 5. EGGS AND *SALMONELLA*

Among salmonellosis outbreaks reported to Korea Centers for Disease Control and Prevention (KCDC), those caused by *S.* Enteritidis have been especially frequent, and foods associated with outbreaks of *S.* Enteritidis infection have often included eggs or egg products (Hong et al., 2015). Salmonellosis caused by contaminated eggs has been consistently reported. Between 2008 and 2012, 23 outbreaks of disease caused by *Salmonella* spp. that were traceable to contaminated livestock or livestock products were reported. Among these outbreaks, food-borne outbreaks by *S.* Enteritidis were most frequent (15 outbreaks/cases) and egg-related products were identified as the major causative agents (Table 3.5). Eggs have been considered as one of the most important food sources because of their high nutritional value and reasonable price. Egg and egg products, however, have been implicated in the outbreak of human salmonellosis. The risk of developing *Salmonella* poisoning can be especially higher in consumers with the consumption habits of eating raw eggs. Traditionally, consumers in Korea habitually eating raw or lightly cooked egg should especially be aware of the risks via raw eggs, which represents a possible route of acquiring food-borne infections.

**Table 3.5** Food-Borne Pathogens and Vehicles Related to Food-Borne *Salmonella* Outbreak due to Livestock or Livestock Products Between 2007 and 2012 in Korea

| Pathogen | Number of Outbreaks | Vehicle | Outbreak Place |
|---|---|---|---|
| *Salmonella* Enteritidis | 15 | Meatball, Minister's head, Jajang-myeon (meat), egg, stir-fried glass noodles and vegetables (meat, egg), Gimbap (egg), rolled omelet, Gimbap, potato salad (egg), steamed egg, broiled quail eggs | Gyeonggi, Gyeongbuk, Seoul, Jeonnam, Chungnam, Chungbuk, Jeju |
| *Salmonella* Thompson | 2 | Gimbap (egg) | Gyeonggi |
| *Salmonella* Typhimurium | 2 | Meatball (egg), Gimbap (egg) | Seoul, Chungnam |
| *Salmonella* D group | 2 | Pork | Jeonnam |
| *Salmonella* B group | 1 | Egg | Gangwon |
| *Salmonella* Schleissheim | 1 | Rolled omelet | Chungnam |
| *Salmonella* Montevideo | 1 | Sweet and sour pork | Seoul |
| *Salmonella* Newport | 1 | Sliced raw beef bibimbap | Jeonnam |
| Total | 23 | | |

*Data from KCDC (Korea Centers for Disease Control and Prevention), 2015.*

Cases of *S*. Enteritidis isolation have been reported in countries other than Korea, including the United States (Baker, 1980; Humphrey et al., 1989); however, in Korea, no such cases were reported from eggs or egg products between 2000 and 2011 (Chun and Hong, 2009; Lee et al., 2002; Woo, 2005). The study of Cho and Shin (1985) showed that *Salmonella* Mississippi was isolated from 1 of 260 eggs (0.38%) collected between November and December 1983 at three poultry farms in Gyeonggi-do. Kwon and Ko (1997) reported that *Salmonella* serogroup E was isolated from only one eggshell (0.26%) of 390 eggs sampled from a wholesaler in Cheon-an. Strikingly, the isolated strain was found in all parts of the same egg including eggshell, albumen, and yolk materials indicating internalization of *Salmonella* in egg. In the study conducted in 1998 by Chang (2000), none of the egg yolks were found to contain *Salmonella* among 135 packs of one dozen shell eggs purchased from supermarkets in Seoul and Gyeonggi-do areas in Korea. In the study of Lee et al. (2002), *S*. Enteritidis and antibodies directed against *S*. Enteritidis were not detected in the yolks of eggs available in the market. The study was evaluated on the egg pools composed of 171 eggs of 57 commercial brands (3 eggs per brand) collected from food stores in a department store located in the Incheon metropolitan city. Lee et al. (2004) carried out an investigation to analyze biological hazards, microbial contamination of egg (normal, dirty, and cracked), water, feed, manure, and equipment associated with laying farms, where one isolate of *S.* Enteritidis originated from the manure of egg-laying hens ($n=32$ samples) among a total of 196 environmental samples and one isolate of *Salmonella* Bardo was detected from dirty egg shells ($n=34$) among a total of 192 egg samples. In the study by Chun and Hong (2009), *Salmonella* was not identified in four different brands of eggs collected from hypermarkets in the northern Gyeonggi area. In the study by Woo (2005), a single isolate of *Salmonella* Gallinarum was isolated only on the eggshell part of one egg among 446 eggs (0.2%) collected from conventional markets and department stores located in Seoul and Gyeonggi regions in 1996. However, *S.* Enteritidis had rarely been isolated from eggs in Korea until the first isolation and identification in the surveys conducted by Kim et al. (2013), in which *S.* Enteritidis was first isolated from eggshells and the contents of eggs distributed at the grocery stores in Korea. Two strains of *S.* Enteritidis from the eggshells and one strain from egg contents collected from the Eumseong city region were first identified in the fall of 2011.

As the results of previous studies indicated, *Salmonella* was not prevalent among shell eggs in South Korea. However, this low frequency of detection, compared with those reported in previous studies in other countries, might partly have resulted from the use of a detection method that did not adopt the pooling criterion based on the US Food and Drug Administration (FDA) Bacteriological Analytical Manual, which has been proved to be the most effective detection method for *Salmonella* in egg contents (US Food and Drug Administration, 2012). Furthermore, most of the previous studies in Korea tested for *Salmonella* in a narrow geographic region, and the bulk pooling method was not applied. In most of these studies, either the sample size for the assessment of *Salmonella* was small (one to five eggs per farm) or the survey was conducted only once without consecutive testing. When the number of eggs for *Salmonella* test is increased and a nationwide study conducted in Korea, the prevalence of *Salmonella* from eggs might be increased. As neither the method of pooling the contents of 20 shell egg samples nor that of sampling by bulk pooling had been used, no *Salmonella* seemed to be detected in the contents of shell egg samples. Since previous reports have estimated that only 1 in 20,000 eggs is positive for *Salmonella*, more than 20,000 eggs should be tested to estimate the prevalence of *Salmonella* in eggs in any given region.

In Korea, the method used for the isolation of *Salmonella* from shell eggs was revised by the government in 2013 (QIA Notice, 2012-162). In a survey by Kim et al. (2015) conducted in bulk, no eggs were positive for *Salmonella* among a total of 2400 shell eggs (120 pooled samples consisting of 20 eggs per pool) between March 2011 and December 2012. Meanwhile in the same microbiological survey conducted at the egg-processing plants (i.e., plants that produce pasteurized and unpasteurized liquid egg), four *Salmonella*-positive samples from 120 unpasteurized liquid egg samples (3.3%) and five positive samples from 75 pasteurized liquid egg samples (6.7%) were identified at eight egg-breaking plants, which appears to prove that the pasteurization processes conducted in egg processing plants failed to adequately eliminate *Salmonella* in liquid egg products. Hygiene and sanitation standards in liquid egg products in South Korea need to be improved. Also, in the study by Lee et al. (2013), conducted according to the FDA method to evaluate the extent and species of *Salmonella* contamination in shell eggs in South Korea, a total of 26 *S.* Gallinarum isolates were obtained from the contents of 7000 shell eggs, but *S.* Enteritidis was not detected. In this study, *S.* Gallinarum was found to be more common in eggs from organic farms: 20.0% from organic and 5.3% from conventional farms.

Although cases of *S.* Enteritidis isolation from eggs have been reported in other countries, such cases have rarely been reported in Korea. This may be attributed to the hypothesis by Bäumler et al. (2000) who suggested that *S.* Enteritidis filled the ecologic niche vacated by the eradication of *S.* Gallinarum from poultry, leading to an epidemic increase in human infections. This hypothesis was also supported by the study by Rabsch et al. (2000).

Even though *S.* Gallinarum has adapted to its avian host and rarely induces food poisoning in humans, it causes a significant poultry disease, fowl typhoid (FT), which has been responsible for considerable economic losses to the poultry industry. In Australia, North America, and most European countries, FT has almost disappeared as a result of improved surveillance and slaughter practices (Silva et al., 1981; Barrow, 1990; Kim et al., 1991; Wigley et al., 2005; Basnet et al., 2008). However, as the results show, FT has become one of the most serious bacterial diseases of poultry in Korea since the first case was reported in the field in 1992 (Kim et al., 2007). After experiencing severe damage from the disease since the massive outbreak in 1992, the Korean authorities decided to introduce a nationwide vaccination program with a live attenuated strain (SG 9R) for commercial layer chickens in 2001 instead of an eradication policy. Despite the introduction of live *S.* Gallinarum vaccines, a total of 928 national FT outbreaks were identified between 2000 and 2008 in Korea, and the annual number of outbreaks reached a peak (206 farms affected) in 2002 (Kwon et al., 2010).

## 6. THE FUTURE OF THE EGG-PRODUCING POULTRY INDUSTRY IN KOREA

At present, tests are conducted annually on eggs from farms in Korea to detect *S.* Enteritidis (MIFAFF Notice, 2009-169). However, confidently assessing the risk of *Salmonella* contamination is difficult due to the complexity of egg production and distribution in Korea. The performance of a national survey to obtain reliable data on the distribution of *Salmonella* contamination among eggs and egg products in Korea is strongly recommended as a preventive measure for egg safety.

In Korea, policies for the control and eradication of FT and Pullorum disease have been instituted since the 1970s. According to these policies, all breeder chickens older than 120 days should be serologically tested for FT and Pullorum disease using the whole blood or serum agglutination test with *Salmonella* Pullorum antigen, and the eggs produced by seropositive flocks should not be used as hatchery eggs. Based on the prevalence of FT over the past 15 years, this policy seems to no longer be sufficiently effective. Therefore, to minimize economic losses in the poultry industry from *S.* Gallinarum infection and to eradicate FT, a new nationwide policy for the control of FT, similar to the National Poultry Improvement Plan of the United States, should be developed that permits the vaccination of certain flocks that are reared in highly contaminated areas. Another option would be to modify the established control program and strictly apply it to all the breeder farms in the country with a combination of stringent management procedures (Kwon et al., 2010).

## REFERENCES

Bajk, G.J., Roh, W.S., 1998. Estimates of cases and social economic costs of foodborne salmonellosis in Korea. J. Food Hyg. Saf. 13, 299–304.

Baker, R.C., Golf, J.P., Timoney, J.P., 1980. Prevalence of *Salmonella* on eggs from poultry farms in New York State. Poult. Sci 59, 289–292.

Barrow, P.A., 1990. Immunity to experimental fowl typhoid in chickens induced by a virulence plasmid-cured derivative of *Salmonella* gallinarum. Infect. Immun. 58, 2283–2288.

Basnet, H.B., Kwon, H.J., Cho, S.H., Kim, S.J., Yoo, H.S., Park, Y.H., Yoon, S.I., Shin, N.S., Youn, H.J., 2008. Reproduction of fowl typhoid by respiratory challenge with *Salmonella* Gallinarum. Avian Dis. 52, 156–159.

Bäumler, A.J., Hargis, B.M., Tsolis, R.M., 2000. Tracing the origins of *Salmonella* outbreaks. Science 287, 50–52.

Chang, Y.H., 2000. Prevalence of *Salmonella* spp. in poultry broilers and shell eggs in Korea. J. Food Prot. 63 (5), 655–658.

Cho, D.I., Shin, K.S., 1985. Isolation and identification of *Salmonella* and *Escherichia coli* from chicken eggs. Korean J. Vet. Public Health 9 (2), 13–18.

Chun, M.S., Hong, S.H., 2009. Identification of microorganisms from eggs in hypermarket in the Northern Gyeonggi area. Korean J. Food Nutr. 22, 394–401.

Codex Alimentarius Commission, 2001. Food Hygiene Basic Texts. Codex Alimentarius – Joint FAO/WHO Food Standards Programs, 3rd. Bernan Association, Lanham, MD.

Foley, S.L., Zhao, S., Walker, R.D., 2007. Comparison of molecular typing methods for the differentiation of *Salmonella* foodborne pathogens. Foodborne Pathog. Dis 4 (3), 253–276.

Food Safety Korea, 2015. Food Poisoning Statistics. Available at http://www.foodsafetykorea.go.kr/portal/healthyfoodlife/foodPoisoningStat.do?menu_no=519&menu_grp=MENU_GRP02 accessed July 2015.

GFN, 2015. Country Databank of the World Health Organisation. Global Foodborne Infections Network. Available at http://www.who.int/gfn/activities/en. accessed July 2015.

Hong, S.H., Park, N.Y., Jo, H.J., Ro, E.Y., Ko, Y.M., Na, Y.J., Park, K.C., Choi, B.G., Min, K.J., Lee, J.K., Moon, J.S., Yoon, K.S., 2015. Risk ranking determination of combination of foodborne pathogens and livestock or livestock products. J. Food Hyg. Saf. 30 (1), 1–12.

Humphrey, T.J., Greenwood, M., Gilbert, R.J., Rowe, B., Chapman, P.A., 1989. The survival of *Salmonellas* in shell eggs cooked under simulated domestic conditions. Epiderm. Infect 103, 35–45.

KAPE, 2015. Animal products grading statistical year book. Korea Institute for Animal Products Quality Evaluation. Available at http://www.ekapepia.com/758.su. accessed July 2016.

KCDC, 2011. Public Health Weekly Report 4 (43), 790–792. Korea Centers for Disease Control and Prevention.

KCDC, 2012. Public Health Weekly Report 5 (25), 461–466. Korea Centers for Disease Control and Prevention.

KCDC, 2013. Public Health Weekly Report 6 (6), 105–110. Korea Centers for Disease Control and Prevention.

KCDC, 2014. Public Health Weekly Report 7 (18), 385–390. Korea Centers for Disease Control and Prevention.

KCDC, 2015. Public Health Weekly Report 8 (21), 470–476. Korea Centers for Disease Control and Prevention.

Kekeocha, C.C., 1985. Introduction to Poultry Keeping. Poultry Production Hand Book. Pfizer Corporation, Nairobi, pp. 1–15.

Kim, C.J., Nagaraja, K.V., Pomeroy, B.S., 1991. Enzyme-linked immunosorbent assay for the detection of *Salmonella* enteritidis infection in chickens. Am. J. Vet. Res. 52, 1069–1074.

Kim, A., Lee, Y.J., Kang, M.S., Kwag, S.I., Cho, J.K., 2007. Dissemination and tracking of *Salmonella* spp. in integrated broiler operation. J. Vet. Sci 8, 155–161.

Kim, Y.Y., Kil, D.Y., Oh, H.K., Han, I.K., 2005. Acidifier as alternative material to antibiotics in animal feed. Asian-Aust. J. Anim. Sci 18, 1048–1060.

Kim, Y.J., Song, B.R., Lim, J.S., Heo, E.J., Park, H.J., Wee, S.H., Oh, S.M., Moon, J.S., 2013. First report on isolation of *Salmonella* enteritidis from eggs at grocery stores in Korea. Korean J. Food Sci. Anim. Res. 33, 239–243.

Kim, Y.J., Chon, J.W., Lim, J.S., Song, B.R., Seo, K.H., Heo, E.J., Park, H.J., Wee, S.H., Oh, D.H., Moon, J.S., 2015. Traceback investigation for *Salmonella* contamination at egg processing plants in South Korea: prevalence, antibiotic resistance, and epidemiological tracing by rep-PCR fingerprinting. J. Food Sci. 80, M759–M764.

KOSIS, 2015. Korean Statistical Information Service. Available at http://kosis.kr/statHtml/statHtml.do?orgId=101&tblId=DT_1EO061&vw_cd=MT_ZTITLE&list_id=F1A&seqNo=&lang_mode=ko&language=kor&obj_var_id=&itm_id=&conn_path=E1 accessed July 2016.

Kwon, O.P., Ko, K.K., 1997. Studies of the human pathogenic bacteria from the chicken eggs. J. Soonchunhyang Med. Coll. 3, 509–516.

Kwon, Y.K., Kim, A., Kang, M.S., Her, M., Jung, B.Y., Lee, K.M., Jeong, W., An, B.K., Kwon, J.H., 2010. Prevalence and characterization of *Salmonella* Gallinarum in the chicken in Korea during 2000 to 2008. Poult. Sci. 89, 236–242.

Lee, S.K., Chon, J.W., Song, K.Y., Hyeon, J.Y., Moon, J.S., Seo, K.H., 2013. Prevalence, characterization, and antimicrobial susceptibility of *Salmonella* Gallinarum isolated from eggs produced in conventional or organic farms in South Korea. Poult. Sci. 92, 2789–2797.

Lee, S.M., Kim, K.H., Lee, J.G., Park, E.J., Lee, S.W., Hong, C.H., 2002. Hygienic quality of eggs in the department food stores in the Incheon Metropolitan area. J. Food Hyg. Saf. 17, 129–136.

Lee, S.M., Yoo, H.S., Hong, C.H., 2004. Analysis of biological hazards and control points in layer houses. Korean J. Vet. Res. 44, 593–605.

MAFRA, 1970. Agriculture, Food and Rural Affairs Statistics Yearbook, 1970. Ministry of Agriculture, Food and Rural Affairs, South Korea.

MAFRA, 1975. Agriculture, Food and Rural Affairs Statistics Yearbook, 1975. Ministry of Agriculture, Food and Rural Affairs, South Korea.

MAFRA, 1980. Agriculture, Food and Rural Affairs Statistics Yearbook, 1980. Ministry of Agriculture, Food and Rural Affairs, South Korea.

MAFRA, 1985. Agriculture, Food and Rural Affairs Statistics Yearbook, 1985. Ministry of Agriculture, Food and Rural Affairs, South Korea.

MAFRA, 1990. Agriculture, Food and Rural Affairs Statistics Yearbook, 1990. Ministry of Agriculture, Food and Rural Affairs, South Korea.

MAFRA, 1995. Agriculture, Food and Rural Affairs Statistics Yearbook, 1995. Ministry of Agriculture, Food and Rural Affairs, South Korea.

MAFRA, 2000. Agriculture, Food and Rural Affairs Statistics Yearbook, 2000. Ministry of Agriculture, Food and Rural Affairs, South Korea.

MAFRA, 2005. Agriculture, Food and Rural Affairs Statistics Yearbook, 2005. Ministry of Agriculture, Food and Rural Affairs, South Korea.

MAFRA, 2008. Application of HACCP System on Laying Hens and Broilers Farms. Ministry of Agriculture, Food and Rural Affairs, South Korea. Available at http://www.ihaccp.or.kr/eng/haccp.jsp#eng_hitory. accessed July 2016.

MAFRA, 2010. Agriculture, Food and Rural Affairs Statistics Yearbook, 2010. Ministry of Agriculture, Food and Rural Affairs, South Korea.

MAFRA, 2011. Application of HACCP System on Gathering Edible Egg Centers. Ministry of Agriculture, Food and Rural Affairs, South Korea. Available at http://www.ihaccp.or.kr/eng/haccp.jsp#eng_hitory. accessed July 2016.

MAFRA, 2014. Agriculture, Food and Rural Affairs Statistics Yearbook, 2014. Ministry of Agriculture, Food and Rural Affairs, South Korea.

QIA, Animal and Plant Quarantine Agency, 2012. Standards for Processing and Ingredients Specifications of Livestock Products (QIA Notice 2012-162).

Rabsch, W., Hargis, B.M., Tsolis, R.M., Kingsley, R.A., Hinz, K.H., Tschape, H., Baumler, A.J., 2000. Competitive exclusion of *Salmonella* enteritidis by *Salmonella* gallinarum in poultry. Emerg. Infect. Dis. 6, 443–448.

Silva, E.N., Snoeyenbos, G.H., Weinack, O.M., Smyser, C.F., 1981. Studies on the use of 9R strain of *Salmonella* gallinarum as a vaccine in chickens. Avian Dis. 25, 38–52.

US Food and Drug Administration, 2012. Bacteriological Analytical Manual, Chapter 5. *Salmonella*. http://www.fda.gov/Food/FoodScienceResearch/LaboratoryMethods/ucm070149.htm.

Wigley, P., Hulme, S., Powers, C., Beal, R., Smith, A., Barrow, P., 2005. Oral infection with the *Salmonella enterica* serovar Gallinarum 9R attenuated live vaccine as a model to characterize immunity to fowl typhoid in the chicken. BMC Vet. Res. 1, 2.

Woo, Y.K., 2005. Microbial hygienic status of poultry meats and eggs collected at the public markets in Seoul and Kyung-gi regions in 1996. Korean J. Microbiol. 41, 38–46.

CHAPTER

# Egg Production Systems and *Salmonella* in Canada

4

**Doug Korver, Lynn McMullen**

*University of Alberta, Edmonton, AB, Canada*

## 1. EGG PRODUCTION SYSTEMS IN CANADA

Egg production in Canada is nationally regulated (supply management) as are the dairy, broiler hatching egg, chicken, and turkey industries. The Egg Farmers of Canada (EFC), which is the national egg marketing board, determines the amount of production allocated to each province, and each provincial marketing board allocates production quota to individual producers (AAFC, 2015). Quota holders receive a guaranteed price for their product based on cost of production. Tariffs and import controls applied by the federal government limit the importation of eggs from outside of Canada (AAFC, 2015). Within each province, layer flocks under a specific number of hens are exempt from the supply management system, and thus from the regulations followed by quota-holding producers, including the national food safety program for the table egg industry, Start Clean-Stay Clean. The maximum number of hens allowed before a producer is required to be part of the supply management system is between 99 and 300, although some provinces allow certain religious communities to have up to 500 hens. The focus of this chapter is therefore the production of eggs under supply management, which represents approximately 97% of all eggs consumed in Canada (Statistics Canada, 2007). However, surplus hatching eggs used for human consumption, farm gate sales, and farmers' market sales are not included in this estimate, and so the actual percentage may be lower. Unregulated flocks, and the entry of surplus broiler hatching eggs into the food supply represent greater potential risk to consumers because of the absence of testing (small flocks), or less frequent testing (broiler breeder flocks) for *Salmonella enterica* serovar Enteritidis.

In 2015, the 1021 regulated Canadian producers, with an average flock size of 20,811 hens, produced approximately 610.5 million dozen eggs (Egg Farmers of Canada, 2015). Because production quota is allocated nationally to each of the provinces, the production of eggs within each province is generally proportional to the population across the country (Statistics Canada, 2015; Egg Farmers of Canada, 2015). Two provinces (Ontario and Quebec) are allocated additional quota for the production of egg-sourced vaccines; surplus table eggs are sold to the liquid egg market and are called Eggs for Processing (Egg Farmers of Canada, 2014).

**Producing Safe Eggs. http://dx.doi.org/10.1016/B978-0-12-802582-6.00004-5**

## 1.1 *SALMONELLA* IN THE CANADIAN FOOD SUPPLY

*S. enterica* is the most common bacterial pathogen reported to cause food-borne outbreaks in Canada, and *Salmonella* Enteritidis is the most often reported serotype associated with the consumption of eggs (Bélanger et al., 2015; DeWinter et al., 2011). Before 1990, other serotypes including *Salmonella* Bareilly, Typhimurium, Infantis, Schwarzengrund, and Java were implicated in cases of salmonellosis linked to the consumption of eggs (Todd, 1996); however, since then, *Salmonella* Enteritidis has accounted for the majority of egg-related cases of human illness in Canada. In Canada, layer flocks are not force molted, which reduces the risk of *S.* Enteritidis contamination of eggs (Denagamage et al., 2015). Past and current data on the incidence of *Salmonella* in eggs is extremely limited and current surveillance programs do not specifically monitor eggs or egg products in any jurisdiction in Canada. This includes the FoodNet (formerly C-Enternet; facilitated by the Public Health Agency of Canada) sentinel site surveillance, and the Canadian Integrated Program for Antimicrobial Resistance Surveillance (CIPARS). However, from 2009 to 2011, CIPARS did obtain 15 isolates of *S.* Enteritidis from spent layer hens at slaughter and as part of targeted research, reported that 5/300 samples of eggs at breaking stations were positive for *S.* Enteritidis (Parmley et al., 2013). In addition to the lack of available data from surveillance programs, there is a paucity of data for the attribution of salmonellosis to the consumption of eggs based on outbreak data.

From 2003 to 2009 there was a substantial increase in human cases of *S.* Enteritidis infection from all sources reported in Canada. In 2003, the national annual incident rate of *S.* Enteritidis was 2.13/100,000 person-years but this increased to a high of 6.72/100,000 person-years in 2008 (Nesbitt et al., 2012). The Provinces of British Columbia (BC) (Galanis et al., 2012; Taylor et al., 2012) and Ontario (Anonymous, 2015) reported substantial increases in the incidence rate of *S.* Enteritidis infections, although not all of these cases were linked to the consumption of eggs (Tighe et al., 2012). However, data from disease surveillance in BC established a link between consumption of eggs and salmonellosis due to *S.* Enteritidis (Taylor et al., 2015). Ultimately, although it is clear that *S.* Enteritidis remains a serious concern for the egg industry in Canada, quantifying the relative risk of egg consumption is difficult because of the lack of available data.

## 2. ROLES AND RESPONSIBILITIES

In Canada, prevention of *S.* Enteritidis is viewed as being a joint responsibility of the industry, government, and consumers (Health Canada, 2013). Several provincial, federal, and local authorities are involved in maintaining food safety. Health Canada sets food safety standards and develops guidelines to minimize food safety risks, as well as working with the Canadian Food Inspection Agency (CFIA) and provincial and territorial governments to ensure effective responses to outbreaks of food-borne illness. The CFIA and provincial and territorial governments oversee the food industry to ensure it meets food safety responsibilities. The Public Health Agency of Canada (PHAC) plays a role in food safety surveillance nationally. Health Canada, the CFIA, and the

PHAC also provide reference laboratory service and health risk assessments, and work together when a recall action is necessary. Agriculture and Agri-Food Canada (AAFC) plays an indirect role, providing assistance to industry in the development of food safety biosecurity and traceability risk management programs (Health Canada, 2015).

The egg industry is required to comply with the Food and Drugs Act (Government of Canada, 2016c), federal legislation regulating food, drugs, cosmetics, and therapeutic devices in Canada. Section 4 of the Food and Drugs Act (Government of Canada, 2016c) states that "no person shall sell an article of food that has in or on it any poisonous or harmful substance." Anyone producing food in Canada must comply with this regulation. The egg industry must also comply with the Egg Regulations (Government of Canada, 2016b), which set standards for egg grading, grading stations, packaging, and inspection. This legislation also covers the international and interprovincial trade in eggs.

Consumers are expected to play an important role in food safety by adopting safe handling and food preparation practices. Health Canada, the CFIA and PHAC, as well as industry trade organizations and provincial and territorial governments develop and deliver science-based educational material to consumers (Health Canada, 2015).

## 3. CONTROL OF *SALMONELLA* IN THE CANADIAN EGG SUPPLY

The primary breeders supplying Canada's egg industry are not considered to be a risk for introducing *Salmonella* into Canada, as these flocks are tested every 3 weeks under the EU Export Protocol, and are required to be *S.* Enteritidis-free (Health Canada, 2013). The results of this testing are also shared with CFIA. Canadian egg-type multiplier flocks are also required to be free of *S.* Enteritidis. Therefore the primary breeders and multiplier flocks producing pullets for commercial egg production are not considered to be a risk for transmission of *S.* Enteritidis in Canada. Under the nonintegrated poultry production system in Canada, hatcheries may produce both layer pullets and broiler chicks; therefore eggs from *S.* Enteritidis-infected broiler breeder flocks could cross-contaminate layer pullets. In addition, surplus broiler hatching eggs can enter the human food supply, and may represent an increased risk of *S.* Enteritidis infection to consumers because until recently, the hatching egg industry in Canada has had less stringent monitoring for *S.* Enteritidis. The broiler hatching egg industry in several Canadian provinces has initiated expanded testing programs for *Salmonella.*

## 4. NATIONAL *SALMONELLA* CONTROL PROGRAM—START CLEAN-STAY CLEAN

In 1990, the Canadian egg industry implemented an *S.* Enteritidis control program called "Safe from *Salmonella*," which introduced biosecurity measures in primary egg production on layer farms (Health Canada, 2013). In 1998, Hazard Analysis and Critical Control Point principles were incorporated into the program, which was renamed "Start Clean-Stay Clean" (SC-SC). Although the program was developed nationally, implementation of the program has been the responsibility of the provincial boards. As a result, although the overall objectives are consistent across the country, the implementation of the national program can vary by province. As of 2013, over 90% of regulated egg producers (i.e., part of supply management) participate in the full SC-SC program. Regardless of individual provincial egg board requirements regarding how the SC-SC program is implemented, any egg producer with a regulated flock (i.e., part of supply management) must perform regular environmental testing for *S.* Enteritidis (Health Canada, 2013). Pullet flocks are subject to a similar program, Start Clean-Stay Clean: Pullets.

The SC-SC manual for producers (Canadian Egg Marketing Agency, 2004) identifies *Salmonella*, *Campylobacter*, *Escherichia coli*, and mycotoxins as relevant biological hazards, but focuses on control of *S.* Enteritidis. Chemical hazards such as water treatment chemicals, disinfectants, pharmaceuticals, and lubricants have also been identified as potential risks for contamination. The SC-SC program identifies hazards and critical control points for all types of egg production, and identifies particular biological and chemical hazards relevant to free-run, free-range, and organic production systems.

Canadian *S.* Enteritidis control and prevention programs involve testing throughout the supply chain—breeder flocks, hatcheries, pullet barns, and layer barns to ensure that *S.* Enteritidis-positive flocks are quickly identified and the appropriate actions taken. At a producer level, *S.* Enteritidis testing is carried out in the barn, rather than samples collected directly from the birds. Environmental swabs are collected from each flock once in the pullet barn (between 3 and 15 weeks of age), early in lay (between 19 and 35 weeks of age), and late in lay (between 36 and 60 weeks of age) by qualified egg board staff. At least 60 different sites within the barn must be collected, with a focus on dust and egg moving equipment and composites. Samples from up to 15 different sites may be pooled for analysis, with a minimum of four pooled samples to be tested. If rodent droppings or dead insects are found, they must also be sampled. Swabs are taken from walls, floors, fans, egg belts, manure belts, coolers, etc. The samples are tested at an accredited laboratory (typically, the provincial government microbiology laboratory) using a culture-based method approved by the Chief Veterinary Officer of each province. Simultaneously, the samples are also tested for other pathogens. Currently, the CFIA is developing standardized testing procedures for serotypes of *Salmonella* (Health Canada, 2015).

If the pullet barn tests positive for *S.* Enteritidis, the flock is humanely euthanized and the carcasses disposed of in an approved manner (composting, burial, or incineration) according to provincial rules and regulations. If the layer barn environment tests positive, all eggs are diverted to a breaking plant, and the product pasteurized until a decision is made regarding the disposition of the flock. Canadian breaking plants have specific protocols for handling and processing eggs coming from flocks known to be infected with *S.* Enteritidis. Depending on market circumstances, a flock testing positive for *S.* Enteritidis will be humanely euthanized, and the carcasses disposed of in an approved manner (composting, burial, or incineration) according to provincial rules and regulations. If the flock is allowed to remain in production, all eggs are diverted to a breaking facility and the product pasteurized until the end of life of the affected flock. Once the flock has been depopulated, the barn is cleaned and disinfected according to standard procedures. The barn must be tested for *S.* Enteritidis (i.e., *S.* Enteritidis below detection limits) after cleaning and disinfection. The EFC recommends that the possibility of probiotics or *S.* Enteritidis vaccination in subsequent flocks in a facility that had previously been positive for *S.* Enteritidis be discussed with the attending veterinarian.

Individual provincial boards may recommend additional actions, including having a provincial veterinarian complete a disease investigation, in which the veterinarian works with the producer to attempt to identify the potential sources of the infection, as well as to offer further guidance on cleaning and disinfection procedures.

Producers that have been impacted by a positive *S.* Enteritidis test can receive insurance compensation through a program such as the Canadian Egg Industry Reciprocal Alliance (CIERA, 2016) or the Poultry Insurance Exchange Reciprocal of Canada. To be eligible for this insurance coverage, a producer must use facilities only for conducting business for the regulated egg industry. For example, producers with single buildings housing both pullets and laying hens may not be eligible for coverage.

Yearly SC-SC audits are conducted by inspectors authorized by EFC within each province. For example, Alberta producers are expected to score a minimum of 92%, based on adherence to the good management practices. Here, producers that score below 92% must provide a 6-month action plan to the board to rectify the shortfalls. Financial penalties of $0.20–$0.40 CAD per dozen eggs are collected by the Alberta provincial board until the standards are met. The funds collected as penalties are returned to the producer when compliance is achieved (Egg Farmers of Alberta, personal communication). Other provinces have different thresholds, expectations, and remedial actions required.

## 5. EGG GRADING STATIONS

Eggs may be stored on-farm at 10–13°C for no more than 7 days under approved conditions before delivery to grading station or further processing. Although some farms have a grading station on site, eggs are typically picked up from the farm once a week by their egg grader and transported to the egg grading station. Egg grading stations wash, candle, and (based on weight) sort eggs received from farms. After grading, the eggs are stored under refrigeration until being shipped to the retail market. Grading stations must be registered with the CFIA, and are subject to the Egg Regulations under the Canada Agricultural Products Act (Government of Canada, 2016a). The Egg Regulations specify storage conditions and limits, hygienic requirements, and require environmental sampling for *Salmonella* twice per year. Individual egg producers are not permitted to wash eggs on-farm, unless they are registered as an egg grading station. According to the federal Egg Regulations, egg wash water must be a minimum of 40°C, at least 11°C warmer than the eggs, must have a pH above 10, and must contain a safe and effective shell cleaning compound (Government of Canada, 2016a). Washing reduces the microbial load on the surface of eggshells (Musgrove et al., 2005). In conjunction with egg washing to remove surface contamination, refrigeration decreases the proliferation of any salmonellae that may be present in the interior of the egg (Gast and Holt, 2000).

## 6. OTHER SOURCES OF TABLE EGGS—SURPLUS BROILER HATCHING EGGS

Until recently, the broiler chicken supply chain has not conducted intensive monitoring for *S.* Enteritidis because of the relatively lower risk for transmission from chicken meat to humans (Keery, 2010), likely because chicken meat is normally cooked thoroughly before being consumed, whereas eggs are often used raw or undercooked in many dishes (Mumma et al., 2004; Patrick et al., 2004). However, hatching eggs entering the table egg food chain have been linked to an outbreak of *S.* Enteritidis in BC (Table 4.1). As a result, surplus hatching eggs are no longer permitted to be sold as table eggs in BC. The on-farm food safety assurance program of the Canadian hatching egg industry referred to as the

**Table 4.1** Summary of Outbreaks of *Salmonella* Enteritidis Linked to the Consumption of Eggs or Egg-Containing Foods in Canada and the Risk Factors Associated With the Outbreaks

| Year | Cases Confirmed | Province | Source and Risk Factors | References |
|---|---|---|---|---|
| 2000 | 62 | British Columbia (BC) | Grade B eggs; poor handling and food safety practices, ill baker | Strauss et al. (2005) |
| 2007 | 3 | BC | Raw egg noodles (eggs and layers linked with same strain) | Taylor et al. (2012) |
| 2008–10 | 584 | BC | Ungraded eggs from hatching and table egg flocks; poor hygiene, handling, and preparation; graded eggs; temperature abuse of eggs; consumption of mayonnaise prepared with raw egg; potential link to consumption of chicken meat | Taylor et al. (2012). |
| 2010–11 | 91 | Alberta | Illegally sourced eggs; unsanitary handling during preparation and use of eggs | Honish et al. (2013) |
| 2013 | 3; 9 (two outbreaks) | BC | Poor-quality ungraded table eggs | CDPCS (2014) |

"Canadian Hatching Egg Quality" program, is a Hazard Analysis Critical Control Point-based on-farm food safety program covering biosecurity, pest control, monitoring of incoming materials, and egg quality and handling. The hatching egg industry has also implemented a more rigorous *Salmonella* testing program. The program includes testing of chick tray papers at the farm, but before broiler breeder chicks are placed in the barn, environmental swabs collected between 15 and 17 weeks of age, and sampling of fluff from broiler chicks at the hatchery at 6-week intervals. If a positive fluff sample is detected, the barn is swabbed again to determine if the breeder farm is the source of contamination. If *S.* Enteritidis is detected, the breeder flock is depopulated.

## 7. OTHER SOURCES OF TABLE EGGS—UNREGULATED FLOCKS

Table egg production outside of the regulated industry represents an unknown level of risk for transmission of *S.* Enteritidis to humans, although ungraded eggs have been linked to several outbreaks in Canada (Table 4.1). The production of these eggs occurs on smaller farms where producers are not required to adhere to the on-farm food safety systems that would reduce the risks for consumers. These eggs can be sold at the farm gate or at farmers' markets. However, consumers may not be aware that these eggs are produced under circumstances that do not require the producers to adhere to these programs.

## 8. IS THE SYSTEM WORKING?

Since 1990, the incidence of human cases of *Salmonella* food-borne infections in Canada has increased substantially. To establish if control programs have an impact on the safety of eggs in Canada, it would be important to evaluate the incidence of *Salmonella* in eggs and food-borne disease related to eggs and egg products. The lack of publically available data from surveillance programs makes it extremely difficult to say with certainty that on-farm control programs have impacted the public health burden from *Salmonella* in eggs. In addition, changes in how surveillance is done, changes in sample collection, and changes in detection methodology in laboratories could each account for changes in prevalence data for *S.* Enteritidis in eggs in Canada. In spite of the lack of publically available data, large outbreaks of salmonellosis due to *S.* Enteritidis in shell eggs were reported in 2005 to be rare in Canada (Strauss et al., 2005).

In Canada, eggs are considered by some as one of the primary sources of *S.* Enteritidis and according to the latest risk assessment from Health Canada, the incidence of *S.* Enteritidis in table eggs sourced from regulated flocks in Canada is 1.7 per million eggs (DeWinter et al., 2011). However, other poultry products (Anonymous, 2015), almonds (Isaacs et al., 2005), cheese (Ahmed et al., 2000), and mung bean sprouts (Honish and Nguyen, 2001; Nesbitt et al., 2012) have each been implicated in outbreaks of salmonellosis in Canada. An outbreak of salmonellosis in Canada in 2015 was linked to handling of baby poultry sourced from hatcheries in Alberta (Public Health Agency of Canada, 2015). Animal contact (specifically contact with dogs) was implicated as a risk factor for the increased number of cases of infection caused by *S.* Enteritidis PT8 that occurred in Ontario in 2011 (Varga et al., 2012). Another case–control study completed in 2011 in Ontario, Canada, found that other risk factors including consumption of poultry meat or processed chicken and not washing hands following handling of raw eggs were associated with *S.* Enteritidis infections (Middleton et al., 2014). The risk of an *S.* Enteritidis infection increased threefold in those who did not wash hands after handling raw eggs compared with those who reported washing hands. Consumption of undercooked eggs or consumption of eggs away from home was not identified as significant risk factors. Middleton et al. (2014) concluded that, although egg consumption was historically thought to be responsible for *S.* Enteritidis infections, consumption of poultry is a more important risk factor. There is no doubt that attribution of salmonellosis to specific foods can be a challenge as many foods have been implicated in outbreaks and it is often very difficult to link illness to a specific etiological agent, especially when multiple food sources may be implicated.

Data from publically available sources provide some insight as to the nature of the outbreaks of salmonellosis that have been linked to consumption of eggs. These outbreaks are summarized in Table 4.1. In 2000, an outbreak occurred in BC, where Grade B eggs were implicated as the source of *S.* Enteritidis but poor handling practices, use of a raw egg wash on a bakery product, and an ill person preparing food were also risk factors associated with the outbreak (Taylor et al., 2012). Ungraded hatching eggs or illegally sourced eggs were responsible for three other documented outbreaks. In two of these cases, unsanitary handling and poor food safety practices in a food service facility were identified as risk factors. In one documented outbreak, the consumption of raw egg noodles was implicated in cases of salmonellosis and the outbreak strain was also found in samples taken from the layer farm that supplied the eggs (Taylor et al., 2012). In cases where ungraded eggs were implicated as the source of the outbreaks, the eggs were not always refrigerated (Taylor et al., 2012), which would increase the risk of growth of *S.* Enteritidis. There have been other outbreaks of salmonellosis that have been potentially linked to the consumption of eggs but often other poultry products were implicated in the same outbreak.

According to Taylor et al. (2012), who investigated the increase in human infections in BC between 2007 and 2010, the increase in human cases of *S.* Enteritidis may be linked to an increase in *S.* Enteritidis in hatcheries; however, at the same time, table egg monitoring programs did not detect any change in the number of samples positive for *S.* Enteritidis. The authors speculated that the increase in the hatcheries may have been linked to repopulation of large numbers of barns after a massive depopulation and restocking of poultry flocks after an outbreak of avian influenza. The increased number of outbreaks related to hatching eggs, graded eggs, and the lack of safe food handling practices resulted in action by the BC Hatching

Egg Commission to stop the sale of these eggs directly from farms to consumers or to food service companies.

Based on the source and risk factors associated with reported outbreaks of *S.* Enteritidis, there is little evidence to say either way if the on-farm food safety programs have made an impact on reducing the burden of salmonellosis linked to the consumption of table eggs in Canada. Most outbreaks have been linked to eggs that are not part of the regulated supply system and where they were linked, food safety systems throughout the food chain were not adequate to protect consumers. Based on this, one can speculate that the regulated egg supply is of lower risk for attribution of human salmonellosis but there is little evidence to support this statement. There is a need for additional investigation and surveillance to provide evidence of efficacy of on-farm food safety programs.

Reducing the disease burden from the presence of *S.* Enteritidis in eggs in Canada requires an integrated approach to ensure that all parties involved from the primary breeders to the consumer understand the risks. There is definitely a need for more attention to handling practices in food service to improve food safety standards related to general hygiene and controls to reduce risks across food systems.

## REFERENCES


Agriculture and Agri-Food Canada, 2015. Supply Management 101. Retrieved from: http://www.agr.gc.ca/eng/industry-markets-and-trade/agri-food-trade-policy/current-negotiations/bilateral-and-regional-negotiations-and-initiatives/trans-pacific-partnership/supply-management-101/?id=1443191400607.

Ahmed, R., Soule, G., Demczuk, W.H., Clark, C., Khakhria, R., Ratnam, S., Marshall, S., Ng, L.-K., Woodward, D.L., Johnson, W.M., Rodgers, F.G., 2000. Epidemiologic typing of *Salmonella enterica* serotype enteritidis in a Canada-wide outbreak of gastroenteritis due to contaminated cheese. J. Clin. Microbiol. 38, 2403–2406.

Anonymous, 2015. Nuggets of wisdom: trends in *Salmonella* and outbreaks associated with frozen, processed chicken products. Public Health Ontario Monthly Infect. Dis. Surveill. Rep. 4 (15), 1–8. Retrieved from: https://www.publichealthontario.ca/en/DataAndAnalytics/Documents/PHO_Monthly_Infectious_Diseases_Surveillance_Report_-_October_2015.pdf.

Bélanger, P., Tanguay, F., Hamel, M., Phypers, M., 2015. An overview of foodborne outbreaks in Canada reported through outbreak summaries: 2008–2014. Can. Comm. Dis. Rep. 41, 254–262.

Canadian Egg Industry Reciprocal Alliance, 2016. Canadian Egg Industry Reciprocal Alliance. Retrieved from: http://www.ceira.ca/.

Canadian Egg Marketing Agency (CEMA), 2004. Start Clean Stay Clean™. HACCP – Based On-Farm Food Safety Program for Shell Egg Production and Spent Hens in Canada. Canadian Egg Marketing Agency, Ottawa, Ontario.

Communicable Disease Prevention and Control Services (CDPCS), 2014. British Columbia Annual Summary of Reportable Diseases. Retrieved from: www.bccdc.ca/resource-gallery/Documents/Statistics%20and%20Research/Statistics%20and%20Reports/Epid/Annual%20Reports/2013CDAnnualReportFinal.pdf.

Denagamage, T., Jayarao, B., Patterson, P., Wallner-Pendleton, E., Kariyawasam, S., 2015. Risk factors associated with *Salmonella* in laying hen farms: systematic review of observational studies. Avian Dis. 59, 291–302.

DeWinter, L.M., Ross, W.H., Couture, H., Farber, J.F., 2011. Risk assessment of shell eggs internally contaminated with *Salmonella* Enteritidis. Int. Food Risk Anal. J. 1, 40–81.

Egg Farmers of Canada, 2014. http://www.eggfarmers.ca/wp-content/uploads/2015/03/Egg-Farmers-of-Canada_Annual-Report_2014.pdf.

Egg Farmers of Canada, 2015. Egg Farmers of Canada Annual Report 2015. Retrieved from: http://www.eggfarmers.ca/wp-content/uploads/2016/03/Egg-Farmers-of-Canada_Annual-Report-20151.pdf.

Galanis, E., Parmley, J., De With, N., 2012. Integrated surveillance of *Salmonella* along the food chain using existing data and resources in British Columbia, Canada. Food Res. Int. 45, 792–801.

Gast, R.K., Holt, P.S., 2000. Influence of the level and location of contamination on the multiplication of *Salmonella enteritidis* at different storage temperatures in experimentally inoculated eggs. Poult. Sci. 79, 559–563.

Government of Canada, 2016a. Canada Agricultural Products Act (R.S.C., 1985, c. 20 [4th Supp.]). Retrieved from: http://laws-lois.justice.gc.ca/eng/acts/C-0.4//.

Government of Canada, 2016b. Egg Regulations (C.R.C., c. 284). Retrieved from: http://laws-lois.justice.gc.ca/eng/regulations/C.R.C.,_c._284/.

Government of Canada, 2016c. The Food and Drugs Act (R.S.C., 1985, c.F-27). Retrieved from: http://laws-lois.justice.gc.ca/eng/acts/F-27/page-1.html#h-5.

Health Canada, 2013. Health Canada Guidance on Reducing the Risk of *Salmonella* Enteritidis in Shell Eggs Produced in Canada. Retrieved from: http://www.hc-sc.gc.ca/fn-an/legislation/guide-ld/salmonella-enteritidis-eng.php.

Health Canada, 2015. National Strategy for the Control of Poultry-related Human *Salmonella* Enteritidis Illness in Canada. Available from: www.hc-sc.gc.ca/fn-an/legislation/guide-ld/salmonella-enteritidis-illness-maladie-eng.php.

Honish, L., Nguyen, Q., 2001. Outbreak of *Salmonella* enteritidis phage type 913 gastroenteritis associated with mung bean sprouts – Edmonton, 2001. Can. Commun. Dis. Rep. 27, 151–156.

Honish, L., Greenwald, D., McIntyre, K., Lau, W., Nunn, S., Nelson, D., McDonald, J., Keegan, V., Wilkinson, K., 2013. *Salmonella* Enteritidis infections associated with foods purchased from mobile lunch trucks – Alberta, Canada, October 2010–February 2011. Morb. Mortal.Wkly. Rep. Surveill. Summ. 62, 567–569.

Isaacs, S., Aramini, J., Ciebin, G., Farrar, J.A., Ahmed, R., Middleton, D., Chandran, A.U., Harris, L.J., Howes, M., Chan, E., Pichette, A.S., Campbell, K., Gupta, A., Lior, L.Y., Pearce, M., Clark, C., Rodgers, F., Jamieson, F., Brophy, I., Ellis, A., *Salmonella* Enteritidis PT30 Outbreak Working Group, 2005. An international outbreak of salmonellosis associated with raw almonds contaminated with a rare phage type of *Salmonella* Enteritidis. J. Food Prot. 68, 191–198.

Keery, I., 2010. *Salmonella* Enteritidis Control Programs in the Canadian Poultry Industry. Prepared for: Surveillance and Epidemiology Advisory Committee. Retrieved from: http://www.agf.gov.bc.ca/lhmr/pubs/se_control_programs0910.pdf.

Middleton, D., Savage, R., Tighe, M.K., Vrbova, L., Walton, R., Whitefield, Y., Varga, C., Lee, B., Rosella, L., Dhar, B., Johnson, C., Ahmed, R., Allen, V.G., Crowcroft, N.S., 2014. Risk factors for sporadic domestically acquired *Salmonella* serovar Enteritidis infections: a case-control study in Ontario, Canada, 2011. Epidemiol. Infect. 142, 1411–1421.

Mumma, G.A., Griffin, P.M., Meltzer, M.I., Braden, C.R., Tauxe, R.V., 2004. Egg quality assurance programs and egg associated *Salmonella enteritidis* infections, United States. Emerg. Infect. Dis. 10, 1782–1789.

Musgrove, M.T., Jones, D.R., Northcutt, J.K., Harrison, M.A., Cox, N.A., 2005. Impact of commercial processing on the microbiology of shell eggs. J. Food Prot. 68, 2367–2375.

Nesbitt, A., Ravel, A., Murray, R., McCormick, R., Savelli, C., Finley, R., Parmley, J., Agunos, A., Majowicz, S.E., Gilmour, M., The Canadian Integrated Program for Antimicrobial Resistance Surveillance Public Health Partnership, the Canadian Public Health Laboratory Network, 2012. Integrated surveillance and potential sources of *Salmonella* Enteritidis in human cases in Canada from 2003 to 2009. Epidemiol. Infect. 140, 1–16.

Parmley, E.J., Pintar, K., Majowicz, S., Avery, B., Cook, A., Jokien, C., Gannon, V., Lapen, D.R., Topp, E., Edge, T.A., Gilmour, M., Pollari, F., Reid-Smith, R., Irwin, R., 2013. A Canadian application of one health: integration of *Salmonella* data from various surveillance programs (2005–2010). Foodborne Pathog. Dis. 9, 747–756.

Patrick, M.E., Adcock, P.M., Gomez, T.M., Alterkruse, S.F., Holland, B.H., Tauxe, R.V., Swerdlow, D.L., 2004. *Salmonella enteritidis* infections, United States: 1985–1999. Emerg. Infect. Dis. 10, 1–7.

Public Health Agency of Canada, 2015. Public Health Notice – Outbreak of *Salmonella* Infections Related to Contact with Live Baby Poultry. Available from: http://www.phac-aspc.gc.ca/phn-asp/2015/salmonella-eng.php.

Statistics Canada, 2007. Poultry and Egg Statistics, October to December 2006. Catalogue No. 23-015-XIE.

Statistics Canada, 2015. Population by Year, by Province and Territory (Number). Retrieved from: http://www.statcan.gc.ca/tables-tableaux/sum-som/l01/cst01/demo02a-eng.htm.

Strauss, B., Fyfe, M., Higo, K., Sisler, M., Paccagnella, A., Trinidad, A., Louie, K., Kurzac, C., Zaharia, B., 2005. *Salmonella enteritidis* outbreak linked to baked goods from a local bakery in lower mainland, British Columbia. Can. Commun. Dis. Rep. 31, 73–82.

Taylor, M., Leslie, M., Ritson, M., Stone, J., Cox, W., Hoang, L., Galanis, E., Outbreak Investigation Team, 2012. Investigation of the concurrent emergence of *Salmonella enteritidis* in humans and poultry in British Columbia, Canada, 2008–2010. Zoonoses Public Health 59, 584–592.

Taylor, M., Galanis, E., Forsting, S., Gustafson, L., Ip, J., Jeyes, J., Lem, M., Murti, M., Nowakowski, C., Ritson, M., Stone, J., Tone, G., 2015. Enteric outbreak surveillance in British Columbia 2009–2013. Can. Commun. Dis. Rep. 41, 263–271.

Tighe, M.K., Savage, R., Vrbova, L., Toolan, M., Whitfield, Y., Varga, C., Lee, B., Allen, V., Maki, A., Walton, R., Johnson, C., Dhar, B., Ahmed, R., Crowcroft, N.S., Middleton, D., 2012. The epidemiology of travel-related *Salmonella* Enteritidis in Ontario, Canada, 2010–2011. BMC Public Health 12, 310.

Todd, C.D., 1996. Risk assessment of use of cracked eggs in Canada. Int. J. Food Microbiol. 30, 123–143.

Varga, C., Middleton, D., Walton, R., Savage, R., Tighe, M.K., Allen, V., Ahmed, R., Rosella, L., 2012. Evaluating risk factors for endemic human *Salmonella* Enteritidis infections with different phage types in Ontario, Canada using multinomial logistic regression and a case-case study approach. BMC Public Health 12, 866–875.

CHAPTER

# Egg Production Systems and *Salmonella* in Australia

# 5

**Kapil K. Chousalkar, Andrea McWhorter**

*The University of Adelaide, Roseworthy, SA, Australia*

The globalization of food production processes means that food-borne gastrointestinal disease is an important global public health issue. Consumption of food items contaminated with enteric bacteria species are among the most common causes of food-related diarrheal disease, with *Campylobacter* spp. and *Salmonella* spp. being the most frequently identified bacterial species in food-related disease outbreaks. A wide variety of food items may become contaminated with *Salmonella* spp. but the consumption of raw or undercooked eggs is often associated with cases of salmonellosis. Infection with *Salmonella* spp. typically results in self-limiting diarrhea but can be more severe in children, the elderly, or immunocompromised individuals (Fabrega and Vila, 2013). In the *Salmonella enterica* type I group, there are over 2500 different serovars but only a limited number of them are consistently isolated following food-borne human salmonellosis (Fabrega and Vila, 2013).

*Salmonella enterica* serovar Enteritidis (*S.* Enteritidis) is a major concern for most egg industries around the world and is associated with the majority of egg-related outbreaks of human salmonellosis. *S.* Enteritidis is, however, not endemic to Australian commercial layer flocks (Arzey, 2002; NSWDPI, 2015). The serovar *Salmonella* Typhimurium (*S.* Typhimurium) has filled this niche and is commonly isolated from Australian layer farms (Cuttell et al., 2014; NSWDPI, 2015). To date, *S.* Typhimurium has been the most frequently reported serovar during egg product-related food poisoning outbreaks Australia-wide (OzFoodnet, 2002, 2005, 2003, 2006, 2012). In this chapter, we present an Australian picture of egg-associated *Salmonella* epidemiology and discuss the overall disease potential of different serovars found in Australia and current on-farm control measures.

## 1. EGG PRODUCTION SYSTEMS IN AUSTRALIA

The egg industry is an important component of Australian intensive food animal production industries. Australian consumers prefer brown-shelled eggs; hence brown egg layer breeds are popular in the Australian egg industry. In 2013–14,

Producing Safe Eggs. http://dx.doi.org/10.1016/B978-0-12-802582-6.00005-7

the Australian commercial egg industry produced 397.4 million dozen eggs from 16.55 million commercial laying hens. Egg production occurs in all Australian states, including Western Australia, South Australia, New South Wales, Victoria, Queensland, the Northern Territory, the Australian Capital Territory, and Tasmania but large-scale production is limited to fewer locations. Production data of 2015 show that New South Wales was the largest egg producer (33%) followed by Queensland (26%) and Victoria (25%) (AECL, 2015).

Under the Primary Industries (Excise) Levies Act 1999 and the Primary Industries Levies and Charges and Collections Act 1991, the Australian Government collects and appropriates levies from egg producers who produce eggs for human consumption (NSWDPI, 2015). Funds raised from these levies are distributed to the Australian Egg Corporation Limited (AECL), Animal Health Australia, and the Department of Agriculture and water resources. These funds are invested in research, development and extension, marketing, residue testing, animal health programs, and disease management (NSWDPI, 2015).

In Australia, there are three major types of commercial egg production systems, cage, barn, and free range. In 2013–14, cage production systems contributed the largest volume of eggs (53% of total volume) followed by free range (38%) and barn (8%) with per capita egg consumption of 213.3 (AECL, 2015). Egg producers can perform all stages of egg production, including collection, grading, packing, and marketing. Primary breeders (mostly commercial companies) manage the elite genetic stock of layer hens that have been imported to Australia as fertile eggs. These birds (elite genetic stock often referred to as Great grandparents) are hatched in quarantine facilities and are tested for *Salmonella* Enteritidis. Egg producers either rear day-old chicks purchased from hatchery or buy point of lay pullets. Commercial egg farms raise multiage flocks housed in separate sheds permitting a continuous egg production cycle. Some cage producers also house multiage flocks in the same shed. Medium to large-scale egg producers have their own egg grading and packing floors. Small-scale egg producers, however, pack eggs and depending on marketing arrangements, transport them to egg grading facilities operated by egg distributors (NSWDPI, 2015).

Most egg producers are represented by the AECL. The AECL has developed a voluntary egg quality program that provides guidelines for food safety, biosecurity, environmental use, hen health, welfare, and labeling for the national egg industry. In addition to State and Territory legislation, the AECL has developed voluntary Codes of Practice for assisting egg producers (AECL). These codes provide guidance on hygienic egg production, storage, packaging, and distribution of shell eggs and egg products for human consumption. It has been reported that few outbreaks of food-borne salmonellosis can be attributed to the consumption of eggs that are produced under an authoritative quality control system (Thomas et al., 2006); however, consumption of noncommercially produced ungraded eggs sold directly off the farm has also been responsible for food-borne salmonellosis (Thomas et al., 2006).

## 2. *SALMONELLA* OUTBREAKS ASSOCIATED WITH EGGS

In the United States and Europe, *S.* Enteritidis has been most frequently associated with egg-related outbreaks of salmonellosis. In Australia, however, definitive types of *S.* Typhimurium are of primary concern. In 2015, the Department of Health recorded 16,952 cases of human salmonellosis. Between 2009 and 2015, cases of *Salmonella* infection have increased, with *S.* Typhimurium being the most predominant serovar reported (Health, 2016; OzFoodnet, 2002, 2003; OzFoonet, 2005, 2006, 2012) (Fig. 5.1). During this period, eggs or consumption of raw egg–based foods were frequently implicated as the source of infection. Australian outbreaks of foodborne salmonellosis have been attributed to the consumption of eggs produced under both authoritative quality control systems and noncommercially produced ungraded eggs sold directly off the farm (Thomas et al., 2006).

The documented number of *Salmonella* cases from 2001 to 2013 is presented in Fig. 5.1. During this time, there has been an increase in the number of *Salmonella* notifications. It should be noted, however, that reporting systems and *Salmonella* diagnostic methodologies have improved. The OzFoodnet annual and quarterly reports from 2001 to 2013 reported 440 *S.* Typhimurium outbreaks, of which 171 (38.8%) were linked to the consumption of egg-based products (excluding bakery items). Bakery products and chicken meat products were also linked with *S.* Typhimurium cases, although some outbreaks were caused by unknown sources.

In 2001, the overall rate of *Salmonella* notification in Australia was 36.2 cases per 100,000 individuals (Fig. 5.1). In 2002, separate salmonellosis outbreaks linked to the consumption of cream-based pastries and a raw egg dish resulted in two mortalities (OzFoodnet, 2003). Cake and other desserts prepared from presumably contaminated eggs were identified during three 2004 outbreaks. During one outbreak, *S.* Typhimurium PT 126 was traced to an organic egg brand in Victoria. Investigations into other outbreaks did not link contaminated eggs to a single source (OzFoonet, 2005). In 2005, six egg-related *Salmonella* food poisoning outbreaks were reported (OzFoonet, 2006).

Between 2006 and 2010, 92 *Salmonella* food poisoning outbreaks resulted in 1740 cases with a hospitalization rate of 23% (400/1740). During this period, New South Wales (37 cases) and Victoria (22 cases) recorded the highest number of outbreaks. Outbreaks were more frequent in warmer months (Oct. to Mar.). Of 92 outbreaks, most (91%) were due to different phage types (PT) of *S.* Typhimurium. *S.* Typhimurium PT 170 (31.5%) and *S.* Typhimurium PT 193 (19.6%) were the most frequently recorded PTs in egg-implicated outbreaks followed by *S.* Typhimurium PT 9 (14%) and PT 135a (8.7%). Despite being implicated during *Salmonella* outbreaks, it has been difficult to accurately identify the source of egg contamination (Stephens et al., 2008). For example, five *S.* Typhimurium PT 135 outbreaks were reported in Tasmania (Jun. to Dec. 2005) involving 125 laboratory confirmed cases. These outbreaks were investigated through personal interviews, cohort studies, microbiological testing, environmental health investigation of food business, trace

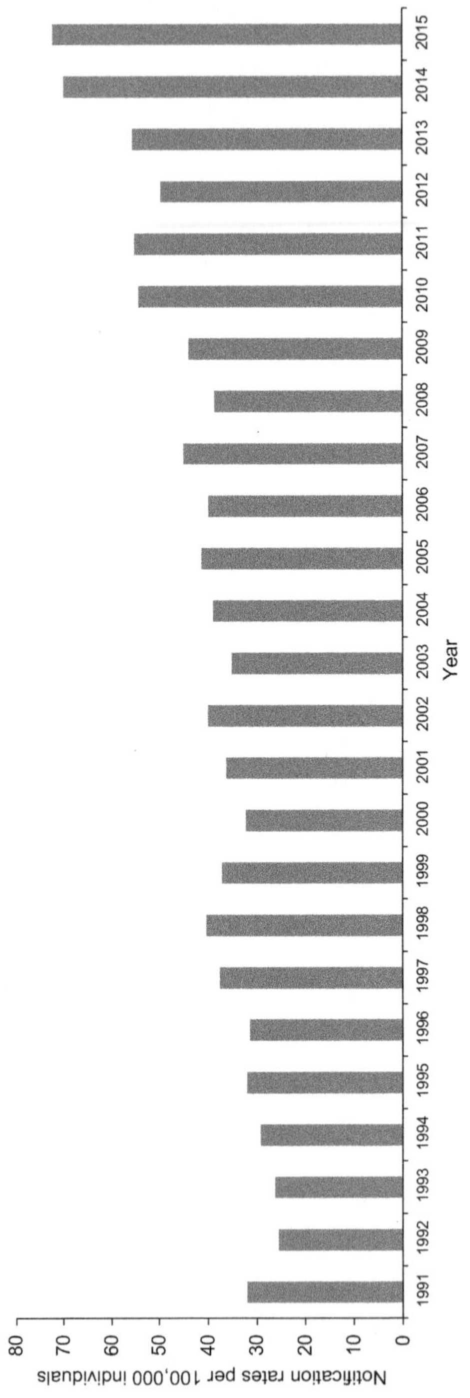


**FIGURE 5.1**

Number of notifications of human salmonellosis in Australia (data from 1991 to 2015).

*Adapted from Health Department, 2016. National Notifiable Disease Surveillance System, Department of Health, Australia.*

back, inspection, and drag swabbing of egg farms. This investigation revealed that food containing raw eggs or cross-contamination of food items due to improper handling and storage were the cause of the outbreaks. Eggs and packing containers contaminated with feces were identified as sources of raw ingredient contamination (Stephens et al., 2008). The same egg farm was implicated in two additional egg-associated *S.* Typhimurium PT 135 outbreaks (Mar. 2007 and Jan. 2008), affecting 66 people (Stephens et al., 2008). For all outbreaks, investigation was limited to phage typing, which cannot differentiate isolates of the same phage type. Hence, it is difficult to definitively prove that the *S.* Typhimurium PT 135 isolated from the egg farm was the same strain detected in the human cases.

Since 2010, the number of egg-related *Salmonella* outbreaks has continued to increase. In 2011, Victoria (12 cases) and New South Wales (9 cases) recorded the highest number of outbreaks (OzFoonet, 2012). In 2012, the role of eggs in *Salmonella* outbreaks was publically highlighted in the media during one of the largest outbreaks in the Australian Capital Territory. This outbreak affected 140 people and resulted in the hospitalization of 15 individuals. Mayonnaise prepared with raw eggs was suspected as a source of infection in this outbreak.

## 3. EPIDEMIOLOGY OF *SALMONELLA* ON AUSTRALIAN COMMERCIAL EGG FARMS

Layer hens have a common opening for the intestinal, urinary, and reproductive tracts. Thus, external egg shell contamination with fecal material is often unavoidable. Although vertical transmission of *Salmonella* from bird to egg has been demonstrated (as has been shown for *S.* Enteritidis), it is generally accepted that horizontal transmission is also the most likely source of contamination of shell eggs (Gantois et al., 2009). Small defects in the egg shell may provide the means for bacteria on the egg shell surface to penetrate and move into the egg contents (De Reu et al., 2006).

An Australian investigation found that *S.* Typhimurium PT 9 was not detectable from eggs laid by infected hens in a field environment or hens infected at the onset of lay (Gole et al., 2014c). A *Salmonella* survey of the Queensland egg industry revealed that *S.* Infantis was the most prevalent among egg layer flocks (Cox et al., 2002). Another microbiological survey conducted by New South Wales Food Authority on 49 egg farms in New South Wales showed that 20% of the farms were positive for *S.* Typhimurium, whereas a survey conducted on 21 egg farms by Safe Food Queensland reported that 13.5% of farms were positive for *S.* Typhimurium (Cuttell et al., 2014).

Currently there is no nationwide prevalence data of *S.* Typhimurium on egg farms. A longitudinal study of cage farms indicated that fecal samples were the best indicator of *S.* Typhimurium egg contamination on the farm and that over time, the multiple loci variable tandem repeats analysis (MLVA) pattern of *S.* Typhimurium was unstable (Gole et al., 2014c). The odds of an eggshell testing positive for *Salmonella* were 91.8, 61.5, and 18.2 times higher when fecal,

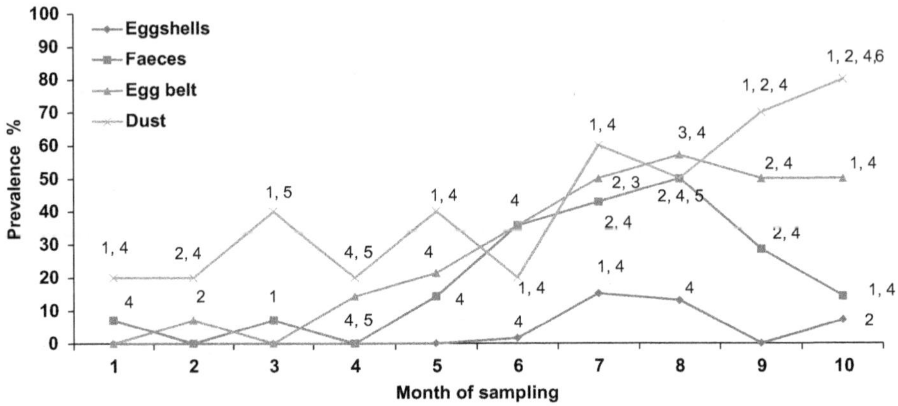


1: *Salmonella* Typhimurium phage type 9
2: *Salmonella* Worthington
3: *Salmonella* subsp.1 ser 4,5,12:-:-
4: *Salmonella* Oranienburg
5: *Salmonella* Agona
6: *Salmonella* subsp.1 ser rough g,s,t:-

**FIGURE 5.2**

Percentage prevalence of *Salmonella* in different types of samples over a period of 10 months in caged flock.

*Adapted from Gole, V.C., Torok, V., Sexton, M., Caraguel, C.G., Chousalkar, K.K., 2014c. Association between indoor environmental contamination by* Salmonella enterica *and contamination of eggs on layer farms. J. Clin. Microbiol. 52, 3250–3258.*

egg belt, and dust samples on egg farms tested *Salmonella* positive (Gole et al., 2014c). One of the major challenges in establishing prevalence is that shedding of *S.* Typhimurium from known positive hens is highly variable (Fig. 5.2) and can be influenced by stress experienced by hens on the farm (Gole et al., 2014a). Single time point sampling may not be sufficient to determine true prevalence; therefore longitudinal sampling of flocks is essential. Due to increasing consumer demand for free range eggs and the likelihood of environmental contamination, it is particularly important to monitor the long-term shedding of *S.* Typhimurium on free range farms. Birds raised in free range production systems are potentially exposed to more environmental stressors than caged birds, including social stress and aggression, predation, or thermal challenges. The current Australian free range egg production system standards require all birds to have access to the range for a minimum of 8 h per day once they are reasonably feathered (i.e., by onset of lay). The only exception to this is during extreme weather conditions (e.g., exceptionally hot or cold weather, high humidity, very strong winds, or heavy rain) or under veterinary advice (e.g., due to a disease outbreak).

The effect environmental stressors have on layer hens and how they contribute to *Salmonella* shedding patterns have been studied in both controlled and field trials (Nakamura et al., 1994; Seo et al., 2000). Both studies reported the short-term increase in shedding of *S.* Enteritids with short-term exposure to stress. In addition, chronic stress can have an immunosuppressive effect on laying hens, which could further influence *Salmonella* infection and shedding (El-Lethey et al., 2003; Humphrey, 2006). Over the lifespan of a layer hen, a bird may be subjected to a wide range of stressors, such as transport of day-old birds, vaccinations during rearing, moving from the rearing site to the egg producing plant (Hughes et al., 1989), the onset of lay (Humphrey, 2006), final stages of the production period, thermal extremes (Mashaly et al., 2004), or transportation (Beuving and Vonder, 1978). Molting-related stress in hens has also been shown to enhance *Salmonella* shedding in eggs and feces (Holt, 2003) and could also cause higher levels of colonization in internal organs (Holt et al., 1995). Further studies are necessary to establish links between environmental stressors and *Salmonella* shedding in free range production systems.

Human *Salmonella* infections can also be attributed to direct human contact with animals. Backyard chicken ownership has increased in recent years (Behravesh et al., 2014), increasing the likelihood of direct hen to human transmission of *Salmonella*. A small-scale Australian survey of backyard layer hens showed that the serovars *S.* Agona and *S.* Bovismorbificans were the most prevalent (Manning et al., 2015). Historically, *S.* Bovismorbificans has been associated with nine outbreaks of salmonellosis in Australia (Stafford et al., 2002), but were not linked with eggs or egg products. Flock owners involved in this study were reluctant to permit multiple samplings so the prevalence data represent only a single sampling. Longitudinal sampling on a wider scale is essential to characterize the prevalence of *S.* Typhimurium in backyard hens and establish potential risks associated with direct transmission of bacteria.

## 4. PREVALENCE AND EGGSHELL PENETRATION BY *SALMONELLA*

According to Australian food safety authorities, *Salmonella* contamination of eggs and egg products is a major public health issue. The actual number of commercially produced eggs contaminated with *Salmonella* is unclear. Multiple surveys have been conducted using both large and small sample sizes. A survey conducted in 1986 sampled 360 eggs sourced from both wholesale and retail markets in Queensland and all were negative for *Salmonella* (Douglas, 2004). In 1989, a similar survey involving 199 eggs also did not detect *Salmonella* (Cozens, 2010; Douglas, 2004). A large-scale survey of commercial eggs examined the external surface of 10,000 eggs and the internal contents of 20,000 eggs (Daughtry et al., 2005). Consistent with previous small-scale surveys, all samples in Daughtry's study were *Salmonella* negative. A study conducted on isolation of *Salmonella* from egg shell wash, egg shell pores and internal contents using a relatively small number of samples revealed that the egg shell and egg internal contents were negative (Chousalkar et al., 2010). However, in a later study performed in 2012, *S.* Infantis was isolated from egg shell wash (Chousalkar and Roberts, 2012). Fearnley et al. (2011) reported that prevalence of *Salmonella* on retail eggs on the shelf was 0.30%. It is important to note that *S.* Typhimurium was not isolated from eggs during any of the egg-based surveys, although eggs and raw egg products have still been associated with *Salmonella* outbreaks. Egg-based surveys are important but laborious due to the large number of eggs required for *Salmonella* testing. In Australia, 4.7 billion eggs are produced annually; hence reliance on surveying of a small number of eggs may not be an accurate or true reflection of *Salmonella* prevalence. *S.* Typhimurium-positive hens could shed up to $10^6$ colony forming units (CFU) of bacteria on egg shells (Gole et al., 2014c), but egg shell contamination could also occur further down the supply chain or egg handling, on or off the egg farm. These factors are yet to be investigated thoroughly in Australia.

The presence of *S.* Typhimurium on egg shells, egg shell quality, and storage temperature can influence the penetration ability of bacteria across the egg shell (Gole et al., 2014b). There are three physical barriers that protect the egg from bacterial penetration, the cuticle, the crystalline eggshell, and the membranes, which divide the albumen and the eggshell (Galiş et al., 2013). Even though bacteria penetrate through shell membranes, egg internal contents are protected by a number of antimicrobial chemical components in the albumen. *S.* Typhimurium can survive on dry egg shell for up to 3 weeks (Gole et al., 2014b). Eggshell penetration by *S.* Typhimurium can also be influenced by various eggshell ultrastructural features including cap quality, alignment, erosion, confluence, Type B bodies, and cuticle cover. It is important to note that egg penetration studies were performed after infecting eggs with *Salmonella* in laboratory conditions followed by incubation at room temperature (Gole et al., 2014b). *Salmonella* penetration across the egg shell, however, is low at refrigeration temperature (4°C) (Chousalkar et al., 2010), underscoring the importance of proper egg storage in both commercial and domestic environments.

## 5. HUMAN FACTOR IN *SALMONELLA* VIRULENCE FACTORS

Nontyphoidal *Salmonella* (NTS) serovars such as *S.* Typhimurium generally cause self-limiting diarrhea and exhibit a low fatality rate of 0.1% (Fabrega and Vila, 2013). In 2010, there was a twofold increase in human cases of egg-related food poisoning outbreaks due to serovars other than *S.* Typhimurium. Other NTS including *S.* Infantis, *S.* Singapore, *S.* Anatum, *S.* Postdam, *S.* Saintpaul, *S.* Virchow, and *S.* Montevideo continue to be reported sporadically (OzFoodnet, 2002, 2003, 2005, 2006, 2012). Not all *Salmonella* isolated from egg or egg products, however, are equally invasive and/or pathogenic (McWhorter and Chousalkar, 2015; McWhorter et al., 2015).

Variation in pathogenicity between *Salmonella* serovars is likely due to genomic variation within *Salmonella* pathogenicity islands (SPI) (McWhorter and Chousalkar, 2015). Furthermore, a favorable growth environment is required for certain serovars (McWhorter and Chousalkar, 2015; McWhorter et al., 2015). This is in part due to the upregulation of invasion genes encoded within SPI 1 in a nutritive environment. *S.* Typhimurium PT9 has been frequently reported during Australian egg-related outbreaks. In vitro, PT 9 is highly invasive following

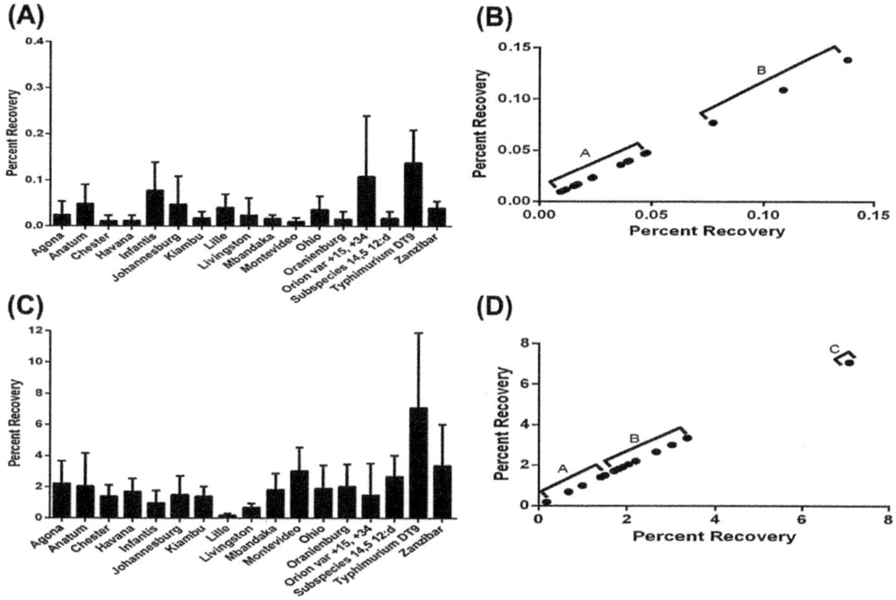


**FIGURE 5.3**

The in vitro invasive capacity of 17 nontyphoidal *Salmonella* serovars before and after enrichment in media. The in vitro invasive capacity of 17 nontyphoidal *Salmonella* serovars was assessed using the gentamicin protection assay with the human intestinal epithelial cell line, Caco2. Bacteria were either suspended in 0.9% saline (A) or grown to stationary phase in LB (C) and then added to cell monolayers at an MOI of 100. Data are represented as mean percent recovery. Assays were repeated five times. Statistical analysis was performed by Kruskal–Wallis analysis of variance with post hoc analysis utilizing Dunn's multiple comparisons test. A significant effect of serotype for both treatment groups ($p<0.006$, saline; $p<0.0002$ LB) was detected but no significant differences were observed between individual serotypes. *k-means* Cluster analysis was performed to identify invasion groups. Two invasive types were identified for serovars suspended in 0.9% saline and are identified as Group A (low) and Group B (moderate) (B). Following growth in an enriched environment (LB broth), substantial increases in percent recoveries were observed (C). Three invasion types were identified by cluster analysis and were classified as low (Group A), moderate (Group B), and high (Group C) (D). *LB*, Luria-Bertani; *MOI*, Multiplicity of infection.

*Taken from McWhorter, A.R., Davos, D., Chousalkar, K.K., 2015. Pathogenicity of* Salmonella *strains isolated from egg shells and the layer farm environment in Australia. Appl. Environ. Microbiol. 81, 405–414.*

growth in media (Fig. 5.3) (McWhorter et al., 2015) yet retains invasive potential in a nonnutritive environment, an ability not observed for all serovars. As such, it may have constitutively active virulence gene(s) that facilitate invasion under any conditions. Therefore, if there is any point during food preparation or storage that encourages the growth or enrichment of *Salmonella* within the food item, the risk of potential food poisoning increases if these serovars are present.

Although proper food or egg handling procedures coupled with kitchen cleanliness limits the possibility of infection, only $10^2$ CFU of pathogenic strains of *Salmonella* are required to cause disease in humans (Fabrega and Vila, 2013). Therefore, targeted *Salmonella* control strategies are not only required for egg farms but through the entire egg supply chain to food preparation and consumption.

## 6. CURRENT ON-FARM CONTROL MEASURES WITH FOCUS ON EGG WASHING

Various methods have been used to control *Salmonella* in layer flocks, which can be either preharvest or postharvest (Galiş et al., 2013). Preharvest methods include genetic selection of hens for resistance to *Salmonella*, flock management involving sanitation, regular flock testing, biosecurity, vaccination, as well as the use of natural antimicrobial products such as prebiotics, probiotics, or organic acids (Galiş et al., 2013). Postharvest methods involving eggshell decontamination using egg washing by sanitizers, ozone, electrolyzed water, and irradiation of egg shell by ultraviolet light, gas plasma technology, or use of biological methods (plant extracts) have also been discussed in the literature; however, their commercial or wide-scale use in the industry is not known. Vaccination of pullets, to either prevent infection or reduce the duration of shedding of *Salmonella* in exposed flocks, is one measure of control (Kilroy et al., 2015). Both killed virulent and live attenuated *Salmonella* vaccines are used, which can result in varying degrees of protection in chickens (Holt et al., 2003; Pavic et al., 2010). Vaxsafe ST (Bioproperties Pty Ltd, Australia) is the only live attenuated vaccine registered for use in poultry in Australia. The long-term efficacy of the vaccine in commercial flocks that are actively shedding *S.* Typhimurium remains unclear.

Egg washing with sanitizers is one of the most common methods of reducing eggshell contamination in Australia. This technique is adopted in many countries including Australia, Japan, and the United States but is banned in the European Union (Messens, 2013). Egg washing protocols involve prewashing, sanitizer wash, and drying. During egg washing, the maintenance of rinse water temperature is important. If an egg is placed in a cool environment, egg internal contents contract causing the membranes to pull away from the eggshell creating negative pressure and movement of air or contamination across the eggshell (Messens, 2013). The temperature of wash water should be maintained 6.7–12°C higher than egg shell temperature to avoid cracking during washing (Galiş et al., 2013). The major advantage of egg washing is the removal of fecal debris thereby reducing the overall bacterial load on the eggshell surface (Galiş et al., 2013); this in turn minimizes the chance of cross-contamination in the kitchen environment.

It should be noted that egg washing chemicals have the potential to alter the eggshell surface and damage the cuticle layer (Gole et al., 2014b). Gole et al. (2014b) reported that egg penetration by *S.* Typhimurium was higher in washed eggs than in unwashed eggs, which may be due to cuticle damage sustained by egg washing chemicals. Therefore, eggs should be kept under appropriate storage and drying conditions to prevent *Salmonella* contamination post washing. Regular cleaning of the egg washing and grading equipment is also necessary to avoid recontamination of eggs after washing. Proper hand washing is also important after egg handling to avoid cross-contamination of food items in private and commercial kitchen environments.

## 7. *SALMONELLA* CONTROL MEASURES AND LIMITATIONS

Trends in the United States, United Kingdom, Europe, and Australia indicate that the incidence of food-borne illness is increasing, and is likely to remain a threat to public health for the foreseeable future. In the modern Australian society, people cook fewer meals at home and are more reliant on ready-to-cook, ready-to-eat foods and takeaway meals. Furthermore, cultural diversity within Australia has contributed to a greater selection of food, incorporating a greater range of raw foods of animal origin into diets. In addition to the egg farm environment, eggs are also likely to become contaminated within the commercial supply chain. Contamination prevention therefore requires control measures during processing in commercial and domestic settings.

Treatment of *S.* Typhimurium-related human illness is symptomatic in focus. Intestinal limited NTS infections in immunocompetent patients are treated with fluid or rehydration therapy. The use of antibiotics is reserved for high-risk individuals such as the elderly, children younger than 1 year, or immunocompromised patients (Hohmann, 2001). The use of antibiotic therapy has unfortunately led to increasing rates of antibiotic resistance overseas (Fabrega and Vila, 2013). In Australia, however, low levels of β lactamase activity have been reported among human *S.* Typhimurium isolates (Micalizzi, 2013). A low level of antimicrobial resistance was also recorded in *S.* Typhimurium isolates from commercial egg layer flocks (Pande et al., 2015). Both of these reports reflect a prudent use of antimicrobials in human and poultry medicine in Australia.

In Australia, human *S.* Typhimurium is notifiable and the structure of reporting is explained in Fig. 5.4 (Chousalkar et al., 2015). Food Standards Australia New Zealand (FSANZ) develops and administers the Australia New Zealand Food Standards Code (including egg standard). Production of eggs and egg products is currently regulated by various departments of health and primary industries from individual states and codes are enforced by state authorities (FSANZ, 2009). All producers and processors are required to comply with codes including prohibition of selling cracked, dirty eggs and egg stamping/labeling for traceability. According to FSANZ standard code 4.2.5, all egg processors are required to pasteurize egg products prepared from cracked and dirty eggs. The AECL strategy is focused on *Salmonella* risk management through the supply chain, which includes producing public documentation designed to educate people on safe egg handling procedures as well as farm/production specific issues. Standards or codes developed by FSANZ have been adopted by various state authorities; however, there are inconsistencies in

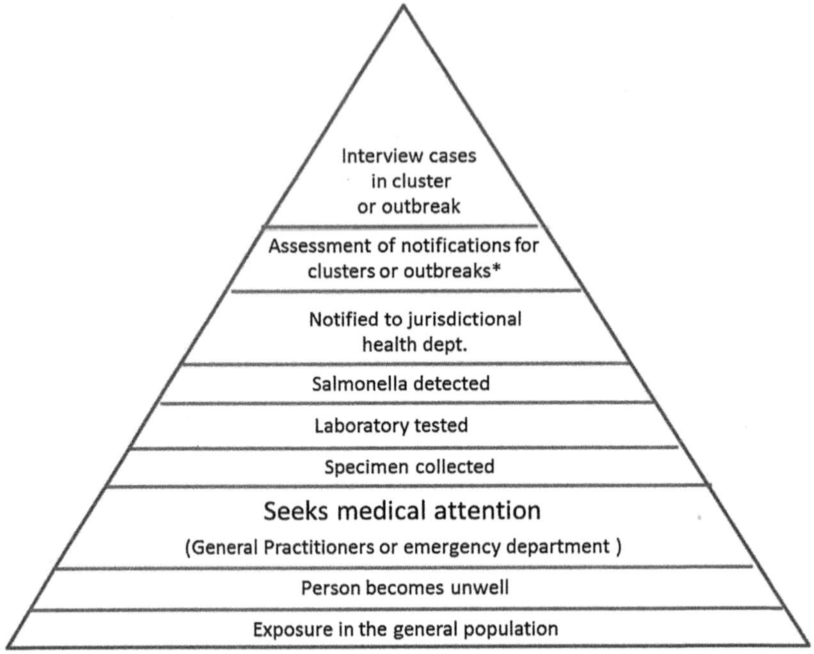


**FIGURE 5.4**

Current *Salmonella* infection notification process in Australia. * Individual jurisdictions interview cases depending on the number of notifications and available resources.

*Taken from Chousalkar, K.K., Sexton, M., McWhorter, A., Hewson, K., Martin, G., Shadbolt, C., Goldsmith, P., 2015.* Salmonella typhimurium *in the Australian egg industry: multidisciplinary approach to addressing the public health challenge and future directions. Crit. Rev. Food Sci. Nutr. (Accepted, in press). http://dx.doi.org/10.1080/10408398.2015.1113928.*

the understating or implementation of these food safety codes. Although there are difficulties in implementing these standards from a national perspective, it would enable a more uniform national food safety code. However, different systems within the states could pose a challenge in implementing a uniform code.

In Australia, MLVA is widely used for the epidemiological/trace back investigation of *S.* Typhimurium outbreaks. MLVA is a useful epidemiological tool for tracing the source of infection, but it does not provide information regarding the virulence or predominant pathotype of *Salmonella* strain involved in human cases. *S.* Typhimurium MLVA types have traced back to the egg farm on multiple occasions (Slinko et al., 2009). However, in some outbreaks, multiple MLVA types that have caused human illness unrelated to egg farms have also been detected from a commercial kitchen environment (Slinko et al., 2009). Egg-borne human illness is not necessarily associated only with poor farm management. An understanding of the ecology of *Salmonella* on layer farms, the supply chain, and the mechanisms that influence association with egg-borne human cases is also required (Arzey, 2002). Focusing only on farm-specific

*Salmonella* intervention strategies is not sufficient and joint efforts from Australian primary industries, health departments, communicable disease, and food branches are essential to the control of *S. Typhimurium* through the entire egg supply chain.

## REFERENCES


AECL, 2015. Australian Egg Corporation Limited Annual Report. AECL, Sydney, p. 3.

Arzey, G., 2002. Salmonella Enteritidis in Australia – Facts and fiction. In: Proceedings of Poult. Inf. Exchange, Gold Coast.

Behravesh, C.B., Brinson, D., Hopkins, B.A., Gomez, T.M., 2014. Backyard poultry flocks and salmonellosis: a recurring, yet preventable public health challenge. Clin. Infect Dis. 58, 1432–1438.

Beuving, G., Vonder, G.M., 1978. Effect of stressing factors on corticosterone levels in the plasma of laying hens. Gen. Comp. Endocrinol. 35, 153–159.

Chousalkar, K.K., Flynn, P., Sutherland, M., Roberts, J.R., Cheetham, B.F., 2010. Recovery of *Salmonella* and *Escherichia coli* from commercial egg shells and effect of translucency on bacterial penetration in eggs. Int. J. Food Microbiol. 142, 207–213.

Chousalkar, K.K., Roberts, J.R., 2012. Recovery of *Salmonella* from table eggs in Australia. Poult. Sci. 91, 1739–1741.

Chousalkar, K.K., Sexton, M., McWhorter, A., Hewson, K., Martin, G., Shadbolt, C., Goldsmith, P., 2015. *Salmonella* Typhimurium in the Australian egg industry: multidisciplinary approach to addressing the public health challenge and future directions. Crit. Rev. Food Sci. Nutr. (Accepted, in press). http://dx.doi.org/10.1080/10408398.2015.1113928.

Cox, J.M., Woolcock, J.B., Sartor, A.L., 2002. The Significance of *Salmonella*, Particularly *S. Infantis*, to the Australian Egg Industry. Corporation, Rural Industries Research Development Corporation.

Cozens, M., 2010. *Salmonella* Enteritidis Monitoring and Certification Program, Queensland. Department of Primary Industries, Queensland, Brisbane, pp. 7–8.

Cuttell, L., Groves, M., Wilson, A., 2014. Microbiological Baseline Survey of the Queensland Egg Production Environment. Queensland, S.F., Brisbane, p. 2014.

Daughtry, B., Sumner, J., Hooper, G., Thomas, C., Grimes, T., Horn, R., Moses, A., Pointon, A., 2005. National Food Safety Risk Profile of Eggs and Egg Products. Australian Egg Corporation Limited, Sydney, Australia.

De Reu, K., Grijspeerdt, K., Messens, W., Heyndrickx, M., Uyttendaele, M., Debevere, J., Herman, L., 2006. Eggshell factors influencing eggshell penetration and whole egg contamination by different bacteria, including *Salmonella* Enteritidis. Int. J. Food Microbiol. 112, 253–260.

Douglas, I., 2004. *Salmonella* Enteritidis – Situation in Queensland. Fisheries, Q.D.o.P.I.a, Brisbane.

El-Lethey, H., Huber-Eicher, B., Jungi, T.W., 2003. Exploration of stress-induced immunosuppression in chickens reveals both stress-resistant and stress-susceptible antigen responses. Vet. Immunol. Immunopathol. 95, 91–101.

Fabrega, A., Vila, J., 2013. *Salmonella enterica* serovar Typhimurium skills to succeed in the host: virulence and regulation. Clin. Microbiol. Rev. 26, 308–341.

Fearnley, E., Raupach, J., Lagala, F., Cameron, S., 2011. *Salmonella* in chicken meat, eggs and humans; Adelaide, South Australia. Int. J. Food Microbiol. 146, 219–227.

Food Standards Australia and New Zealand, 2009. Primary Production and Processing Standard for Egg and Egg Products. Food Standards Australia and New Zealand, Canberra.

Galiş, A.M., Marcq, C., Marlier, D., Portetelle, D., Van, I., Beckers, Y., Théwis, A., 2013. Control of *Salmonella* contamination of shell eggs—preharvest and postharvest methods: a review. Compr. Rev. Food Sci. Food Saf. 12, 155–182.

Gantois, I., Ducatelle, R., Pasmans, F., Haesebrouck, F., Gast, R., Humphrey, T.J., Van Immerseel, F., 2009. Mechanisms of egg contamination by *Salmonella* Enteritidis. FEMS Microbiol. Rev. 33, 718–738.

Gole, V.C., Caraguel, C.G., Sexton, M., Fowler, C., Chousalkar, K.K., 2014a. Shedding of *Salmonella* in single age caged commercial layer flock at an early stage of lay. Int. J. Food Microbiol. 189, 61–66.

Gole, V.C., Chousalkar, K.K., Roberts, J.R., Sexton, M., May, D., Tan, J., Kiermeier, A., 2014b. Effect of egg washing and correlation between eggshell characteristics and egg penetration by various *Salmonella* Typhimurium strains. PLoS One 9, e90987.

Gole, V.C., Torok, V., Sexton, M., Caraguel, C.G., Chousalkar, K.K., 2014c. Association between indoor environmental contamination by *Salmonella enterica* and contamination of eggs on layer farms. J. Clin. Microbiol. 52, 3250–3258.

Health Department, 2016. National Notifiable Disease Surveillance System. Australian Government Department of Health, Canberra.

Hohmann, E.L., 2001. Nontyphoidal salmonellosis. Clin. Infect. Dis. 32, 263–269.

Holt, P.S., 2003. Molting and *Salmonella enterica* serovar enteritidis infection: the problem and some solutions. Poult. Sci. 82, 1008–1010.

Holt, P.S., Gast, R.K., Kelly-Aehle, S., 2003. Use of a live attenuated *Salmonella typhimurium* vaccine to protect hens against *Salmonella* Enteritidis infection while undergoing molt. Avian Dis. 47, 656–661.

Holt, P.S., Macri, N.P., Porter Jr., R.E., 1995. Microbiological analysis of the early Salmonella enteritidis infection in molted and unmolted hens. Avian Dis. 39, 55–63.

Hughes, C.S., Gaskell, R.M., Jones, R.C., Bradbury, J.M., Jordan, F.T., 1989. Effects of certain stress factors on the re-excretion of infectious laryngotracheitis virus from latently infected carrier birds. Res. Vet. Sci. 46, 274–276.

Humphrey, T., 2006. Are happy chickens safer chickens? Poultry welfare and disease susceptibility. Br. Poult. Sci. 47, 379–391.

Kilroy, S., Raspoet, R., Devloo, R., Haesebrouck, F., Ducatelle, R., Van Immerseel, F., 2015. Oral administration of the *Salmonella* Typhimurium vaccine strain Nal2/Rif9/Rtt to laying hens at day of hatch reduces shedding and caecal colonization of *Salmonella* 4,12:i:-, the monophasic variant of *Salmonella* Typhimurium. Poult. Sci. 94, 1122–1127.

Manning, J., Gole, V., Chousalkar, K., 2015. Screening for *Salmonella* in backyard chickens. Prev. Vet. Med. 120, 241–245.

Mashaly, M.M., Hendricks 3rd, G.L., Kalama, M.A., Gehad, A.E., Abbas, A.O., Patterson, P.H., 2004. Effect of heat stress on production parameters and immune responses of commercial laying hens. Poult. Sci. 83, 889–894.

McWhorter, A.R., Chousalkar, K.K., 2015. Comparative phenotypic and genotypic virulence of *Salmonella* strains isolated from Australian layer farms. Front. Microbiol. 6, 12.

McWhorter, A.R., Davos, D., Chousalkar, K.K., 2015. Pathogenicity of *Salmonella* strains isolated from egg shells and the layer farm environment in Australia. Appl. Environ. Microbiol. 81, 405–414.

Messens, W., 2013. Egg Decontamination by Washing. In: Van Immerseel, F., Nys, Y., Bain, M. (Eds.), Improving the Safety and Quality of Eggs and Egg Products, Woodhead Publishing Limited, Cambridge, pp. 163–177.

Micalizzi, G., 2013. ss-Lactamases in *Salmonella enterica* isolated in Australia. Commun. Dis. Intell. Q. Rep. 37, E47–E51.

Nakamura, M., Nagamine, N., Takahashi, T., Suzuki, S., Kijima, M., Tamura, Y., Sato, S., 1994. Horizontal transmission of *Salmonella* Enteritidis and effect of stress on shedding in laying hens. Avian Dis. 38, 282–288.

New South Wales Department of Primary Industries, 2015. NSW poultry egg industry overview 2015. In: Industries, New South Wales Department of Primary Industries. NSW Department of Primary Industries, Paterson.

OzFoodnet, 2002. In: Branch, H.P.P. (Ed.), Enhancing Foodborne Disease Surveillance Across Australia in 2001. Communicable Diseases Intelligence, Canberra, ACT, pp. 375–406.

OzFoodnet, 2003. Foodborne Disease in Australia: Incidence, Notifications and Outbreaks. Annual Report of the OzFoodNet Network, 2002. In: Branch, H.P.P. (Ed.). Communicable Diseases Intelligence, Canberra, ACT, pp. 164–190.

OzFoonet, 2005. Reported Foodborne Illness and Gastroenteritis in Australia: Annual Report of the OzFoodNet Network, 2004. Communicable Diseases Intelligence, Canberra, ACT, pp. 164–190.

OzFoonet, 2006. Burden and Causes of Foodborne Disease in Australia: Annual Report of the OzFoodNet Network, 2005, Branch. Communicable Diseases Intelligence, Canberra, ACT, pp. 278–300.

OzFoonet, 2012. Monitoring the Incidence and Causes of Diseases Potentially Transmitted by Food in Australia: Annual Report of the Ozfoodnet Network, 2010. Communicable Diseases Intelligence, Canberra, ACT, pp. E213–E241.

Pande, V.V., Gole, V.C., McWhorter, A.R., Abraham, S., Chousalkar, K.K., 2015. Antimicrobial resistance of non-typhoidal *Salmonella* isolates from egg layer flocks and egg shells. Int. J. Food Microbiol. 203, 23–26.

Pavic, A., Groves, P.J., Cox, J.M., 2010. Utilization of a novel autologous killed tri-vaccine (serogroups B [Typhimurium], C [Mbandaka] and E [Orion]) for *Salmonella* control in commercial poultry breeders. Avian Pathol. 39, 31–39.

Seo, K.H., Holt, P.S., Gast, R.K., Hofacre, C.L., 2000. Combined effect of antibiotic and competitive exclusion treatment on *Salmonella* Enteritidis fecal shedding in molted laying hens. J. Food Prot. 63, 545–548.

Slinko, V.G., McCall, B.J., Stafford, R.J., Bell, R.J., Hiley, L.A., Sandberg, S.M., White, S.A., Bell, K.M., 2009. Outbreaks of *Salmonella* Typhimurium phage type 197 of multiple genotypes linked to an egg producer. Commun. Dis. Intell. Q. Rep. 33, 419–425.

Stafford, R.J., McCall, B.J., Neill, A.S., Leon, D.S., Dorricott, G.J., Towner, C.D., Micalizz, G.R., 2002. A statewide outbreak of *Salmonella* Bovismorbificans phage type 32 infection in Queensland. In: OzFoodnet. Communicable Disease Intelligence, Brisbane, pp. 568–573.

Stephens, N., Coleman, D., Shaw, K., 2008. Recurring outbreaks of *Salmonella typhimurium* phage type 135 associated with the consumption of products containing raw egg in Tasmania. Commun. Dis. Intell. Q. Rep. 32, 466–468.

Thomas, C., Daughtry, B., Padula, D., Jordan, D., Arzey, G., Davey, K., Holds, G., Slade, J., Pointon, A., 2006. An Egg: Salmonella Quantitative Risk Assessment Model for the Australian Egg Industry. Australian Egg Corporation Limited., Sydney.

CHAPTER

# Egg Production Systems and *Salmonella* in South America

# 6

**Dante Javier Bueno[1], María Cecilia Soria[2], Mario Alberto Soria[1], Francisco Procura[1,3], Francisco Isabelino Rodriguez[1,3], Eduardo Ignacio Godano[4]**

*[1]National Agricultural Technology Institute EAA Conception of Uruguay, Conception of Uruguay, Entre Rios, Argentina; [2]INIQUI-CONICET, National University of Salta, Salta, Argentina; [3]National Council of Scientific and Technical Research (CONICET), Argentina; [4]Tecnovo S.A., Industrial Park, Crespo, Entre Ríos, Argentina*

## 1. INTRODUCTION

South America is a continent located in the Western Hemisphere, mostly in the Southern Hemisphere, with a relatively small portion in the Northern Hemisphere. It is also considered as a subcontinent of the Americas. It is bordered on the west by the Pacific Ocean and on the north and east by the Atlantic Ocean; North America and the Caribbean Sea lie to the northwest. It includes 12 sovereign countries: Argentina, Bolivia, Brazil, Chile, Colombia, Ecuador, Guyana, Paraguay, Peru, Suriname, Uruguay, and Venezuela (Dorst and Winkel, 2015).

South America is a historically unstable region because of continuing changes in monetary policy, which has led to constant internal and external conflicts with different outcomes. Despite this, in recent years, several countries made enormous development strides in recent decades, from the consolidation of democratic governments and continued advances in health and education to more recent progress in protecting the environment and reducing inequality. By 2015, the region met the majority of the Millennium Development Goals (MDGs) – a historic achievement (Thorpe and Aguilar Ibarra, 2010; United Nations Development Programme, 2016). South America has an area of 1,780,800 thousands of hectares and its population has been estimated at 413,651,000 in 2015. The most populous countries in South America are Brazil, Colombia, Argentina, Venezuela, and Peru (Table 6.1; Economic Commission for Latin America and the Caribbean, 2014).

There are some competitive advantages in the South American countries for poultry production. Some of the countries are large grain producers (Argentina, Brazil, and Paraguay); others import grains, but have a good production structure (Colombia and Peru). There are abundant and qualified hand labor, modern industry and dynamic entrepreneurs, low production costs, good animal health (Brazil, Argentina, Chile, and Uruguay), and a large internal market in Brazil. However, egg layer production and export

**Producing Safe Eggs. http://dx.doi.org/10.1016/B978-0-12-802582-6.00006-9**

**Table 6.1** Capital, Population, Total Area, Poverty, and Total Gross Domestic Product at Constant Market Prices of South American Countries (Economic Commission for Latin America and the Caribbean, 2014)

| Country | Capital | Population (Thousands of Persons Estimated in 2015) | Total Area, 2012 (Land Area/Area of Inland Waters in Thousands of Hectares) | National Poverty (Percentages of the Total Population) | Total Gross Domestic Product, at Constant Market Prices in 2013 (Millions of Dollars at Constant 2010 Prices) |
|---|---|---|---|---|---|
| Argentina | Buenos Aires | 42,119 | 278,040 (273,669/4,371) | – | 524,029.9 |
| Bolivia (Plurinational State of) | Sucre | 10,746 | 109,858 (108,330/1528) | 36.3[a] | 23,208.8 |
| Brazil | Brasilia | 202,956 | (861,577 (845,814/15,763) | 18.0[b] | 2,279,748.1 |
| Chile | Santiago | 17,889 | 75,610 (74,353/1256) | 7.8[b] | 252,538.5 |
| Colombia | Bogotá | 49,633 | 114,175 (110,950/3225) | 30.7[b] | 333,209.5 |
| Ecuador | Quito | 16,268 | 25,637 (24,836/801) | 33.6[b] | 82,609.1 |
| Guyana | Georgetown | 808 | 21,497 (19,685/1812) | – | 2672.2 |
| Paraguay | Asunción | 6,993 | 40,675 (39,730/945) | 40.7[b] | 23,597.5 |
| Peru | Lima | 30,994 | 128,522 (128,000/522) | 23.9[b] | 175,425.1 |
| Suriname | Paramaribo | 548 | 16,382 (15,600/782) | – | 4873.2 |
| Uruguay | Montevideo | 3,430 | 17,622 (17,502/120) | 5.6[b] | 45,172.1 |
| Venezuela (Bolivarian Republic of) | Caracas | 31,267 | 91,205 (88,205/3000) | 32.1[b] | 267,213.3 |
| Total | - | 413,651 | 1,780,800 | – | – |

[a] *In 2011.*
[b] *In 2013.*

**Table 6.2** Number of Hens and Per Capita Egg Consumption in South America During 2014 (Ruiz, 2015)

| Country | No. of Hens (×1000) | Per Capita Egg Consumption (Units) |
|---|---|---|
| Argentina | 41,200 | 257 |
| Bolivia (Plurinational State of) | 3965 | 138 |
| Brazil | 94,000 | 169 |
| Chile | 12,000 | 191 |
| Colombia | 34,700 | 242 |
| Ecuador | 9500 | 140 |
| Paraguay | 2500 | 135 |
| Peru | 16,000 | 171 |
| Uruguay | 3100 | 262 |
| Venezuela (Bolivarian Republic of) | 21,212 | 190 |

challenges are increasing the production scale, adapting egg farms and plants to new animal welfare requirements, and implementing quality programs (Mendes, 2011).

In general, the production of eggs for consumption in South American countries has increased to a greater or lesser extent. This is explained in part by genetic improvement in commercial lines, the quality and availability of raw materials for the manufacture of feed, and the increasing demand for animal protein worldwide. Also, the incorporation of technology in housing systems for laying hens has certainly favored the growth of the egg industry in several South American countries (Dreyer and Windhorst, 2011; Evans, 2014).

Although statistics for hen egg production can vary greatly depending on the source, of approximately 69.7 million metric tons of eggs produced worldwide in 2014, South America produced 4.7 million metric tons (about 6.75%). Among the top 20 countries for egg production in 2013, Brazil and Colombia ranked 7th and 18th, respectively (Conmay, 2015). Brazilian companies account for 2 of the 25 largest egg producers worldwide when ranked by size of layer flock (Clements, 2015). Table 6.2 indicates the number of hens and per capita egg consumption in South America during 2014; Guyana and Suriname data are not included in this chapter due to the lack of information from these countries for 2014. According to the Latin-American Top Poultry Companies database (Table 6.3), 5 of 10 of the biggest table egg producers in South America, measured by the number of layers, are located in Brazil, 2 in Venezuela and Colombia, and 1 in Peru (Ruiz, 2015). The genetic lines used for laying hens in South American countries (Ruiz, 2015) are Hy-Line (Argentina, Bolivia, Brazil, Chile, Colombia, Ecuador, Paraguay, Peru, Uruguay, Venezuela), Lohmann (Argentina, Brazil, Chile, Colombia, Paraguay, Peru, Venezuela), ISA (Bolivia, Brazil, Chile, Colombia, Paraguay, Peru, Venezuela), Hysex (Brazil, Peru), Dekalb (Brazil, Uruguay), Lhom (Brazil), H&N (Brazil, Colombia, Ecuador, Uruguay), Shaver (Chile), and Bovans (Peru, Venezuela).

**Table 6.3** South America's Top 10 Largest Egg Companies in 2014 (Ruiz, 2015)

| Company | Country | Number of Hens (×1000) |
|---|---|---|
| Grupo Mantiqueira | Brazil | 11,000 |
| Granja Yabuta | Brazil | 9000 |
| Avícola La Calera | Peru | 4500 |
| Granja La Caridad | Venezuela | 4485 |
| Ademar Kerkoff | Brazil | 4000 |
| Incubadora Santander | Colombia | 3500 |
| Avicultura Josidith | Brazil | 3000 |
| Nutriavícola S.A. | Colombia | 2500 |
| Ovomar | Venezuela | 2480 |
| Somai Nordeste | Brazil | 2460 |

Nonenriched cage systems are not prohibited in South America and most of the laying hens are located in battery cage farms, which include automatic and conventional (manual) battery cage systems. In general, the incorporation of modern technology is evident in South America with the migration of a simple pyramidal cage system with no conveyors for feed or eggs to industrial systems with manure or egg conveyors (Rodriguez, 2006; Scheurer, 2014). It is not common that producers and entrepreneurs use enriched cages (furnished or modified cages), noncage, "barn egg," or "free range" systems, as required by regulations of the European Union.

The prevalence data about *Salmonella* spp. pertaining to layer hens and egg production in South America is scarce, limited to some countries and probably underestimated. Also, the data from the literature are diverse in kind and number of samples, number of flocks/farms tested, and the methodology used to isolate the bacteria. *S.* Gallinarum biovar Gallinarum (SG) has not been eradicated in breeder flocks and commercial layers in South American countries. This can be explained by the use of vaccines that are ineffective, misused, or not always controlled, and the abandonment of hygienic prophylaxis measures (Servicio Nacional de Sanidad y Calidad Agroalimentaria, 2002).

In general, live and inactivated vaccines are approved for *Salmonella* control in poultry from South American countries, but there are some differences in licensed vaccines. The live vaccine is based on SG, *Salmonella* Enteritidis (SE), or *Salmonella* Typhimurium (ST) strains. *S.* Gallinarum 9R, using a rough strain of SG, is a vaccine produced by different companies to protect poultry against *S.* Gallinarum and SE, especially in layers (Servicio Nacional de Sanidad y Calidad Agroalimentaria, 2002). Live attenuated vaccines for SE (AviPro *Salmonella* Vac E and Gallivac Se) and ST (AviPro *Salmonella* Vac T) by oral administration via drinking water are approved in many South American countries like Bolivia, Brazil, Chile, Ecuador, Paraguay, Peru, Uruguay, and Colombia (Lohmann Animal Health, 2013, 2014; Merial, 2016).

An inactivated vaccine for SE, ST, and/or *S.* Gallinarum protection is produced by different companies (Servicio Nacional de Sanidad y Calidad Agroalimentaria, 2002). However, an Argentinean company started to produce an oral inactive subunit vaccine against non–host-specific *Salmonella* serotypes. The antigen is neither live attenuated nor injectable and it is recommended for different poultry categories (Vetanco, 2013).

In this chapter, based on the above-mentioned information, we describe egg production, the main *Salmonella* serotypes, and antimicrobial resistance for these serotypes found in layer hens and eggs, as well as the public health implications, and the *Salmonella* spp. National Poultry Improvement Plan for each South American country. In reference to Guyana, Paraguay, and Surinam, we did not find any information about prevalence and antimicrobial resistance in layer hens or eggs.

## 2. ARGENTINA

In Argentina, a population of 37 million laying hens is estimated, showing a growth of 54% from 2003 to 2013. Egg production in Argentina grew about 55% in the years 2000–2010 and has risen from 327,000 tons to over 500,000 tons. With an average annual growth rate of 5% in 2012, egg production in Argentina almost doubled to reach 600,000 tons. However, the industry went through a difficult period during 2012/2013, when production dropped considerably. But now a recovery backed by an increase in domestic consumption and the expected expansion of the shell egg exports are forecast. About 85% of the birds are in cages, 10% in floor systems, and 5% outdoors. The ratio of white eggs to brown eggs is 60:40 (Evans, 2015a).

Egg production is mainly concentrated in the provinces of Buenos Aires (41%) and Entre Rios (24%). Shell egg production totaled 11,770 million eggs in 2014. This means an increase of 8.4% over the previous year. The per capita egg consumption reached 256 per year, and it has been increasing year after year. Shell eggs go to the domestic market and the industrialized eggs are used for export. Processing plants industrialize 9% of total egg production (Mair et al., 2015).

After the introduction of intensive production of birds from 1960 in Argentina, there were cases of Avian Typhoid and Pullorum disease, caused by SG and *S.* Gallinarum biovar Pullorum (SP), respectively, with a significant mortality rate among affected birds. Vaccination lead to a decrease of these two biovars but a subsequent increase beginning in the 1980s of a new infection that started in breeding chicken flocks mainly caused by SE. This was then isolated in egg breeder flocks (Sandoval et al., 1989; Sandoval and Terzolo, 1989).

There are few research studies about *Salmonella* contamination on laying hen farms, eggs, their packaging, etc. in Argentina. Viora et al. (1993) found SE in 50% of fiber egg cartons. On the other hand, Soria (2012) found 1.8% (29/1643) positive for *Salmonella* spp. in packaging, shell and egg content samples from eggs sold in supermarkets in the center and east of Entre Rios province, Argentina. In egg packaging, the serotypes isolated were SE, *S.* Agona, *S.* Westhampton, and *S.* Muenchen.

Meanwhile, the serotypes found in shell egg samples were SE, *S.* Montevideo, *S.* Brandenburg, and ST. Tests of the egg contents by this author found *S.* Gallinarum, SE, and ST. However, De Franceschi et al. (1998) did not find *Salmonella* spp. in 7760 eggs from farms in the province of Buenos Aires.

Soria (2013), using multiple samples from 30 commercial layer farms in Entre Ríos, found that *Salmonella* spp. was most frequently detected in boot swab samples (30%). Only 4.8% (56/1167) of samples were positive for *Salmonella* spp., whereas egg, water, and air samples were negative for these bacteria. The *Salmonella* serotypes most frequently observed in the layer houses (Soria et al., data not published) were *S.* Schwarzengrund (17.5%), SE (15%), *S.* Mbandaka (7.5%), and *S.* Newport (7.5%).

In Argentina, in terms of public health, a significant increase in the number of SE isolates from foods associated with food-borne outbreaks has been observed since 1987 (Caffer and Eiguer, 1994). During the period 2000–2005, the National Reference Laboratory studied 443 *Salmonella* isolates derived from these foods and reported SE (31.8%), ST (19.6%), *S.* Newport (6.9%), and *S.* Agona (6.1%). The other serotypes were isolated at a lower frequency. On the other hand, from 2006 ST was the most frequent serotype isolated from humans, animals, and foods (Caffer et al., 2010). In addition, Favier et al. (2012) determined the prevalence of *Salmonella* spp. in foods of animal origin sold at retail stores over the period 2005–2011 in San Luis, Argentina, and only found two ST strains from liquid eggs (2/60).

Melo et al. (2007) evaluated antimicrobial susceptibility of 28 strains of SG, isolated from outbreaks of fowl typhoid from commercial laying hens in Argentina during 2006. Only two isolates had sensitivity greater than or equal to 90%. Gentamicin, norfloxacin, and tetracycline resistance was greater than 30%. Fosfomycin and sulfadiazine + trimethoprim combination appear to be the drugs of choice for the treatment of fowl typhoid in Argentina. On the other hand, Soria (2012) observed that 100% of ST strains tested were sensitive to gentamicin, norfloxacin, ciprofloxacin, imipenem, and amoxicillin/clavulanic acid for different strains of ST isolated from commercial eggs in Entre Ríos, Argentina. The resistance to doxycycline, tetracycline, and streptomycin was observed to be 68.1%, 93.6%, and 97.9%, respectively. Overall, 98% of the strains were resistant to two or more drugs belonging to the same or different groups of antibiotics. Resistance to quinolone and phenicol groups was not observed.

The National Poultry Health Plan defines a *Salmonella* control program for breeder flocks to prevent the transmission of SG, SP, SE, ST, and *S.* Heidelberg in Argentina (Servicio Nacional de Sanidad y Calidad Agroalimentaria, 2002). It is applied by the National Service of Agricultural and Food Safety. The samples include unhatched eggs (chopped) from incubators, cloacal and boot swabs from breeder flocks, and breeders in the rearing period, respectively (Programa de Aves, 2015). Testing for the four *Salmonella* serotypes involves only bacteriologic monitoring. Sampling should be done every 4 and 9 weeks for grandparent and parent breeder flocks, respectively, and at 9 and 18 weeks old for breeders in the rearing period. Live and inactivated vaccines are not recommended for breeder flocks by Argentina *Salmonella* control program (Servicio Nacional de Sanidad y Calidad Agroalimentaria, 2002), but are recommended for laying hens. *Salmonella* spp. monitoring and control program in commercial poultry farms was approved in 2016. It is included in the National Poultry Health Plan and is focused on SE, ST and S. Heidelberg. Laying hen farms should be sampled twice per year. In cage flocks, 2 × 150 grams of naturally pooled feces shall be taken from all belts or scrapers in the house after running the manure removal system; and 150 grams of dust shall be collected from 20 prolific sources of dust throughout the house. However, in the case of step cage houses without scrapers or belts, 2 × 150 grams of mixed fresh feces must be collected from 60 different places beneath the cages in the dropping pits (Servicio Nacional de Sanidad y Calidad Agroalimentaria, 2016).

## 3. BOLIVIA

According to the News Agency of Bolivia, the per capita consumption reported by the National Institute of Statistics is 166 eggs/person/year. Between 2008 and 2014, egg production has increased about 6%. Also, the total egg production for 2014 was 1591.75 million units, 8.5% more than in 2008. Geographically, the population of laying hens in Bolivia is well concentrated, 29% in Cochabamba and 66% in Santa Cruz (Asociación de avicultores Cochabamba-Bolivia, 2015).

According to the National Health Information Systems in 2001, 47% of the reported cases of food-borne disease were due to *Salmonella* spp. (SEDES, 2001). In reference to egg production, Espinoza et al. (2007) analyzed 40 eggs, collected from different markets in La Paz city. *Salmonella* spp. was identified in 17.5% of the samples studied.

In Bolivia, the Control and Eradication National Program for *Salmonella* spp. is obligatory for all poultry establishments (Ministerio de Agricultura, Ganadería y Desarrollo Rural, 2002). The general objective of this program is to eradicate SG and SP and control paratyphoid salmonellae in Bolivia, which is applied by the National Agricultural Health and Food Safety Service. In farms, monitoring of lots will be held during the breeding and rearing stages, at 4 and 12 weeks old, which includes 1% of the population, using serological tests accompanied by bacteriological analysis from the lots tested. During the production stage, a control is performed in an official laboratory using the rapid serum plate agglutination (SPA) test, confirmed by a microagglutination (MA) test. The first serological control should be done at 17 weeks old to 10% of the breeder population. Also, 1% of the poultry population should undergo a bimonthly serological and bacteriological test during the productive life of the hen. Poultry establishments that have negative results to the first tests will be classified as "suitable," keeping this category if consecutive negative results continue. When birds test positive for SP, SG, SE, or ST in breeding, rearing, or production, the birds will be quarantined and assigned the status of "lot in control." Confirmation should be done through serological and bacteriological tests within no more than 15 calendar days. If these birds test negative, the lot will be categorized as "apt." However, if they test positive, the lot will be classified as "infected." If the birds are "infected" with SP or SG, they will be immediately separated, proceeding to sacrifice on the same farm or in a poultry slaughterhouse. Disinfection of the slaughterhouse occurs immediately after sacrifice. Also, ST- and SE-positive lots will be subject to a control program, where no sacrifice will be necessary. On the other hand, serological and bacteriological study is contemplated during the different production stages of hens. In this case, the study will be conducted in 1% of the population.

## 4. BRAZIL

Egg production in Brazil grew at an annual rate of 2.8% between 2000 and 2013, from 1.51 million tons to 2.17 million tons. All commercial egg-laying hens are in cages, and about three-quarters of production are white eggs (Evans, 2015a). The states of São Paulo (34%) and Mina Gerais (11%) have the most concentrated amount of egg production for consumption. Per capita egg consumption in Brazil grew from 148 in 2012 to 182 eggs/person/year in 2014, representing an increase of over 20% in just 2 years. At present, the production of eggs is 37,245,133,103 units. Also, breeder flock housings have increased by 25% from 2008 to 2014. On the other hand, almost all egg production in Brazil is sold in the domestic market (Associcäo Brasileira de Proteina Animal, 2015). Brazilian exports of eggs (fresh and processed) achieved 7000 tons in the first half of 2015, 41.9% more than the total exported in the same period of 2014 (El Sitio Avícola, 2015b).

Gama et al. (2003) investigated the presence of *Salmonella* spp. in the transport boxes of 12 flocks. The positive flocks were subsequently monitored at every 4 weeks, up to 52 weeks, using bacteriological examination of cecal fresh feces. Also, samples of eggs were studied at 52 and 76 weeks. About 33.3% of flocks were positive and several serotypes of *Salmonella* were found in fecal samples: SE, *S.* Infantis, *S.* Mbandaka, and *S.* Javiana. The authors also reported the presence of a rough strain of *Salmonella*. In reference to egg samples, SE was found in one egg from a total of 500 eggs sampled. Kotwizz et al. (2008) evaluated the prevalence of *Salmonella* spp. in 30 farms of laying hens in the state of Paraná, taking samples of eggs, feces, and cloacal swabs. They found that 23% of farms were positive to *Salmonella* spp. without isolation from egg samples. The serotypes isolated were *S.* Mbandaka, *S.* Infantis, and *S.* Newport (11% each). Furthermore, Kotwizz et al. (2013) found that *Salmonella* spp. was isolated from 52.0% of discarded hatching eggs, in which the predominant serotype was SE (84.6%). On the other hand, in the region of San Pablo, Freitas Neto et al. (2014) reported that 25% of eight laying hen flocks were *Salmonella* spp. positive, recording the presence of *S.* Havana and SE in cecal feces. They also found five samples of eggs (shell and content, total of 340 samples) positive with the presence of *S.* Mbandaka, *Salmonella enterica* subspecies enterica 6, 7: z10, and *S.* Braenderup. From the serotypes found in the different works reviewed earlier, SE, *S.* Infantis, *S.* Javiana, *S.* Newport, and *S.* Braenderup are listed among the 20 most frequent *Salmonella* serotypes found in humans in Brazil from 2001 to 2007 by the World Health Organization (Hendriksen et al., 2011).

Dias de Oliveira et al. (2005) studied antimicrobial susceptibility of 91 SE isolates from broiler carcasses, food, human, and poultry-related samples (25 isolates from meals with eggs or other poultry product samples), which originated from the south of Brazil. A great proportion of resistant strains were found, 90.1% showing resistance to at least one antimicrobial drug. There was a high resistance to sulfonamides (75.8%) and nitrofurantoin (52.8%) and low resistance to tetracycline (15.4%), streptomycin (7.7%), nalidixic acid (7.7%), gentamicin (5.5%), norfloxacin (3.3%), trimethoprim (3.3%), cefalotin (2.2%), ampicillin (1.1%), and chloramphenicol (1.1%). Resistance to ciprofloxacin was not detected. A total of 51.6% of SE strains were multidrug resistant with 18 resistance patterns found. The highest resistance was found in strains isolated from poultry-related samples, where all strains were resistant to at least one antimicrobial agent.

In another study in Brazil, Silva et al. (2013) described the presence of extended-spectrum beta-lactamase (ESBL) CTX-M-2-producing *S. enterica* isolates belonging to serotypes Schwarzengrund and Agona in poultry farms for the first time. From 2008 to 2009, 93 *Salmonella* spp. strains were isolated from commercial poultry and related sources in farms of five Brazilian states. Thirteen *S. enterica* isolates that were grouped into two major pulsed-field gel electrophoresis clusters (A and B), belonging to serotypes Schwarzengrund and Agona, respectively, were found to produce ESBL CTX-M-2. The blaCTX-M-2 genotype was associated with the presence of an IncP plasmid of approximately 40 kb. CTX-M lactamases have been widely distributed in South America at least since 1989 and possibly before appearing in Europe (Radice et al., 2002). It is known that Enteropathogens such as *Salmonella* spp. were among the first microorganisms found to carry the blaCTX-M-2 gene (Rossi et al., 1995). Previous studies have shown the production of CTX-M-8 and CTX-M-9 by *S. enterica* isolated from human and animal samples (Peirano el al. 2006), whereas the production of CTX-M-2 has been documented, so far, in ST isolated from pediatric patients and poultry (Fernandes et al., 2009).

In Brazil, the National Poultry Health Plan is applied for *Salmonella* spp. by the Ministry of Agriculture, Livestock and Supply. The program testing includes bacteriologic monitoring of the environment (boot swabs, shaver transport boxes) and animals (meconium, feces, cloacal swabs), and serologic monitoring of birds (enzyme immunoassay methods, SPA, slow serum tube agglutination, and MA). The health certification program for salmonellosis in breeder flocks objective is to prevent, detect early, and control the presence of SG, SP, SE, and ST, by conducting surveillance with routine laboratory tests in poultry establishments. The establishment will be certified as free of these serotypes after three consecutive tests with negative results. On the other hand, the monitoring program for *Salmonella* in commercial layer farms has the objective to prevent, detect early, and control the presence of SE and ST in poultry, with sampling for laboratory testing. These birds are under monitoring or surveillance and the farmer is not allowed to use a vaccine of any kind against *Salmonella* in breeder flocks or vaccines prepared with oil adjuvant during the 4 weeks before testing (Ministério da Agricultura, Pecuária e Abastecimiento, 2010).

In breeder flocks, a *Salmonella* control should be done at 1–7 days old (bacteriologic monitoring), 12 weeks old, in the beginning of egg production, and every 3 months until the end of the lot life (bacteriologic and serologic monitoring). In layer hens, a *Salmonella* control should be done in the beginning of egg production and every 4 months up to the end of the lot life (bacteriologic monitoring). When great-grandparent and grandparent flocks are positive for SG, SP, SE, or ST, birds will be sanitary sacrificed or immediately slaughtered and there will be destruction of all eggs from the affected lots. In layer hens and parent breeder flocks, a positive diagnosis for SE and ST will change the lot from "monitored" to "under surveillance." In this case, antibiotic therapy is allowed in parent breeder flocks. However, in general, the sale of eggs from breeder flocks infected with these serotypes cannot be made for human consumption (Ministério da Agricultura, Pecuária e Abastecimiento, 2010).

## 5. CHILE

Domestic production of eggs in Chile consists of a large industrial sector, which contributes around 3200 million eggs a year. Small egg producers, whose production is less relevant to the national economy, play an important socioeconomic role in the domestic economy. The egg industry is in the hands of about 300 producers; 60 farms account for 90% of production and are located mainly in the central region. Egg production is mainly concentrated in Santiago, Valparaíso, and O'Higgins, comprising 70% of the national production overall. It is estimated that there were almost 12 million layer hens in 2014, of which 9 million were producers of white eggs and nearly 3 million were reported producing colored eggs. The growth in production in the period 2002–2013 showed an increase of 35.7%, which means an average growth rate of 2.8% annually. The factors behind this growth are incorporating new technologies and infrastructure, increasing productivity by specializing in poultry and better feed conversion, genetic advancements, and the continuous improvement of the health of the birds. Furthermore, a strong and growing demand for animal protein of high quality at low cost in the country has contributed to the sustained development of the sector over the years (Giacomozzi Carrasco, 2014; Instituto Nacional de Estadísticas Chile, 2015).

The frequency of acquiring SE in Chile is unknown. Alexandre et al. (2000) evaluated this serotype contamination in eggs offered in retail markets in the Metropolitan Area during two consecutive years (1998–1999). *S.* Enteritidis was found in 0.09% of egg samples (1/1081) and the contaminated sample was offered in a supermarket. On the other hand, Prat et al. (2001) studied phage typing of SE isolates from clinical, food, and poultry samples in Chile. According to 47 food isolates obtained during SE outbreaks, they found that one and four phage types were identified. Most of the foods involved in these outbreaks (38.3%) were products made from eggs such as mayonnaise or meringue cakes. Furthermore, from a total of 27 strains isolated in surveillance studies of poultry-raising establishments, these authors recognized four phage types, including 14 of type 1 (51.9%) and 10 of type 4 (37%). Some of these isolate samples came from hen cloaca or ovaries. They also identified strains that belong to phage types 2 and 7 in one and two cases, respectively.

Fica et al. (2012) studied salmonellosis outbreaks and public policies role in terms of food security control. In the case of SE, rates have increased twofold since 1998 (5.3–10.7 per 100,000 habitants) with an important increase in the number of outbreaks linked to this agent (7–31 annual outbreaks) since year 2005. Persistence of this problem is probably associated to the low surveillance of poultry farms made by the Chilean state; to the absence of a cold chain during collection, distribution, and selling of eggs; and to the lack of an educational program directed to the population. The regulation that bans homemade mayonnaise in restaurants or fast food stores is an important advance that requires further evaluation.

Ninety four strains were isolated from feces samples of 30 poultry farms from the central zone in Chile between Mar. and Jun. 2004 (broiler and layer hens). Thirty nine of these strains were resistant to flumequine, nalidixic acid, and oxolinic acid simultaneously, whereas only two strains were resistant to nalidixic acid and oxolinic acid. All the strains were sensitive to ciprofloxacin and enrofloxacin. A total of 3, 20, and 16 strains phenotypically resistant to quinolones presented patterns I (137-43-22-15 bp), II (152-43-37 bp), and III (137-58-22 bp), respectively. Mutations in codons 83 and 87 of the QRDR region of *gyr*A gene were observed in SE and *S.* Heidelberg, which were the most frequent serotypes identified. No patterns were found that suggested mutation in codon 81 or restriction patterns for double mutations of the QRDR region of *gyr*A gene (San Martín et al., 2005).

The Chile National Plan to control *Salmonella* spp. in poultry was published in 1998 by González Díaz and Correo Munida due to an SE human outbreak (González Díaz, and Correo Munida, 1998). In this plan, they used bacteriologic and serologic (300 birds by SPA test) monitoring of birds in breeder flocks and bacteriologic monitoring in layer hens for SG, SP, SE, and ST. For hatcheries with breeder flocks, eggs, discard chicks, and meconium were recommended samples. Birds that were positive to SPA test are sent to the laboratory for isolation. The measures recommended for *Salmonella*-positive birds were removing SPA-positive birds, lot drug treatment, application of competitive exclusion, and/or SPA testing for 10% of birds and culture for SPA-positive birds. The Agriculture and Livestock Service from the Agriculture Ministry executes a *Salmonella* spp. control program to facilitate export certification processes for poultry, hatching eggs, and poultry meat. This targets serotypes established by the European Union: SE and ST for all levels of commercial poultry. It also includes SG and SP (Servicio Agrícola y Ganadero, 2015).

## 6. COLOMBIA

Commercial egg production has a tradition of over 50 years in Colombia. The poultry business represents 10% of gross domestic product (GDP) and employs 450,000 people across the country. The egg industry had a 4.3% annual growth between 2000 and 2013. Per capita consumption is 251 eggs/person/year. It is estimated that the current population is 35,000,000 laying hens, housed in floor production systems (70%) and battery cage confinement (30%). About 92% of eggs produced in Colombia are brown and they are classified by weight. The egg production is scattered among a large number of producers distributed in different regions, with scales ranging from 500 to 500,000 layers hens. In Colombia, the egg industry developed sliced eggs, which are unique in the world (Aguilera Díaz, 2014; Avila Cortez, 2015; Evans, 2015a).

Despite the fact that SE detection in eggs is not regulated in Colombia, according to Rodriguez et al. (2015) active surveillance has currently evaluated the health status of the flocks in the country both by governmental and academic institutions in addition to the poultry and egg industry. Also, studies contributed to know the *Salmonella* spp. prevalence in laying hens from Colombia. Ramirez et al. (2014) found that 1.7% of 230 eggs were positive for SE in samples of shell, egg white, and yolks. On the other hand, Rodriguez et al. (2015) reported that 5 of 15 laying hen farms were *Salmonella* positive in the Tolima region of Colombia. The pathogen was isolated most frequently from the egg surface (57.15%, $n=8$), followed by the feed (28.57%, $n=4$) and environmental samples (14.29%, $n=2$), whereas the cloacal swab samples were negative. Eight and six strains were identified as *S.* Shannon, and SE, respectively. All strains were resistant to amikacin, cephalothin, cefoxitin, cefuroxime, and gentamicin; however, five SE isolates presented intermediate resistance to nitrofurantoin. All isolates were sensitive to amoxicillin clavulanate, ampicillin, aztreonam, cefepime, ceftazidime, ceftriaxone, ciprofloxacin, imipenem, levofloxacin, meropenem, piperacillin-tazobactam, and tigecycline.

Pulido-Landínez et al. (2014) studied 35 cases of high mortality and severe egg drops occurring in commercial brown layer farms in Colombia, located in the states of Cundinamarca, Santander, Bolivar, and San Andres from 2008 to 2012. Samples studied were liver, spleen, and ovarian follicles. They showed that SE was identified in 45.71% of samples from farms suspected to be suffering outbreaks of fowl typhoid based on clinical signs. In 37.14% of the cases, SG was identified. Their results prove to be a threat to public and animal health, since SE is the main serotype implied in human outbreaks and SG compromises the economic viability of the industry.

In Colombia, the only focus of the National Poultry Plan for *Salmonella* control, coordinated by the Colombian Agricultural Institute, is SG (Instituto Colombiano Agropecuario, 1976). The official test for monitoring is SPA. In breeder flocks, 100% of birds from the lot should be sampled. The first sampling will be taken at 16 and 20 weeks old. If the first test is negative, a second check of 10% of the birds from the farm will be taken at 21–30 days after the first test. If all results are negative, a certificate, free of *Salmonella*, will be issued, which will be valid for 6 months. The certificate will be renewed if after testing 10% of birds all results are negative. If suspicious or positive results are found, 100% of the birds will be tested 21–30 days after the last testing has been done and so on. If an increase in the number of positive birds is observed in two successive samplings, the lot is considered infected and their removal by sacrifice is ordered. Any bird that tests positive in any of the tests must be removed immediately by sacrifice. On the other hand, in commercial laying hens, a representative of all farm bird percentages is taken and serological testing is done at 16 and 20 weeks old. If all results are negative in the first test, a certificate, free of SG, will be issued, which is valid for 6 months. If positive birds are found, a new test of 100% of birds will be taken, 21–30 days after the first checks have been performed and all positive birds are slaughtered. Hundred percent of all birds with suspicious or positive serological and cultural tests will be slaughtered and incinerated. Eggs from poultry farms positive for SG cannot be used for incubation, and they should be properly disinfected to be used for other commercial purposes.

## 7. ECUADOR

According to the latest census of poultry, Ecuador has more than 1600 producers dedicated to the operation of commercial egg laying hens. This includes the participation of small, medium, and large companies that together account for about 14% of GDP, according to its own agriculture industry estimates. The egg producer sector is the main ally of farmers as it consumes 100% of the production of yellow corn and soybean expeller. As for the genetic lines of the existing layers in the country, Lohmann represents 56%, Hy Line 25%, and ISA Brown 19% of market share (Corporación Nacional de Avicultores de Ecuador, personal communication).

As for the bird population, it is estimated that there are 12,500,000 laying hens in Ecuador. Of this, 9,400,000 are in production. Therefore a production of 2800 million eggs/year is estimated. Given that domestic consumption of egg reaches 2169 million per year, it is determined that there is a surplus of 657 million eggs (23% of production). Per capita consumption is 140 eggs/person/year. The egg sector has grown from 6 to 12.5 million layers from 2008 to 2015. Production of eggs for consumption is highly diversified; two provinces in the central area produced about 60% with the participation of several medium and small companies. Large companies are involved in about 15% of the national production. Poultry production in the country has sufficient technical infrastructure to meet all domestic demand in terms of table eggs. Within areas, 49% of egg production is in Tungurahua, 22% in Manabi, 15% in Pichincha, 11% in Cotopaxi, and the remaining 3% in other provinces (El Sitio Avícola, 2015a; Evans, 2015b).

In Ecuador, there are no data or study about the level of contamination of eggs that could affect humans and the social and economic impact in Public Health (Sanchez Mora, 2013). In this country, food-borne diseases are transmitted by water and food, with salmonellosis being one of the most important causes of an outbreak (Estrada Aguila and Valencia Bustamante, 2012). From 150 egg samples, taken from 50 laying farms in Tungurahua, Ecuador, 0.0133% (2/150) were positive for SE (Sanchez Mora, 2013). On the other hand, Estrada Aguila and Valencia Bustamante (2012) did not find any sample positive for *Salmonella* spp. in eggs for consumption (0/282) in a similar study in Pichincha, Quito.

The Ecuadorian Agency for Quality Assurance in Agriculture (Ministry of Agriculture, Livestock, Aquaculture and Fisheries) adopted the National Poultry Health Program (Agencia ecuatoriana de aseguramiento de la calidad del agro, 2013), which includes a salmonellosis control project in breeding flocks and commercial laying hens. This would take into account the monitoring of birds for the detection of *Salmonella*, the laboratory diagnosis, vaccinations, competitive exclusion, quarantine control, and health education (biosecurity) and communication.

## 8. PARAGUAY

The current situation of egg production in Paraguay shows several positive aspects, including a sustained increase in egg consumption and the campaigns about the benefits of eggs as food. In this country, the egg industry has been favored with a large investment of technology introduced to the sector from the 1990s. Implementing production quality assurance allows some industries to be certified to expand into the international market. There are 4.5 million laying hens in industrial production, of which 2.5 million are egg-producing hens and 1.5 million are replacement pullets and cocks. About 2,500,000 eggs are produced and 75% of them are sold. Per capita consumption is 135 eggs/person/year. Paraguay is among the countries with the lowest consumption of eggs even though nongovernmental organizations have made several awareness campaigns and it is one of the cheapest food products. The outlook for egg production is positive, but consumption remains low. However, the producers of laying hens in Paraguay look at it as an excellent opportunity for growth (Molinas, 2013; Ruiz, 2015).

In Paraguay, the National Poultry Health and Quality Plan includes a *Salmonella* control program for SG, SP, ST, and SE, especially in breeder flocks (Spaini and Ydoyaga, 2010). It also includes laying hens and broilers. It is applied by the National Quality and Animal Health Service. This Program involves bacteriologic monitoring of the environment and animals (feces, meconium), and serologic monitoring of birds (enzyme immunoassay methods and rapid hemagglutination plate and tube). Sampling should be done in breeder flocks at different times in life: 1 day old, 4 weeks old, 2 weeks before entering egg phase, and each 2 weeks in this phase. In laying hens, the sampling should be done in the animals at one day old, 2 weeks before entering egg phase, and each 10 weeks in this phase. A positive result in serologic monitoring for SE or ST implies the isolation of positive or suspect birds and the lot remains interdicted for providing eggs for consumption and/or incubation. On the other hand, when a positive result for SG or SP in serologic monitoring is found, all poultry and hatching eggs from the lot should be removed and destroyed.

## 9. PERU

Peru has increased its egg production an average of 6% a year from 2000 to 2013. It is estimated that about 99% of production is brown shell, whereas the proportion of hens in cages is estimated at 84%, 10% in floor systems, and 6% outdoors (Evans, 2015a). There is currently a population of 16 million laying hens in Peru and the per capita consumption has increased to 171 eggs/person/year (Ruiz, 2015). The production is concentrated in Ica, Lima, La Libertad, Lambayeque, and Arequipa. The most important egg producers are in San Martin and Loreto, and Cajamarca and Puno in the jungle and in the mountain range, respectively. Raising hens is done mostly in the coastal and jungle area; however, they can also be grown in the mountains up to 3000 m (Cumpa Gavidia, 2014).

Egg production totaled 359,000 tons in 2014 due to local demand, but it is not so high compared with other countries in the region. The industry of laying hens in Peru is growing faster than its economy. In 2005, egg consumption was only 110 eggs/person/year. Due to the impressive economic growth in the last decade, mainly driven by the mining industry, egg consumption has been increasing steadily. Peru is a market for middle-sized brown eggs, but lately white eggs are also demanded by consumers with rising purchasing power and by foreigners who come to work from countries where white eggs are usually consumed. In the past, all layer farms had a very simple structure consisting of open houses built with wooden poles, wire, and, in some areas where it never rains, not even a hard roof; they used canvas soaked in asphalt or straw. Labor was cheap and available, but it has become expensive and scarce with the rise of the economy. Today, the egg industry is undergoing rapid modernization in the production process and automation is playing an important role (Cousinet, 2015).

Lévano-Muñante and Lopez-Flores (2001) investigated the presence of *Salmonella* spp. in 680 pools of four eggs from farms ($n=4$) and markets ($n=4$) in different cities of Peru. They found that only eggs from markets were contaminated with the pathogen in shell and egg content, 2.4% and 0.6%, respectively. The serotypes isolated were *S.* Djugu (90.5%) and *S.* Mbandaka (9.5%). On the other hand, Sanchez-Ingunza et al. (2013) isolated *S.* Infantis, *S.* Kentucky, *S.* Mbandaka, and *S.* Senftenberg from poultry carcasses, drag swabs, and chicken feed. Although this information is not related to samples of eggs or layer hens, it has becomes important with the report of Zamudio et al. (2011) who observed a genetic association between *S.* Infantis strains from poultry meat and human clinical isolation. With these results, further works are required to generate more information about *Salmonella* prevalence in Peru, particularly related to the poultry egg industry.

There is a Control Health Law for fowl typhoid and Pullorum disease in poultry and hatcheries in Peru (García Perez and Morales Bermudez, 1987). The official tests for monitoring are SPA or rapid whole blood plate agglutination. Breeder flocks should not receive any treatment with antibiotics at least 15 days before the tests are done. Rapid whole blood plate agglutination/SPA test is conducted in the whole lot of birds when 5% of hens start laying eggs (first test) and repeated in 5% of the negative tested birds 21 days after (second test). All birds that are positive and/or suspicious should be separated into a house apart within the same lot and MA test will be done. In parallel, up to four birds are sent to an authorized laboratory for SG and SP isolation. Poultry establishments that remain negative for two consecutive tests of the batch production will be classified as Free. If there is no bird positive with MA test or isolation of *Salmonella*, the lot will be classified in the category Free. On the other hand, if positive birds are found, the lot will be separated for immediate slaughter. It is recommended to perform a retest throughout the lot within 4 weeks. If one or more of the retest birds are positive for MA and/or *Salmonella* isolation, the lot will be categorized as infected and will go into quarantine, preparing poultry for immediate slaughter. It is prohibited to use a vaccine, with a rough strain of SG, against typhoid in poultry establishments. On the other hand, there is a Draft plan for Free Certification *Salmonella* Poultry Farms, which aims to reduce the prevalence of important *Salmonella* serotypes for public and animal health in all stages of the primary types of poultry production (Separovic, personal communication).

## 10. GUYANA AND SURINAME

In Guyana, the number of laying hens and egg production decreased from 300,000 in 2000 to 185,000 in 2013 and from 1506 tons in 2000 to 898 tons in 2013, respectively, whereas the per capita consumption was reduced from 2 kg egg/person/year in 2000 to 0.7 kg egg/person/year in 2009 (Evans, 2014; Food and Agriculture Organization of the United Nations, 2015). On the other hand, the exportation of egg was 2 tons in 2011 with a peak of 90 tons in 2007, whereas importation of table eggs increased from 872 in 2000 to 1692 tons in 2011 (Evans, 2015b).

In Suriname, the number of laying hens and egg production decreased from 750,000 in 2000 to 670,000 in 2013, and from 3000 tons in 2000 to 2721 tons in 2013, respectively, whereas the per capita consumption was reduced from 5.9 kg egg/person/year in 2000 to 4.1 kg egg/person/year in 2009 (Evans, 2014; Food and Agriculture Organization of the United Nations, 2015). However, the importation of table eggs increased from 110 tons in 2000 to 316 tons in 2011 (Evans, 2015b). Most eggs consumed in Suriname are domestically produced, and this country has been self-sufficient in egg production. The main challenge for the production of eggs remains the high cost of feed for hens, as most feed components for poultry are imported (Derlagen et al., 2013).

No information about the National Poultry Improvement Plan was found for Guyana and Suriname. However, it is known that the veterinary legislative framework of Suriname is relatively outdated; several areas of animal health are not regulated or included in the legislation. The Animal Diseases Act is not comprehensive enough to deal with the risks relating to many diseases, which could potentially impact animal and public health in Suriname (Bessy et al., 2013).

## 11. URUGUAY

There is currently a population of 3.1 million laying hens in Uruguay, with 2,200,000 replacement pullet chicks annually in most genetic lines. There are not any grandparent flocks in this country, but two complete integrations exist (from breeder flocks to the sale of table eggs). Furthermore, there are six hatcheries and two shell egg processing plants; the production of egg products is for the domestic market. The per capita consumption is 262 eggs/person/year, of which 80% are brown eggs. Also, 85% of the birds are housed in cages from the first day of their life (Trenchi, personal communication, Ruiz, 2015). According to the Agricultural Statistical Yearbook 2014 (DIEA, 2014), production was 921.6 million eggs in Uruguay in 2013 with 36,360 eggs exported.

From 1997 to 2004, SE was the most frequently identified serotype in Uruguay, accounting for more than 50% of the strains received each year at the National *Salmonella* Center and for more than 85% of the strains isolated from humans (Betancor et al., 2004). For this reason, a countrywide serological and microbiological survey of chicken flocks and commercially available eggs from 2000 to 2002 was carried out by Betancor et al. (2010). They found that at least 1 of every 214 eggs (58 of 12,400) in the country was contaminated with *S. enterica*, an estimated prevalence of 0.0049. *S.* Derby was the most frequently isolated, followed by SG, SE, and *S.* Panama.

Despite the highest prevalence in eggs, *S.* Derby was not isolated from humans during the analysis period, suggesting a low capacity to infect humans. On the other hand, all SE and *S.* Derby isolates were susceptible to ampicillin, nalidixic acid, gentamicin, trimethoprim-sulfamethoxazole, tetracycline, and chloramphenicol. Two *Salmonella* Panama isolates were resistant to ampicillin, cephalothin, cefuroxime, and ceftazidime, but susceptible to all of the other antibiotics tested. Among SG isolates, three antibiotic susceptibility patterns were found.

The occurrence of ESBL-producing *Salmonella* isolates is an extremely rare event in Uruguay. In this sense, Vignoli et al. (2006) reported the presence of bla-TEM-144 carried in a transferable plasmid in *S.* Derby obtained from eggs, but the clinical relevance of such a finding is still unknown. Nevertheless, so far in Uruguay, there have been no reports of ESBLs in SE of human origin. In this country, this serotype and ST are the most frequent agents of food-borne diseases (Betancor et al., 2004, 2010; Macedo-Viñas et al., 2009), and historically they have remained susceptible to oxyiminocephalosporins (Betancor et al., 2010; Macedo-Viñas et al., 2009).

*Salmonella* control is only obligatory for SG and SP in breeder flocks from Uruguay by rapid hemagglutination test after starting the egg production and before the incubation of fertile eggs. The retest shall be determined by the Poultry Health Department (Ministry of Livestock, Agriculture and Fisheries), when deemed pertinent to lot control. Any positive or suspect bird to this serological test should be removed immediately by sacrifice (Ministerio de Agricultura y Pesca, 2001).

## 12. VENEZUELA

The Venezuelan Poultry Federation (FENAVI) estimates the average table egg production per month to be 1,180,000 boxes of 360 units in 2011. About 85% of Venezuela's egg production is concentrated in the central region with 10% in the western region and 5% in the eastern region (USDA, 2011). The per capita consumption was 180 eggs/person/year in 2012 and the goal was to achieve 234 eggs/person/year in 2015 (Villa, 2012). It was 190 eggs/person/year in 2014. The number of hens and the egg production increased from 13 million in 2010 to 21 million in 2014, and from 174,600 tons in 2000 to 280,900 tons in 2013, respectively (Evans, 2015a; Ruiz, 2015). However, the number of laying hens was 14 million in 2015, slightly lower than existing in 2011, due to the severe shortage of supplies. Today, many farms have gone bankrupt and closed because they are unable to remain active under these conditions, where shortage of food and medicine for birds has become a commonplace for producers. Furthermore, there is a lack of foreign exchange to import raw materials (Herráez Martinez, 2016).

There is little information about *Salmonella* spp. in layer hens or eggs for Venezuela. The available data are related to other kinds of samples from the poultry industry where the serotypes found were *S.* Heidelberg, SE, ST, and *S.* Meleagridis (Molina et al., 2010). In terms of public health, the frequency of serotypes from food samples is not exactly known in Venezuela (Molina et al., 2010). On the other hand, Infante et al. (1986) evaluated the serotypes isolated from poultry samples from 1976 to 1986.

Although it is not clear if the authors were referring to organs or eggs, they found that 22.8% of the samples related to laying hens were positive for *Salmonella* spp. The serotypes identified were ST, *S.* Agona, *S.* Ohio, *S.* Infantis, *S.* Heardt, *S.* Albany, *S.* Isangi, *S.* Fresno, *S.* Orion, *S.* Senftenberg, *S.* Whortington, and *S.* Johannesburg.

The National Program against Avian Salmonellosis is applied by the National Institute of Integral Agricultural Health (formerly Autonomous Agricultural Health Service) in Venezuela, which belongs to the Ministry of Popular Power for Agriculture and Lands (formerly Ministry of Agricultural and Breeding). This program was created to eradicate SG, SP, and SE in parent flocks, to control and eradicate SG and SP in breeder flocks, and to control these two *Salmonella* biovars in layer farms (Ministerio de Agricultura y Cría, República de Venezuela. 1995). Bacteriological and serological monitoring of birds (SPA and MA test) are obligatory to test for SG and SP. The use of vaccines or bacterins from homologous strains to control SG and SP is forbidden. Also, the application of drugs in birds 3 weeks before the field trial that can cover up the reactions or *Salmonella* isolation, and other procedures are prohibited that mask the free classification of lot for these biovars. When all poultry houses or animals from a farm are negative for the isolate or SPA detection for SG and SP, the farm will be considered free of these bacteria. These diagnostic tests should be conducted at least once a year. On the other hand, when the SPA test is positive in one or more birds of a house, a representative number of positive birds will be sent to the official laboratory so a confirmation, using serological and bacteriological tests can be done. If there is a bird positive for some of these biovars by bacteriological test, the poultry house where this bird is will be in quarantine, the lot will be declared in sanitation, and birds will be sacrificed entirely in the closest slaughter plant. After the necessary sanitary measures have been implemented, a sample of 10% of the remaining animals from the farm should be taken. If any animals are positive, the measures described earlier will be applied.

There is a draft resolution on the National Program of prevention, control, and eradication of salmonellosis in Venezuela (Ministerio del Poder Popular para la Agricultura y Tierras, 2011). Some objectives are to control and eradicate SG, SP, SE, and ST in breeder flocks and layer farms and control these Salmonellas in balanced food-processing plants.

## 13. CONCLUSIONS

The growth of the poultry industry in South America has been explosive in recent years. Some countries are large grain producers (Argentina, Brazil, and Paraguay), others import grain but have a good production structure (Colombia and Peru), which impacts the cost of production positively. More than 214 million layers are located in South America, producing more than 4 million tons of eggs, with Brazil, Colombia, and Argentina being the most important egg producers in this region.

Nonenriched cage systems are not prohibited in South America and most of the laying hens are located in battery cage farms. Nonconventional systems, including free range, pasture, and organic, are rare in this region.

The prevalence data about *Salmonella* spp. and antimicrobial resistance for *Salmonella* isolated from eggs, layer hens, and their environment are scarce in South America. Increased international interest in controlling SE and ST in poultry has led to the development and implementation of monitoring programs in South American countries by different National Health Services. However, since *S*. Gallinarum has not been eradicated in breeder flocks in these countries, the National Poultry Improvement Plan of each country defines stringent sanitation and testing standards, especially to prevent the transmission of this serotype to the egg-laying stock.

Although vaccines against SG have been administered for many years in most of the countries of this region, especially using a rough strain of SG (9R), this biovar has not been eradicated in commercial layers. Furthermore, SE can be coisolated from farms with SG in commercial layer hens, indicating that the latter biovar does not necessary exclude SE in poultry.

## ACKNOWLEDGMENTS

We thank Prof. Lucille St. Cyr de Bueno (Tucuman, Argentina), MLS, SBB (ASCP)[CM] Marina U. Bueno (National Institute of Health, USA), and MT (ASCP) Carol Reid (USA) for reading the manuscript.

## REFERENCES

Agencia Ecuatoriana de aseguramiento de la calidad del agro, 2013. Programa Nacional Sanitario Avícola. http://www.agrocalidad.gob.ec/wp-content/uploads/2014/11/Programa-Nacional-Sanitario-Avicola-PNSA1.pdf.

Aguilera Díaz, M., 2014. Determinantes del desarrollo en la Avicultura en Colombia; instituciones, organizaciones y tecnologías. http://www.banrep.gov.co/docum/Lectura_finanzas/pdf/dtser_214.pdf.

Alexandre, S.M., Pozo, M.C., González, G.V., Martínez, H.M.C., Prat, M.S., Fernández, R.A., Fica, C.A., Fernández, O.J., Heitmann, G.I., 2000. Detection of *Salmonella* Enteritidis in samples of poultry products for human consumption in the Chilean metropolitan area. Rev. Méd. Chile 128, 1075–1083.

Asociación de avicultores Cochabamba- Bolivia, 2015. Boletín Estadístico Anual, Gestión 2014, Cochabamba, Boliva. http://adacochabamba.com/web/wp-content/uploads/2013/08/BOLETIN-ANUAL-2014-FORMATO-DIGITAL.pdf.

Associcão Brasileira de Proteina Animal, 2015. Annual Report 2015. http://abpa-br.com.br/files/publicacoes/c59411a243d6dab1da8e605be58348ac.pdf.

Avila Cortez, F., 2015. 2014 Avicultura: con buena clasificación. Avicultores 223, 18–37.

Bessy, C., Lazarus, C., Romero, J., Sosa, O., Wijngaarde, J., 2013. Project: Suriname: FAO/Idb Agriculture Policy Loan: Agricultural Services Study Inter-alia Analysis of Suriname's Agricultural Health and Food Safety System. www.iadb.org/Document.cfm?id=38149589.

Betancor, L., Schelotto, F., Martinez, A., Pereira, M., Algorta, G., Rodriguez, M.A., Vignoli, R., Chabalgoity, J.A., 2004. Random amplified polymorphic DNA and phenotyping analysis of *Salmonella* enteric serovar Enteritidis isolates collected from humans and poultry in Uruguay from 1995 to 2002. J. Clin. Microbiol. 42, 1155–1162.

Betancor, L., Pereira, M., Martinez, A., Giossa, G., Fookes, M., Flores, K., Barrios, P., Repiso, V., Vignoli, R., Cordeiro, N., Algorta, G., Thomson, N., Maskell, D., Schelotto, F., Chabalgoity, J.A., 2010. Prevalence of *Salmonella enterica* in poultry and eggs in Uruguay during an epidemic due to *Salmonella enterica* serovar Enteritidis. J. Clin. Microbiol. 48, 2413–2423.

Caffer, M.I., Eiguer, T., 1994. *Salmonella* enteritidis in Argentina. Int. J. Food Microbiol. 21, 15–19.

Caffer, M.I., Alcain, A., Panagopulo, M., Moroni, M., Brengi, S., Terragno, R., 2010. Serovariedades de *Salmonella* spp.en Argentina, 2007–2009. XII Congreso Argentino de Microbiología. Rev. Arg. Microbiol. 42, 80.

Clements, M., 2015. Who Are the World's Largest Egg Producers? http://www.wattagnet.com/Who_are_the_world%E2%80%99s_largest_egg_producers_.html.

Conmay, A., 2015. World egg output grows despite disease impact. Poult. Trends 28–35.

Cousinet, L., 2015. Lohmann Tierzucht, market leader in Peru. Poult. News 1, 10–11.

Cumpa Gavidia, M., 2014. Producción de gallinas ponedoras en el Perú. Agro Enfoque. http://agroenf.com/2014/01/02/produccion-de-gallinas-ponedoras/.

De Franceschi, M., Viora, S., Anselmo, R., Barrios, H., 1998. *Salmonella* Enteritidis en huevos frescos. Rev. Med. Vet. 79, 5–7.

Derlagen, C., Barreiro-Hurlé, J., Shik, O., 2013. Agricultural Sector Support in Suriname. IDB/FAO, Rome, Italy.

Dias de Oliveira, S., Flores, F.S., dos Santos, L.R., Brandelli, A., 2005. Antimicrobial resistance in *Salmonella* Enteritidis strains isolated from broiler carcasses, food, human and poultry-related samples. Int. J. Food Microbiol. 97, 297–305.

DIEA, 2014. Anuario Estadístico Agropecuario 2014. Ministerio de Agricultura, Ganadería y Pesca-Uruguay. http://www.mgap.gub.uy/portal/page.aspx?2,diea,diea-anuario-2014,O,es,0.

Dorst, J.P., Winkel, C.M., 2015. South America. Encyclopaedia Britannica. http://www.britannica.com/place/South-America#toc41759.

Dreyer, J., Windhorst, H.-W., 2011. Análisis del mercado mundial del huevo y ovoproductos. www.elsitioavicola.com/articles/2044/analisis-del-mercado-mundial-del-huevo-y-ovoproductos/#sthash.Oxd8O8wi.dpuf.

Economic Commission for Latin America and the Caribbean (ECLAC), 2014. Statistical Yearbook for Latin America and the Caribbean, 2014. LC/G.2634-P, Santiago, Chile.

El Sitio Avicola, 2015a. Cada Ecuatoriano Come 12 Huevos Al Mes. http://www.elsitioavicola.com/poultrynews/22925/cada-ecuatoriano-come-12-huevos-al-mes/.

El Sitio Avícola, 2015b. Volumen récord de exportación de carne de pollo en junio de 2015. www.elsitioavicola.com/poultrynews/30456/volumen-racord-de-exportacian-de-carne-de-pollo-en-junio-de-2015/.

Espinoza, E., Revollo, S., Espada, A., 2007. Identificación de *Salmonella* sp. mediante la técnica de reacción en cadena de la polimerasa anidado (nested PCR) y técnicas convencionales en huevos recolectados en los principales mercados de la ciudad de La Paz. Vis. Cienti 2, 10–16.

Estrada Aguila, J.P., Valencia Bustamante, B.A., 2012. Determinación de *Salmonella* spp. en huevos frescos de gallina en los principales mercados de la ciudad de Quito (Veterinarian and Zootecnist thesis). Facultad de Medicina Veterinaria y Zootecnia, Universidad Central del Ecuador, Quito, Ecuador.

Evans, T., 2014. Tendencias Avícolas Mundiales: El Sector Avícola Latinoamericano. http://www.elsitioavicola.com/articles/2558/tendencias-avacolas-mundiales-el-sector-avacola-latinoamericano/.

Evans, T., 2015a. Tendencias Avícolas Mundiales 2014: América abastece la quinta parte del huevo en el mundo. http://www.elsitioavicola.com/articles/2698/tendencias-avacolas-mundiales-2014-amarica-abastece-la-quinta-parte-del-huevo-en-el-mundo/#sthash.k6sETcat.dpuf.

Evans, T., 2015b. Tendencias Avícolas Mundiales 2014: Se duplicó la exportación de huevo de América. http://www.elsitioavicola.com/articles/2699/tendencias-avacolas-mundiales-2014-se-duplica-la-exportacian-de-huevo-de-amarica/#sthash.TurZZjC0.dpuf.

Favier, G.I., Lucero Estrada, C.M.S., Lazarte Otero, V., Escudero, M.E., 2012. Prevalence, antimicrobial susceptibility, and molecular characterization by PCR and pulsed field gel electrophoresis (PFGE) of *Salmonella* spp. isolated from foods of animal origin in San Luis, Argentina. Food Control 29, 49–54.

Fernandes, S.A., Paterson, D.L., Ghilardi-Rodrigues, A.C., Adams Haduch, J.M., Tavechio, A.T., Doi, Y., 2009. CTX-M-2-producing *Salmonella Typhymurium* isolated from pediatric patients and poultry in Brazil. Microb. Drug Resist. 15, 317–321.

Fica, A., Acosta, G., Dabanch, J., Perret, C., Torres, M., Lopez, J., Jofre, L., Weitzel, T., 2012. Brotes de salmonelosis y el tamaño y rol del Estado en Chile. Rev. Chil. Infect. 29, 207–214.

Food and Agriculture Organization of the United Nations, 2015. Statistic Divisions. FAOSTAT. http://faostat3.fao.org/download/Q/QA/E.

Freitas Neto, O.C., Galdino, V., Campello, P.L., Almeida, A.d., Fernandes, S., Berchieri Júnior, A., 2014. *Salmonella* serovars in laying hen flocks and commercial table eggs from a region of São Paulo state, Brazil. Rev. Bras. Ciênc. Avíc. 16, 57–61.

Gama, N., Berchieri Jr., A., Fernandes, S., 2003. Occurrence of *Salmonella* sp. in laying hens. Rev. Bras. Ciênc.Avíc. 5, 15–21.

García Pérez, A., Morales Bermúdez, P.R., 1987. Reglamento sanitario de control de la pulorosis y tifosis en planteles avícolas y plantas de incubación a nivel nacional. Decreto Supremo No. 022-87-AG.

Giacomozzi Carrasco, J., December 2014. Panorama y actualización del mercado del huevo. Oficina de estudios y políticas agrarias. http://www.odepa.cl/wp-content/files_mf/1418835552huevos.pdf.

González Díaz, N., y Correo Munida, J., 1998. Plan nacional de control de salmonella en la avicultura chilena. vol. 4. TecnoVet. http://www.tecnovet.uchile.cl/index.php/RT/article/view/6246/6102.

Hendriksen, R.S., Vieira, A.R., Karlsmose, S., Lo Fo Wong, D.M., Jensen, A.B., Wegener, H.C., Aarestrup, F.M., 2011. Global monitoring of *Salmonella* serovar distribution from the world health organization global foodborne infections network country data bank: results of quality assured laboratories from 2001 to 2007. Foodborne Pathog. Dis. 8, 887–900.

Infante, D.M., León, A.J., de Noguera, C., Quiroz, C.A., Herrera, A.J., Valdillo, P., 1986. Serovariedades de Salmonelas Paratifoides aisladas de aves durante 1976–1986. Vet. Trop. 11, 3–8.

Instituto Colombiano Agropecuario, 1976. Control de la Salmonelosis en las aves de corral. Resolución Número 1476 del 10 de septiembre de 1976.

Instituto Nacional de Estadísticas Chile, 2015. Criadero de aves y cerdos. www.ine.cl/canales/chile_estadistico/estadisticas_agropecuarias/estadisticas_pecuarias/pecuarias.php.

Kottwitz, L.B.M., Back, A., Leão, J.A., Alcocer, I., Karan, M., Oliveira, T.C.R.M., 2008. Contaminação por *Salmonella* spp. em uma cadeia de produção de ovos de uma integração de postura comercial. Arq. Bras. Med. Vet. Zootec. 60, 496–498.

Kottwitz, L.B.M., Leão, J.A., Back, A., dos P.Rodriguez, D., Magnani, M., de Oliveira, T.C.R.M., 2013. Commercially laid eggs vs. discarded hatching eggs: contamination by *Salmonella* spp. Braz. J. Microbiol. 44 (2), 367–370.

Lévano Muñante, G.C., López Flores, C., 2001. Evaluación de la presencia de *Salmonella* en huevos frescos, utilizando el medio xilosa-lisina-tergitol 4 (XLT4). Ciencia e Investigación 4, 50–56.

Lohmann Animal Health, 2013. Regulatory Affairs.

Lohmann Animal Health, 2014. Colombia: Gold Standard Vaccine AviPro Salmonella Vac E Has Been Recently Approved. https://www.salmonella360.com/news/2014/sep/24/colombia-gold-standard-vaccine-avipro-salmonella-vac-e-has-been-recently-approved.

Macedo-Viñas, M., Cordeiro, N.F., Bado, I., Herrera-Leon, S., Vola, M., Robino, L., Gonzalez-Sanz, R., Mateos, S., Schelotto, F., Algorta, G., Ayala, J.A., Echeita, A., Vignoli, R., 2009. Surveillance of antibiotic resistance evolution and detection of class 1 and 2 integrons in human isolates of multi-resistant *Salmonella* Typhimurium obtained in Uruguay between 1976 and 2000. Int. J. Infect. Dis. 13, 342–348.

Mair, G., Beczkowski, G.y, Lamelas, K., 2015. Boletín Avícola 2014. Ministerio de Agricultura, Ganadería y Pesca. http://www.minagri.gob.ar/site/ganaderia/aves/02-informes/_archivos/000001-Boletines/000072_Nro72%20Junio%202014.pdf.

Martínez Herráez, N., 2016. Situación crítica para la avicultura venezolana. http://www.elsitioavicola.com/poultrynews/31407/situacian-cratica-para-la-avicultura-venezolana/.

Melo, J., Harkes, R., Elías, J., 2007. Sensibilidad de aislamientos de salmonella gallinarum de Argentina a diferentes antimicrobianos. http://www.engormix.com/MA-avicultura/sanidad/articulos/sensibilidad-aislamientos-salmonella-gallinarum-t1552/165-p0.htm.

Mendes, A.A., 2011. Poutry and Egg Production in South America – Consumer Behavior, Trends and Perspectives. http://www.poultryclub.com/fileadmin/european-clubs/poultryclub/binary/activities/Presentation-A-Mendes-Arg-2011.pdf.

Merial, 2016. Gallivac Se. http://www.merial.com.br/avicultores/produtos/Pages/gallivac_se.aspx.

Ministério da Agricultura, Pecuária e Abastecimiento, 2010. Norma Técnica para certificaçãosanitária de estabelecimentos avícolas de reprodução e monitoramento dos estabelecimentos avícolas comerciais. Portaria N° 297 www.avisite.com.br/legislacao/anexos/portaria297_20100621.pdf.

Ministerio de Agricultura y Cría, República de Venezuela, 1995. Resolución Oficial del Programa Nacional Contra la Salmonelosis Aviar. N° DM/96 Servicio Autónomo de Sanidad Agropecuaria- SASA. www.pgr.gob.ve/dmdocuments/1995/35812.pdf.

Ministerio de Agricultura y Pesca, 2001. Sanidades especiales, Sanidad avícola. In: Infante, G., Chans, L.E., Costa, R. (Eds.), Legislación Sanitaria Animal, tomo I, pp. 178–183.

Ministerio de Agricultura, Ganadería y Desarrollo Rural, 2002. Resolución administrativa. N°19/2002.

Ministerio del Poder Popular para la Agricultura y Tierras, 2011. Proyecto de Resolución sobre el Programa de Prevención, Control y Erradicación de la Salmonelosis Aviar.

Molina, N., Millán, B., Araque, M., 2010. Indicadores de calidad sanitaria y fenotipificación de *Salmonella* enterica aislada de pollo crudo comercializado en el área urbana de Mérida, Venezuela. Rev. Infect 14, 174–185.

Molinas, A., 2013. Producción, Comercialización y consumo de huevo en Paraguay. http://ing-alfredo-molinas.blogspot.com.ar/2013/12/produccion-de-huevo-en-paraguay.html.

Peirano, G., Agersø, Y., Aarestrup, F.M., dos Reis, E.M., dos Prazeres Rodrigues, D., 2006. Occurrence of integrons and antimicrobial resistance genes among *Salmonella enterica* from Brazil. J. Antimicrob. Chemother. 58, 305–309.

Prat, S., Fernandez, A., Fica, A., Fernandez, J., Alexandre, M., Heitmann, I., 2001. Phage typing of *Salmonella* enteritidis isolates from clinical, food, and poultry samples in Chile. Pan. Am. J. Public Health 9, 7–12.

Programa de Aves, 2015. Manual de procedimientos operativos. http://www.senasa.gov.ar/prensa/DNSA/dir_programacion_sanitaria/Programa_Enfermedades_de_Aves_y_Animales_de_Granja%20/Plan_Nacional_Sanidad_Avicola/Manual_procedimientos-PNSA-2015-V2.pdf.

Pulido-Landinez, M., Sanchez-Ingunza, R., Guard, J., do Nascimento, V.P., 2014. Presence of *Salmonella* enteritidis and *Salmonella* gallinarum in commercial laying hens diagnosed with fowl typhoid disease in Colombia. Avian Dis. 58, 165–170.

Radice, M., Power, P., Di Conza, J., Gutkind, G., 2002. Early dissemination of CTX-M-derived enzymes in South America. Antimicrob. AgentsChem. 46, 602–604.

Ramírez, R.Y., Rincón, D.P., Vargas, J., 2014. *Salmonella* Enteritidis en huevos de gallina comercializados en Tunja (Colombia). Salud Soc. Uptc 1, 22–27.

Rodriguez, R., Fandino, C., Donado, P., Guzman, L., Verjan, N., 2015. Characterization of *Salmonella* from commercial egg-laying hen farms in a central region of Colombia. Avian Dis. 59, 57–63.

Rodríguez, G.A., 2006. El manejo de las pollitas en recría para futuras ponedoras. In: Hemisferio Sur, Buenos Aires.

Rossi, A., Lopardo, H., Woloj, M., Picandet, A.M., Mariño, M., Galds, M., Radice, M., Gutkind, G., 1995. Non-typhoid *Salmonella* spp. resistant to cefotaxime. J. Antimicrob. Chemother. 36, 697–702.

Ruiz, B., March 2015. Retoma su camino la industria avícola latinoamericana. Industria avícola. 12–30. http://www.industriaavicola-digital.com/201503/Default/3/0#&pageSet=11.

San Martín, B., Lapierre, L., Toro, C., Bravo, V., Cornejo, J., Hormazabal, J.C., Borie, C., 2005. Isolation and molecular characterization of quinolone resistant *Salmonella* spp. from poultry farms. Vet. Microbiol. 110, 239–244.

Sanchez Mora, M.J., 2013. Determinación de la prevalencia de Enterobacterias del género *Salmonella* spp. en huevos frescos de gallina de empresas avícolas de la provincia de Tungurahua (Veterinarian and Zootecnist thesis). Facultad de Medicina Veterinaria y Zootecnia, Universidad Central del Ecuador, Quito, Ecuador.

Sanchez-Ingunza, R., Chirinos-Zegarra, B., Guard, J., 2013. Evidence of *Salmonella* Infantis infection in humans linked to contamination of poultry products in Peru, South America. Int. Poult. Scientific forum. http://www.southernpoultrysciencesociety.org/pdfs/13ProgramGuide.pdf.

Sandoval, V.E., Terzolo, H.R., 1989. Las infecciones por *Salmonella* entérica subesp. Entérica serovar Enteritidis afectan seriamente a la producción avícola argentina y constituyen un problema de salud pública. Rev. Arg. Prod. Anim. 9, 132–134.

Sandoval, V.E., Terzolo, H.R., González Rivero, C.V., 1989. Reproducción de paratifosis en pollos parrilleros por inoculación oral de *Salmonella* entérica subesp. Entérica serovar Enteritidis. Rev. Arg. Prod. Anim. 9, 309–314.

Scheurer, G., 2014. Manual de buenas prácticas en aves de postura comerciales. In: Dunken, Buenos Aires.

SEDES (Ministerio de Salud), 2001. Informe mensual de Laboratorio. Salmonelosis, ámbito urbano/rural. Servicio Agrícola y Ganadero- Chile. 2015. Informe Sanidad Animal Chile, Año 2014.

Servicio Agrícola y Ganadero- Chile, 2015. Informe Sanidad Animal Chile, Año 2014.

Servicio Nacional de Sanidad y Calidad Agroalimentaria, 2002. Programa de control de las Micoplasmosis y Salmonelosis de las aves y prevención y vigilancia de enfermedades exóticas y de alto riesgo en planteles de reproducción. Resolución 882/2002. http://infoleg.mecon.gov.ar/infolegInternet/anexos/80000-84999/80362/norma.htm.

Servicio Nacional de Sanidad y Calidad Agroalimentaria, 2016. Programa de vigilancia y control de la contaminación por Salmonella sp. en granjas avícolas comerciales. Resolución 86/2016 http://www.cira.org.ar/index.php?option=com_content&view=article&id=6600:resolucion-86-2016&catid=112&Itemid=500.

Silva, K.C., Fontes, L.C., Moreno, A.M., Astolfi-Ferreira, C.S., Ferreira, A.J.P., Lincopan, N., 2013. Emergence of extended-spectrum–lactamase CTXM-2-producing *Salmonella enterica* serovars Schwarzengrund and Agona in poultry farms. Antimicrob. Ag. Chem. 57, 3458–3459.

Soria, M.A., 2012. Presencia de *Salmonella* y características físicas de huevos destinados a consumo humano (M.Sc. thesis). Facultad de Ciencias Veterinarias, Universidad Nacional del Litoral, Esperanza, Santa Fe. Argentina.

Soria, M.C., 2013. *Salmonella* y aflatoxinas en granjas de gallinas ponedoras comerciales (Ph.D. thesis). Facultad de Ciencias Exactas de la Universidad Nacional de La Plata, La Plata, Buenos Aires. Argentina.

Spaini, G., Ydoyaga, N., 2010. Plan Nacional de Sanidad y Calidad Avícola. (Paraguay).

Thorpe, A., Aguilar Ibarra, A., 2010. Los modelos económicos de utilización de recursos naturales en Latinoamérica. Ensayos de economía. 20, 121–143. http://www10.iadb.org/intal/intalcdi/PE/2011/08042.pdf.

United Nations Development Programme, 2016. About Latin America and the Caribbean. http://www.latinamerica.undp.org/content/rblac/en/home/regioninfo.html.

USDA, 2011. Venezuela. International Egg and Poultry Review. 14 N°42.

Vetanco, 2013. Vetanco Inaugurated Their Third Industrial Plant in Argentina. Vetanco News. http://www.vetanco.com/news/opening_plant_gralrodriguez.html.

Vignoli, R., Cordeiro, N.F., García, V., Mota, M.I., Betancor, L., Power, P., Chabalgoity, J.A., Schelotto, F., Gutkind, G., Ayala, J.A., 2006. New TEM-derived extended-spectrum lactamase and its genomic context in plasmids from *Salmonella enterica* serovar Derby isolates from Uruguay. Antimicrob. Agents Chem. 50, 781–784.

Villa, F., 2012. Venezuela se mantiene como el país con menos consumo de huevos. http://www.elmundo.com.ve/noticias/negocios/consumo/venezuela-se-mantiene-como-el-pais-con-menos-consu.aspx.

Viora, S., Anselmo, R., Barrios, H., y De Franceschi, M., 1993. *Salmonella* enteritidis en maples. Rev. Med. Vet. 74, 96–99.

Zamudio, M.L., Meza, A., Bailón, H., Martinez-Urtaza, J., Campos, J., 2011. Experiencias en la vigilancia epidemiológica de agentes Patógenos transmitidos por alimentos a través de Electroforesis en Campo Pulsado (PFGE) en el Perú. Rev. Peru Med. Exp. Salud Publica 28, 128–135.

SECTION 2

# *Salmonella* Contamination in Layer Flocks: Pathogenesis, Dissemination, and Current Control Strategies

CHAPTER

# Overview of Salmonellosis and Food-borne *Salmonella*: Historical and Current Perspectives

7

**Turki M. Dawoud[1,2], Zhaohao Shi[1], Young Min Kwon[1], Steven C. Ricke[1]**

*[1]University of Arkansas, Fayetteville, AR, United States;*
*[2]King Saud University, Riyadh, Saudi Arabia*

## 1. INTRODUCTION

Food-borne *Salmonella* and salmonellosis continues to be a major issue in the public health sector and reports of outbreaks and recalls still occur in news releases. Certainly *Salmonella* is considered one of the prominent food-borne pathogens that still warrants ongoing investigation and extensive research efforts. Consequently, there is a considerable body of scientific literature that covers every aspect of this organism's biology, epidemiology, and ecology. As scientific methodology has advanced with approaches such as transcriptional profiling and whole genome sequencing becoming more routine, a better understanding has emerged of *Salmonella's* physiological versatility, both in the environment and in the host, and also of the considerable diversity that occurs among serovars and individual isolates.

Although much knowledge has been gained in fairly recent times, *Salmonella* and salmonellosis are an age old phenomenon. Certainly as a human disease, variants of *Salmonella* as causative agents have probably existed beyond recorded history. However, appreciation of the current widespread dissemination of *Salmonella* in food products requires an overview of how *Salmonella* species came to become so prominent in food production systems, in particular certain serovars such as *S.* Enteritidis that appear to be so closely associated with egg production. In this chapter, an overview of the historical milestones of the awareness of *Salmonella,* first as a human systemic disease causing pathogen, then as a food-borne pathogen will be discussed. This will be followed by a description of current patterns and trends of *Salmonella* occurrence worldwide. Finally, this chapter will feature an introduction to food-borne *Salmonella* appearance in poultry and egg production, with the current perspectives and details being left to be discussed in other chapters of this book.

**Producing Safe Eggs. http://dx.doi.org/10.1016/B978-0-12-802582-6.00007-0**

## 2. HISTORICAL DEVELOPMENT OF SALMONELLOSIS

*Salmonella* infection has been encountered and described frequently since the beginning of the 19th century, but typhoid fever disease certainly existed even before the 18th century with ancient physicians such as Hippocrates and Galen characterizing the disease with common symptoms of prolonged fever and partial unconsciousness (Lancaster, 1990). Symptoms of the disease reported at that time were only fever without any other visible symptoms. As a result, those diseases were classified under one large group with several subgroups according to the length of fever time as short, long, or continuous. The *Hippocratic typhus* was classified under those characterized fevers and from this classification emerged an ongoing confusion of typhoid fever being associated with other diseases (Paul, 1930).

Historically, there has been considerable uncertainty linked to the typhoid fever that is caused by *Salmonella* Typhi. The first source of dispute occurred between the two types of infection, namely, typhus and typhoid. Initial observations of typhoid fever were made by Thomas Willis in 1659 (Newsom, 2007; White, 2008). An early attempt to distinguish these two infections was attempted by John Huxham in 1737 who referred to typhoid fever as a slow nervous fever, and the typhus fever as a putrid malignant fever (Bechah et al., 2008; McCrae, 1907). In 1810, von Hildenbrand distinguished between typhus and typhoid fever as simple regular typhus and irregular typhus. By 1824, Nathan Smith provided a clear and accurate description of Typhous (typhoid) fever (Smith, 1914, 1981). During the 1820s, Pierre Bretonneau along with other French physicians made two observations. The first observation originated from those who died from typhoid fever being characterized as swelling, inflamed victims, with ulcerated Peyer patches and enlarged lymph nodes, and the second occurred in people who recovered from typhoid infection who were subsequently immunized against further typhoid infections (Paterson, 1949; Shulman, 2004). In 1829, Pierre Charles Alexandre Louis, a French physician, was the first to propose the name "typhoid fever." Although numerous observations were reported to distinguish between typhus and typhoid fever, some confusion remained. However, two reports made a significant contribution toward differentiating the typhus and typhoid forms of the disease and resolving the ensuing confusion associated with these two diseases. In 1837, William Wood Gerhard made a clear distinction between typhus and typhoid fever (Cirillo, 2000; Ober, 1976). By the late 1840s, William Jenner delineated the differences between typhus and typhoid fever based on the distinctive infection symptoms and epidemiology he personally experienced after infection by both diseases (Bechah et al., 2008; Cook, 2001; Seddon and Queen, 2004). William Budd drew some key conclusions from his observations on several outbreaks of typhoid fever occurring between 1839 and 1873 that were first reported in a series of published papers and ultimately his comprehensive work on "Typhoid Fever." The most important observation by Budd was the conclusion that typhoid fever was spread by an unknown agent mostly transmitted through contaminated water thus introducing the concept of a fecal-oral route of dissemination (Moorhead, 2002; Parry, 2006).

The second primary point of confusion occurred between malaria and typhoid fever by associating them together with the name "typhomalarial fever." This name was given by the US Army pathologist Joseph J. Woodward in 1862 during the American Civil War (Bynum, 2002; Sulaiman, 2006; Cunha and Cunha, 2008; Cunha et al., 2013). This association was because of the difficulty faced by many physicians at that time to draw a clear diagnostic conclusion on the causative agent of both infections. In 1892, William Osler was able to distinguish between typhoid and malaria based on their clinical features and did so without conducting any laboratory tests (Bryan, 1996). Interestingly, those historical observations on the symptoms of both diseases are still applicable and practical to modern day physicians (Cunha, 2007; Cunha and Cunha, 2008; Pradhan, 2011). One of the main clinical signs was the acute fever that Osler noted, and it clearly exhibited a differential pattern between the two diseases. In malaria, fever becomes elevated rapidly and remains high. However, in typhoid, the fever gradually rises through the early few days post infection up to the second or third week (Cunha, 2004, 2005; Uneke, 2008). One commonly used method, the Widal reaction, was a specific agglutination test used in the diagnosis of typhoid bacilli; however, Jhaveri et al. (1995) concluded that the Widal reaction may yield misleading results as a false positive with cross-reactions occurring between typhoidal *Salmonella*, nontyphoidal *Salmonella* (NTS), and malaria (Jhaveri et al., 1995; Shanthi et al., 2012).

## 2.1 ISOLATION AND CHARACTERIZATION OF *SALMONELLA*

The first reported observation of this bacillus was in 1880 by Karl (Carl) Joseph Eberth from specimens of typhoid victims. This bacterium was called *Eberthella typhosa*, the typhoid bacillus, now known as *Salmonella* Typhi, and was first isolated by Georg Gaffky in the mid-1880s (Collins and Petts, 2011). A year later, Theobald Smith who worked under Daniel E. Salmon's supervision, isolated *Salmonella choleraesuis*, now known as the species *Salmonella enterica* from pigs and named it "the hog-cholera" as it was originally thought to be the cause of swine fever, but was later proved to be a secondary infection agent (Schultz, 2008).

In 1888, Gäertner isolated *Bacterium enteritidis*, known now as *S.* Enteritidis, and this was considered the first laboratory-confirmed case connecting a human salmonellosis outbreak to a consumed food. In 1896, four studies reached the same conclusion from two different sources. A study by Pfeiffer and Koller (Bazin, 2011; Gröschel and Hornick, 1981) and another study by Gruber and Durham (Wright and Semple, 1897; Wilson, 1909; Bensted, 1951; Sansonetti, 2011) recognized that serological responses to typhoid bacillus could be demonstrated when using the serum from animals that had been immunized by typhoid bacillus. Likewise, a study conducted by G. F. Widal, a French physician, and another one reported by A. S. F. Grünbaum and H. E. Durham reached the same conclusion that typhoid bacillus could be agglutinated by typhoid patient serum (Fison, 1897; Waller, 2002; Chart et al., 2007; Shanthi et al., 2012; Tan and Linskey, 2012). Widal called this test the "sero-diagnosis" test, and subsequently used this approach on samples isolated from patients with typhoid symptoms (Gupta and Rao, 1981). Although some results were negative, these were later identified as a new species named *Bacille paratyphique*, now known as *Salmonella* Paratyphoid (Wray and Wray, 2000). In 1897, Smith and Stewart came to the conclusion that all previously identified organisms could be classified under one large group since they shared numerous morphological and biochemical relatedness (Hornick, 1974).

In 1900, Joseph Léon Marcel Lignieres, a French bacteriologist, proposed the genus name *Salmonella* after D. E. Salmon, which became generally acknowledged by 1933. In 1925, a new classification was initiated by P. B. White using a serological diagnostics method through the H (flagellar) antigen. By 1930, Kauffman set up the basis for the existing serological analysis now known as Kauffman–White classification system using O (somatic or lipopolysaccharide) and H antigens. A few years later, 44 serotypes were listed in the first publication using the Kauffman–White serotyping scheme. Kauffman had been leading the *Salmonella* International Centre for three decades and by 1964, a total of 958 serovars had been identified. In 1965, L. Le Minor took over responsibility of supervising the *Salmonella* Centre until retiring in 1989. When he left the office, there were 2267 serovars identified as *Salmonella*. Grimont and Weill in a World Health Organization (WHO) publication "Antigenic Formulae of *Salmonella* Serovars" proposed renaming the serological scheme that had historically been known as the Kauffman–White Scheme to the White–Kauffman–e Minor Scheme in honor of Le Minor's efforts to characterize the numerous *Salmonella* serotypes (Grimont and Weill, 2007; Guibourdenche et al., 2010; Wattiau et al., 2011).

In the early 1960s, the identified serotypes were considered as species and were commonly given names based on their geographical sites of isolation, such as *Salmonella panama* and *Salmonella london*. Because of the confusion caused by this classification and lack of differentiation of the serovars with biochemical analysis, numerous proposals were suggested to resolve this matter. In 1944, Borman and colleagues proposed three species: *S. choleraesuis*, *Salmonella typhosa*, and *Salmonella kauffmannii*. In 1952, F. Kauffmann and P. R. Edwards also suggested having three species: *S. choleraesuis*, *S. typhosa*, and *S. enterica*. Ten years later, W. H. Ewing suggested *S. choleraesuis*, *Salmonella typhi*, and *Salmonella enteritidis* (Agbaje et al., 2011; Evangelopoulou et al., 2010; Ewing, 1972; Su and Chiu, 2007). In 1966, Kauffmann divided *Salmonella* into four subgenera designated by Roman numbers (I–IV), which were later considered as species. These subgenera were *S. kauffmannii* (I), *Salmonella salamae* (II), *Salmonella arizonae* (III), and *Salmonella houtenae* (IV) (Boyd et al., 1996; Brenner et al., 2000; Le Minor and Popoff, 1987). A historically significant publication regarding this bacterium's nomenclature was generated by Skerman et al., in 1980 for the Approved Lists of Bacterial Names that included five *Salmonella* species: *S. arizonae*, *S. choleraesuis*, *S. enteritidis*, *S. typhi*, and *S. typhimurium* (Euzéby, 1999; Ezaki et al., 2000).

Currently, the genus *Salmonella* is grouped in the Gammaproteobacteria class under the Enterobacteriaceae family. *Salmonella* consists of two species, *S. enterica* and *Salmonella bongori*, as developed in 1987 by Le Minor and Popoff and proposed by Reeves et al., in 1989 (Tindall et al., 2005; Su and Chiu, 2007). *Salmonella enterica* has been further divided into six subspecies, *enterica* (subspecies I), *salamae* (subspecies II), *arizonae* (subspecies IIIa) *diarizonae* (subspecies IIIb), *houtenae* (subspecies IV), and *indica* (subspecies VI). These two species are now divided into more than 2579 serotypes (serovars) using serological methods introduced by White and Kauffmann in the late 1920s with classification depending on the cell antigens, mainly lipopolysaccharide (O antigens) and flagella (H antigens) that were discovered in 1918 by Weil and Felix and less frequently the capsular antigens (Vi) (Grimont and Weill, 2007; Guibourdenche et al., 2010; Rodríguez and Hardy, 2015; Mølbak et al., 2006). The Vi antigen is uniquely produced by *Salmonella* Typhi, Paratyphi C, and some strains of *S.* Dublin. In 2004, a new species was proposed, *Salmonella subterranean* (Shelobolina et al., 2004). With all the controversy and complexity associated with *Salmonella* nomenclature over the years, this proposed new species name was announced to be effectively valid, but not for taxonomic purposes (Euzéby, 2005); however, in 2010 it was confirmed that this species is most closely related to *Escherichia hermannii* (Canals et al., 2011; Parte, 2014; http://www.bacterio.net/salmonella.html#subterranea).

## 2.2 *SALMONELLA* BACTERIOPHAGE AND FURTHER CLASSIFICATION

*Salmonella* isolates are placed into serogroups based on similarities for one or more of the somatic (O) antigens, and most of these serotypes belong to the species *S. enterica*. *Salmonella* serovars were first considered as species and were given their names according to the associated diseases, the animals from which they were isolated, or in a few cases, the person who first isolated them or the place where they were initially isolated (Su and Chiu, 2007).

During the 20th century, *S.* Typhimurium received considerable attention compared with other *Salmonella* species, both generally and for foundational genetic studies in bacterial nitrogen regulation and glutamine synthetase ammonia metabolism (Kustu et al., 1979, 1984; Merrick, 1982; Humphrey, 1999; Rabsch et al., 2001). To distinguish within this species, phage typing (PT) was introduced by Lilleengen in 1948 to characterize as many as 24 isolated strains based on their recognition by specific bacteriophages (Zinder and Lederberg, 1952; Baggesen et al., 2010). This Lilleengen typing (LT) was designated in recognition of the most studied Typhimurium serovar, Typhimurium LT2 (Stanley et al., 1993). In the late 1950s, another PT system was reported by B. R. Callow after modifying the original system by Felix and Callow to differentiate with 34 typing phages; a later effort by Anderson discriminated 195 phage types, and eventually at least 300 definitive phage types have been distinguished using this system (Ghilardi et al., 2006; Rabsch et al., 2002). For additional differentiation beyond PT, biotyping can be performed that depends on the fermentative characteristics originally applied by Edwards, tests developed by Kristensen and others, and improvements made by Hansen and Harhoff to recognize 21 biotypes (Duguid et al., 1975). In the early 1970s, Alfredsson et al. (1972) made a modification in the Kristensen fermentation test that improved results. This approach was further improved by Duguid et al. (1975) to contain 32 primary biotypes and 144 subtypes, thus achieving full biotyping (Duguid et al., 1975; Pang et al., 2012; Platt et al., 1987; Rabsch, 2007; Rabsch et al., 2011).

## 2.3 *SALMONELLA* AND SALMONELLOSIS

*Salmonella* and salmonellosis first became known as being associated with each other in the late 1870s and the species now known as *S.* Typhi was identified as the cause of the enteric fever transmitted by water and milk. By the 1930s, the importance of isolating environmental sources and human carriers was recognized by public health officials (Leavitt, 1996). With this, water supplies were secured with the rise of water chlorination and filtration as well as sewage schemes, whereas milk supplies were improved with heat treatments and pasteurization (Hardy, 2015). For direct protection of humans, the implementation of vaccination programs, education programs on proper food handling and protocols, and proper disposal of wastes helped to decrease the incidence of typhoid (Hathcock, 1982).

Salmonellosis is now mostly known as a food-borne illness and was first linked to the consumption of beef in the late 1880s by Gärtner who isolated this bacillus. Early in the second half of the 20th century, *Salmonella* Agona appeared in association with Latin American fishmeal formulated as an ingredient in poultry feeds (Clark et al., 1973; Fleming et al., 2000; Wright et al., 2011) and in the mid-1970s the appearance of *Salmonella* Hadar in turkey flocks and subsequently in chicken broiler flocks was reported (Rowe et al., 1980; O'Brien, 2013). *S.* Enteritidis has been a major problem with a high frequency of infections associated with eggs for over three decades (St. Louis et al., 1988; Threlfall et al., 2014; Velge et al., 2005). It has been suggested that the appearance of *S.* Enteritidis in chicken flocks by the 1960s was a direct consequence of the control measures taken toward eradicating *S.* Gallinarum and *S.* Pullorum from poultry (Bäumler et al., 2000; Hitchner, 2004).

In general, *Salmonella* infection develops into one of the following: the systemic infection referred to as enteric fever, an intestinal infection such as gastroenteritis, or a blood infection in humans referred to as bacteremia. The process of salmonellosis starts with the intake of *Salmonella* cells that must resist the acidity of the stomach with a pH range from 1 to 2 during digestion (Foster and Spector, 1995; Smith, 2003; Bearson et al., 2006; Álvarez-Ordóñez et al., 2009, 2011) and subsequently the colonization of the small intestine with a possibility of causing systemic infection. Numerous host factors play roles in susceptibility to *Salmonella* infections such as age, health conditions (immunocompromised patients), and other circumstances. The transmission of *Salmonella* species is usually through a fecal to oral route with the consumption of contaminated food or water with feces and/or urine from infected humans or animals (Chen et al., 2013; Monack, 2012; Raffatellu et al., 2008; Sterzenbach et al., 2013; Tsolis et al., 2011).

*Salmonella* Typhimurium and *S.* Enteritidis are the most common serotypes of *Salmonella* that cause food-borne illness in humans worldwide as well as in the United States (Galanis et al., 2006). They cause gastroenteritis in humans and a typhoid-like disease in mice in contrast to symptoms attributed to *Salmonella* Typhi infections. These two serotypes are typically associated with the consumption of poultry, beef, lamb, seafood, vegetables, fruits, and their food products (Brands et al., 2005; Davies et al., 2004; de Freitas et al., 2010; Heinitz et al., 2000; Rajashekara et al., 2000; Martinez-Urtaza et al., 2004; Heaton and Jones, 2008; Perelle et al., 2004; Mor-Mur and Yuste, 2010; St. Louis et al., 1988; White et al., 2001; Lynch et al., 2009; Hanning et al., 2009). *Salmonella* Enteritidis has historically been associated with chicken eggs, with 82% of the reported human infections of *S.* Enteritidis in the United States between the mid- and late 1980s being linked to chicken egg consumption (Dhillon et al., 2001; Seo et al., 2003; Kimura et al., 2004). Humans can become ill with salmonellosis by consuming raw or lightly cooked eggs showing the need for proper consumer cooking and handling techniques (Braden, 2006; Hope et al., 2002; Kimura et al., 2004; Patrick et al., 2004; St. Louis et al., 1988). It is often possible to reduce *Salmonella* infections by better hygiene practices and strategies (Cox et al., 2011). However, the rapid emergence of antibiotic resistance in some *Salmonella* strains has complicated the clinical treatment of their infection (Chen et al., 2004; Van den Bogaard and Stobberingh, 2000; Fey et al., 2000; Randall et al., 2004; Threlfall, 2002; Threlfall et al., 2000; Velge et al., 2005; White et al., 2001).

NTS species are responsible for human gastroenteritis and bacteremia cases with annual estimated global illnesses occurring in 93.8 million individuals along with 155,000 deaths (Majowicz et al., 2010). Also, *Salmonella* Typhi causes typhoid fever with an estimated 21.7 million cases and 217,000 deaths occurring annually (Crump et al., 2004). This serovar is mostly transmitted through the oral route from food or water contaminated by human feces or urine. An annual average of 9.4 million incidents was caused by 31 major food-borne pathogens. NTS species were responsible for 11% of those incidents as the second most common pathogenic agent (Scallan et al., 2011). In the United States, *Salmonella* as a predominant bacterial agent was responsible for 44% of laboratory-confirmed bacterial food-borne infections in 2009. A more recent report by Centers for Disease Control and Prevention (CDC) in 2014 revealed that the incidence of laboratory-confirmed NTS infections in the United States for 2013 decreased by 9% in comparison with the period 2010–2012 and exhibited no changes when compared with the period 2006–2008. In both comparisons, the incidence rate of 15.19 in 2013 is still above the national health objectives of 11.4 cases per 100,000 persons for 2020 (Crim et al., 2014).

## 2.4 CURRENT EPIDEMIOLOGY ISSUES

The infections associated with *Salmonella* cost the US economy annually an estimated average of 11.39 billion US dollars, exceeding the annual cost of other bacterial food-borne infections and making NTS the costliest food-borne pathogen in terms of health outcomes, with losses of approximately 3.7 billion dollars (Batz et al., 2014; ERS-USDA, 2014; Hoffmann and Anekwe, 2013; Scharff, 2012; Byrd-Bredbenner et al., 2013). In 2013, a 9% decrease in *Salmonella* infection incidences was reported in comparison with the period 2010–2012 of other bacterial food-borne pathogens;

of those, 90% were serotyped isolates. Among them were *S.* Enteritidis with 19%, *S.* Typhimurium with 14%, and *S.* Newport with 10%. Concurrently, it was reported that between 1998 and 2008 approximately 34% of all single serotype *Salmonella* outbreaks were linked to a particular food, of those 66% were caused by the top four serotypes, *S.* Enteritidis with 36%, *S.* Typhimurium with 14%, *S.* Newport with 10%, and *S.* Heidelberg with 6% (Jackson et al., 2013). Poultry and poultry products have accounted for 54% of *Salmonella* outbreaks connected to a single identified product with eggs, chickens, and turkeys as the most frequent foods of origin (Gould et al., 2013; Jackson et al., 2013; Painter et al., 2013). In addition, outbreaks of *Salmonella* in association with fresh produce accounted for approximately 50% of the total (Strawn et al., 2014).

A program initiated by WHO in 2000 as a part of the global *Salmonella* surveillance (Salm-Surv) system "External Quality Assurance System" (EQAS) for *Salmonella* species serotyping is considered the largest surveillance system worldwide and is operated through a web-based self-evaluating system. To enhance and facilitate the capability of the central laboratories in many countries to serotype *Salmonella* species with high accuracy, WHO conducted an EQAS *Salmonella* serotyping annual comparative series of studies between 2000 and 2007 with 249 participating laboratories from 97 nations using serological titration. The average for all processes among the participating laboratories was 76% with the accuracy of detecting all serotypes at 82% and 94% for *S.* Enteritidis. The accuracy for reporting *Salmonella* serotypes increased by approximately 20% from 2000 to 2007, although the incorrect reporting of some serotypes was observed during this same period with the lowest percentage of 3.6% in 2007 and the highest reported in 2006 of 41% (Chaitram et al., 2003; Petersen et al., 2002; Hendriksen et al., 2009a).

The sensitivity of identifying *Salmonella* serovars using a phase two flagellar antigen (H-antigen) detection approach accounted for the high percentage error when reporting serotype results mostly due to the loss of this antigen (Hopkins et al., 2010). More importantly, the occurrence of laboratory errors was reported to be influenced by several issues. Approximately 30% of the participating laboratories lacked high-quality antigenic detection materials, which caused difficulties in detecting uncommon serotypes in 26% of the overall laboratories due to false positives or false negatives (Galanis et al., 2006; Hendriksen et al., 2009a). Difficulties in identifying unusual strains and in getting access to high-quality antisera appeared to be more likely to occur in regions of Africa, Central Asia, the Middle East, Russia, and the Caribbean (Hendriksen et al., 2009a). Problems with identifying the correct serotype can in turn lead to major delays in tracing back to the main source of numerous *Salmonella* infections (Galanis et al., 2006; Hendriksen et al., 2009a).

The WHO Global Salm-Surv system had been focusing solely on the surveillance of *Salmonella* species and has been expanded to include food-borne and other enteric pathogens. This system is now known as Global Foodborne Infections Network (GFN). This system has two subsystems, EQAS and Country Databank (CDB); the first system was established to assess the quality of *Salmonella* serotyping and currently is employed for food-borne pathogens serotyping and antimicrobial susceptibility worldwide. The CDB has been established to report the data annually for the 15 most frequently identified *Salmonella* serotypes as a global passive surveillance system from members of national laboratories (Hendriksen et al., 2009b, 2011).

A study by Hendriksen and others (2011) reported on the global distribution of the 15 most consistently prevalent *Salmonella* serotypes identified from human clinical specimens as a part of the WHO GFN system between 2001 and 2007. Two specific serovars out of all the serovars that had been reported were predominant in most regions: *S.* Enteritidis ranked first and *S.* Typhimurium second. This pattern of predominance is in contrast to the regions of North America and Oceania (Australia and New Zealand). These reported serotypes were isolated from both human and nonhuman (animals, food, feed, and environment) sources (Hendriksen et al., 2011; Vieira et al., 2009). The data collected from the CDB were uploaded by public health laboratories of 37 countries within six geographical regions (18 subregions). The European region was divided into five subregions: Northern Europe, Western Europe, Central Europe, Southern Europe, and Eastern Europe. *Salmonella* Enteritidis was the most common serovar isolated from humans in all subregions, ranging from 56% to 89.9%, followed by *S.* Typhimurium, which ranged from 5.4% to 21.6%. For nonhuman sources, *S.* Enteritidis was the most commonly isolated serovar in Central Europe, Southern Europe, and Eastern Europe, ranging from 30.9% to 49.6%, followed by *S.* Typhimurium in Central Europe with 27.7% and Eastern Europe with 8.1%. Other serovars isolated from nonhuman sources that ranked second were *S.* Gallinarum in Southern Europe with 13.8% and *S.* Infantis in Northern Europe with 9.8%. The data for nonhuman sources were not reported from Western Europe.

In North America, the most frequently isolated serovar from both human and nonhuman sources was *S.* Typhimurium with 29.4% and 23.1%, respectively, followed by *S.* Enteritidis isolated from humans with 23.6% and *S.* Heidelberg originating from nonhuman sources at 14%. In more recent years, the most frequently reported *Salmonella* serovars from US human infections reported to CDC (2012) were *S.* Enteritidis followed by *S.* Typhimurium. In addition, the most frequent serovars from samples analyzed by FSIS-USDA (2013) for young broiler chicken were *S.* Kentucky and *S.* Enteritidis. Latin America has been partitioned into three subregions for analysis: Central America, South America, and the Caribbean. The most frequently isolated serovars from humans were *S.* Typhimurium with 31.5% and 24.4% followed by *S.* Enteritidis with 30.5% and 17.1% in Central America and the Caribbean, respectively. In Central America, *Salmonella* II 1,4,12,27 at 25.3% and *S.* Heidelberg at 16.8% were the predominant serotypes isolated. On the other hand, in the Caribbean, *S.* Kentucky and *S.* Typhimurium with 42.7% and 9.8% frequencies were identified as the leading serovars, respectively. In South America, from both human and nonhuman sources, *S.* Enteritidis occurred at 48.3% and 34.1% and *S.* Typhimurium at 18% and 9.8%, respectively.

The African continental region has been divided into two subregions for reporting purposes, North Africa and sub-Saharan Africa. *Salmonella* Enteritidis was the major serovar isolated from humans with 30% occurring in North Africa, 28.3% in sub-Saharan Africa, and 32.9% from nonhuman sources in North Africa. *Salmonella* Typhi was the second most frequently isolated serovar from humans in sub-Saharan Africa. *S.* Kentucky with 13% was the most frequently isolated serovar from nonhumans in sub-Saharan Africa. Rare serovars have also been isolated and reported as the second most common serovar: *S.* Livingstone from humans with a frequency of 15.6%, *S.* Anatum with 17.9% from nonhuman sources in North Africa, and *S.* Bredeney with 8.5% from nonhuman sources in sub-Saharan Africa. In the Middle East, East Asia, Central Asia, South Asia and Southwest Asia, *S.* Enteritidis was the predominant serovar isolated from humans, whereas in Australia and New Zealand, *S.* Typhimurium was leading with 59.3% of all isolated serovars. The data were collected from participating countries using the WHO GFN system. Eighty-three countries reported data with 1.5 million human and 360,000 nonhuman *Salmonella* isolations and 307 distinct serovars.

The general conclusion based on all reported data is that *S.* Enteritidis and *S.* Typhimurium accounted for 78.8% of human isolates and 37.9% of nonhuman isolates (Hendriksen et al., 2011; Vieira et al., 2009). In the period from 2009 to 2010, *S.* Enteritidis accounted for 34% of all *Salmonella* serotype reported outbreaks with a total of 39 outbreaks in 2009 and 37 outbreaks in 2010 (CDC, 2013). According to the US FoodNet surveillance system, *S.* Enteritidis has led other serovars with laboratory-confirmed *Salmonella* infections from 2007 to 2012 followed by *S.* Typhimurium. *Salmonella* Enteritidis and *S.* Typhimurium accounted for approximately 16% and 12%, respectively, in 2012 (CDC, 2014). In 2013, *S.* Enteritidis was accountable for 19% of *Salmonella* cases and 14% for *S.* Typhimurium. In comparison with the 2010–2012 period, *Salmonella* incidences for some serotypes were substantially lowered with a 14% decrease for *S.* Enteritidis while incidence rates for *S.* Typhimurium remained steady (Crim et al., 2014).

*Salmonella* infections result in a considerable cost to the United States and other countries both in terms of human health and in terms of their negative impact on the economy. Because of this, several surveillance programs exist to monitor outbreaks and identify outbreak strains by cooperating with the WHO to keep track of illnesses. *Salmonella* Enteritidis and *S.* Typhimurium rank as the top two disease-causing serotypes associated with outbreaks throughout the world with isolates taken from human and nonhuman sources. However, it is important to continue identifying and monitoring *Salmonella* serovars to lessen the impact of outbreaks and watch for changes in patterns or trends in infections.

## 3. POULTRY- AND POULTRY PRODUCTS–ASSOCIATED *SALMONELLA*

The sources of *Salmonella* infection are relatively diverse, but one of the primary sources is poultry and poultry products. The association of poultry and poultry products has long been documented (Buncic and Sofos, 2012; Chittick et al., 2006;

Currie et al., 2005; Finstad et al., 2012; Foley et al., 2011, 2013; Howard et al., 2012; Liljebjelke et al., 2005; Singh et al., 2010; Smith et al., 2008; Vandeplas et al., 2010; Yildirim et al., 2011). Throughout the last decade, approximately 80% of all *Salmonella* disease outbreaks were linked to a specific *Salmonella* serovar; of those, 34% (403 outbreaks) were traced to a particular food product (Jackson et al., 2013). In those outbreaks, *S.* Enteritidis caused 35.7% with 65% of these linked to egg-associated outbreaks and 13% to chicken-associated outbreaks. *Salmonella* Typhimurium was associated with 14.4% of all outbreaks; 26% of those were chicken-related infections, whereas only 7% were connected to eggs. Another serotype, *S.* Newport, was responsible for 10% of all outbreaks caused by a single food product. This serotype had a similar percentage to *S.* Enteritidis (13%) in association with chicken outbreaks, but was not associated with eggs. *S.* Heidelberg was associated with a particular food product at roughly 6%, with 42% of these outbreaks associated with eggs and 33% with chicken (Jackson et al., 2013). These serotypes have been the primary serovars isolated from human and other nonhuman sources (Vieira et al., 2009; Painter et al., 2013). During the last decade, the food products and processing commercial entities regulated by the US Department of Agriculture (USDA) and Food and Drug Administration (FDA) that have been most often linked with *Salmonella* infections were those involving poultry and poultry products, with 458 outbreaks connected to meat and 125 outbreaks connected to eggs and egg dishes (DeWaal and Glassman, 2013). The following sections describe the growth in poultry product markets and the relationship between *Salmonella* and various poultry products.

## 3.1 POULTRY PRODUCT MARKET TRENDS

Certainly one factor in *Salmonella* in poultry being a prominent source is simply due to the considerable market growth in the production and consumption of poultry products. Eggs and meat products from broiler chickens and other poultry along with turkey products are some of the most highly consumed foods, internationally (Harmon, 2013; Kearney, 2010; Rask and Rask, 2011). On an annual basis, US per capita consumption of chicken is over 80 pounds (MacDonald, 2008), overtaking beef in 2010 (Bentley, 2012). The United States is one of the largest producers of poultry products and these products consist mostly of chickens, turkey meats, and eggs (Harvey, 2012). The production of broiler meat from Jan. through Oct. for the year 2015 was around 33.6 billion pounds, an increase of about 4.1% from the previous year with most of the gains due to higher average bird slaughter weights (Mathews and Haley, 2015b). The US broiler amount held in domestic cold storage was approximately 851.7 million pounds at the end of the year of 2015, showing an increase in all broiler product categories and a total increase of 28% from 2014 (Mathews and Haley, 2015a). For the year 2016, broiler meat is forecast to total 41 billion pounds with stocks also estimated to increase (Mathews and Haley, 2015b).

Turkey meat is the second most prevalent avian meat product in the United States, with 1.35 billion pounds produced for the third quarter of 2015 (Mathews and Haley, 2015a). A total of 356 million pounds of turkey meat products were in cold storage holdings with whole bird stocks totaling 194 million pounds at the end of September (Mathews and Haley, 2015a). The US table egg production was forecasted to be approximately 20.28 billion eggs for the fourth quarter of 2015 and around 81.52 billion eggs for the year (Mathews and Haley, 2015a). In the month of Oct., approximately 276 million eggs were shipped, reaching a total of 4.8 billion for the year (Mathews and Haley, 2015a).

In European Union countries, 26.3 billion pounds of poultry meat were produced in 2011; of those, approximately 80.7% consisted of broiler chicken meat and 15.7% turkey meat (AVEC, 2012). In 2012, the production of poultry meat increased by 259.4 million pounds. Moreover, broiler chicken meat per capita consumption was approximately 38.8 pounds in 2011, and 7.5 pounds per person of turkey meat was consumed. The total consumption of poultry meat was 25.9 billion pounds, with 51.6 pounds of poultry meat consumed per capita in 2011. Approximately 135.5 million additional pounds of poultry meat were consumed in 2012. Furthermore, world production of poultry meat was approximately 225.31 billion pounds and the total poultry meat consumed was 225.25 billion pounds (AVEC, 2012). In 2014, the United States accounted for 20% of the world broiler meat production and 17% of the world broiler meat consumption. For turkey meat, the United States accounted for world production and consumption of 49% and 45%, respectively, in 2014 (AVEC, 2014).

This high quantity of demand and production will continue to require advanced food safety standards since several food-borne pathogens such as *Salmonella* can contaminate poultry meat and eggs at any stage of processing and/or storage of food products (Carrasco et al., 2012; Luber, 2009; Mor-Mur and Yuste, 2010; Newell et al., 2010; Park et al., 2008; Todd et al., 2010). In the following sections the association of *Salmonella* with poultry and eggs will be discussed in terms of serovar specificity and sources.

## 3.2 *SALMONELLA* SEROVARS COMMONLY ASSOCIATED WITH POULTRY AND ITS PRODUCTS

Most common NTS isolated from humans are also associated with poultry, including chicken, turkey, their meat products, eggs, and eggs products (EFSA, 2010; de Freitas Neto et al., 2010; Hoelzer et al., 2011; Sandt et al., 2013). According to a study by CSPI (2013) for food-borne outbreaks solely caused by meat, poultry, or their products in the United States between 1998 and 2010, chicken was listed in the top category of the risk pyramid with 452 outbreaks and recalls of 127 million pounds of chicken and corresponding products. Turkey was the second highest category with 130 outbreaks and 33 million pounds of turkey meat recalled. In general, *Salmonella* spp. were associated with animal products, meat, and poultry as one of two pathogens responsible for 30% of infections and 27.2% of hospitalization cases (DeWaal and Glassman, 2013; CSPI, 2013). In another report about the top 10 riskiest foods regulated by the FDA, eggs and egg-related products were ranked the second highest food source with 352 outbreaks and 11,163 cases of infection (Klein et al., 2009). In European countries, approximately 1426 reported outbreaks were food-borne enteric diseases; of those 20% were associated with poultry consumption (Kessel et al., 2001). Chickens were linked in approximately 75% of the outbreaks, whereas turkey accounted for about 20%. The prevalence of *Salmonella* spp. in those outbreaks was approximately 30%.

The association of *Salmonella* with fowl species has a long history of documentation. In the late 19th century, the two avian-adapted *S. enterica* subspecies *enterica* serotypes Gallinarum and Pullorum were first identified as being responsible for fowl typhoid and pullorum disease, respectively. During the early part of the last century, these two serovars were widespread in the poultry flocks of the United States and Europe, mainly chickens and turkeys (Bullis, 1977; Rabsch et al., 2000; Shivaprasad, 2000; Barrow and Freitas Neto, 2011). Because of the increased mortality associated with poultry, and the significant economic losses from these serovars, monitoring programs were initiated to control and eradicate these diseases (Bullis, 1977; Bäumler et al., 2000; Hitchner, 2004; Shivaprasad, 2000). In the United States, two serovars were prevalent causing pullorum disease and fowl typhoid, *Salmonella* Pullorum and *S.* Gallinarum, respectively, and the diseases identified with these species were controlled and eventually eradicated by the mid-1960s (Bäumler et al., 2000; Barrow and Freitas Neto, 2011; Shivaprasad, 2000).

## 3.3 *SALMONELLA* ASSOCIATION WITH LAYER HENS AND EGGS

The eradication of *S.* Gallinarum and *S.* Pullorum was followed by the emergence of *S.* Enteritidis in chickens and this serotype became the food-borne pathogen most associated with eggs between the early 1980s and mid-1990s (Bäumler et al., 2000). *S.* Enteritidis infections increased steadily from the sixth most commonly isolated serovar in the 1960s until it peaked in 1995, being associated mainly with foods containing raw eggs, lightly cooked egg dishes, as well as less frequently with meat products and sometimes raw produce in cases of possible cross-contamination (St. Louis et al., 1988; Rabsch et al., 2001; Patrick et al., 2004). In more recent years, this serovar has declined as a pathogen associated with the consumption of chicken, but still remains linked to chicken eggs (Braden, 2006; Olsen et al., 2000, 2001; Howard et al., 2012; Foley et al., 2013). Another serovar, *S.* Heidelberg, has become more predominant since 1997, with the highest percentage reported at the beginning of the last decade, comprising over 50% of all *Salmonella* isolations from chickens (Foley et al., 2008). The increase of *S.* Heidelberg in chickens has been connected to the measurements taken to control and reduce *S.* Enteritidis by the National Poultry Improvement Plan–USDA (Foley et al., 2008, 2011).

Before its more recent association with poultry and eggs, *S.* Enteritidis was historically linked to rodents with reports from the earlier part of the 20th century finding 30% of rodent isolates harboring *S.* Enteritidis and only a 0.5% incidence among other animals (Edwards and Bruner, 1943). This also led to the current problems with eradicating *S.* Enteritidis in poultry flocks because unlike *S.* Gallinarum and *S.* Pullorum, which could be controlled through careful monitoring and culling of infected birds from breeder flocks, rodent populations near poultry farms could act as a reservoir and reinfect the birds (Henzler and Opitz, 1992). In addition, although *S.* Gallinarum and *S.* Pullorum cause fowl typhoid and pullorum disease, which are deadly to the bird, *S.* Enteritidis can colonize the bird without causing any negative physical effects, making detecting and eliminating carriers more difficult (Guard-Petter, 2001). It has also been hypothesized that stress-induced survival mechanisms allow *S.* Enteritidis to survive in eggs where other *Salmonella* cannot (Van Immerseel, 2010).

Eggs, as a leading vehicle of food-borne *Salmonella* infections, can be contaminated externally by penetration through the shell, which is known as horizontal transmission, or internally during the formation of the egg, also known as vertical transmission (Gantois et al., 2009). Horizontal transmission can occur during and after oviposition as the egg comes into contact with *Salmonella* from contaminated surfaces and depends on the condition of the shell as well as external conditions such as environmental cleanliness, temperature, relative humidity, and housing conditions (Messens et al., 2005; De Reu et al., 2008). Vertical transmission, on the other hand, occurs if the bird's reproductive tract has been colonized by *Salmonella*, which contaminates the egg contents as it travels through the oviduct (Humphrey et al., 1993; Humphrey, 1994, 1999; Howard et al., 2012). Studies have shown that *S.* Enteritidis is able to evade destruction by the host immune system and infect the ovaries and the reproductive tract of laying hens and penetrate into the internal contents of eggs (Okamura et al., 2001; Gast et al., 2004, 2007; Gantois et al., 2009).

The exact site within the reproductive tissues where *Salmonella* reside is not known. However, it is known that certain factors that increase stress can facilitate the colonization process and increase the spreading of *Salmonella.* Historically, a common industry practice to increase productivity was to induce molting by placing the birds in an altered light–dark cycle and withdrawing feed as this acted to "recycle" the flock, causing it to produce higher quality eggs at an increased rate after the molt period (Ricke, 2003; Norberg et al., 2010; Ricke et al., 2013). However, this has been associated with a depressed immune system, and altered gut microbiota, allowing for easier infection and colonization by *S.* Enteritidis as well as increased shedding for increased horizontal transmission (Holt, 1992, 1999, 2003; Ricke, 2003). As the induction of molting in layer flocks was considered an economic necessity for the US layer industry, research into alternative molt diets has been looked into and there continues to be evidence for new developments that may decrease colonization and infection (Holt, 2003; Norberg et al., 2010; Ricke et al., 2013)

## 4. SUMMARY AND CONCLUSIONS

*Salmonella* has a long history of causing illness in humans. Known since antiquity for causing typhoid and paratyphoid fever, it is now mostly known as a food-borne pathogen manifesting itself with general gastroenteritis symptoms. Techniques developed throughout the 19th and early 20th centuries have greatly improved the ability to isolate and identify *Salmonella* from patients and nonhuman sources, giving a better understanding of the organism. It is now known to consist of two species, six subspecies, and over 2500 serovars through the identification of different antigens found on the surface of the cells using techniques such as PT. Both typhoidal and non-typhoidal *Salmonella* species are associated with causing millions of illness and hundreds of thousands of deaths worldwide, greatly impacting the health and economies of nations across the world. Accordingly, surveillance programs and public health policies have been put in place, including some monitored by the WHO to keep tabs on illness and respond to outbreaks. In the case of NTS serotypes, there has been an association with several different meats and produce, but the closest association has been with poultry and poultry products, especially eggs. *Salmonella* continues to be an issue for public health and continued research and monitoring is necessary to minimize its impact.

## REFERENCES


Agbaje, M., Begum, R.H., Oyekunle, M.A., Ojo, O.E., Adenubi, O.T., 2011. Evolution of *Salmonella* nomenclature: a critical note. Folia Microbiol. 56 (6), 497–503.

Alfredsson, G.A., Barker, R.M., Old, D.C., Duguid, J.P., 1972. Use of tartaric acid isomers and citric acid in the biotyping of *Salmonella typhimurium*. J. Hyg. 70 (04), 651–666.

Álvarez-Ordóñez, A., Fernández, A., Bernardo, A., López, M., 2009. Comparison of acids on the induction of an acid tolerance response in *Salmonella* Typhimurium, consequences for food safety. Meat Sci. 81 (1), 65–70.

Álvarez-Ordóñez, A., Begley, M., Prieto, M., Messens, W., López, M., Bernardo, A., Hill, C., 2011. *Salmonella* spp. survival strategies within the host gastrointestinal tract. Microbiology 157 (12), 3268–3281.

AVEC, 2012. Annual Report. Retrieved from: http://www.avec-poultry.eu/annual-reports-overview/annual-report-2012.

AVEC, 2014. Annual Report. Retrieved from: http://www.avec-poultry.eu/annual-reports-overview/annual-report-2014.

Baggesen, D.L., Sorensen, G., Nielsen, E.M., Wegener, H.C., 2010. Phage typing of *Salmonella* Typhimurium – Is it still a useful tool for surveillance and outbreak investigation? Euro Surveillance 15 (4), 19471.

Barrow, P.A., Freitas Neto, O.C., 2011. Pullorum disease and fowl typhoid–New thoughts on old diseases: a review. Avian Pathol. 40 (1), 1–13.

Batz, M., Hoffmann, S., Morris Jr., J.G., 2014. Disease-outcome trees, EQ-5D scores, and estimated annual losses of quality-adjusted life years (QALYs) for 14 foodborne pathogens in the United States. Foodborne Pathog. Dis. 11 (5), 395–402.

Bäumler, A.J., Hargis, B.M., Tsolis, R.M., 2000. Tracing the origins of *Salmonella* outbreaks. Science 287 (5450), 50–52.

Bazin, H., 2011. Vaccination: A History from Lady Montagu to Genetic Engineering. John Libbey Eurotext, France.

Bearson, S.M.D., Bearson, B.L., Rasmussen, M.A., 2006. Identification of *Salmonella enterica* serovar Typhimurium genes important for survival in the swine gastric environment. Appl. Environ. Microbiol. 72 (4), 2829–2836.

Bechah, Y., Capo, C., Mege, J.L., Raoult, D., 2008. Epidemic typhus. Lancet Infect. Dis. 8 (7), 417–426.

Bensted, H.J., 1951. Standardization of serological reagents. Br. Med. Bull. 7 (3), 178–180.

Bentley, J., 2012. U.S. Per Capita Availability of Chicken Surpasses That of Beef. Retrieved from: http://www.ers.usda.gov/amber-waves/2012-september/us-consumption-of-chicken.aspx#.VMMsW9K_dfQ.

Boyd, E.F., Wang, F.S., Whittam, T.S., Selander, R.K., 1996. Molecular genetic relationships of the salmonellae. Appl. Environ. Microbiol. 62 (3), 804–808.

Braden, C.R., 2006. *Salmonella enterica* serotype Enteritidis and eggs: a national epidemic in the United States. Clin. Infect. Dis. 43 (4), 512–517.

Brands, D.A., Inman, A.E., Gerba, C.P., Maré, C.J., Billington, S.J., Saif, L.A., Levine, J.F., Joens, L.A., 2005. Prevalence of *Salmonella* spp. in oysters in the United States. Appl. Environ. Microbiol. 71 (2), 893–897.

Brenner, F.W., Villar, R.G., Angulo, F.J., Tauxe, R., Swaminathan, B., 2000. *Salmonella* nomenclature. J. Clin. Microbiol. 38 (7), 2465–2467.

Bryan, C.S., 1996. Fever, famine, and war: William Osler as an infectious diseases specialist. Clin. Infect. Dis. 23 (5), 1139–1149.

Bullis, K.L., 1977. The history of avian medicine in the U.S. II. Pullorum disease and fowl typhoid. Avian Dis. 21 (3), 422–429.

Buncic, S., Sofos, J., 2012. Interventions to control *Salmonella* contamination during poultry, cattle and pig slaughter. Food Res. Int. 45 (2), 641–655.

Bynum, B., 2002. Typhomalaria. Lancet 360 (9342), 1339.

Byrd-Bredbenner, C., Berning, J., Martin-Biggers, J., Quick, V., 2013. Food safety in home kitchens: a synthesis of the literature. Int. J. Environ. Res. Public Health 10 (9), 4060–4085.

Canals, R., McClelland, M., Santiviago, C.A., Andrews-Polymenis, H., 2011. Chapter 7: genomics of *Salmonella* species. In: Wiedmann, M., Zhang, W. (Eds.), Genomics of Foodborne Bacterial Pathogens. Springer Books, New York, pp. 171–235.

Carrasco, E., Morales-Rueda, A., García-Gimeno, R.M., 2012. Cross-contamination and recontamination by *Salmonella* in foods: a review. Food Res. Int. 45 (2), 545–556.

Centers for Disease Control and Prevention (CDC), 2012. Foodborne Diseases Active Surveillance Network (FoodNet). FoodNet Surveillance Report for 2011 (Final Report) U.S. Department of Health and Human Services, Atlanta, Georgia. Retrieved from: http://www.cdc.gov/foodnet/pdfs/2011_annual_report_508c.pdf.

Centers for Disease Control and Prevention (CDC), 2013. Incidence and trends of infection with pathogens transmitted commonly through food – foodborne diseases active surveillance network, 10 US sites, 1996–2012. MMWR. Morbid. Mortal. Weekly Rep. 62 (15), 283–287.

Centers for Disease Control and Prevention (CDC), 2014. Foodborne Diseases Active Surveillance Network (FoodNet). FoodNet Surveillance Report for 2012 (Final Report) U.S. Department of Health and Human Services, Atlanta, Georgia. Retrieved from: http://www.cdc.gov/foodnet/PDFs/2012_annual_report_508c.pdf.

Center for Science in the Public Interest (CSPI), 2013. Risky Meat: A CSPI Field Guide to Meat and Poultry Safety. Retrieved from: https://www.cspinet.org/foodsafety/riskymeat.html.

Chaitram, J.M., Jevitt, L.A., Lary, S., Tenover, F.C., 2003. The World Health Organization's external quality assurance system proficiency testing program has improved the accuracy of antimicrobial susceptibility testing and reporting among participating laboratories using NCCLS methods. J. Clin. Microbiol. 41 (6), 2372–2377.

Chart, H., Cheasty, T., de Pinna, E., Siorvanes, L., Wain, J., Alam, D., Nizami, Q., Bhutta, Z., Threlfall, E.J., 2007. Serodiagnosis of *Salmonella enterica* serovar Typhi and *S. enterica* serovars Paratyphi A, B and C human infections. J. Med. Microbiol. 56 (9), 1161–1166.

Chen, S., Zhao, S., White, D.G., Schroeder, C.M., Lu, R., Yang, H., McDermott, P.F., Ayers, S., Meng, J., 2004. Characterization of multiple-antimicrobial-resistant *Salmonella* serovars isolated from retail meats. Appl. Environ. Microbiol. 70 (1), 1–7.

Chen, S., Sanderson, M.W., White, B.J., Amrine, D.E., Lanzas, C., 2013. Temporal-spatial heterogeneity in animal-environment contact: implications for the exposure and transmission of pathogens. Sci. Rep. 3, 3112.

Chittick, P., Sulka, A., Tauxe, R.V., Fry, A.M., 2006. A summary of national reports of foodborne outbreaks of *Salmonella* Heidelberg infections in the United States: clues for disease prevention. J. Food Protect. 69 (5), 1150–1153.

Cirillo, V.J., 2000. Fever and reform: the typhoid epidemic in the Spanish-American War. J. Hist. Med. Allied Sci. 55 (4), 363–397.

Clark, G.M., Kaufmann, A.F., Gangarosa, E.J., Thompson, M.A., 1973. Epidemiology of an international outbreak of *Salmonella* Agona. The Lancet 2 (7872), 490–493.

Collins, C., Petts, D., 2011. Early masters: a story of bacteriology from A to G. Biomed. Sci. 530–533. Retrieved from: https://www.ibms.org/includes/act_download.php?download=pdf/2011-Aug-Bacteriology.pdf.

Cook, G.C., 2001. The college of physicians in the nineteenth century. Clin. Med. 1 (3), 234–239.

Cox, N.A., Cason, J.A., Richardson, L.J., 2011. Minimization of *Salmonella* contamination on raw poultry. Ann. Rev. Food Sci. Technol. 2, 75–95.

Crim, S.M., Iwamoto, M., Huang, J.Y., Griffin, P.M., Gilliss, D., Cronquist, A.B., Cartter, M., Tobin-D'Angelo, M., Blythe, D., Smith, K., Lathrop, S., Zansky, S., Cieslak, P.R., Dunn, J., Holt, K.G., Lance, S., Tauxe, R., Henao, O.L., 2014. Incidence and trends of infection with pathogens transmitted commonly through food–foodborne diseases active surveillance network, 10 US Sites, 2006–2013. MMWR. Morbid. Mortal. Weekly Rep. 63 (15), 328–332.

Crump, J.A., Luby, S.P., Mintz, E.D., 2004. The global burden of typhoid fever. Bull. World Health Organ. 82 (5), 346–353.

Cunha, C.B., Cunha, B.A., 2008. Brief history of the clinical diagnosis of malaria: from Hippocrates to Osler. J. Vector Borne Dis. 45 (3), 194–199.

Cunha, B.A., Gran, A., Munoz-Gomez, S., 2013. Typhoid fever vs. malaria in a febrile returning traveler: Typhomalaria revisited–an Oslerian perspective. Travel Med. Infect. Dis. 11 (1), 66–69.

Cunha, B.A., 2004. Osler on typhoid fever: differentiating typhoid from typhus and malaria. Infect. Dis. Clin. North Am. 18 (1), 111–125.

Cunha, B.A., 2005. Malaria or typhoid fever: a diagnostic dilemma? Am. J. Med. 118 (12), 1442–1443.

Cunha, B.A., 2007. Fever of unknown origin: clinical overview of classic and current concepts. Infect. Dis. Clin. North Am. 21 (4), 867–915.

Currie, A., MacDougall, L., Aramini, J., Gaulin, C., Ahmed, R., Isaacs, S., 2005. Frozen chicken nuggets and strips and eggs are leading risk factors for *Salmonella* Heidelberg infections in Canada. Epidemiol. Infect. 133 (05), 809–816.

Davies, R.H., Dalziel, R., Gibbens, J.C., Wilesmith, J.W., Ryan, J.M.B., Evans, S.J., Byrne, C., Paiba, G.A., Pascoe, S.J.S., Teale, C.J., 2004. National survey for *Salmonella* in pigs, cattle and sheep at slaughter in Great Britain (1999–2000). J. Appl. Microbiol. 96 (4), 750–760.

de Freitas, C.G., Santana, A.P., da Silva, P.H., Goncalves, V.S., Barros Mde, A., Torres, F.A., Murata, L.S., Perecmanis, S., 2010. PCR multiplex for detection of *Salmonella* Enteritidis, Typhi and Typhimurium and occurrence in poultry meat. Int. J. Food Microbiol. 139 (1), 15–22.

de Freitas Neto, O.C., Penha Filho, R.A.C., Barrow, P., Berchieri Junior, A., 2010. Sources of human non-typhoid salmonellosis: a review. Revista Brasileira de Ciência Avícola 12 (1), 01–11.

De Reu, K., Messens, W., Heyndrickx, M., Rodenburg, T.B., Uyttendaele, M., Herman, L., 2008. Bacterial contamination of table eggs and the influence of housing systems. World's Poult. Sci. J. 64 (01), 5–19.

DeWaal, C.S., Glassman, M., 2013. Outbreak Alert! 2001–2010: A Review of Foodborne Illness in America. Center for Science in the Public Interest (CSPI), Washington, DC. Retrieved from: http://cspinet.org/new/pdf/outbreak_alert_2013_final.pdf.

Dhillon, A.S., Shivaprasad, H.L., Roy, P., Alisantosa, B., Schaberg, D., Bandli, D., Johnson, S., 2001. Pathogenicity of environmental origin *Salmonellas* in specific pathogen-free chicks. Poult. Sci. 80 (9), 1323–1328.

Duguid, J.P., Anderson, E.S., Alfredsson, G.A., Barker, R., Old, D.C., 1975. A new biotyping scheme for *Salmonella* Typhimurium and its phylogenetic significance. J. Med. Microbiol. 8 (1), 149–166.

Edwards, P.R., Bruner, D.W., 1943. The occurrence and distribution of *Salmonella* types in the United States. J. Infect. Dis. 72 (1), 58–67.

ERS-USDA, 2014. Cost Estimates of Foodborne Illnesses. Retrieved from: http://ers.usda.gov/data-products/cost-estimates-of-foodborne-illnesses.aspx.

European Food Safety Authority (EFSA), 2010. Analysis of the baseline survey on the prevalence of *Campylobacter* in broiler batches and of *Campylobacter* and *Salmonella* on broiler carcasses, in the EU, 2008-Part A: *Campylobacter* and *Salmonella* prevalence estimates. EFSA J. 8 (03). http://dx.doi.org/10.2903/j.efsa.2010.1503. 1503 (100 pp.) Available online: http://www.efsa.europa.eu/en/efsajournal/pub/1503.htm#.

Euzéby, J.P., 1999. Revised *Salmonella* nomenclature: designation of *Salmonella enterica* (ex Kauffmann and Edwards 1952) Le Minor and Popoff 1987 sp. nov., nom. rev. as the neotype species of the genus *Salmonella* Lignieres 1900 (approved lists 1980), rejection of the name *Salmonella choleraesuis* (Smith 1894) Weldin 1927 (approved lists 1980), and conservation of the name *Salmonella* Typhi (Schroeter 1886) Warren and Scott 1930 (approved lists 1980). Request for an opinion. Int. J. Syst. Evol. Bacteriol. 49 (2), 927–930.

Euzéby, J., 2005. Validation of publication of new names and new combinations previously effectively published outside the IJSEM. Int. J. Syst. Evol. Microbiol. 55 (3), 547–549.

Evangelopoulou, G.D., Burriel, A., Spyrou, V., 2010. A concise history of *Salmonella* spp. nomenclature. J. Hellenic Vet. Med. Soc. 61 (4), 323–329.

Ewing, W.H., 1972. The nomenclature of *Salmonella*, its usage, and definitions for the three species. Can. J. Microbiol. 18 (11), 1629–1637.

Ezaki, T., Kawamura, Y., Yabuuchi, E., 2000. Recognition of nomenclatural standing of *Salmonella* Typhi (approved lists 1980), *Salmonella* enteritidis (approved lists 1980) and *Salmonella* Typhimurium (approved lists 1980), and conservation of the specific epithets enteritidis and typhimurium. Request for an opinion. Int. J. Syst. Evol. Bacteriol. 50 (2), 945–947.

Fey, P.D., Safranek, T.J., Rupp, M.E., Dunne, E.F., Ribot, E., Iwen, P.C., Bradford, P.A., Angulo, F.J., Hinrichs, S.H., 2000. Ceftriaxone-resistant *Salmonella* infection acquired by a child from cattle. N. Engl. J. Med. 342 (17), 1242–1249.

Finstad, S., O'Bryan, C.A., Marcy, J.A., Crandall, P.G., Ricke, S.C., 2012. *Salmonella* and broiler processing in the United States: relationship to foodborne salmonellosis. Food Res. Int. 45 (2), 789–794.

Fison, E.T., 1897. Widal's sero-diagnosis of typhoid fever. Br. Med. J. 2 (1909), 266–269.

Fleming, L.E., Bean, J.A., Katz, D., Hammond, R., 2000. The epidemiology of seafood poisoning. In: Hui, Y.H., Kitts, D., Stanfield, P.S. (Eds.), Foodborne Disease Handbook: Seafood and Environmental Toxins. Marcel Dekker, Inc., New York, pp. 287–310.

Foley, S.L., Lynne, A.M., Nayak, R., 2008. *Salmonella* challenges: prevalence in swine and poultry and potential pathogenicity of such isolates. J. Animal Sci. 86 (Suppl. 14), E149–E162.

Foley, S.L., Nayak, R., Hanning, I.B., Johnson, T.J., Han, J., Ricke, S.C., 2011. Population dynamics of *Salmonella enterica* serotypes in commercial egg and poultry production. Appl. Environ. Microbiol. 77 (13), 4273–4279.

Foley, S.L., Johnson, T.J., Ricke, S.C., Nayak, R., Danzeisen, J., 2013. *Salmonella* pathogenicity and host adaptation in chicken-associated serovars. Microbiol. Mol. Biol. Rev. 77, 582–607.

Food Safety and Inspection Service, United States Department of Agriculture (FSIS-USDA), 2013. Progress Report on *Salmonella* and *Campylobacter* Testing of Raw Meat and Poultry Products, 1998–2013. Retrieved from: http://www.fsis.usda.gov/wps/wcm/connect/885647f4-2568-48bf-ae5c-4a0d8279f435/Progress-Report-Salmonella-Campylobacter-CY2013.pdf?MOD=AJPERES.

Foster, J.W., Spector, M.P., 1995. How *Salmonella* survive against the odds. Annual Rev. Microbiol. 49 (1), 145–174.

Galanis, E., Wong, D.M.A.L.F., Patrick, M.E., Binsztein, N., Cieslik, A., Chalermchaikit, T., Aidara-Kane, A., Ellis, A., Angulo, F.J., Wegener, H.C., 2006. Web-based surveillance and global *Salmonella* distribution, 2000–2002. Emerg. Infect. Dis. 12 (3), 381–388.

Gantois, I., Ducatelle, R., Pasmans, F., Haesebrouck, F., Gast, R., Humphrey, T.J., Van Immerseel, F., 2009. Mechanisms of egg contamination by *Salmonella* Enteritidis. FEMS Microbiol. Rev. 33 (4), 718–738.

Gast, R.K., Guard-Bouldin, J., Holt, P.S., 2004. Colonization of reproductive organs and internal contamination of eggs after experimental infection of laying hens with *Salmonella* Heidelberg and *Salmonella* Enteritidis. Avian Dis. 48 (4), 863–869.

Gast, R.K., Guraya, R., Guard-Bouldin, J., Holt, P.S., Moore, R.W., 2007. Colonization of specific regions of the reproductive tract and deposition at different locations inside eggs laid by hens infected with *Salmonella* Enteritidis or *Salmonella* Heidelberg. Avian Dis. 51 (1), 40–44.

Ghilardi, Â.C.R., Tavechio, A.T., Fernandes, S.A., 2006. Antimicrobial susceptibility, phage types, and pulse types of *Salmonella* Typhimurium, in São Paulo, Brazil. Memórias do Instituto Oswaldo Cruz 101 (3), 281–286.

Gould, L.H., Walsh, K.A., Vieira, A.R., Herman, K., Williams, I.T., Hall, A.J., Cole, D., 2013. Surveillance for foodborne disease outbreaks – United States, 1998–2008. MMWR Surveillance Summary 62 (2), 1–34.

Grimont, P.A.D., Weill, F.X., 2007. Antigenic Formulae of the *Salmonella* Serovars. WHO Collaborating Center for Reference and Research on *Salmonella*, Paris, France (Institut Pasteur).

Gröschel, D.H., Hornick, R.B., 1981. Who introduced typhoid vaccination: Almroth Wright or Richard Pfeiffer? Rev. Infect. Dis. 3 (6), 1251–1254.

Guard-Petter, J., 2001. The chicken, the egg and *Salmonella enteritidis*. Environ. Microbiol. 3 (7), 421–430.

Guibourdenche, M., Roggentin, P., Mikoleit, M., Fields, P.I., Bockemuhl, J., Grimont, P.A., Weill, F.X., 2010. Supplement 2003–2007 (No. 47) to the White-Kauffmann-Le Minor scheme. Res. Microbiol. 161 (1), 26–29.

Gupta, A.K., Rao, K.M., 1981. A simple bactericidal antibody test for sero-diagnosis of typhoid fever. J. Immunol. Methods 42 (1), 121–125.

Hanning, I.B., Nutt, J.D., Ricke, S.C., 2009. Salmonellosis outbreaks in the United States due to fresh produce: sources and potential intervention measures. Foodborne Pathog. Dis. 6 (6), 635–648.

Hardy, A., 2015. *Salmonella* infections, Networks of Knowledge, and Public Health in Britain, 1880–1975. Oxford University Press, New York.

Harmon, D.E., 2013. Poultry: From the Farm to Your Table, first ed. The Rosen Publishing Group, Inc., New York.

Harvey, D., 2012. Poultry & Eggs. Retrieved from: http://www.ers.usda.gov/topics/animal-products/poultry-eggs/.aspx#.Uu9K1z2Qy1J.

Hathcock, J. (Ed.), 1982. Nutritional Toxicology, vol. 1. Academic Press, Inc, New York.

Heaton, J.C., Jones, K., 2008. Microbial contamination of fruit and vegetables and the behaviour of enteropathogens in the phyllosphere: a review. J. Appl. Microbiol. 104 (3), 613–626.

Heinitz, M.L., Ruble, R.D., Wagner, D.E., Tatini, S.R., 2000. Incidence of *Salmonella* in fish and seafood. J. Food Protect. 63 (5), 579–592.

Hendriksen, R.S., Mikoleit, M., Carlson, V.P., Karlsmose, S., Vieira, A.R., Jensen, A.B., Seyfarth, A.M., DeLong, S.M., Weill, F.X., Wong, D.M.A.L.F., Angulo, F.J., Wegener, H.C., Aarestrup, F.M., 2009a. WHO Global Salm-Surv external quality assurance system for serotyping of *Salmonella* isolates from 2000 to 2007. J. Clin. Microbiol. 47 (9), 2729–2736.

Hendriksen, R.S., Seyfarth, A.M., Jensen, A.B., Whichard, J., Karlsmose, S., Joyce, K., Mikoleit, M., Delong, S.M., Weill, F.X., Aidara-Kane, A., Wong, D.M.A.L.F., Angulo, F., Wegener, H.C., Aarestrup, F.M., 2009b. Results of use of WHO Global Salm-Surv external quality assurance system for antimicrobial susceptibility testing of *Salmonella* isolates from 2000 to 2007. J. Clin. Microbiol. 47 (1), 79–85.

Hendriksen, R.S., Vieira, A.R., Karlsmose, S., Lo Fo Wong, D.M., Jensen, A.B., Wegener, H.C., Aarestrup, F.M., 2011. Global monitoring of *Salmonella* serovar distribution from the world health organization global foodborne infections network country data bank: results of quality assured laboratories from 2001 to 2007. Foodborne Pathog. Dis. 8 (8), 887–900.

Henzler, D.J., Opitz, H.M., 1992. The role of mice in the epizootiology of *Salmonella enteritidis* infection on chicken layer farms. Avian Dis. 625–631.

Hitchner, S.B., 2004. History of biological control of poultry diseases in the USA. Avian Dis. 48 (1), 1–8.

Hoelzer, K., Moreno Switt, A., Wiedmann, M., 2011. Animal contact as a source of human non-typhoidal salmonellosis. Vet. Res. 42 (1), 34–61.

Hoffmann, S., Anekwe, T.D., 2013. Making Sense of Recent Cost-of-foodborne-illness Estimates. (Economic Information Bulletin No. 118). Retrieved from: http://www.ers.usda.gov/media/1204379/eib118.pdf.

Holt, P.S., 1992. Effects of induced moulting on immune responses of hens. Br. Poult. Sci. 33, 165–175.

Holt, P.S., 1999. Chapter 33: impact of induced molting on immunity and *Salmonella enterica* serovar Enteritidis infection in laying hens. In: Saeed, A.M., Gast, R.K., Potter, M.E., Wall, P.G. (Eds.), *Salmonella enterica* Serovar Enteritidis in Humans and Animals: Epidemiology, Pathogenesis and Control. Iowa State University Press, Ames, pp. 183–192.

Holt, P.S., 2003. Molting and *Salmonella enterica* serovar enteritidis infection: the problem and some solutions. Poult. Sci. 82 (6), 1008–1010.

Hope, B.K., Baker, A.R., Edel, E.D., Hogue, A.T., Schlosser, W.D., Whiting, R., McDowell, R.M., Morales, R.A., 2002. An overview of the *Salmonella* Enteritidis risk assessment for shell eggs and egg products. Risk Anal. 22 (2), 203–218.

Hopkins, K.L., Kirchner, M., Guerra, B., Granier, S.A., Lucarelli, C., Porrero, M.C., Jakubczak, A., Threlfall, E.J., Mevius, D.J., 2010. Multiresistant *Salmonella enterica* serovar 4,[5],12:i:- in Europe: a new pandemic strain? Euro Surveillance 15 (22), 19580. Available online: http://www.eurosurveillance.org/ViewArticle.aspx?ArticleId=19580.

Hornick, R.B., 1974. Jeremiah Metzger lecture: *Salmonella* infections–newer perspectives of an old infection. Trans. Am. Clin. Climatol. Assoc. 85, 164–174.

Howard, Z.R., O'Bryan, C.A., Crandall, P.G., Ricke, S.C., 2012. *Salmonella* Enteritidis in shell eggs: current issues and prospects for control. Food Res. Int. 45 (2), 755–764.

Humphrey, T.J., Baskerville, A., Whitehead, A., Rowe, B., Henley, A., 1993. Influence of feeding patterns on the artificial infection of laying hens with *Salmonella enteritidis* phage type 4. Vet. Record 132 (16), 407–409.

Humphrey, T.J., 1994. Contamination of egg shell and contents with *Salmonella enteritidis*: a review. Int. J. Food Microbiol. 21 (1), 31–40.

Humphrey, T.J., 1999. Chapter 18: contamination of eggs and poultry meat with *Salmonella enterica* serovar Enteritidis. In: Saeed, A.M., Gast, R.K., Potter, M.E., Wall, P.G. (Eds.), *Salmonella enterica* Serovar Enteritidis in Humans and Animals: Epidemiology, Pathogenesis and Control. Iowa State University Press, Ames, pp. 183–192.

Jackson, B.R., Griffin, P.M., Cole, D., Walsh, K.A., Chai, S.J., 2013. Outbreak-associated *Salmonella enterica* serotypes and food commodities, United States, 1998-2008. Emerg. Infect. Dis. 19 (8), 1239–1244.

Jhaveri, K.N., Nandwani, S.K., Mehta, P.K., Surati, R.R., Parmar, B.D., 1995. False positive modified Widal test in acute malaria. J. Assoc. Physicians India 43 (11), 754–755.

Kearney, J., 2010. Food consumption trends and drivers. Philos. Trans. Royal Soc. B 365 (1554), 2793–2807.

Kessel, A.S., Gillespie, I.A., O'Brien, S.J., Adak, G.K., Humphrey, T.J., Ward, L.R., 2001. General outbreaks of infectious intestinal disease linked with poultry, England and Wales, 1992-1999. Commun. Dis. Public Health/PHLS 4 (3), 171–177.

Kimura, A.C., Reddy, V., Marcus, R., Cieslak, P.R., Mohle-Boetani, J.C., Kassenborg, H.D., Segler, S.D., Hardnett, F.P., Barrett, T., Swerdlow, D.L., 2004. Chicken consumption is a newly identified risk factor for sporadic *Salmonella enterica* serotype Enteritidis infections in the United States: a case-control study in FoodNet sites. Clin. Infect. Dis. 38 (Suppl. 3), S244–S252.

Klein, S., Witmer, J., Tian, A., DeWaal, C.S., 2009. The Ten Riskiest Foods Regulated by the US Food and Drug Administration. Center for Science in the Public Interest (CSPI). Retrieved from: http://www.cspinet.org/new/pdf/cspi_top_10_fda.pdf.

Kustu, S., Burton, D., Garcia, E., McCarter, L., McFarland, N., 1979. Nitrogen control in *Salmonella*: regulation by the *glnR* and *glnF* gene products. Proc. Nat. Acad. Sci. USA 76 (9), 4576–4580.

Kustu, S., Hirschman, J., Burton, D., Jelesko, J., Meeks, J.C., 1984. Covalent modification of bacterial glutamine synthetase: physiological significance. Mol. Gen. Genet. 197 (2), 309–317.

Lancaster, H.O., 1990. Expectations of Life: A Study in the Demography, Statistics, and History of World Mortality. Springer Science + Business Media, New York.

Le Minor, L., Popoff, M.Y., 1987. Designation of *Salmonella enterica* sp. nov., nom. rev., as the type and only species of the genus *Salmonella*: request for an opinion. Int J. Syst. Bacteriol. 37 (4), 465–468.

Leavitt, J.W., 1996. Typhoid Mary: Captive to the Public's Health. Beacon Press, Boston, MA.

Liljebjelke, K.A., Hofacre, C.L., Liu, T., White, D.G., Ayers, S., Young, S., Maurer, J.J., 2005. Vertical and horizontal transmission of *Salmonella* within integrated broiler production system. Foodborne Pathog. Dis. 2 (1), 90–102.

Luber, P., 2009. Cross-contamination versus undercooking of poultry meat or eggs – which risks need to be managed first? Int. J. Food Microbiol. 134 (1), 21–28.

Lynch, M.F., Tauxe, R.V., Hedberg, C.W., 2009. The growing burden of foodborne outbreaks due to contaminated fresh produce: risks and opportunities. Epidemiol. Infect. 137 (03), 307–315.

MacDonald, J.M., 2008. The Economic Organization of US Broiler Production. Economic Information Bulletin No. (EIB-38) ERS-USDA. Retrieved from: http://www.ers.usda.gov/publications/eib-economic-information-bulletin/eib38.aspx.

Majowicz, S.E., Musto, J., Scallan, E., Angulo, F.J., Kirk, M., O'Brien, S.J., Jones, T.F., Fazil, A., Hoekstra, R.M., 2010. The global burden of nontyphoidal *Salmonella* gastroenteritis. Clin. Infect. Dis. 50 (6), 882–889.

Martinez-Urtaza, J., Peiteado, J., Lozano-Leon, A., Garcia-Martin, O., 2004. Detection of *Salmonella* Senftenberg associated with high saline environments in mussel processing facilities. J. Food Protect. 67 (2), 256–263.

Mathews, K., Haley, M., 2015a. Lower Livestock, Dairy and Poultry Prices Expected in 2016. Livestock, Dairy, and Poultry Outlook No. (LDPM-258) ERS-USDA. Retrieved from: http://www.ers.usda.gov/media/1959920/ldp-m-258.pdf.

Mathews, K., Haley, M., 2015b. High-valued U.S. Dollar Driving U.S. Exports Lower, Imports Higher. Livestock, Dairy, and Poultry Outlook No. (LDPM-257) ERS-USDA. Retrieved from: http://www.ers.usda.gov/media/1942630/ldpm257.pdf.

McCrae, T., 1907. Typhoid fever. In: Osler, W., McCrae, T. (Eds.), Modern Medicine, its Theory and Practice: In Original Contributions by American and Foreign Authors, vol. 3. Lea Brothers & Company, Philadelphia, pp. 70–92.

Merrick, M.J., 1982. A new model for nitrogen control. Nature 297 (5865), 362–363.

Messens, W., Grijspeerdt, K., Herman, L., 2005. Eggshell penetration by *Salmonella*: a review. World's Poult. Sci. J. 61 (01), 71–86.

Mølbak, K., Olsen, J.E., Wegener, H.C., 2006. Chapter 3: *Salmonella* infections. In: Riemann, H.P., Cliver, D.O. (Eds.), Foodborne Infections and Intoxications. Academic Press, Cambridge, pp. 57–136.

Monack, D.M., 2012. *Salmonella* persistence and transmission strategies. Curr. Opin. Microbiol. 15 (1), 100–107.

Moorhead, R., 2002. William Budd and typhoid fever. J. Royal Soc. Med. 95 (11), 561–564.

Mor-Mur, M., Yuste, J., 2010. Emerging bacterial pathogens in meat and poultry: an overview. Food Bioprocess Technol. 3 (1), 24–35.

Newell, D.G., Koopmans, M., Verhoef, L., Duizer, E., Aidara-Kane, A., Sprong, H., Opsteeghb, M., Langelaarb, M., Threfalld, J., Scheutze, F., van der Giessenb, J., Kruse, H., 2010. Food-borne diseases—the challenges of 20 years ago still persist while new ones continue to emerge. Int. J. Food Microbiol. 139, S3–S15.

Newsom, S.W.B., 2007. Typhoid fever: a continuing problem. Br. J. Infect. Control 8 (2), 8–11.

Norberg, L.M., McReynolds, J.L., Kim, W.K., Chalova, V.I., Nisbet, D.J., Ricke, S.C., 2010. Chapter 6: Colonization and pathogenesis of food-borne *Salmonella* in egg-laying hens. In: Ricke, S.C., Jones, F.T. (Eds.), Perspectives on Food-Safety Issues of Animal-Derived Foods. University of Arkansas Press, Fayetteville, pp. 63–84.

Ober, W.B., 1976. American pathology in the 19th century: notes for the definition of a specialty. Bull. N. Y. Acad. Med. 52 (3), 326–347.

O'Brien, S.J., 2013. The "decline and fall" of nontyphoidal *Salmonella* in the United Kingdom. Clin. Infect. Dis. 56, 705–710.

Okamura, M., Kamijima, Y., Miyamoto, T., Tani, H., Sasai, K., Baba, E., 2001. Differences among six *Salmonella* serovars in abilities to colonize reproductive organs and to contaminate eggs in laying hens. Avian Dis. 45 (1), 61–69.

Olsen, S.J., MacKinnon, L.C., Goulding, J.S., Bean, N.H., Slutsker, L., 2000. Surveillance for foodborne-disease outbreaks–United States, 1993–1997. MMWR CDC Surveillance Summaries 49 (1), 1–62.

Olsen, S.J., Bishop, R., Brenner, F.W., Roels, T.H., Bean, N., Tauxe, R.V., Slutsker, L., 2001. The changing epidemiology of *Salmonella*: trends in serotypes isolated from humans in the United States, 1987–1997. J. Infect. Dis. 183 (5), 753–761.

Painter, J.A., Hoekstra, R.M., Ayers, T., Tauxe, R.V., Braden, C.R., Angulo, F.J., Griffin, P.M., 2013. Attribution of foodborne illnesses, hospitalizations, and deaths to food commodities by using outbreak data, United States, 1998–2008. Emerg. Infect. Dis. 19 (3), 407–415.

Pang, S., Octavia, S., Reeves, P.R., Wang, Q., Gilbert, G.L., Sintchenko, V., Lan, R., 2012. Genetic relationships of phage types and single nucleotide polymorphism typing of *Salmonella enterica* serovar Typhimurium. J. Clin. Microbiol. 50 (3), 727–734.

Park, S.Y., Woodward, C.L., Kubena, L.F., Nisbet, D.J., Birkhold, S.G., Ricke, S.C., 2008. Environmental dissemination of foodborne *Salmonella* in preharvest poultry production: reservoirs, critical factors, and research strategies. Crit. Rev. Environ. Sci. Technol. 38 (2), 73–111.

Parry, C.M., 2006. Chapter 1: Epidemiological and clinical aspects of human typhoid fever. In: Mastroeni, P., Maskell, D. (Eds.), *Salmonella* Infections: Clinical, Immunological and Molecular Aspects. Cambridge University Press, Cambridge, pp. 1–24. http://dx.doi.org/10.1017/CBO9780511525360.002. Available from: Cambridge Books Online.

Parte, A.C., 2014. LPSN—list of prokaryotic names with standing in nomenclature. Nucleic Acids Res. 42 (D1), D613–D616.

Paterson, J.C., 1949. Typhoid and the paratyphoids: a review. Postgrad. Med. J. 25 (287), 413–420.

Patrick, M.E., Adcock, P.M., Gomez, T.M., Altekruse, S.F., Holland, B.H., Tauxe, R.V., Swerdlow, D.L., 2004. *Salmonella* Enteritidis infections, United States, 1985–1999. Emerg. Infect. Dis. 10 (1), 1–7.

Paul, J.R., 1930. Nathan Smith and typhoid fever. Yale J. Biol. Med. 2 (3), 169. b1–181.

Perelle, S., Dilasser, F., Malorny, B., Grout, J., Hoorfar, J., Fach, P., 2004. Comparison of PCR-ELISA and LightCycler real-time PCR assays for detecting *Salmonella* spp. in milk and meat samples. Mol. Cell. Probes 18 (6), 409–420.

Petersen, A., Aarestrup, F.M., Angulo, F.J., Wong, S., Stohr, K., Wegener, H.C., 2002. WHO global salm-surv external quality assurance system (EQAS): an important step toward improving the quality of *Salmonella* serotyping and antimicrobial susceptibility testing worldwide. Microb. Drug Resist. 8 (4), 345–353.

Platt, D.J., Brown, D.J., Old, D.C., Barker, R.M., Munro, D.S., Taylor, J., 1987. Old and new techniques together resolve a problem of infection by *Salmonella* Typhimurium. Epidemiol. Infect 99 (01), 137–142.

Pradhan, P., 2011. Coinfection of typhoid and malaria. J. Med. Lab. Diag. 2 (3), 22–26.

Rabsch, W., Hargis, B.M., Tsolis, R.M., Kingsley, R.A., Hinz, K.H., Tschäpe, H., Bäumler, A.J., 2000. Competitive exclusion of *Salmonella* Enteritidis by *Salmonella* Gallinarum in poultry. Emerg. Infect. Dis. 6 (5), 443–448.

Rabsch, W., Tschäpe, H., Bäumler, A.J., 2001. Non-typhoidal salmonellosis: emerging problems. Microb. Infect. 3 (3), 237–247.

Rabsch, W., Andrews, H.L., Kingsley, R.A., Prager, R., Tschäpe, H., Adams, L.G., Bäumler, A.J., 2002. *Salmonella enterica* serotype Typhimurium and its host-adapted variants. Infect. Immun. 70 (5), 2249–2255.

Rabsch, W., Truepschuch, S., Windhorst, D., Gerlach, R.G., 2011. Chapter 2: typing phages and prophages of *Salmonella*. In: Porwollik, S. (Ed.), *Salmonella*: From Genome to Function. Caister Academic Press, Norfolk, pp. 25–48.

Rabsch, W., 2007. Chapter 10: *Salmonella* Typhimurium phage typing for pathogens. In: Schatten, H., Eisenstark, A. (Eds.), *Salmonella*: Methods and Protocols. Humana Press Inc., Totowa, pp. 177–211.

Raffatellu, M., Wilson, R.P., Winter, S.E., Bäumler, A.J., 2008. Clinical pathogenesis of typhoid fever. J. Infect. Dev. Countries 2 (04), 260–266.

Rajashekara, G., Haverly, E., Halvorson, D.A., Ferris, K.E., Lauer, D.C., Nagaraja, K.V., 2000. Multidrug-resistant *Salmonella* Typhimurium DT104 in poultry. J. Food Protect. 63 (2), 155–161.

Randall, L.P., Cooles, S.W., Osborn, M.K., Piddock, L.J., Woodward, M.J., 2004. Antibiotic resistance genes, integrons and multiple antibiotic resistance in thirty-five serotypes of *Salmonella enterica* isolated from humans and animals in the UK. J. Antimicrob. Chemother. 53 (2), 208–216.

Rask, K.J., Rask, N., 2011. Economic development and food production–consumption balance: a growing global challenge. Food Policy 36 (2), 186–196.

Ricke, S.C., Dunkley, C.S., Durant, J.A., 2013. A review on development of novel strategies for controlling *Salmonella* Enteritidis colonization in laying hens: fiber-based molt diets. Poult. Sci. 92 (2), 502–525.

Ricke, S.C., 2003. The gastrointestinal tract ecology of *Salmonella* Enteritidis colonization in molting hens. Poult. Sci. 82 (6), 1003–1007.

Rodríguez, C., Hardy, E., 2015. Lipopolysaccharide aggregates in native agarose gels detected by reverse negative staining with imidazole and zinc salts. Anal. Biochem. 485, 72–80.

Rowe, B., Hall, M.L., Ward, L.R., de Sa, J.D., 1980. Epidemic spread of *Salmonella* Hadar in England and Wales. Br. Med. J. 280 (6221), 1065–1066.

Sandt, C.H., Fedorka-Cray, P.J., Tewari, D., Ostroff, S., Joyce, K., M'Ikanatha, N.,M., 2013. A comparison of non-typhoidal *Salmonella* from humans and food animals using pulsed-field gel electrophoresis and antimicrobial susceptibility patterns. PLoS One 8 (10), e77836.

Sansonetti, P., 2011. Chapter 9: Vaccination against typhoid fever: a century of research. End of the beginning or beginning of the end? In: Plotkin, S.A. (Ed.), History of Vaccine Development. Springer Science + Business Media, New York, pp. 65–72.

Scallan, E., Hoekstra, R.M., Angulo, F.J., Tauxe, R.V., Widdowson, M.A., Roy, S.L., Jones, J.L., Griffin, P.M., 2011. Foodborne illness acquired in the United States–major pathogens. Emerg. Infect. Dis. 17 (1), 7–15.

Scharff, R.L., 2012. Economic burden from health losses due to foodborne illness in the United States. J. Food Protect. 75 (1), 123–131.

Schultz, M., 2008. Theobald Smith. Emerg. Infect. Dis. 14 (12), 1940.

Seddon, D., Queen, V., 2004. Sir William Jenner (1815-1898) and Queen Victoria. J. Med. Biogr. 12 (3), 161–163.

Seo, K.H., Holt, P.S., Stone, H.D., Gast, R.K., 2003. Simple and rapid methods for detecting *Salmonella enteritidis* in raw eggs. Int. J. Food Microbiol. 87 (1), 139–144.

Shanthi, J., Usha Rani, R., Balagurunathan, R., 2012. A brief study of diagnosis and frequency of typhoid fever incidence by Widal test. Ann. Biol. Res. 3 (4), 1847–1851.

Shelobolina, E.S., Sullivan, S.A., O'Neill, K.R., Nevin, K.P., Lovley, D.R., 2004. Isolation, characterization, and U (VI)-reducing potential of a facultatively anaerobic, acid-resistant bacterium from low-pH, nitrate- and U (VI)-contaminated subsurface sediment and description of *Salmonella* subterranea sp. nov. Appl. Environ. Microbiol. 70 (5), 2959–2965.

Shivaprasad, H.L., 2000. Fowl typhoid and pullorum disease. Revue scientifique et technique 19 (2), 405–424.

Shulman, S.T., 2004. The history of pediatric infectious diseases. Pediat. Res. 55 (1), 163–176.

Singh, S., Yadav, A.S., Singh, S.M., Bharti, P., 2010. Prevalence of *Salmonella* in chicken eggs collected from poultry farms and marketing channels and their antimicrobial resistance. Food Res. Int. 43 (8), 2027–2030.

Smith, K.E., Medus, C., Meyer, S.D., Boxrud, D.J., Leano, F., Hedberg, C.W., Elfering, K., Braymen, C., Bender, J., Danila, R.N., 2008. Outbreaks of salmonellosis in Minnesota (1998 through 2006) associated with frozen, microwaveable, breaded, stuffed chicken products. J. Food Protect. 71 (10), 2153–2160.

Smith, E.J., 1914. The Life and Letters of Nathan Smith, M.B., M.D. Yale University Press, New Haven.

Smith, D.C., 1981. Medical science, medical practice, and the emerging concept of typhus in mid-eighteenth-century Britain. Med. Hist. 25 (S1), 121–134.

Smith, J.L., 2003. The role of gastric acid in preventing foodborne disease and how bacteria overcome acid conditions. J. Food Protect. 66 (7), 1292–1303.

St Louis, M.E., Morse, D.L., Potter, M.E., DeMelfi, T.M., Guzewich, J.J., Tauxe, R.V., Blake, and the working group, 1988. The emergence of grade A eggs as a major source of *Salmonella* Enteritidis infections: new implications for the control of salmonellosis. J. Am. Med. Assoc. 259 (14), 2103–2107.

Stanley, J., Baquar, N., Threlfall, E.J., 1993. Genotypes and phylogenetic relationships of *Salmonella* Typhimurium are defined by molecular fingerprinting of IS*200* and 16S *rrn* loci. J. Gen. Microbiol. 139 (6), 1133–1140.

Sterzenbach, T., Crawford, R.W., Winter, S.E., Bäumler, A.J., 2013. Chapter 5: *Salmonella* virulence mechanisms and their genetic basis. In: Barrow, P.A., Methner, U. (Eds.), *Salmonella* in Domestic Animals. CABI Publishing, Wallingford, pp. 80–103.

Strawn, L.K., Danyluk, M.D., Worobo, R.W., Wiedmann, M., 2014. Distributions of *Salmonella* subtypes differ between two U.S. produce-growing regions. Appl. Environ. Microbiol. 80 (13), 3982–3991.

Su, L.H., Chiu, C.H., 2007. *Salmonella*: clinical importance and evolution of nomenclature. Chang Gung Med. J. 30 (3), 210–219.

Sulaiman, W., 2006. Typhoid and malaria co-infection - an interesting finding in the investigation of a tropical fever. Malaysian J. Med. Sci. 13 (2), 64–65.

Tan, S.Y., Linskey, K., 2012. George Fernand Widal (1862-1929): serologist and clinician-scientist. Singapore Med. J. 53 (5), 297–298.

Threlfall, E.J., Ward, L.R., Frost, J.A., Willshaw, G.A., 2000. The emergence and spread of antibiotic resistance in food-borne bacteria. Int. J. Food Microbiol. 62 (1), 1–5.

Threlfall, E.J., Wain, J., Peters, T., Lane, C., De Pinna, E., Little, C.L., Wales, A.D., Davies, R.H., 2014. Egg-borne infections of humans with *Salmonella*: not only an *S.* Enteritidis problem. World's Poult. Sci. J. 70 (01), 15–26.

Threlfall, E.J., 2002. Antimicrobial drug resistance in *Salmonella*: problems and perspectives in food- and water-borne infections. FEMS Microbiol. Rev. 26 (2), 141–148.

Tindall, B.J., Grimont, P.A., Garrity, G.M., Euzeby, J.P., 2005. Nomenclature and taxonomy of the genus *Salmonella*. Int. J. Syst. Evol. Microbiol. 55 (1), 521–524.

Todd, E.C., Michaels, B.S., Greig, J.D., Smith, D., Holah, J., Bartleson, C.A., 2010. Outbreaks where food workers have been implicated in the spread of foodborne disease. Part 7. Barriers to reduce contamination of food by workers. J. Food Protect. 73 (8), 1552–1565.

Tsolis, R.M., Xavier, M.N., Santos, R.L., Bäumler, A.J., 2011. How to become a top model: impact of animal experimentation on human *Salmonella* disease research. Infect. Immun. 79 (5), 1806–1814.

Uneke, C.J., 2008. Concurrent malaria and typhoid fever in the tropics: the diagnostic challenges and public health implications. J. Vector Borne Dis. 45 (2), 133–142.

Van den Bogaard, A.E., Stobberingh, E.E., 2000. Epidemiology of resistance to antibiotics. Links between animals and humans. Int. J. Antimicrob. Agents 14 (4), 327–335.

Van Immerseel, F., 2010. Stress-induced survival strategies enable *Salmonella* Enteritidis to persistently colonize the chicken oviduct tissue and cope with antimicrobial factors in egg white: a hypothesis to explain a pandemic. Gut Pathog. 2, 23.

Vandeplas, S., Dubois Dauphin, R., Beckers, Y., Thonart, P., Thewis, A., 2010. *Salmonella* in chicken: current and developing strategies to reduce contamination at farm level. J. Food Protect. 73 (4), 774–785.

Velge, P., Cloeckaert, A., Barrow, P., 2005. Emergence of *Salmonella* epidemics: the problems related to *Salmonella enterica* serotype Enteritidis and multiple antibiotic resistances in other major serotypes. Vet. Res. 36 (3), 267–288.

Vieira, A., Jensen, A.R., Pires, S.M., Karlsmose, S., Wegener, H.C., Wong, D.L.F., 2009. WHO global foodborne infections network country databank – a resource to link human and non-human sources of *Salmonella*. In: Proceeding of the 12th Symposium of the International Society for Veterinary Epidemiology and Economics. Durban, South Africa.

Waller, J., 2002. The Discovery of the Germ: Twenty Years that Transformed the Way We Think About Disease. Columbia University Press, New York.

Wattiau, P., Boland, C., Bertrand, S., 2011. Methodologies for *Salmonella enterica* subsp. *enterica* subtyping: gold standards and alternatives. Appl. Environ. Microbiol. 77 (22), 7877–7885.

White, D.G., Zhao, S., Sudler, R., Ayers, S., Friedman, S., Chen, S., McDermott, P.F., McDermott, S., Wagner, D.D., Meng, J., 2001. The isolation of antibiotic-resistant *Salmonella* from retail ground meats. N. Engl. J. Med. 345 (16), 1147–1154.

White, J.B., 2008. Infection in a village community in the 19th century and the development of the dispensary system: presidential address to the Ulster medical society, Thursday 11 October 2007. Ulster Med. J. 77 (3), 153–159.

Wilson, W.J., 1909. On heterologous agglutinins more particularly those present in the blood serum of cerebro-spinal fever and typhus fever cases. J. Hyg. 9 (03), 316–340.

Wray, C., Wray, A. (Eds.), 2000. *Salmonella* in Domestic Animals. CABI Publishing, New York.

Wright, A.E., Semple, D., 1897. Remarks on vaccination against typhoid fever. Br. Med. J. 1 (1883), 256–259.

Wright, A.P., Gould, L.H., Mahon, B., Sotir, M.J., Tauxe, R.V., 2011. Overview of the impact of epidemic-assistance investigations of foodborne and other enteric disease outbreaks, 1946–2005. Am. J. Epidemiol. 174 (Suppl. 11), S23–S35.

Yildirim, Y., Gonulalan, Z., Pamuk, S., Ertas, N., 2011. Incidence and antibiotic resistance of *Salmonella* spp. on raw chicken carcasses. Food Res. Int. 44 (3), 725–728.

Zinder, N.D., Lederberg, J., 1952. Genetic exchange in *Salmonella*. J. Bacteriol. 64 (5), 679.

CHAPTER

# *Salmonella* in Preharvest Chickens: Current Understanding and Approaches to Control

8

**Paul Wigley[1], Paul Barrow[2]**

[1]*University of Liverpool, Neston, Cheshire, United Kingdom;*
[2]*University of Nottingham, Loughborough, Leicestershire, United Kingdom*

## 1. INTRODUCTION: GENERAL *SALMONELLA* INFECTION BIOLOGY

*Salmonella enterica* subsp. *enterica* comprises more than 2500 separate serovars that differ genetically and in their pathogenicity. From the point of view of infection biology they may be divided into two groups or pathovars.

The first group comprises a small number of serovars that typically produce typhoid-like diseases in a narrow range of host species. These infections generally affect animals of all ages and that are healthy, not immunologically compromised, and not pregnant. Examples of these include *S. enterica* subsp. *enterica* serovar Typhi (*S.* Typhi) causing typhoid in man, *S.* Gallinarum affecting domestic fowl and other avian species, *S.* Dublin in cattle, *S.* Abortusovis affecting sheep, *S.* Abortusequi affecting horses, and *S.* Choleraesuis most frequently associated with pigs. Many of these serovars are also characterized by persistent infections in convalescent animals and by localization in the reproductive tract. In some cases this results in abortion (*S.* Dublin and the *S.* Abortus serovars) or transmission via the egg to the progeny (*S.* Gallinarum) such that both vertical and horizontal transmission become major parts of the infection cycle. Some of these serovars are, in all probability, becoming increasingly adapted to the host and as with this characteristic in other bacterial pathogens such as *Mycobacterium leprae*, genome shrinkage occurs characterized by increasing auxotrophy. They generally colonize the intestine poorly in the absence of clinical disease and therefore enter the human food chain rarely.

The vast majority of the remaining serovars do not normally produce systemic disease in healthy, adult animals. However, most are prototrophic and colonize the alimentary tract well and as a result of this are shed in the feces, contaminate carcasses at slaughter, and enter the human food chain.

**Producing Safe Eggs. http://dx.doi.org/10.1016/B978-0-12-802582-6.00008-2**

The two currently most frequently isolated serovars associated with human food poisoning, *S.* Typhimurium and *S.* Enteritidis are unusual in that they belong to both these pathogroups. They both colonize the gut of several animal species well and are major causes of food poisoning and they also produce typical typhoid, in this case in mice. This fact raises the question of whether NRAMPs mice are unusual since *S.* Dublin, *S.* Abortusovis, and *S.* Choleraesuis also produce typical murine typhoid.

*S. enterica* subsp. *enterica* serovars are classified within the Kauffmann White system based on the combination of lipopolysaccharide (O) and flagella (H) antigens, serovars possessing several of each creating a mosaic of surface antigens that contributes toward defining the specific serovar. Those serovars possessing the O-12 antigen are grouped in serogroup D, which contains *S.* Enteritidis, *S.* Gallinarum, and *S.* Pullorum, which are all associated with infection of laying hens and some form of vertical transmission via the egg, and much more than occurs with most other serovars and serogroups. The related group D serovar, *S.* Dublin, shows similar behavior in cattle. Enzyme and electrophoresis analysis of the nonmotile avian serovars Gallinarum and Pullorum indicated that these had evolved from a nonmotile progenitor that had itself evolved from an *S.* Enteritidis-like ancestor. It was thought that *S.* Pullorum had evolved further from the progenitor than had S. Gallinarum, which correlated with earlier findings of heterogeneity within this serovar (Crichton and Old, 1990). Thompson et al. (2008) showed by whole genome sequencing that this was a reasonable scenario and it also involved horizontal gene transmission following divergence from the *S.* Enteritidis-like progenitor. Analysis of more strains (Langridge et al., 2015) indicated that at least two clades of *S.* Enteritidis strains exist in which one more closely resembles *S.* Gallinarum and also colonizes the chicken intestine less well. These studies show that these taxa are closely related, probably undergoing a process of evolution and combining characteristics associated with host specificity/generalism including invasiveness and gut colonization and thereby association with carcass contamination.

# 2. BIOLOGY OF LAYER INFECTIONS

## 2.1 INVASIVENESS AND SYSTEMIC INFECTION

The extent to which *Salmonella* serovars enter the human food chain is reflected by the ability both to colonize the alimentary tract and to invade the tissues following intestinal colonization. Both are relevant since both lead to contamination of the egg albeit by different means.

Infection in the hatchery can result in extensive horizontal transmission. However, the susceptibility to invasion by *Salmonella* is also the greatest within the first few days of hatching, which can lead to extensive systemic disease.

Some authors have reported that strains of *S.* Enteritidis PT4 were more invasive for young chicks than strains of PT7, 8, and 13a, and they suggested that this may be one of the factors that contributed to the establishment of *S.* Enteritidis

PT4 in the United Kingdom (Hinton et al., 1990). The same authors also found that more recent isolates of *S.* Enteritidis PT4 were more invasive than strains isolated in previous years and suggested that recent isolates of PT4 may have an enhanced virulence for chickens (Hinton et al., 1990). Other authors have found no difference in invasiveness or colonization ability between different phage types of *S.* Typhimurium (Barrow et al., 1987) or *S.* Enteritidis but it rather seems to be strain related (Timoney et al., 1989; Poppe et al., 1993b; Gast and Benson, 1996). The antigenic structure does not seem to be inherently a major factor in virulence, although strains with a wrinkled colony and greater amounts of high-molecular-weight lipopolysaccharides (LPS) are more virulent for chickens, when inoculated parenterally, in terms of bacterial counts in the spleen, localization in the reproductive tract, and percentage of contaminated eggs (Guard-Petter et al., 1996); they are also more tolerant to heat, acid, and hydrogen peroxide than nonwrinkled colonies (Humphrey et al., 1996).

There does also appear to be a degree of organ specificity so that otherwise identical *S.* Enteritidis PT13 strains originally isolated from the ovary or blood showed differences in their isolation from liver, spleen, and ceca after experimental oral inoculation (Poppe et al., 1993a). However, the ovarian isolate formed an entire and smooth colony, whereas the blood isolate developed a corrugated colony appearance after 2 days of growth at room temperature on Luria Bertani agar (C. Poppe, unpublished data).

Random mutagenesis studies have indicated the involvement of genes associated with host interaction, metabolism, and stress responses resulting from survival in an environment to which *Salmonella* is not, in all probability, ideally adapted (Turner et al., 1998; Morgan et al., 2004). Similar studies using S. Gallinarum also identified well-known virulence genes (Shah et al., 2005).

Type three secretion system (TTSS)-1, encoded by *Salmonella* Pathogenicity Island (SPI)-1, is responsible for invasion of epithelial cells whether in vitro or in vivo (Galán and Curtiss, 1989). The genes involved in SPI1 mediating invasion are highly conserved among the genus *Salmonella* and absent from the genomes of close relatives, such as *Escherichia coli*.

The biology of the invasion process is complex and involves not only SPI1 but also SPI4 (Gerlach et al., 2008). Adhesion is a vital initial process, although the role of the <13 different fimbriae expressed by *Salmonella* serovars remains to be defined fully. The main function of the SPI1-encoded T3SS-1 apparatus is to translocate <15 effector proteins into the host cell (Ibarra and Steele-Mortimer, 2009). These effector proteins are encoded by genes located within SPI1 on SPI5, on pathogenicity islets, or on bacteriophages. A subset of these, SipA, SipC, SopA, SopB, SopD, SopE, and SopE2, rearrange intracellular actin to promote bacterial entry into epithelial cells. Much of the work has been done in vitro or in vivo using mice or ligated intestinal loops in calves. The little work that has been done with chickens indicates that SPI1 is less important for systemic disease (Jones et al., 2001) than it is for intestinal gastroenteritis indicating the importance of nonprofessional phagocytic cells in the latter, whereas uptake from the intestine as the first stage of systemic disease involves phagocytic cells in the Peyer patch, cecal tonsil, and other cell clusters more than epithelial cells (Barrow et al., 2000).

The role of flagella is unclear. That they induce inflammation following recognition via TLR5 is clear both in mammals (Schmitt et al., 2001) and chickens and this explains to some extent the difference between the intestinal response to *S.* Typhimurium and *S.* Enteritidis, which results in a strong inflammatory response, and the response to the nonflagellate *S.* Gallinarum and *S.* Pullorum, where it is thought that invasion takes place by stealth in the absence of an inflammatory response (Kaiser et al., 2000). It might be significant that monophasic strains of *S.* Typhimurium have been appearing more recently in several countries in pigs and poultry (Parsons et al., 2013) and nonmotile derivatives of *S.* Dublin have also appeared in the United States.

If *Salmonella* bacteria are injected intravenously into chickens, they are taken up rapidly by macrophages in the spleen and liver. How the bacteria reach these organs following intestinal colonization is unclear, although there is an indication that with *S.* Dublin cell-free bacteria are involved.

Once the bacteria become localized within macrophages, SPI1 genes are normally downregulated (Eriksson et al., 2003), although this does not occur with serovars such as *S.* Infantis and *S.* Montevideo (Imre et al., 2013), which may explain in part at least the reduced virulence of such serovars. Macrophages are the preferred intracellular niche for persistence of *Salmonella* serotypes in tissue (Dunlap et al., 1992; Santos and Bäumler, 2004). A key virulence factor required for survival in macrophages is the type III secretion system encoded by SPI2 (T3SS-2) (Ochman et al., 1996).

The ability to resist the intracellular antibacterial effects of reactive oxygen and nitrogen species and multiply is important. Key to this is the expression of genes on the TTSS-2 encoded by SPI2 on the genome, which is present in all members of the species *S. enterica*, but are absent from *Salmonella bongori* or *E. coli* (Ochman and Groisman, 1996). The T3SS-2 translocates at least 16 effector proteins into the host cell cytosol, including SpiC, SseF, SseG, SlrP, SspH1, SspH2, SifA, SifB, SseI, SseJ, PipB, PipB2, SseK1, SseK2, GogB, and SopD2 (Abraham and Hensel, 2006). Although the molecular functions are known for some of these effector proteins, in most cases it remains unclear how they contribute to T3SS-2-mediated macrophage survival. One purpose of the T3SS-2 seems to be altering the properties of the *Salmonella*-containing vacuole by manipulating vesicular trafficking events (Uchiya et al., 1999; Vazquez-Torres et al., 2000).

Some evidence suggests that the *spvRABCD* operon is also involved in the interaction of *Salmonella* serovars with macrophages (Libby et al., 2000). The *spv* operon is located on virulence plasmids present in a small number of *S. enterica* subsp. *enterica* serotypes, generally those that cause systemic disease (Gulig, 1990), or on the chromosome of *S. enterica* subsp. *arizonae* serotypes (Libby et al., 2002).

Invasiveness and systemic infection are likely to be important to infection of the reproductive tract. The association between serovars and reproductive infection leading to contaminated eggs is poorly understood, although there is an association particularly with certain group D serovars, namely, the serotype cluster involving *S.* Enteritidis, *S.* Gallinarum, and *S.* Pullorum. Considerable experimental work with *S.* Enteritidis has shown that a proportion of infected eggs arise from infections of the oviduct and ovary. In the case of *S.* Pullorum this is a clear association with persistent *S.* Pullorum infection and infection of the ovary and oviduct resulting in <10% eggs being infected (Wigley et al., 2001). Both this serovar and *S.* Gallinarum are rarely associated these days with food poisoning but nevertheless are models of vertical transmission. The situation with *S.* Gallinarum is not as clear as with *S.* Pullorum (Barrow & Neto, 2011) because, although there is considerable epidemiological evidence of vertical transmission, it is more difficult to demonstrate this experimentally and it seems likely that the genetic background of the birds is an important factor in whether this happens or not.

Infection of *S.* Enteritidis during the laying period results in the production of infected eggs, which, if these are fertile and are hatched, results in extensive infection of the progeny, which themselves continue to excrete *S.* Enteritidis until they also come into lay.

## 2.2 INTESTINAL COLONIZATION

Experimental infection of chickens with either *S.* Enteritidis or *S.* Typhimurium results in extensive colonization of the intestine and fecal shedding, whereas vertical transmission seems to occur much less with the latter serovar than with the former. A higher proportion of eggs laid by *S.* Enteritidis are surface contaminated, as a result of cloacal infection, than a result from systemic infection, which itself results in infection of the egg contents (Barrow and Lovell, 1991), although it may nevertheless be the latter that contributes in a greater way to transmission to humans via the food chain.

The exact mechanism whereby *Salmonella* strains colonize the intestine remains unclear, although this virulence attribute is likely to be multifactorial.

Following the realization in the 1970s that some intestinal pathogens attach to the mucosa this was thought to be an important feature for colonization of the ceca, where the highest numbers of *Salmonella* bacteria occur, although this seemed unlikely since the flow rate of contents through this organ is very low with the ceca being emptied two to four times a day. Initial screening of transposon mutant libraries for colonization ability identified a range of genes associated with metabolism, attachment, and also invasion (Turner et al., 1998; Morgan et al., 2004). These genes included LPS, fimbrial genes, and a small number of SPI1, SPI3, and SPI5 genes in addition to a large number of transport genes. These studies were complemented by gene expression studies harvesting *S.* Typhimurium or *S.* Enteritidis from cecal contents of newly hatched chickens (Harvey et al., 2011; Dhawi et al., 2011), which indicated that during colonization the bacteria were subjected to heat stress and were using a set of carbon sources different from those in laboratory media, including propane-2-diol and ethanolamine, both of which may be degradation products from host epithelial cells. These require tetrathionate as electron acceptor and cobalamin as cofactor. More recent studies indicate that tetrathionate and the carbon sources are now thought to be the results of the inflammatory reaction against invasive *Salmonella* (Winter et al., 2010). Most bacterial growth takes place close to the mucosa (Harvey et al., 2011) where nutrient concentrations and oxygen levels are likely to be the highest, although host antibacterial factors such as neutrophil leukocytes and defensins are also likely to be active. In addition to respiration whereby electron acceptors are available, there is also some evidence that substrate level phosphorylation (fermentation) is likely to be a major way of generating ATP (Barrow et al., 2015).

## 2.3 IMMUNITY

The close association between *Salmonella* and the host intestinal mucosa during colonization and invasion suggest that the innate response to infection will be an important component defining the course of infection. Interaction between the bacterial surface pattern-associated molecular patterns such as LPS and flagella and the pattern recognition receptors, in this case TLR 4 and 5, results in the intestinal inflammatory response expressed by neutrophil diapedesis and antimicrobial beta-defensin peptides produced by Paneth cells at the base of the intestinal crypts.

The immune responses to the two pathovars of *S. enterica*, namely, the typhoid and gastroenteritis groups are different as indicated earlier. The response to the typhoid group in chickens, namely, *S.* Gallinarum and *S.* Pullorum, has been described by Chappell et al. (2009). The absence of an inflammatory response within the intestinal mucosa during infection by these serovars is associated with no significant influx of heterophil granulocytes (Henderson et al., 1999). Chemoattractants for granulocytes and monocytes, chCXCLi1 and chCXCLi2, are downregulated after infection (Chappell et al., 2009). Following internalization by macrophages and dendritic cells underlying the mucosal epithelium they are transported to the spleen, liver, ovaries, and bone marrow (Wigley et al., 2001). Infection is characterized by high amounts of specific antibodies and T-cell proliferation (Wigley et al., 2005). Reduced levels of inflammatory cytokines interleukin (IL)-18 and interferon (IFN)-γ are detected in the spleens of birds infected with *S.* Pullorum when compared with *S.* Enteritidis-infected chickens. Unlike *S.* Enteritidis, *S.* Pullorum appears to induce a Th2-type response that drives antibody production but is poor in clearing intracellular infection (Chappell et al., 2009). Surviving chickens remain *Salmonella* carriers (Wigley et al., 2001) with the bacteria persisting within macrophages in the spleen and liver (Wigley et al., 2001).

The immune response to the more proinflammatory serovars such as *S.* Enteritidis and also *S.* Typhimurium is characterized by an influx of immune cells such as granulocytes, macrophages, and T and B cells in cecum, spleen, and bursa of Fabricius accompanied by upregulation in transcription of genes leading to the production of the relevant chemokines and cytokines, including IL-1, IL-6, and IFN-γ (Berndt and Methner, 2001; Beal et al., 2004a; Berndt et al., 2006). Following infection with *S.* Enteritidis, changes in immune cell composition occur in the ovary and oviduct

(Withanage et al., 1998; Barua and Yoshimura, 2004). There is experimental evidence that older chickens show a more competent T-cell immunity in gut and stronger T-cell responses after *S.* Typhimurium infection compared with very young birds (Beal et al., 2004b, 2005). Interestingly, T cells appear to be essential to clearance of *Salmonella* from the intestine, whereas B cells are not an absolute requirement, although the actual mechanisms of immune clearance are not known (Beal et al., 2006).

## 3. APPROACHES TO CONTROL

The entry of infected poultry products, meat, or eggs, into the food chain is exacerbated by the gross spread and cross-infection that occurs during slaughter (Mead, 1989). The World Health Organization has long recognized that the three areas where infection control may be sensibly exerted are by education of the public, by improvements in slaughter hygiene and technology, and by control of infection in the birds themselves (World Health Organisation, 1980, 1990). There are, however, limitations in the extent to which public education can be effective in this. The economics of abattoir processing indicate that there may also be financial limitations in the extent to which improvements may be made at this stage. It seems likely therefore that, as with past control of other bacterial zoonoses, such as bovine tuberculosis and brucellosis, control in the animals themselves must be central to infection control policy. However, the economics of poultry production must be an important factor in introducing control measures. Profit margins are small and the apparent absolute requirement for free trade will result in those countries that introduce such measures to control infection, such as salmonellosis, being placed at a financial disadvantage in contrast to countries where no measures are taken. Introduction into international free-trade legislation of a public health component and conditions would seem an important measure. This would mean that those countries wishing to improve the bacterial zoonosis status of their national flocks will not be penalized as a consequence.

It is possible to rear poultry totally in the absence of *Salmonella*. Large breeding companies and research establishments do it albeit at a high cost. This is done through the introduction of high-quality housing and diet, together with employment of skilled staff and efficient management structures. In addition, introduction of thorough hygiene and disinfection measures and other schemes to reduce the chances of cross-infection, such as "all in–all out" rearing are required. In some cases, these things will be possible and, as existing housing degenerates, requiring replacement, improvements can be made slowly. In countries with high ambient temperatures, open-sided housing may limit the extent to which such improvements may contribute to reduce environmental sources of infection. However, the financial incentives to eliminate a food-poisoning pathogen from stock, which has very little direct impact on productivity and for which financial incentives are not available, pose considerable imponderable problems for governments and poultry companies. It seems likely, therefore, for the foreseeable future, that biological control measures will be an increasingly attractive option.

### 3.1 VACCINES

Vaccination against host-specific *Salmonella* serotypes that cause severe systemic disease in a particular host species (e.g., *S.* Gallinarum in poultry), induces a strong serotype-specific protective immunity against infection and disease (Smith, 1956; Barrow and Wallis, 2000). In contrast, vaccination against host nonspecific *Salmonella* serotypes has yielded variable success rates. The two infection types display very different epidemiological pictures and patterns of pathogenicity, which, together with the nature of the immune response to systemic and intestinal infections, may account for these differences. As a result of public health interest this has been a fruitful area for research over several years. A number of reviews have appeared that summarize our knowledge and understanding up to 5–6 years ago.

Killed vaccines have been used to control host nonspecific *Salmonella* infections in poultry with very varying success. Autologous vaccines have been used extensively and little information is available on their efficacy. Some work (Timms et al., 1994; Liu et al., 2001) supports earlier observations that they may be used to reduce mortality, although this is of little practical significance in the field.

The work of McCapes et al. (1967) and Truscott and Friars (1972) supports earlier contentions that maternal vaccination with bacterins does not reduce significantly excretion of *Salmonella* in the progeny, although mortality can be reduced. A vaccine containing both *S.* Enteritidis and *S.* Typhimurium, both grown under iron-restricted conditions, is commercially available in some European countries (Clifton-Hadley et al., 2002). Iron restriction is known to upregulate bacterial factors that stimulate virulence and thus may stimulate important immunogens. However, given that many other relevant genes are also upregulated in macrophages (Eriksson et al., 2003) it might be more appropriate to produce the vaccines under the conditions experienced in that environment. The inactivated *S.* Enteritidis vaccine was efficient at decreasing egg contamination after intravenous challenge with *S.* Enteritidis (Woodward et al., 2002). This work is difficult to evaluate since oral or respiratory challenge would have been more relevant. However, the combined *S.* Enteritidis and *S.* Typhimurium vaccine, when given intramuscularly at day 1 and week 4, did decrease shedding after oral challenge with *S.* Typhimurium in a seeder-bird challenge model (Clifton-Hadley et al., 2002).

Attention has been paid to the development of live, attenuated vaccine strains of *Salmonella* because of the accumulation of evidence that such strains of *Salmonella* are more immunogenic in mice and in poultry than are killed or subunit vaccines (Collins, 1974; Zhang-Barber et al., 1999). Live vaccines have been tested extensively in mice and also in poultry. Although a number of different live *Salmonella* strains have been tested for their efficacy in experimental or semi-field studies, only a few are registered and commercially available for use in poultry in Europe. The commercially available live *S.* Typhimurium and *S.* Enteritidis vaccine strains are either auxotrophic double-marker mutants derived through chemical mutagenesis or developed on the basis of the principle of metabolic drift mutations (Vielitz et al., 1992; Meyer et al., 1993; Linde et al., 1990). These are negative mutations in essential enzymes and metabolic regulatory centers as a consequence of which the resulting metabolic processes lead to prolonged generation times and corresponding reductions in virulence (Linde et al., 1990). Some of these *Salmonella* live vaccines have been further characterized by molecular methods (Schwarz and Liebisch, 1994).

Another live vaccine registered for prophylactic use against *S.* Enteritidis, which was developed initially for immunization against *S.* Gallinarum, is the rough strain *S.* Gallinarum 9R (Smith, 1956). This vaccine strain has been tested more extensively in recent years since it has been shown to give cross-protection against *S.* Enteritidis (Barrow et al., 1991), a member of the same serogroup. The extent of cross-protection against other serotypes, from either the same or other serogroups remains unclear. In a large field trial in the Netherlands in which 80 commercial flocks were vaccinated with the *S.* Gallinarum 9R vaccine strain, the flock level occurrence of *S.* Enteritidis infections was 2.5% (2/80 flocks). This was significantly less than the flock level occurrence of *S.* Enteritidis infections in unvaccinated flocks (214 of 1854 flocks or 11.5%) (Feberwee et al., 2001a). In 4500 eggs derived from 5 *S.* Gallinarum 9R vaccinated flocks, no vaccine strain bacteria were detected, whereas no evidence was found in another study for the fecal spread of the vaccine strain (Feberwee et al., 2001b).

*AroA* mutants have been tested extensively in poultry and found to be effective, albeit less protective than the "gold standard" produced in chickens infected with a wild-type strain (Barrow et al., 1990; Cooper et al., 1990). Given the general consensus that there is little cross-protection between serovars, it is not surprising that Parker et al. (2001) found no significant differences in egg or reproductive tract infection when laying hens were vaccinated at day of hatch, and at 4 and 22 weeks with an *aroA* mutant of *S.* Typhimurium and challenged with *S.* Enteritidis 8 weeks after the final immunization.

As stated earlier, most data on the nature of vaccine-induced protection are derived from mice studies and care should be taken in extrapolating these data to poultry. Although killed vaccines can be efficacious in reducing *Salmonella* in poultry, live vaccines are thought to have some advantages over killed vaccines, including stimulation of both cell-mediated and humoral immune arms and expression of all appropriate antigens in vivo, whereas the latter stimulate mainly antibody production and express only the antigens present at the time of in vitro harvesting (Collins, 1974). Killed vaccines may also be destroyed rapidly and eliminated from the host, they may be poorly immunogenic in unprimed hosts and unable to induce cytotoxic T cells. Live vaccines have been shown to be more effective in increasing lymphocyte proliferation in response to *S.* Enteritidis antigens in laying hens (Babu et al., 2003). They also have additional protective effects, particularly when administered orally, which can be exploited during their development and application. These include (1) genus-specific colonization inhibition (competitive exclusion) demonstrated to be primarily an effect of microbial metabolism and (2) the stimulation of primed PMNs in the gut (Van Immerseel et al., 2005). Commercial vaccines that are administered in this way may induce the exclusion effect. The mechanism of this inhibition is poorly understood but is likely to result from competition for available nutrients and electron acceptors under the prevailing redox conditions in the gut.

Killed vaccines are unable to induce these effects. It seems unlikely at the moment that more effective killed or subunit vaccines will be produced in the next few years because many basic questions relating to identification of the major protective immunogens and the nature of the immune response in the chicken remain unanswered. Live vaccines have some disadvantages, including, perhaps most significantly, those associated with public acceptability, particularly where genetic manipulation has been used to produce the vaccine. This is a major issue that should be addressed since the safety requirements are different for live vaccines than for inactivated vaccines.

The criteria for an ideal vaccine have been discussed previously (Pritchard et al., 1978; Barrow, 1991) and they include (1) effective protection against both mucosal and systemic infection, (2) attenuation for animals and man, (3) efficacy in reducing intestinal colonization, and thus reduced environmental contamination, and egg infection, (4) compatibility with other control measures, and (5) cost-effective application. As indicated earlier, it is already possible to attenuate strains in a number of ways but inability to induce gastroenteritis is not always evaluated. It should be possible in the next few years to produce live, attenuated strains that are immunogenic for poultry and other food animals but that maintain attenuation in man and other nontarget species. This will, by necessity, require molecular genetics as a tool. The alternative is that live, attenuated vaccines are produced, as currently, by undefined chemical mutagenesis with strains possessing a combination of uncharacterized lesions, including antibiotic resistance, and whose cumulative effects may also not be completely known. The vaccines currently in use in Europe and elsewhere are highly safe but it is anomalous that it is acceptable to allow their widespread dissemination while being extremely cautious over the use of defined deletion mutants produced by genetic manipulation, but where each deletion is nevertheless known and characterized and where antibiotic resistance genes are not present. The environmental issues associated with the genetic modification of plants and also some food animals that may escape to the wild are very different issues to the use of bacterial deletion mutants, with no additional DNA added. One advantage of the current widespread application of the vaccines that are already in use is that because they are widely distributed data will now accumulate on any reversion and other potential risks to man, target animals, and the environment.

## 3.2 ANTIBIOTICS AND COMPETITIVE EXCLUSION

Antibiotics and other chemotherapeutic agents are used for the treatment and prevention of a number of bacterial diseases of poultry, systemic diseases caused by genera such as *Salmonella* spp., *E. coli*, and *Mycoplasma* spp., and have been used for the reduction of fecal carriage of *Salmonella* and growth promotion/stimulation (sometimes now referred to as digestion enhancement). The list of antibiotics used is long and varies according to country and the extent to which national regulations restrict general use without veterinary prescription. Despite the use of penicillin derivatives and, more recently, fluoroquinolones, the incidence of multiple antibiotic resistance in *Salmonella* has traditionally been very low, in contrast to the situation in the calf rearing industry. Strains such as *S.* Typhimurium DT104 were originally isolated from calves and have been isolated with increasing frequency in the poultry industry. However, in recognition of the extensive use of antibiotics in all aspects of livestock rearing (O'Neill report) there is pressure toward a reduction in use of antibiotics for disease control and elimination of their use for growth promotion.

### 3.2.1 Therapeutic and Prophylactic Use

The use of chemotherapy has at times and in some countries been promoted with *S.* Enteritidis in broiler breeders or layers to attempt to reduce the frequency of egg contamination immediately before stock movement from rearing to laying accommodation, followed by restoration of the gut flora by oral administration of a competitive exclusion preparation. Tetracyclines, furazolidone, and fluoroquinolones have been used for this. The major concern over the use of antibiotic therapy has always been one of selection of resistant clones. Experimental work (Smith and Tucker, 1975) suggests that some antibiotics, including tetracycline, ampicillin, and chloramphenicol, have little or no effect on fecal shedding of the *Salmonella.* After medication was discontinued fecal shedding may increase for a while, presumably as a result of the effect of the antibiotics on the normal intestinal flora that inhibit colonization by pathogens. Plasmid-mediated antibiotic resistance in the *E. coli* flora may also transfer to the *Salmonella* population under such conditions, confirming that the use of antibiotic therapy leads to the encouragement of transfer of resistance in the intestine. The massive increases in the use of quinolones and fluoroquinolones since the mid-1980s have led to the isolation of an increasing number of *Salmonella* strains that are resistant to nalidixic acid (Piddock et al., 1990; Wray et al., 1990). Although resistance is now also found to be transmissible, it is usually chromosomal mutation but may be transferred between certain *Salmonella* strains by transducing bacteriophages (Barrow et al., 1998). Its use in poultry has now been banned in the United States and it has precipitated considerable debate. Even in the absence of clinical salmonellosis, *Salmonella* organisms colonizing the gut are subject to selective pressure from these chemicals when they are used against other bacterial diseases.

### 3.2.2 Growth Promoting Use

In some countries chemotherapeutic antibiotics have been banned for this purpose where they are also used in human or veterinary therapy purposes. Up until 1969 (Anon, 1969), low levels of chemotherapeutic antibiotics, in addition to copper sulfate, were used as growth promoting agents in Europe. The appearance of multiresistant *S.* Typhimurium in calves was perceived to be a very worrying development and the use of such drugs for this purpose was first banned in the United Kingdom and then in the European Union but this has not so far happened in the United States or many other countries. After 1969 the pharmaceutical companies developed new antibiotics that were growth promoting but had no direct effect on *Salmonella* or *E. coli.* Their spectrum of activity was different. However, because they affected members of the gut flora, which were themselves inhibitory to *Salmonella* colonization, they altered the susceptibility of poultry to infection with

*Salmonella*. Some, such as tylosin and nitrovin and the glycopeptide avoparcin considerably increased fecal excretion of *Salmonella* (Smith and Tucker, 1978; Smith et al., 1985). This effect can be reduced by using lower concentrations in the feed. In addition to the effect on *Salmonella*, avoparcin also selects for resistance to glycopeptides, such as vancomycin in *Enterococcus faecium* (Wegener et al., 1999). This is potentially a very great public health threat since the transposon can transfer the resistance to multiresistant strains of *Staphylococcus aureus* (Noble et al., 1992). In addition, since poultry and pig *E. faecium* strains can colonize the human gut (Berchieri, 1999), the development of vancomycin-resistant *S. aureus* concerns those involved in human public health. The use of growth promoting antibiotics, including avoparcin, has now been banned in the European Union. This sort of action, without controlling import of poultry and poultry meat from outside the trading bloc, does not, however, completely solve the problem.

#### *3.2.3 Competitive Exclusion*

Because young poultry, reared intensively, are slow to develop the complex intestinal microflora of older birds, they are particularly prone to colonization with food-poisoning salmonellas. Colonization resistance can be markedly increased, however, by the early establishment of an adult-type flora through oral administration of the requisite organisms. The phenomenon in poultry was first demonstrated by Nurmi and Rantala (1973) and is usually termed "competitive exclusion" (CE). Protection develops rapidly, is apparently unaffected by the breed, sex, or immune status of recipient birds and is active against all host nonspecific serotypes studied so far (Mead, 2000). Commercial CE products contain a wide variety of viable bacteria that are provided by cultures of cecal material from selected donor birds.

In addition to its use in very young chickens it has also been used following oral antibiotic treatment, although this approach is now discouraged for the reasons presented earlier. A third situation where it might be used is in birds that have been stressed by poor handling or management practices that could disrupt the normal gut flora; however, this application has been little studied as yet.

For a number of reasons its effectiveness in the field is less than under experimental conditions. In treated flocks, the proportion of birds becoming *Salmonella* positive is reduced in comparison with untreated controls and for those birds that do become carriers, the numbers of salmonellas being shed are also generally lower.

### 3.3 LEGISLATION

As a consequence of the *S.* Enteritidis epidemic, which resulted in high national publicity and accompanying political fall-out, the United Kingdom and the Netherlands led the way toward European Union–level legislation (Directives 2160/2003 and 1168/2006) to require monitoring for *S.* Enteritidis and *S.* Typhimurium, initially of breeder and layer flocks and eventually also of broilers and turkeys. The intention was to improve feed quality and ultimately to introduce a requirement for national control measures involving surveillance, biosecurity, and vaccination. Each EU member state was required to develop and implement a series of National Control Plans (NCP) for *Salmonella* and to set out targets for its reduction. These plans were produced at different speeds in different countries and the legislation appears to be having a positive impact in reducing *Salmonella* levels within the European Union. *Salmonella* Enteritidis declined from a peak of hundreds of infected flocks and over 20,000 human cases in England and Wales in 1992 to a handful of positive flocks and less than 5000 human cases by 2010. When considering PT4, the main cause of egg-associated infection in the United Kingdom, the decline has been even more marked, with a drop from over 15,000 human cases in 1992 to 459 in 2010. The reduction of *Salmonella* in UK egg production is a clear success for control strategies such as the industry-led voluntary "Lion Mark" scheme introduced in 1998 and which included improved surveillance, hygiene, and biosecurity and perhaps, most significantly, vaccination of laying hens. Most of the measures included in the "Lion Mark" scheme enshrined in legislation within the UK NCP for the commercial egg sector. Furthermore, the NCP lays down specific requirements for surveillance in laying hens. Layers are tested for *Salmonella* at the hatchery, during rearing, at pullet placement, at the start of production, and every 15 weeks during production. Eggs from flocks found to test positive for *Salmonella* cannot be sold as table grade eggs and in practice most positive flocks are culled and houses cleared of *Salmonella* before repopulation.

Despite these successes *S.* Enteritidis remains the most important serovar associated with food-borne salmonellosis worldwide.

Regular baseline surveys carried out by the European Food Safety Authority (EFSA) have given a clear snapshot of *Salmonella* in European poultry with prevalence in egg production falling across Europe and varying between 1% in the United Kingdom and 14% in Spain. In 2008, infection rates in broilers varied from <1% in Scandinavia through 14.9% in Spain to 85% in Hungary. Serovars such as *S.* Infantis and *S.* Hadar are becoming more prevalent in a number of EU countries.

Surveillance of *Salmonell*a in the United States is more fragmentary and it is more difficult to evaluate, although evidence indicates that *Salmonella* infection remains a considerable problem. As much as 23% of US poultry meat is still infected with *Salmonella* (www.fsis.usda.gov). No specific strategy is in place for control, which may reflect state rather than federal decision making. The situation in Australia is more interesting since, although around 13% of carcasses are contaminated with *Salmonella* at slaughter, more than half of these are *S.* Sofia, a serovar with a low potential for virulence in humans.

Countries such as Thailand, which have rapidly expanding industries for both domestic consumption and export, have recognized the problem and are developing strategies for the surveillance and control of both *Salmonella* and *Campylobacter*, although little information is available on *Salmonella* prevalence. Countries such as these are helped by the ability to build housing from zero with the required associated management regimens. In many other countries *Salmonella* remains a problem of secondary animal or human health significance.

No doubt the increasing globalization of trade in poultry meat will increase pressure on producing countries to introduce targets for reducing levels of infection that will comply with the expectations of the consuming countries. This will inevitably lead to reduced levels of infection in an increasing number of countries.

### 3.4 GENETIC RESISTANCE

It is apparent from the earlier sections that the pathogenesis of *Salmonella* involves several discrete stages, perhaps most crudely represented as the intestinal and systemic phases of infection. It is also clear to those working with poultry during a period of just a few years that host genotypes and phenotypes can change rapidly. This is the result of manipulating host genetics to select for particular phenotypes especially those associated with production traits. However, inadvertently, infection resistance traits can also change as a result of this.

Mendelian inheritance of resistance to *Salmonella* has been reported for decades but during the 1980s a systematic investigation was made to identify the gene(s) involved through breeding for F1, F2, and backcrossing with SNP mapping to narrow down the chromosomal region associated with the trait. This was done with resistance to systemic salmonellosis using *S.* Typhimurium, *S.* Enteritidis, *S.* Gallinarum, and *S.* Pullorum. The patterns of resistance between the different inbred lines were the same for all serovars indicating a common mechanism. A similar pattern of resistance was also observed with another intracellular pathogen, *M. avium* infection (Bumstead and Barrow, unpublished results). The resistance, nominated the *SAL1* locus, has been mapped to the long arm on chromosome 5 (Fife et al., 2009). This region includes two candidate genes *AKT1*, a serine threonine/kinase modulating apoptosis, cell proliferation, and development, and *SIVA1*, a proapoptotic factor. The *SAL1* locus accounts for more than 50% of the resistance in the inbred lines with additional contributions from TLR4 and also *NRAMP1* (*Slc11A1*) (Hu et al., 1997).

Additional studies centered on the genetic basis to differences in the level of gut colonization and fecal excretion of *S.* Typhimurium and *S.* Enteritidis in inbred lines. Resistance was not sex linked and a dominant trait not linked to MHC, *SAL1*, or to the gut flora. Interestingly, the resistance to *Salmonella* colonization was mirrored in the patterns of resistance to *Campylobacter jejuni* (Boyd et al., 2005). Mapping this trait has been difficult (Fife et al., 2009; Calenge et al., 2010; Tilquin et al., 2005).

How far this information can be manipulated to increase resistance to infection without other deleterious changes in phenotype remains to be seen. However, simple selection using normal breeding techniques may be done; alternatively genetic manipulation through deriving transgenic animals will also offer the opportunity of making beneficial changes.

## 4. CONCLUSION AND SYNTHESIS

Our understanding of the biology of *Salmonella* infection in layers has improved during the last 10 years by the use of standard microbiology and with the technical improvements in molecular biology, bacterial genetics, and avian immunology. These have led to improvements in diagnosis and will lead to improved efficacy in the vaccines used, which have until now been developed largely empirically. By the same token, other approaches to controlling *Salmonella* infection, such as the use of competitive exclusion, have also been applied without a full understanding of the mechanism involved. Although the application of avian genetics is in its infancy, this technology, perhaps combined with other approaches such as vaccination, would seem to be the most promising in the long run in terms of biological approaches to controlling *Salmonella* infection. However, as with all technologies a note of caution must be introduced. Individual genes are now being identified that are associated with resistance but only with a percentage of the full resistance trait. The complexity of the involvement of individual genes, their interaction with other genes, either also involved in disease resistance or other traits not related to infection, and their effects on production traits remain a long way from realization.

There are huge problems in livestock rearing; the one with the greatest potential to cause disruption in productivity and animal and public health is antimicrobial resistance. This will require huge changes in the longer term in poultry health management. In countries such as Thailand, which are now expanding their poultry industry, involving the introduction of the newest and best quality housing and infrastructure are leading the way. This is built on the understanding that management including biosecurity, food quality, and housing will remain the approaches to infection control with the greatest likelihood of long-term success.

## REFERENCES


Abrahams, G.L., Hensel, M., 2006. Manipulating cellular transport and immune responses: dynamic interactions between intracellular *Salmonella enterica* and its host cells. Cell Microbiol. 8 (5), 728–737.

Babu, U., Scott, M., Myers, M.J., Okamura, M., Gaines, D., Yancy, H.F., Lillehoj, H., Heckert, R.A., Raybourne, R.B., 2003. Effects of live attenuated and killed *Salmonella* vaccine on T-lymphocyte mediated immunity in laying hens. Vet. Immunol. Immunopathol. 91 (1), 39–44.

Barrow, P.A., Freitas Neto, O.C., 2011. Pullorum disease and fowl typhoid–new thoughts on old diseases: a review. Avian Pathol. 40, 1–13.

Barrow, P.A., Lovell, M.A., 1991. Experimental infection of egg-laying hens with *Salmonella enteritidis*. Avian Pathol. 20, 339–352.

Barrow, P.A., Wallis, T.S., 2000. Vaccination against *Salmonella* infections in food animals; rationale, theoretical basis and practical applications. In: Wray, C., Wray, A. (Eds.), *Salmonella* in Domestic Animals. CABI publishing, Wallingford, Oxfordshire, UK, pp. 323–339.

Barrow, P.A., Huggins, M.B., Simpson, J.M., Lovell, M.A., 1987. Observations on the pathogenesis of experimental *Salmonella typhimurium* infection in chickens. Res. Vet. Sci. 42, 194–199.

Barrow, P.A., Hassan, J.O., Berchieri Jr., A., 1990. Reduction in faecal excretion by chickens of *Salmonella typhimurium* by immunization with avirulent mutants of *S. typhimurium*. Epidemiol. Infect. 104, 413–426.

Barrow, P.A., Lovell, M.A., Berchieri Jr., A., 1991. The use of two live attenuated vaccines to immunize egg-laying hens against *Salmonella enteritidis* phage type 4. Avian Pathol. 20, 681–692.

Barrow, P.A., Lovell, M.A., Szmolleny, G., Murphy, C.K., 1998. Effect of enrofloxacin administration on excretion of *Salmonella enterica* serotype Enteritidis and the *Escherichia coli* flora by experimentally infected chickens. Avian Pathol. 27, 586–590.

Barrow, P.A., Lovell, M.A., Stocker, B.A.D., 2000. Protection against experimental fowl typhoid by parenteral administration of live SL5928, an *aroA-serC* (aromatic dependent) mutant of a wild-type *Salmonella* Gallinarum strain made lysogenic for P22*sie*. Avian Pathol. 29, 423–431.

Barrow, P.A., Berchier, A., Neto, O.C.F., Lovell, M.A., 2015. The contribution of aerobic and anaerobic respiration to intestinal colonization and virulence for *Salmonella* Typhimurium in the chicken. Avian Pathol. 44, 401–407.

Barrow, P.A., 1991. Immunological control of Salmonella in poultry. In: Blankenship, L.C. (Ed.), Colonization Control of Human Enteropathogens in Poultry. Academic Press, pp. 199–217.

Barua, A., Yoshimura, Y., 2004. Ovarian cell-mediated immune response to *Salmonella enteritidis* infection in laying hens (*Gallus domesticus*). Poult. Sci. 83 (6), 997–1002.

Beal, R.K., Powers, C., Wigley, P., Barrow, P.A., Smith, A.L., 2004a. Temporal dynamics of the cellular, humoral and cytokine responses in chickens during primary and secondary infection with *Salmonella enterica* serovar Typhimurium. Avian Pathol. 33, 25–33.

Beal, R.K., Wigley, P., Powers, C., Hulme, S.D., Barrow, P.A., Smith, A.L., 2004b. Age at primary infection with *Salmonella enterica* serovar Typhimurium in the chicken influences persistence of infection and subsequent immunity to re-challenge. Vet. Immunol. Immunopathol. 100, 151–164.

Beal, R.K., Powers, C., Wigley, P., Barrow, P.A., Kaiser, P., Smith, A.L., 2005. A strong antigen-specific T-cell response is associated with age and genetically dependent resistance to avian enteric salmonellosis. Infect. Immun. 73, 7509–7516.

Beal, R.K., Powers, C., Davison, T.F., Barrow, P.A., Smith, A.L., 2006. Clearance of *Salmonella enterica* serovar Typhimurium in chickens is independent of B-cell function. Infect. Immun. 74, 1442–1444.

Berchieri, A., 1999. Intestinal colonization of a human subject by vancomycin-resistant *Enterococcus faecium*. Clin. Microbiol. Infect. 5 (2), 97–100.

Berndt, A., Methner, U., 2001. Gamma/delta T cell response of chickens after oral administration of attenuated and non-attenuated *Salmonella typhimurium* strains. Vet. Immunol. Immunopathol. 78 (2), 143–161.

Berndt, A., Pieper, J., Methner, U., 2006. Circulating gamma delta T cells in response to *Salmonella enterica* serovar enteritidis exposure in chickens. Infect Immun. 74 (7), 3967–3978.

Boyd, Y., Herbert, E.G., Marston, K.L., Jones, M.A., Barrow, P.A., 2005. Host genes affect intestinal colonisation of newly-hatched chickens by *Campylobacter jejuni*. Immunogenetics 57, 248–253.

Calenge, F., Kaiser, P., Vignal, A., Beaumont, C., 2010. Genetic control of resistance to salmonellosis and to *Salmonella* carrier-state in fowl: a review. Genet. Sel. Evol. 42, 11. http://dx.doi.org/10.1186/1297-9686-42-11.

Chappell, L., Kasier, P., Barrow, P.A., Jones, M.A., Johnston, C., Wigley, P., 2009. The immunobiology of avian systemic salmonellosis. Vet. Immunol. Immunopathol. 128, 53–59.

Clifton-Hadley, F.A., Breslin, M., Venables, L.M., Sprigings, K.A., Cooles, S.W., Houghton, S., Woodward, M.J., 2002. A laboratory study of an inactivated bivalent iron restricted *Salmonella enterica* serovars Enteritidis and Typhimurium dual vaccine against Typhimurium challenge in chickens. Vet. Microbiol. 89 (2–3), 167–179.

Collins, F.M., 1974. Vaccines and cell-mediated immunity. Bacteriol. Rev. 38 (4), 371–402.

Cooper, G.L., Nicholas, R.A., Cullen, G.A., Hormaeche, C.E., 1990. Vaccination of chickens with a *Salmonella enteritidis* aroA live oral *Salmonella* vaccine. Microb. Pathog. 9 (4), 255–265.

Crichton, P.B., Old, D.C., 1990. Salmonellae of serotypes *gallinarum* and *pullorum* grouped by biotyping and fimbrial-gene probing. J. Med. Microbiol. 32 (3), 145–152.

Dhawi, A.A., Elazomi, A., Jones, M.A., Lovell, M.A., Li, H., Emes, R.D., Barrow, P.A., 2011. Adaptation to the chicken intestine in *Salmonella* Enteritidis PT4 studied by transcriptional analysis. Vet. Microbiol. 153, 198–204.

Dunlap, N.E., Benjamin Jr., W.H., Berry, A.K., Eldridge, J.H., Briles, D.E., 1992. A 'safe-site' for *Salmonella typhimurium* is within splenic polymorphonuclear cells. Microb. Pathog. 13 (3), 181–190.

Eriksson, S., Lucchini, S., Thompson, A., Rhen, M., Hinton, J.C., 2003. Unravelling the biology of macrophage infection by gene expression profiling of intracellular *Salmonella enterica*. Mol. Microbiol. 47 (1), 103–118.

Feberwee, A., de Vries, T.S., Hartman, E.G., de Wit, J.J., Elbers, A.R., de Jong, W.A., 2001a. Vaccination against Salmonella enteritidis in Dutch commercial layer flocks with a vaccine based on a live *Salmonella* gallinarum 9R strain: evaluation of efficacy, safety, and performance of serologic Salmonella tests. Avian Dis. 45 (1), 83–91.

Feberwee, A., Hartman, E.G., de Wit, J.J., de Vries, T.S., 2001b. The spread of *Salmonella* gallinarum 9R vaccine strain under field conditions. Avian Dis. 45 (4), 1024–1029.

Fife, M.S., Salmon, N., Hocking, P.M., Kaiser, P., 2009. Fine mapping of the chicken salmonellosis resistance locus (SAL1). Anim. Genet. 40 (6), 871–877.

Galán, J.E., Curtiss 3rd, R., 1989. Cloning and molecular characterization of genes whose products allow *Salmonella typhimurium* to penetrate tissue culture cells. Proc. Natl. Acad. Sci. U.S.A. 86 (16), 6383–6387.

Gast, R.K., Benson, S.T., 1996. Intestinal colonization and organ invasion in chicks experimentally infected with *Salmonella enteritidis* phage type 4 and other phage types isolated from poultry in the United States. Avian Dis. 40 (4), 853–857.

Gerlach, R.G., Cláudio, N., Rohde, M., Jäckel, D., Wagner, C., Hensel, M., 2008. Cooperation of *Salmonella* pathogenicity islands 1 and 4 is required to breach epithelial barriers. Cell Microbiol. 10 (11), 2364.

Guard-Petter, J., Keller, L.H., Rahman, M.M., Carlson, R.W., Silvers, S., 1996. A novel relationship between O-antigen variation, matrix formation, and invasiveness of *Salmonella enteritidis*. Epidemiol. Infect. 117 (2), 219–231.

Gulig, P.A., 1990. Virulence plasmids of *Salmonella typhimurium* and other salmonellae. Microb. Pathog. 8 (1), 3–11.

Harvey, P.C., Watson, M., Hulme, S., Jones, M.A., Lovell, M., Berchieri Jr., A., Young, J., Bumstead, N., Barrow, P., 2011. *Salmonella enterica* serovar Typhimurium colonizing the lumen of the chicken intestine grows slowly and upregulates a unique set of virulence and metabolism genes. Infect. Immun. 79, 4105–4121.

Henderson, S.C., Bounous, D.I., Lee, M.D., 1999. Early events in the pathogenesis of avian salmonellosis. Infect. Immun. 67 (7), 3580–3586.

Hinton, M., Threlfall, E.J., Rowe, B., 1990. The invasiveness of different strains of *Salmonella enteritidis* phage type 4 for young chickens. FEMS Microbiol. Lett. 58 (2), 193–195.

Hu, J., Bumstead, N., Barrow, P., Sebastiani, G., Olien, L., Morgan, K., Malo, D., 1997. Resistance to salmonellosis in the chicken is linked to NRAMP1 and TNC. Genome Res. 7 (7), 693–704.

Humphrey, T.J., Williams, A., McAlpine, K., Lever, M.S., Guard-Petter, J., Cox, J.M., 1996. Isolates of *Salmonella enterica* Enteritidis PT4 with enhanced heat and acid tolerance are more virulent in mice and more invasive in chickens. Epidemiol. Infect. 117 (1), 79–88.

Ibarra, J.A., Steele-Mortimer, O., 2009. Salmonella–the ultimate insider. *Salmonella* virulence factors that modulate intracellular survival. Cell Microbiol. 11 (11), 1579–1586.

Imre, A., Bukovinszki, A., Lovell, M.A., Li, H., Zhou, X., Barrow, P.A., 2013. Gene expression analysis of *Salmonella enterica* SPI in macrophages indicates differences between serovars that induce systemic disease from those normally causing enteritis. Vet. Microbiol. 167 (3–4), 675–679.

Jones, M.A., Wigley, P., Page, K.L., Hulme, S.D., Barrow, P.A., 2001. *Salmonella enterica* serovar Gallinarum requires the *Salmonella* pathogenicity island 2 type III secretion system but not the *Salmonella* pathogenicity island 1 type III secretion system for virulence in the chicken. Infect. Immun. 69, 5471–5476.

Kaiser, P., Rothwell, L., Galyov, E.E., Barrow, P.A., Burnside, J., Wigley, P., 2000. Differential cytokine expression in avian cells in response to invasion by *Salmonella typhimurium, Salmonella enteritidis* and *Salmonella gallinarum*. Microbiology 146, 3217–3226.

Langridge, G.C., Fookes, M., Connor, T.R., Feltwell, T., Feasey, N., Parsons, B.N., Seth-Smith, H.M.B., Barquista, L., Stedman, A., Humphrey, T., Wigley, P., Peters, S.E., Maskell, D.J., Corander, J., Chabalgoity, J.A., Barrow, P., Parkhill, J., Dougan, G., Thomson, N.R., 2015. Patterns of genome evolution that have accompanied host adaptation in *Salmonella.* PNAS 112, 863–868.

Libby, S.J., Lesnick, M., Hasegawa, P., Weidenhammer, E., Guiney, D.G., 2000. The *Salmonella* virulence plasmid spv genes are required for cytopathology in human monocyte-derived macrophages. Cell Microbiol. 2 (1), 49–58.

Libby, S.J., Lesnick, M., Hasegawa, P., Kurth, M., Belcher, C., Fierer, J., Guiney, D.G., 2002. Characterization of the *spv* locus in *Salmonella enterica* serovar Arizona. Infect Immun. 70 (6), 3290–3294.

Linde, K., Beer, J., Bondarenko, V., 1990. Stable *Salmonella* live vaccine strains with two or more attenuating mutations and any desired level of attenuation. Vaccine 8 (3), 278–282.

Liu, W., Yang, Y., Chung, N., Kwang, J., 2001. Induction of humoral immune response and protective immunity in chickens against *Salmonella enteritidis* after a single dose of killed bacterium-loaded microspheres. Avian Dis. 45 (4), 797–806.

McCapes, R.H., Coffland, R.T., Christie, L.E., 1967. Challenge of turkey poults originating from hens vaccinated with *Salmonella typhimurium* bacterin. Avian Dis. 11 (1), 15–24.

Mead, G.C., 1989. Hygiene problems and control of process contamination. In: Mead, G.C. (Ed.), Processing of Poultry. Elsevier, London, pp. 183–220.

Mead, G.C., 2000. Prospects for 'competitive exclusion' treatment to control salmonellas and other food-borne pathogens in poultry. Vet. J. 159, 111–123.

Meyer, H., Koch, H., Methner, U., Steinbach, G., 1993. Vaccines in salmonellosis control in animals. Zentralbl Bakteriol 278 (2–3), 407–415.

Morgan, E., Campbell, J.D., Rowe, S.C., Bispham, J., Stevens, M.P., Bowen, A., Barrow, P.A., Maskell, D.J., Wallis, T.S., 2004. Identification of host-specific colonization factors of *Salmonella enterica* serovar Typhimurium. Mol. Microbiol. 54, 994–1010.

Noble, W.C., Virani, Z., Cree, R.G., 1992. Co-transfer of vancomycin and other resistance genes from *Enterococcus faecalis* NCTC 12201 to *Staphylococcus aureus.* FEMS Microbiol. Lett. 72 (2), 195–198.

Nurmi, E., Rantala, M., 1973. New aspects of *Salmonella* infection in broiler production. Nature 241 (5386), 210–211.

O'Neill report, 2014. Antimicrobial Resistance: Tackling a Crisis for the Health and Wealth of Nations the Review on Antimicrobial Resistance Chaired by Jim O'Neill.

Ochman, H., Groisman, E.A., 1996. Distribution of pathogenicity islands in *Salmonella* spp. Infect Immun. 64 (12), 5410–5412.

Ochman, H., Soncini, F.C., Solomon, F., Groisman, E.A., 1996. Identification of a pathogenicity island required for *Salmonella* survival in host cells. Proc. Natl. Acad. Sci. U.S.A. 93 (15), 7800–7804.

Parker, C., Asokan, K., Guard-Petter, J., 2001. Egg contamination by *Salmonella* serovar *enteritidis* following vaccination with Delta-aroA *Salmonella* serovar *typhimurium*. FEMS Microbiol. Lett. 195 (1), 73–78.

Parsons, B.N., Crayford, G., Humphrey, T.J., Wigley, P., 2013. Infection of chickens with antimicrobial-resistant *Salmonella enterica* Typhimurium DT193 and monophasic *Salmonella* Typhimurium-like variants: an emerging risk to the poultry industry? Avian Pathol. 42 (5), 443–446.

Piddock, L.J., Wray, C., McClaren, I., Wise, R., 1990. Quinolone resistance in *Salmonella* spp: veterinary pointers. Lancet 336 (8707), 125.

Poppe, C., Demczuk, W., McFadden, K., Johnson, R.P., 1993. Virulence of *Salmonella enteritidis* phage types 4, 8 and 13 and other *Salmonella* spp. for day-old chicks, hens and mice. Can J. Vet. Res. 57 (4), 281–287.

Poppe, C., McFadden, K.A., Brouwer, A.M., Demczuk, W., 1993. Characterization of *Salmonella enteritidis* strains. Can J. Vet. Res. 57 (3), 176–184.

Pritchard, D.G., Nivas, S.C., York, M.D., Pomeroy, B.S., 1978. Effects of Gal-E-mutant of *Salmonella typhimurium* on experimental salmonellosis in chickens. Avian Dis. 22 (4), 562–575.

Report, 1969. The Joint Committee on the Use of Antibiotics in Animal Husbandry and Veterinary Medicine. Cmnd 4190.

Santos, R.L., Bäumler, A.J., 2004. Cell tropism of *Salmonella enterica*. Int. J. Med. Microbiol. 294 (4), 225–233.

Schmitt, C.K., Ikeda, J.S., Darnell, S.C., Watson, P.R., Bispham, J., Wallis, T.S., Weinstein, D.L., Metcalf, E.S., O'Brien, A.D., 2001. Absence of all components of the flagellar export and synthesis machinery differentially alters virulence of *Salmonella enterica* serovar Typhimurium in models of typhoid fever, survival in macrophages, tissue culture invasiveness, and calf enterocolitis. Infect. Immun. 69 (9), 5619–5625.

Schwarz, S., Liebisch, B., 1994. Use of ribotyping, IS200 typing and plasmid analysis for the identification of *Salmonella enterica* subsp. enterica serovar Typhimurium vaccine strain Zoosaloral H and its differentiation from wild type strains of the same serovar. Zentralbl Bakteriol 281 (4), 442–450.

Shah, D.H., Lee, M.J., Park, J.H., Lee, J.H., Eo, S.K., Kwon, J.T., Chae, J.S., 2005. Identification of *Salmonella gallinarum* virulence genes in a chicken infection model using PCR-based signature-tagged mutagenesis. Microbiology 151 (Pt 12), 3957–3968.

Smith, H.W., Tucker, J.F., 1975. The effect of antibiotic therapy on the faecal excretion of *Salmonella typhimurium* by experimentally infected chickens. J. Hyg. 75 (2), 275–292.

Smith, H.W., Tucker, J.F., 1978. The effect of antimicrobial feed additives on the colonization of the alimentary tract of chickens by *Salmonella typhimurium*. J. Hyg. 80 (2), 217–231.

Smith, H.W., Barrow, P.A., Tucker, J.F., 1985. The effect of oral antibiotic administration on the excretion of Salmonellas by chickens. In: Proceedings of the International Symposium on Salmonella, New Orleans, Louisiana, USA. July 19–20 pp. 88–90.

Smith, H.W., 1956. The use of live vaccines in experimental *Salmonella* gallinarum infection in chickens with observations on their interference effect. J. Hyg. 54 (3), 419–432 (No abstract available).

Thomson, N.R., Clayton, D.J., Windhorst, D., Davidson, S., Vernikos, G., Churcher, C., Quail, M.A., Stevens, M., Jones Angela Lord, M.A., Woodward, J., Arrowsmith, C., Norbertczak, H., Rabbinowitch, E., Barrow, P.A., Watson, M., Maskell, D., Humphrey, T., Roberts, M., Parkhill, J., Dougan, G., 2008. Comparative genome analysis of *Salmonella enteritidis* PT4 and *Salmonella gallinarum* 287/91 provides insights into host adaptation in zoonotic bacterial pathogens. Genome Res. 18, 1624–1637.

Tilquin, P., Barrow, P.A., Marly, J., Pitel, F., Plisson-Petit, F., Velge, P., Vignal, A., Baret, P.V., Bumstead, N., Beaumont, C., 2005. A genome scan for quantitative trait loci affecting the *Salmonella* carrier-state in the chicken. Genet. Sel. Evol. 37 (5), 539–561.

Timms, L.M., Marshall, R.N., Breslin, M.F., 1994. Laboratory and field trial assessment of protection given by a *Salmonella* enteritidis PT4 inactivated, adjuvant vaccine. Br. Vet. J. 150 (1), 93–102.

Timoney, J.F., Shivaprasad, H.L., Baker, R.C., Rowe, B., 1989. Egg transmission after infection of hens with *Salmonella enteritidis* phage type 4. Vet. Rec. 125 (24), 600–601.

Truscott, R.B., Friars, G.W., 1972. The transfer of endotoxin induced immunity from hens to poults. Can. J. Comp. Med. 36 (1), 64–68.

Turner, A.K., Lovell, M.A., Hulme, S.D., Zhang-Barber, L., Barrow, P.A., 1998. Identification of *Salmonella typhimurium* genes required for colonization of the alimentary tract and for virulence in newly hatched chickens. Infect. Immun. 66, 2099–2106.

Uchiya, K., Barbieri, M.A., Funato, K., Shah, A.H., Stahl, P.D., Groisman, E.A.A., 1999. Salmonella virulence protein that inhibits cellular trafficking. EMBO J. 18 (14), 3924–3933.

Van Immerseel, F., Methner, U., Rychlik, I., Nagy, B., Velge, P., Martin, G., Foster, N., Ducatelle, R., Barrow, P.A., 2005. Vaccination and early protection against non host-specific *Salmonella* serotypes in poultry; exploitation of innate immunity and microbial activity. Epidemiol. Infect. 133, 959–978.

Vazquez-Torres, A., Xu, Y., Jones-Carson, J., Holden, D.W., Lucia, S.M., Dinauer, M.C., Mastroeni, P., Fang, F.C., 2000. Salmonella pathogenicity island 2-dependent evasion of the phagocyte NADPH oxidase. Science 287 (5458), 1655–1658.

Vielitz, E., Conrad, C., Voss, M., Löhren, U., Bachmeier, J., Hahn, I., 1992. Immunizing against salmonella infections with live and inactivated vaccines. Dtsch Tierarztl Wochenschr 99 (12), 483–485 (Article in German).

Wegener, H.C., Aarestrup, F.M., Jensen, L.B., Hammerum, A.M., Bager, F., 1999. Use of antimicrobial growth promoters in food animals and *Enterococcus faecium* resistance to therapeutic antimicrobial drugs in Europe. Emerg. Infect Dis. 5 (3), 329–335 Review. Erratum in: Emerg Infect Dis, 1999; 5 (6), 844.

Wigley, P., Berchieri Jnr, A., Page, K.L., Smith, A.L., Barrow, P.A., 2001. *Salmonella enterica* serovar Pullorum persists in splenic macrophages and in the reproductive tract during persistent, disease-free carriage in chickens. Infect. Immun. 69, 7873–7879.

Wigley, P., Hulme, S.D., Powers, C., Beal, R.K., Berchieri Jr., A., Smith, A., Barrow, P., 2005. Infection of the reproductive tract and eggs by *Salmonella enterica* serovar Pullorum in the chicken is associated with suppression of cellular immunity at sexual maturity. Infect. Immun. 73, 2986–2990.

Winter, S.E., Thiennimitr, P., Winter, M.G., Butler, B.P., Huseby, D.L., Crawford, R.W., Russell, J.M., Bevins, C.L., Adams, L.G., Tsolis, R.M., Roth, J.R., Bäumler, A.J., 2010. Gut inflammation provides a respiratory electron acceptor for *Salmonella*. Nature 467 (7314), 426–429.

Withanage, G.S., Sasai, K., Fukata, T., Miyamoto, T., Baba, E., Lillehoj, H.S., 1998. T lymphocytes, B lymphocytes, and macrophages in the ovaries and oviducts of laying hens experimentally infected with *Salmonella enteritidis*. Vet. Immunol. Immunopathol. 66 (2), 173–184.

Woodward, M.J., Gettinby, G., Breslin, M.F., Corkish, J.D., Houghton, S., 2002. The efficacy of Salenvac, a *Salmonella enterica* subsp. Enterica serotype Enteritidis iron-restricted bacterin vaccine, in laying chickens. Avian Pathol. 31 (4), 383–392.

World Health Organisation, 1980. Report of the WHO/WAVFH Round Table Conference on the Present Status of the *Salmonella* Problem (Prevention and Control). WHO/VPH/81.27. (Geneva, WHO, Unpublished).

World Health Organisation, 1990. Report of WHO Consultation on Salmonellosis Control in Agriculture. WHO/CDS/VPH/90.94. (Geneva, WHO, Unpublished).

Wray, C., McLaren, I., Wise, R., Piddock, L.J., 1990. Piddock LJ.Nalidixic acid-resistant salmonellae. Vet. Rec. 126 (19), 489.

Zhang-Barber, L., Turner, A.K., Barrow, P.A., 1999. Vaccination for control of *Salmonella* in poultry. Vaccine 17 (20–21), 2538–2545.

CHAPTER

# Developments in Detection Strategies for *Salmonella* Enteritidis in Layer Hen Flocks

9

**Doug Waltman**
*Georgia Poultry Laboratory Network, Gainesville, GA, United States*

## 1. INTRODUCTION

An article was published in 1988 that forever changed the poultry industry. St. Louis et al. (1988) reported an increase in food-borne Salmonellosis in the northeastern United States caused by *Salmonella* Enteritidis (SE) and they implicated grade A shell eggs as the source. This revelation prompted sweeping changes for veterinarians and live production workers, resulting in a dual focus on avian and public health.

The impact of this finding can best be appreciated by looking back to the pre-SE poultry and research community. Before 1988, the industry had successfully dealt with three *Salmonella* challenges. First, pullorum disease (PD) and fowl typhoid (FT) were successfully controlled. Second, egg-associated salmonellosis was resolved with the Egg Products Inspection Act of 1970, which removed cracked and lower quality eggs from the table egg market. And then, the problem of *Salmonella* in feed was reduced by restricting some feed sources (e.g., fish meals) and using better surveillance. The resulting lull in *Salmonella* problems in poultry meant that relatively few laboratories in the United States were actively working with *Salmonella* and most of the work was with processing plants and feed.

The new challenge was SE in eggs. After the initial denial and finger-pointing phase it was obvious that something had to be done. Similar food-borne disease problems were being found and dealt with in Europe and other parts of the world (Barrow, 1992), but in some ways the European problem was different. The particular phage type (PT) of SE was PT4, and it was causing mortality in chickens and was associated with meat-type birds as well. Scientists and others were asking what was different about SE. Why was it contaminating eggs?

A great deal of research was initiated to provide answers related to the epidemiology, infectivity, pathogenesis, detection, and laboratory methodology. Numerous questions needed answers, and perhaps the most critical one was how to detect infected flocks that are laying contaminated eggs so they will not enter the food chain? What do we sample? How do we monitor? How do we test?

Producing Safe Eggs. http://dx.doi.org/10.1016/B978-0-12-802582-6.00009-4

## 2. GENERAL DETECTION STRATEGIES

Not only did the "poultry industry" change, but also its philosophy, purpose, and process of testing. Media and methods had to be developed or modified to meet the new challenges of detecting SE.

### 2.1 SEROLOGICAL TESTING

One of the first detection strategies investigated was serological testing. PD and FT had been detected and brought under control by identifying positive flocks using a simple agglutination test (NPIP, 1989). Both of those *Salmonella* serotypes belong to serogroup D1, which is the same serogroup as SE. When SE infects birds, it should elicit an immune response similar to *S.* Pullorum (SP) and *S.* Gallinarum (SG). Therefore the same agglutination test and antigen that detected antibody responses to SP and SG should detect antibody responses to SE.

The Pullorum-Typhoid test requirement was incorporated into the National Poultry Improvement Plan (NPIP) US Sanitation Monitored classification for multiplier breeders, not to detect SP or SG, but to detect SE (NPIP, 1989), and it is still part of the US *S.* Enteritidis Monitored classification (NPIP, 2014a,b). The primary breeding flocks were already being tested for PD and FT. If any birds were positive, they were submitted (up to 25 birds) to an authorized laboratory and cultured.

The Pullorum-Typhoid agglutination tests, although successful for detecting PD and FT, suffer from sensitivity and specificity problems for detecting SE and other paratyphoids (Weinack et al., 1979; Cooper et al., 1989; Timony et al., 1990; Poppe et al., 1992; Mutalib et al., 1992). Researchers needed to develop a test that would reliably detect infected flocks while maintaining a high level of specificity. The enzyme-linked immunosorbent assay (ELISA) is more sensitive and identifies more infected individual birds than the agglutination tests (Barrow, 1992). What showed great promise under experimental conditions did not do as well under field conditions. The ELISA is, however, being used under specific circumstances with some flocks in some countries.

Several review articles have discussed using serology, especially ELISA, for detecting *Salmonella* or SE-infected flocks (Barrow, 1992, 1994, 2000; Barrow et al., 1996) as well as the development of ELISA using lipopolysaccharide, fimbrial antigens, flagellar antigens, and other cell associated antigens (Chart et al., 1990; Thorns et al., 1990; Kim et al., 1991; Nicholas and Cullen, 1991; van der Heijden and van der Wijngaard, 1992; Wray et al., 1992; Van Zijderveld et al., 1992, 1993; McDonough et al., 1998).

According to the World Organization for Animal Health (OIE), microbiological testing is preferred to serological testing because of its higher sensitivity in broiler flocks and higher specificity in breeder and layer flocks (OIE, 2015).

### 2.2 BIRD TESTING

Another strategy of detecting infected flocks involved testing birds. Birds were submitted for necropsy and cultured for *Salmonella*. Early on, some wanted to differentiate between internal organ positive and intestinal or environmental positive. The lungs and intestines had to be cultured separately (if cultured at all) from the internal organs and reproductive tract, because the former were considered "environmental" isolations. A bird or flock was only considered infected if SE was isolated from the internal organs (Proposed Voluntary Model State Program, Salmonella Committee of the Northeastern Conference on Avian Diseases, August 25, 1988, September 1, 1988, March 1, 1989). Over time the logic of this distinction was not borne out, and today SE isolation from any site in the bird is considered to be SE positive.

SE, at least in the United States, was unlike the other invasive serotypes (SP and SG) in that there was little mortality or morbidity as commonly found with other paratyphoid infections in mature poultry (Gast, 2013). Mortality, morbidity, or production parameters cannot predict if a flock is positive.

The major problem with the strategy of bird testing is the number of birds that have to be tested. Typically the prevalence of SE in a flock is very low and may vary from one flock to another. Not all birds will be positive at one time or ever. Certainly, some may be clearing the infection, whereas others may be acquiring it. In addition, testing birds is laborious and economically not feasible. Studies show that even in infected flocks or flocks living in environmentally positive houses, only a small percentage of birds are positive, and many times it is the intestinal tract and not the organs or reproductive tract that is positive (Poppe et al., 1992; Kinde et al., 1996; Faddoul and Fellows, 1996).

Bird testing, although not recommended as a screening test, is still used as a confirmatory test. Within the NPIP, culturing birds for SE follows the detection of SE in the environment of the house. For egg-type breeders, 60 birds are randomly submitted and cultured for *Salmonella*. If SE is found in the birds, the flock either loses its classification or is depopulated (NPIP, 2014a,b).

Some older *Salmonella* testing programs utilized cloacal swabs as the method of testing birds (Ellis et al., 1976). The practice has largely been discarded as a screening test because cloacal swabs lack sensitivity (Hinton, 1988; Mutalib et al., 1992), fecal shedding is sporadic and *Salmonella* may be shed in low numbers (Carrique-Mas and Davies, 2008), typically low prevalence in the flock, and is labor intensive.

Isolation of SE from individual bird samples or pooling of samples is likely to show low sensitivity compared with indirect environmental monitoring (Van der Giessen et al., 1991; Mutalib et al., 1992; Poppe et al., 1992; Skov et al., 1999; Carrique-Mas and Davies, 2008).

## 2.3 EGG TESTING

Ultimately, the definitive sample would be the egg itself; after all, that is the source of the SE and the human infections. Studies show the number of contaminated eggs is very low even from infected flocks (Humphrey et al., 1989, 1991; Kinde et al., 1996; Schlosser et al., 1999; Humbert, 2000). Two US Department of Agriculture (USDA) Risk Assessments (1998, 2005) reported that industry wide about 1 in 20,000 eggs may be contaminated with SE. In addition, of the SE-positive eggs, the vast majority contain very low numbers of SE (Humphrey et al., 1989, 1991). Studies showed that contaminated eggs were laid intermittently and would be shed for only a limited amount of time. For these reasons, egg testing is not a feasible screening method for detecting SE-positive flocks or positive eggs. However, egg testing may serve as a confirmatory test after SE is detected in commercial layer environmental samples. The US Food and Drug Administration (FDA) Egg Safety Rule (FDA Egg Safety Rule, 2009) established guidelines for sampling and testing layer flocks for SE and the consequences of detection.

Interestingly, there is a linear relationship between the rate of contamination of egg contents and the prevalence of infected chickens (Arnold et al., 2014). They also found a much higher level of egg shell surface contamination than internal contents contamination. This emphasized that some of the contamination may come from the feces present on the shell surface. The presence of *Salmonella* on the surface may result in shell penetration into the interior or serve to directly contaminate individuals, especially outside the United States where eggs are not routinely washed.

### 2.4 ENVIRONMENTAL TESTING

Finally the screening test of choice to detect SE-positive flocks was based on the environmental sample. The environmental test provides indirect evidence that the birds are infected or colonized because samples are tested from the hatchery or house environment and not the birds directly. The rationale was that even though SE was invasive similar to SP/SG and may infect the internal organs and reproductive tract, it also acts as a paratyphoid *Salmonella* in colonizing the intestinal tract, resulting in shedding into the environment. Studies have shown that, although SE may be isolated from organs or reproductive tissues, it may commonly be isolated from the intestine (Poppe et al., 1992; Kinde et al., 1996; Faddoul and Fellows, 1996). Therefore if SE is detected in the environment of the bird it either got there from the bird or its presence in the environment could serve to infect or colonize the bird. An early industry concern was that finding SE in the environment may not be evidence of infection in birds, but may be from feed, rodents, insects, or other sources. It is quite possible that SE's presence in the house is from other sources; however, once it is in the house it can infect the birds.

Environmental sampling has proved to be relatively easy to perform, and provides a good screening test with high sensitivity (Poppe et al., 1992; Davies and Breslin, 2001; Kinde et al., 2005; Gast, 2007; Arnold et al., 2010).

## 3. SAMPLE TYPES AND SAMPLE COLLECTION

Because SE may infect the reproductive system and be deposited in eggs, it presumably could be vertically transmitted to offspring thus spreading the infection. Therefore initially the breeder flocks were targeted for monitoring and control. In the United States the NPIP became the agency directly involved in testing breeding flocks for SE (NPIP, 2014a,b).

## 3.1 BREEDER HOUSES

Typically, birds are raised in a separate pullet facility and are moved into breeder houses before the onset of sexual maturity and lay, which means the birds experience the stresses of moving and coming into sexual maturity and egg production. Stress increases the susceptibility of birds to *Salmonella* infection. In addition, the birds will remain in the house for 40 or more weeks. The extended life span provides adequate time for *Salmonella* to be introduced into the house environment from any of a number of sources. Therefore an ongoing program of testing, which begins in the pullet phase and continues for the life of the flock, is necessary to detect the introduction or presence of *Salmonella* in the house.

For years, the house environmental sample was the pooled litter sample. The effectiveness of the litter sample depended on the collector and the number of areas of the house that were sampled. The greater the number of areas collected, the better the test results. Ellis et al. (1976) recommended that the number of pooled samples collected be based on the size of the house or the number of birds in the house. Olesiuk et al. (1969) had reported that infected flocks were detected more consistently by testing floor litter than by serologic tests or culture of cloacal swabs, nest litter, drinking water, eggs, or embryos.

Several studies (Opara et al., 1992; Juven et al., 1994; Mallinson et al., 2000b; Hayes et al., 2000) showed a correlation between the available water ($A_w$) in the floor litter and the presence or level of *Salmonella*. Mallinson et al. (2000b) suggested a potential intervention strategy is to maintain the house environment as dry as possible. From a sampling perspective, areas with higher $A_w$ values may be better sample sites.

*Salmonella* is not uniformly distributed in the house litter (Riemann et al., 1998; Hayes et al., 2000; Rolfe et al., 2000), which is one reason the drag swab (DS) is more effective than the litter sample (Kingston, 1980; Mallinson et al., 1989; Hayes et al., 2000; de Rezende et al., 2001). The DS sample increases the area of the house floor that is sampled and is easier to collect and work within the laboratory (Mallinson et al., 1989; NPIP, 2014a,b).

Studies have shown a correlation between the number of DS sample sets tested and the detection of *Salmonella*-positive houses (Caldwell et al., 1995; Mahe et al., 2008; Waltman, unpublished data). To detect positive houses with low levels of *Salmonella* a sufficient number and sampling sites of samples need to be collected.

The disadvantage of the DS is its small size, resulting in a small surface area for sample collection and the possibility of getting it stuck in or fall through the slats in the house if present. The boot swab (BS) has evolved from a sponge or piece of gauze attached to a boot, to a piece of fabric tied to the boot, to a hair net covering the boot, to a sock-like material placed over the boot, to its current status of an absorbent material that is placed over a boot or boot cover and held in place by an elastic band. The current BS stays in place, is absorbent, has a much larger surface area than the DS, and allows the collector to walk through the house and do other duties while collecting the sample.

Studies have shown the BS is as good as or better than the DS (Caldwell et al., 1998; McCrea et al., 2005). Skov et al. (1999) found the recovery of *Salmonella* from five pairs of socks and 300 fecal samples (60 pools) were comparable. Gradel et al. (2002) showed that five pairs of socks were better than 60 fecal samples. Two pairs of socks were not as effective as five pairs, but was better than 60 fecal samples. Arnold et al. (2009) compared a pooled pair of BS, two dust samples, and 60 fecal samples. They recommended the use of a pooled pair of BS and two dust samples. Buhr et al. (2007) compared four sampling methods for detecting *Salmonella* and found recovery rates of 59% from socks, 36% from DS, 31% from litter, and 22% from feces.

Carrique-Mas and Davies (2008) recommended that *Salmonella* monitoring programs include sampling for feces/litter and dust. They found that it was commonly easier to isolate *Salmonella* from dust than from feces. They suggested that it may be due to the relative advantage *Salmonella* has in that matrix compared with other bacteria. Ellis et al. (1976) also reported that in some areas dust was more frequently contaminated with *Salmonella* than floor or nest litter. Waltman (unpublished data) found more *Salmonella*-positive houses when both DS and dust were tested.

Other samples types may be of use perhaps not for routine monitoring but in an intensive sampling situation or after cleaning and disinfection. For example, Davies and Wray (1996) found swabs from nest box floors and slave feed hoppers gave a higher isolation rate than litter, dust, and drinker samples.

## 3.2 HATCHERY

Because SE can be vertically spread from parent to offspring, the hatchery becomes a site for testing. Sampling the hatchery not only may serve to identify SE or other *Salmonella*-infected flocks, but also may be used to determine whether the hatchery is contaminated and the effectiveness of the hatchery cleaning and disinfection procedure.

Common hatchery samples include fluff, hatch residue, dead-in-shell embryos, chicks, and chick papers. Fluff and hatch residue are easy to collect and represent a large number of birds. If only one flock was hatched in a particular hatcher, they are good indicators of the *Salmonella* status for the breeder flock and the chicks hatched. However, they are somewhat difficult to work with in the laboratory, because both have high levels of background flora including *Proteus* spp. and *Pseudomonas* spp., fluff tends to go airborne possibly becoming a source of laboratory contamination, and egg shells will cut through plastic bags. Dead-in-shell embryos and chicks are not as representative of the flock status because of the limited number of birds/embryos tested. They also require more labor in sample preparation. The commonly recommended procedure for flock testing is sampling the chick papers. Each paper represents 100 chicks, and the recommended procedure in the United States is to culture 10 papers or the number that represents 10% of the flock (NPIP, 2014a,b). In the European Union (EU), the recommendation is to test one chick box liner for every 500 chicks (European Commission No. 1168/2006).

## 3.3 LAYER HOUSE

Unlike breeder house, which all have a very similar design, layer houses have a variety of sizes and styles. Depending on the type of house, the particular samples that are collected may vary. The challenge of sampling layer houses has increased with the move toward cage-free, free range, or pastured flocks.

In the United States there has been an ongoing debate over whether the manure-type samples or the egg machinery samples are better. The SEPP Progress Report (1995) found 18% of walkways, 17% of egg machinery swabs, 15% of manure DS, 14% of manure scraper swabs, and 12% of fans were positive for SE (essentially no difference in the recovery rates from the different samples). In a subsequent national survey, National Animal Health Monitoring System Layer '99 (USDA, 2000), the reported percentages of SE by sample type were: egg belt (48%), elevator (45%), manure (20%), and walkway (18%). Clearly, the egg machinery samples were more effective. This study also found that 60% of the SE-positive houses only had one SE-positive sample, and no houses had over two SE-positive samples (typically 17 samples were collected from each house).

The USDA conducted another industry survey in 2013 (USDA, 2014) asking what samples were being tested (not the results) on the layer farms. Data from 2008 to 2013, respectively, were: manure (98% and 97%), egg belts (17% and 13%), elevator equipment (11% and 10%), and nests (3% and 7%). The results appear to reflect the position of FDA for manure-based samples.

The European Commission (EC, 2004) baseline study of the prevalence of *Salmonella* in laying flocks required the collection of five naturally pooled samples of feces/litter from each layer house. For cage houses, each fecal sample consisted of 200–300 g from 20 to 40 locations. In floor housing systems, five pairs of BS were collected. Gradel et al. (2002) tested houses using two and five pairs of socks and 60 fecal samples. The two pairs of socks (one sample) were equal to the 60 feces (one pooled sample); however, five pairs of socks (five samples) gave the best recovery.

Studies have concluded that the collection of naturally pooled fecal material is preferable to individual droppings (Davies and Breslin, 2001; Carrique-Mas and Davies, 2008; Arnold et al., 2011). Waltman (unpublished data) extensively sampled 34 layer houses for the isolation of SE and other *Salmonella* and found the following percent positive: manure-type samples (83%), egg machinery samples (55%), walkway DS (59%), dust swabs (44%), and dust (62%). Kinde et al. (2005) environmentally sampled two layer flocks and found litter was more commonly positive than egg belts. Fan dust swabs were also found to be a good sample source.

## 3.4 SAMPLING PROGRAMS

### 3.4.1 NPIP (NPIP, 2014a,b)

The environmental sampling recommendations of the NPIP *Salmonella*-related classifications for breeder flocks include chick paper samples, DS or BS samples, and dust or egg belt samples. Layer pullet houses are tested initially between 2 and 4 weeks of age and every 30 days. After moving to the breeder house, the testing continues at 30-day intervals for the life of the flock. If SE is isolated from the house environment, 60 birds are submitted for necropsy and culture.

### 3.4.2 FDA Egg Safety Rule (FDA, 2009)

The FDA Egg Safety Rule provides guidelines for testing commercial layer flocks. Manure-based samples are the preferred sample type, if possible, and the DS is the preferred sampling device. Flocks are tested at 14–16 weeks, 40–45 weeks, and 4–6 weeks after molt. Generally, two DS are collected for each row/bank. For trace-back testing, manure is tested, plus egg belts and deescalators, fans and walkways.

Because of the variations in housing types and production practices, FDA has published two guidance documents to provide some guidance for designing a sampling procedure (FDA, 2011, 2013).

### 3.4.3 EU Baseline Prevalence Survey

In 2004/2005 all EU member states were required to carry out a standardized survey to establish the baseline prevalence of *Salmonella* in commercial laying flocks. For caged houses, five fecal samples were collected each comprising 200–300 g and two dust samples comprising 50 g each. For noncaged houses, five pairs of BS and two dust samples were collected.

### 3.4.4 National Control Programs

The National Control Programs guidelines require a *Salmonella* test during the rearing stage, between 22 and 26 weeks of age, and then every 15 weeks during lay. In addition, an official sampling by the competent authority is required on one flock per holding per year. In caged houses, two 150-g naturally pooled fecal samples are collected. In noncage houses, the operator collects a pair of BS. For the official test, the competent authority collects two pairs of BS and 100 g of dust.

# 4. CULTURE METHODS

Once it was determined that the environmental sample was the source for the culture, it became clear that the existing media and methodology were not sufficiently sensitive or specific. Methods for the isolation of *Salmonella* were initially developed for clinical specimens in humans and subsequently animals. Later when food became the apparent source of the infection, microbiologists naturally applied methods that were effective in the clinical area to the food area (Galton et al., 1968; Fagerberg and

Avens, 1976). Those methods were not as effective, and then as they were applied to farm or environmental samples, they were severely lacking. Each aspect of the culture process—primary and secondary enrichment, incubation temperature and time, and plating media—had to be optimized.

## 4.1 ENRICHMENT

### 4.1.1 Pre-enrichment

Pre-enrichment (PE), or the use of a nonselective enrichment to resuscitate damaged or "sublethally injured" *Salmonella*, was initially applied to feed and food samples because the processing that occurs with those samples may have resulted in damaged cells. Studies have shown that PE with Buffered Peptone Water (BPW) could be used in conjunction with selective enrichments for feces and environmental samples (Edel and Kampelmacher, 1973; Vassiliadis, 1983; Van Schothrust and Renaud, 1983).

### 4.1.2 Selective Enrichment

Galton et al. (1968) found that it was difficult to isolate *Salmonella* without selective enrichment if the ratio of coliforms to *Salmonella* was as low as 10:1. The background flora in most fecal and environmental samples is extremely high. Therefore selective enrichment is necessary to reduce the level of non-*Salmonella* bacteria while increasing the level of *Salmonella*.

#### 4.1.2.1 Selenite Enrichment

Various formulations of selenite enrichment have been used for a hundred years (Guth, 1916; Leifson, 1936; North and Bartram, 1953; Stokes and Osborne, 1955). A study by Waltman et al. (1995) compared three selenite formulations with three tetrathionate (TT) enrichments and found the recovery rate from the selenite enrichments was substantially less. Other studies have also found selenite enrichment was not as sensitive as TT or Rappaport-Vassiliadis (RV) (Rall et al., 2005; Schonenbrucher et al., 2008).

In addition to the lower recovery, selenite enrichments are reduced to the toxic heavy metal selenium. The Environmental Protection Agency considers selenite a toxin and carcinogen. It is also classified as a hazardous material and must be disposed of under restricted conditions.

#### 4.1.2.2 Tetrathionate Enrichment

TT enrichment was formulated by Mueller and Kauffman (Mueller, 1923; Kauffman, 1930, 1935) and modified by Hajna and Damon (1956) and Jeffries (added novobiocin, 1959). TT enrichment may be used as either a direct selective enrichment or as a secondary enrichment following PE.

Waltman et al. (1995) compared direct enrichment in TT broth base, TT with brilliant green (TBG), and TT Hajna (TTH) incubated at 37°C and 42°C. There was essentially no difference between any of the formulations or the temperatures of incubation, except the TTH incubated at 42°C, which had about a 5% lower recovery rate. They also compared PE in BPW followed by enrichment in TTH, RV, and modified semisolid Rappaport-Vassiliadis (MSRV) media and found them to be comparable or slightly better than the direct TT enrichments. The best recovery was following delayed secondary enrichment (DSE) procedure, which was about 15–20% more sensitive.

Rybolt et al. (2005) found both TT and RV alone were not very sensitive; however, in combination, their collective sensitivity was much better. Some isolation protocols require both selective enrichments be used in combination following PE. Poppe et al. (1992) compared PE followed by selective enrichment in TBG and MSRV. They found that TBG was best for isolating *Salmonella* from tissue samples, but MSRV was better with environmental samples.

#### 4.1.2.3 Rappaport-Vassiliadis Enrichment

RV enrichment was formulated by Rappaport et al. (1956) and Vassiliadis et al. (1970, 1976). The medium has the advantage of being autoclavable and may be stored in the refrigerator. The enrichment was designed to follow PE and be inoculated at a ratio of 1:100. The initial work was done with fecal samples, but has been extended to all sample types. Van der Zee et al. (1990) reported that RV was inhibitory to SE PT4, which may result in the underestimation of the prevalence of SE in samples.

#### 4.1.2.4 Modified Semisolid Rappaport-Vassiliadis Enrichment

Goossens et al. (1984) developed a semisolid media based on the Rappaport enrichment, and DeSmedt et al. (1986) and DeSmedt and Bolderkijk (1987) formulated the current commercial MSRV. Part of the selective ability of MSRV is the semisolid matrix that allows for the selective migration of motile *Salmonella*. Therefore MSRV is not recommended for the isolation of nonmotile *Salmonella* (i.e., SP and SG). Since some paratyphoid *Salmonella* may be nonmotile, it is a good practice to plate MSRV enrichments after incubating 48 h even if there is no migration.

Waltman (unpublished data) compared the isolation of *Salmonella* from naturally contaminated environmental samples by direct TT enrichment, direct TT enrichment followed by DSE, and direct TT followed by MSRV. The TT/MSRV method was more sensitive for *Salmonella* in general and SE in particular (Fig. 9.1). A follow-up study compared additional enrichment combinations (Fig. 9.2). The best recovery was TT/MSRV followed by BPW/MSRV. The BPW/TTH and BPW/RV methods, which are common in many isolation protocols, were not sensitive for *Salmonella*, especially SE.

### 4.2 INCUBATION TEMPERATURE AND TIME

Conventional protocols incubate PE media at 35°C to 37°C. Direct TT enrichment or TT enrichment following PE is incubated at 37°C or 41–43°C, depending on the sample and procedure. RV and MSRV are typically incubated at 41–42°C. Generally, bird tissues or organs are incubated at 37°C, and intestines, intestinal contents, or environmental samples are incubated at 41–43°C. Higher temperatures are usually used with samples with higher bacterial background.

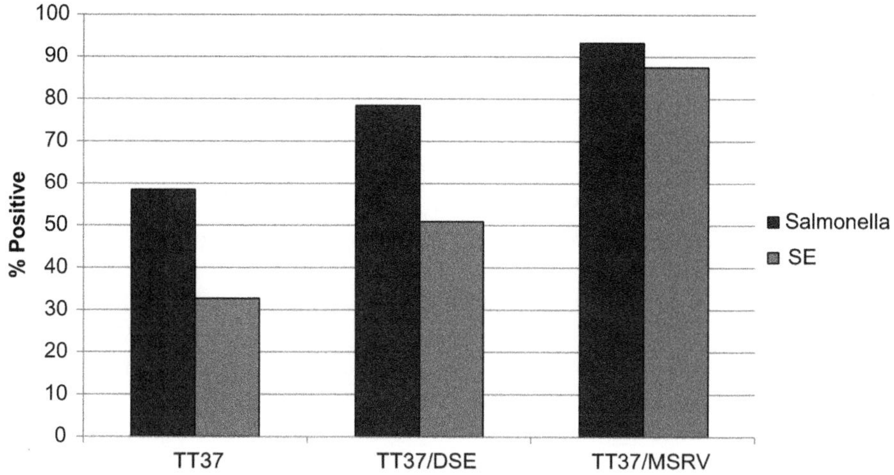


**FIGURE 9.1**

Isolation of *Salmonella* from 2548 naturally contaminated environmental samples using different enrichment procedures. There were 1132 samples positive for *Salmonella* and 336 positive for *Salmonella* Enteritidis. *DSE*, delayed secondary enrichment; *MSRV*, modified semisolid Rappaport-Vassiliadis; *TT37*, tetrathionate enrichment incubated at 37°C

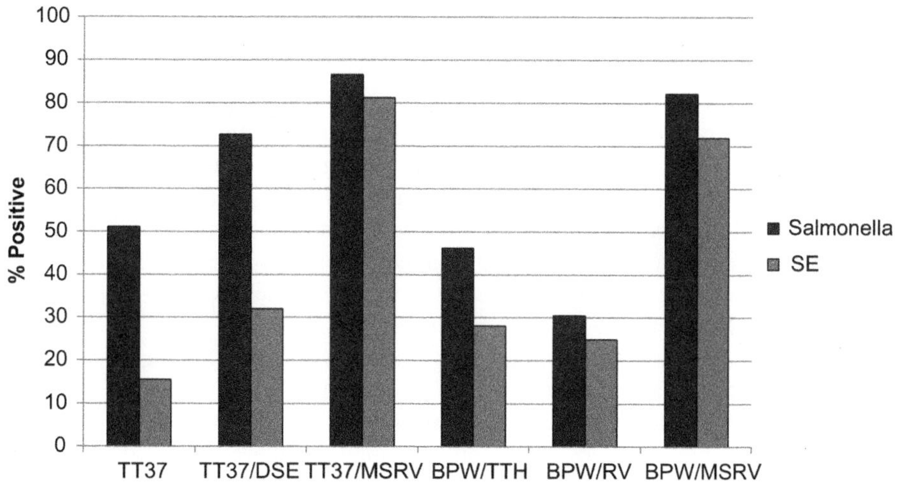


**FIGURE 9.2**

Isolation of *Salmonella* from 867 naturally contaminated environmental samples using different enrichment methods. There were 407 samples positive for *Salmonella* and 128 positive for *Salmonella* Enteritidis. *BPW*, Buffered Peptone Water; *DSE*, delayed secondary enrichment; *MSRV*, modified semisolid Rappaport-Vassiliadis; *RV*, Rappaport-Vassiliadis; *TT37*, tetrathionate enrichment incubated at 37°C; *TTH*, tetrathionate Hajna.

Some *Salmonella* serotypes (e.g., SP and SG) do not tolerate higher temperatures and should be incubated at 37°C. Also it is important to monitor the inside temperature of the incubator, especially with higher incubation temperatures, to make sure there are no spikes in temperature, which would prove lethal to *Salmonella*.

Typically PE media is incubated for 18–24 h, and TT and RV enrichments are incubated for 20–24 h. Some protocols may recommend an additional 24-h incubation if there are no *Salmonella* suspect colonies on the plating media. MSRV is incubated for 20–24 h, and if there is no zone of migration, reincubated another 24 h.

In the United States, NPIP recommended the use of DSE for increasing the isolation of *Salmonella* from clinical and environmental samples until 2010. NPIP continues to recommend DSE for clinical samples, but direct TT enrichment followed by MSRV is now recommended for environmental samples (NPIP, 2014a,b). DSE is the process in which the sample is incubated in TT for 24 h and plated (Pourciau and Springer, 1978; Rigby and Pettit, 1980). The enriched sample is left at room temperature for 5–7 days. If the initial plating was negative for *Salmonella*, 1 mL of the sample is inoculated into a tube of TT, incubated, and plated again. DSE had been shown to be the most sensitive isolation method until the recent MSRV modification (Waltman et al., 1991, 1993, 1995).

### 4.3 PLATING MEDIA

Selective enrichment increases the level of *Salmonella* relative to other bacteria; however, rarely does the enrichment eliminate all other bacteria. Therefore the enrichment is inoculated onto plating media that has two important characteristics. First, the plating media are selective, inhibitory dyes or chemicals, antibiotics, or other conditions further inhibit background flora in favor of *Salmonella*. Second, because other bacteria are still commonly present, the plating media must have some differential characteristic, such as fermentation of sugar, $H_2S$ production, or chromogenic response. Typically, two plating media are used in isolation protocols; each should have different selective and differential characteristics. It is essential that the plating media be able to detect not only typical *Salmonella* colonies, but also atypical ones, such as $H_2S$ negative, lysine negative, and lactose positive.

In the late 1980s, the available plating media (Bismuth Sulfite, MacConkey; Deoxycholate citrate agar; *Salmonella Shigella*; Mannitol lysine crystal violet bile agar; Hektoen Enteric, HE; and Brilliant Green agar, BGA) were being used with various degrees of effectiveness. When these plating media were applied to feces and environmental samples, the sensitivity and especially the specificity were lacking. Most of these media suffered from the presence of *Proteus* spp., which looks like *Salmonella*.

Studies showed that the addition of novobiocin to plating media (e.g., BGA, HE, Modified Lysine Iron Agar, Xylose Lysine Desoxycholate (XLD)) would inhibit *Proteus*, resulting in increased sensitivity and specificity (Hoben et al., 1973; Restaino et al., 1977, 1982; Tate et al., 1990, 1992; Miller et al., 1991).

Miller et al. (1991) developed a new plating media specifically formulated for environmental samples. The media, Xylose lysine tergitol 4 (XLT4), is a modification of XLD agar, but the desoxycholate was replaced by Tergitol 4 (Niaproof 4). Several studies have shown it to be highly sensitive and specific (Miller et al., 1991; Tate et al., 1992; Dusch and Altwegg, 1995; Mallinson et al., 2000a). Mallinson et al. (2000a) modified XLT4 to increase its sensitivity and specificity especially for detecting weak $H_2S$-producing *Salmonella*.

Miller-Mallinson agar was formulated for enhanced detection of weakly $H_2S$-positive *Salmonella* (Mallinson et al., 2000a; Miller and Mallinson, 2000). They incorporated Niaproof 4 as the selective agent and XGAL (5-bromo-4-chloro-3-indolyl-beta-D-galactopyranoside) for its chromogenic characteristics.

Waltman et al. (1995) compared the recovery rates and specificity of several plating media from naturally contaminated environmental samples (Fig. 9.3). XLT4 was the best plating media followed by media that contained novobiocin.

The first of many chromogenic agar media was Rambach (RAM) agar (Rambach, 1990). It was based on the fermentation of propylene glycol and a chromogenic detection of beta-galactosidase. Carrique-Mas et al. (2009) found RAM agar was more sensitive than Brilliant Green Agar supplemented with novobiocin (BGN) and XLD; however, it was important to screen the pale orange colonies on RAM. Davies and Wray (1996) used RAM as the sole plating media from MSRV enrichment.

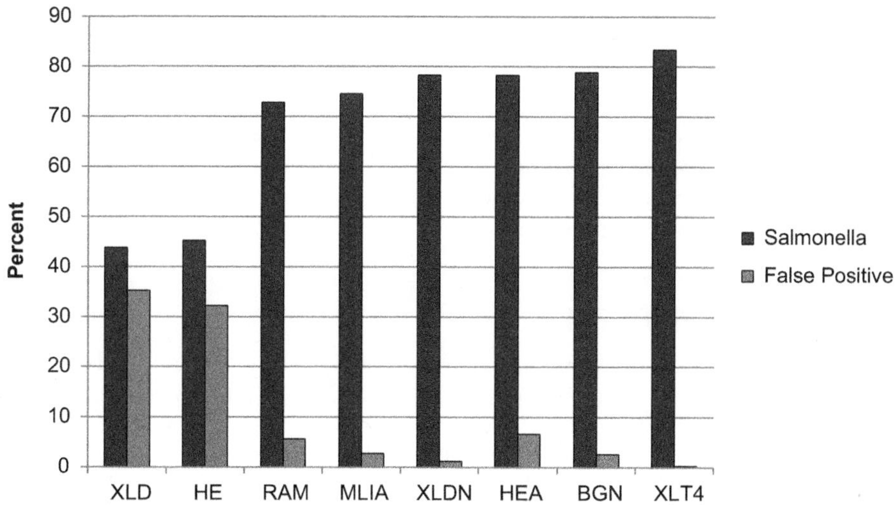


**FIGURE 9.3**

Recovery rates and percentage false-positive rates for various plating media following selective enrichment of naturally contaminated environmental samples. *BGN*, brilliant green agar supplemented with novobiocin; *HE*, hektoen enteric; *HEA*, hektoen enteric agar; *MLIA*, modified lysine iron agar; *RAM*, rambach; *XLD*, xylose lysine desoxycholate agar; *XLDN*, xylose lysine desoxycholate agar supplemented with novobiocin; *XLT4*, xylose lysine tergitol 4.

Many other chromogenic plating media are currently available that typically have good sensitivity and specificity; however, the cost is typically much greater than conventional media.

## 4.4 SCREENING SUSPECT COLONIES

Because the selective and differential characteristics of the plating media are not always definitive for *Salmonella*, the resulting colonies must be confirmed as *Salmonella*. That process typically begins by screening with triple sugar iron (TSI) and lysine iron agar (LIA) slants. Three to five colonies that display morphologies consistent with *Salmonella* are inoculated individually into tubes of TSI and LIA. Based on the resulting reactions in the tubes, the respective isolates are discarded as not *Salmonella* or processed further to biochemically and serologically confirm. When using TSI and LIA, it is imperative that if either tube gives a reaction consistent with *Salmonella* the isolate is processed further.

The enrichment procedures that use MSRV as the secondary enrichment and the more selective plating media have dramatically reduced the number of false-positive colonies on the plates. It was not uncommon in the past to have false-positive rates greater than 30% from some plating media (Dunn and Martin, 1971; Taylor and Shelhart, 1971; Komatsu and Restaino, 1981; Waltman et al., 1995). With the newer methodologies, false-positive rates are routinely less that 5%. This reduction in the number of false-positive colonies has resulted in greater sensitivity and the increased potential to detect multiple serotypes. It is important when picking *Salmonella* suspect colonies to pick various colony morphologies consistent with *Salmonella* to increase the possibility of detecting multiple serotypes. As we are targeting specific serotypes (e.g., SE and ST), they may be present in mixed populations of *Salmonella* and must be detectable.

Because of the reduced number of false-positive colonies some laboratories have instituted different procedures for screening colonies. Some laboratories go directly to biochemical or serological methods.

## 4.5 ISOLATION PROTOCOLS

Depending on the country or region, there are several isolation protocols or programs for the detection of *Salmonella*, including ISO-6579:202 (ISO, 2002a), ISO-6579:2002 Annex D (ISO, 2002b), NMKL 71 (NMKL, 1999), FDA environmental and egg testing procedures (FDA, 2007, 2008), and the NPIP environmental testing procedures (NPIP, 2014a,b).

All of the procedures, except the NPIP TT/MSRV procedure, pre-enrich with BPW. The ISO06579:2002, FDA and NPIP PE procedures use both TT and RV as secondary enrichments. The NMKL 71 procedure uses only RV. The ISO and NMKL 71 procedures specify plating only XLD and either BGN or a second plate of the laboratory's choice. The FDA environmental procedure specifies BGN and XLT4 plates, and the NPIP procedures recommend BGN and XLT4, but they are not required.

## 4.6 CONFIRMATION

### 4.6.1 Biochemical Identification

Suspect isolates should be confirmed biochemically because some non-*Salmonella* may react with serogrouping antisera. Biochemical identification may be done using commercially available automated or manual systems or utilizing tube media (Jones et al., 2000). The biochemical characteristics of *Salmonella* have been detailed by Ewing (1986).

### 4.6.2 Serological Typing

The serological typing of *Salmonella* depends on the detection of both somatic O-antigens and flagellar H-antigens. Classical serotyping utilizes specific antisera that detect the representative antigens. Over the last few years additional molecular methods of serotyping have become available that can completely or partially serotype *Salmonella* based on the Kauffman-White Scheme (Grimont and Weill, 2007).

Although most laboratories serogroup *Salmonella*, they may take different approaches to serotyping. Some laboratories may not serotype, but submit all isolates to a reference laboratory. Others may serotype specific serotypes of major concern such as SE or ST, but send the others to a reference laboratory. A few laboratories may fully serotype and only use a reference laboratory for confirmation. To assign a serotype name to an isolate it must be fully serotyped.

# 5. RAPID DETECTION METHODS

Over the last 20 years, there have been numerous rapid methods that have been introduced and have met with varying degrees of success. Most of these were developed, at least initially, for the clinical or food areas. Typically, for those methods to be effective with environmental samples the test had to be modified in some manner to increase the sensitivity.

## 5.1 ANTISERA BASED

Some of the first tests were immunological tests that used antibodies to detect the presence of *Salmonella* in the samples. These assays range from tube assays, to microtiter-based antigen capture assays, to lateral flow assays. They typically are easy to use and provide fairly good results.

Van der Zee and Huis in't Veld (2000) reviewed the antisera-based assays, including ELISA and enzyme-linked immunosorbent fluorescent assays.

The lateral-flow assays are simple strips impregnated with *Salmonella*-specific antibodies. They may be just strips or may be housed in plastic. The strip is inoculated from the enrichment, or the strip is dipped into the enrichment. A reaction line in the strip denotes a positive test.

Romer's RapidChek *Salmonella* test and Neogen's Reveal SE test have been approved for use by the NPIP (NPIP, 2014a,b). FDA has granted equivalency to Romer's RapidChek SE test and Neogen's Reveal assay (FDA, 2009). Moongkarndi et al. (2011) developed and tested a lateral flow assay that would simultaneously detect SE and ST on one strip.

## 5.2 DNA BASED

Advancements in technology, for example, the polymerase chain reaction (PCR) have resulted in greater sensitivity and specificity of testing. To some extent the usefulness of PCR was slowed by the need to train technicians in the newer technology and the cost of doing the test. As confidence in the technology increased and the cost of the tests decreased, more laboratories have moved from culture to PCR, at least for screening. The Layers 2013 survey in the United States found that only 34% of farms were using PCR or other rapid tests for their environmental testing (USDA, 2014). Currently there is almost a "designer" PCR atmosphere. You can test for all *Salmonella* or for specific serotypes, or multiplex several in one test.

### 5.2.1 *PCR: Generic* Salmonella

There are several commercially available PCR systems for detecting *Salmonella* in general. Studies have compared PCR systems with other PCR systems or with culture (Liu et al., 2002; Kusar et al., 2010; Summer et al., 2012; Soria et al., 2012; Schultz et al., 2012; Tebbs et al., 2012).

### 5.2.2 *PCR: Serotype Specific*

Depending on the program or the specific need of a company, PCR tests can be developed for specific serogroups or serotypes of *Salmonella*. Seo et al. (2004) developed a group D–specific reverse transcription (RT)-PCR for pooled eggs. They used a segment of the gene *sef*A specific to group D *Salmonella*. The assay was sensitive and specific, and was approved by NPIP for environmental samples (NPIP, 2014a,b).

Charlton et al. (2005) developed an SE-specific PCR for use in DS and chick papers. The PCR was approved for use by NPIP for environmental samples (NPIP, 2014a,b).

Lungu et al. (2012) compared the results of an SE-specific RT-PCR to culture using the NPIP methods. The PCR was more sensitive than culture. Adams et al. (2013) tested an SE-specific RT-PCR with pooled DS.

### 5.2.3 *PCR: Multiplex PCR*

If more than one serotype of *Salmonella* is of interest, a multiplex PCR system may be developed that can detect more than one serotype in a single test. McCarthy et al. (2009) developed a multiplex PCR for ST and SH for use in food samples. The assay worked very well with pure cultures and spiked samples. Park and Ricke (2015) developed a multiplex PCR for SE, SH, and ST. Hong et al. (2009) developed an allelotyping PCR to detect SE, SK, and ST from poultry DS.

### 5.2.4 *Isothermal Amplification*

In the last few years, several companies have introduced a variation of the PCR technology that still amplifies the DNA targets, but it does so at one temperature and does not require the temperature cycling of PCR. For that reason the equipment is not as large or expensive, and the test is run in less time.

Ueda and Kuwabara (2009) developed a loop-mediated isothermal amplification (LAMP) assay that has a specificity similar to PCR, but the sensitivity was considered greater. Okamura et al. (2008) developed a LAMP assay to amplify fragments of O9 *Salmonella*-specific insertion element. They reported it was more sensitive and faster than PCR. The NPIP (NPIP, 2014a,b) has approved the commercial LAMP assays produced by Neogen Corporation (ANSR *Salmonella* test) and EnviroLogix (DNAble Salmonella Detection kit).

## 6. CONCLUSION

The detection strategies for SE in layer hen flocks has changed over time as we have learned more about the organism and its epidemiology. Currently the environmental sample is the means for monitoring for the presence of SE in flocks. Confirmation by testing birds or eggs may be used depending on the particular program. Culture methods have been improved or developed to optimize the isolation of SE from these environmental samples and may continue to be improved. The shift toward molecular methods will speed up the detection of SE, but still relies on enrichment of the sample before testing and on subsequent culture if positive.

In the future, as the industry continues to change from the classic caged houses to cage-free, pasture, or free range, the sampling methodology will need to be reevaluated. The molecular methods will continue to be developed and improved as will the experience and the confidence in their results. The development of serotype-specific assays will allow the detection of not only *Salmonella* in general, but specific serotypes, without the necessity of isolation.

## REFERENCES

Adams, D.R., Stensland, W.R., Wang, C.H., O'Conner, A.M., Trampel, D.W., Harmon, K.M., Strait, E.L., Frana, T.S., 2013. Detection of Salmonella enteritidis in pooled poultry environmental samples using a serotype-specific real-time polymerase chain reaction assay. Avian Dis. 57, 22–28.

Arnold, M.E., Mueller-Doblies, D., Carrique-Mas, J.J., Davies, R.H., 2009. The estimation of pooled-sample sensitivity for detection of *Salmonella* in turkey flocks. J. Appl. Microbiol. 107, 936–943.

Arnold, M.E., Carrique-Mas, J.J., Davies, R.H., 2010. Sensitivity of environmental sampling methods for detecting *Salmonella* Enteritidis in commercial laying flocks relative to the within-flock prevalence. Epidemiol. Infect. 138, 330–339.

Arnold, M.E., Carrique-Mas, J.J., McLaren, I., Davies, R.H., 2011. A comparison of pooled and individual bird sampling for detection of *Salmonella* in commercial egg laying flocks. Prev. Vet. Med. 99, 176–184.

Arnold, M.E., Martelli, F., McLaren, I., Davies, R.H., 2014. Estimation of the rate of egg contamination from *Salmonella*-infected chickens. Zoonoses Public Health 61, 18–27.

Barrow, P.A., Desmidt, M., Ducatelle, R., Guittet, M., van der Heijden, H.M.J.F., Holt, P.S., Huis in't Velt, J.H.J., McDonough, P., Nagaraja, K.V., Porter, R.E., Proux, K., Sisak, F., Staak, C., Steinbach, G., Thorns, C.J., Wray, C., van Zijderveld, F., 1996. World Health Organization – supervised interlaboratory comparison of ELISAs for the serological detection of *Salmonella enterica* serotype Enteritidis in chickens. Epidemiol. Infect. 117, 69–77.

Barrow, P.A., 1992. ELISAs and the serological analysis of *Salmonella* infections in poultry: a review. Epidemiol. Infect. 109, 361–369.

Barrow, P.A., 1994. Serological diagnosis of *Salmonella* serotype Enteritidis infections in poultry by ELISA and other tests. Int. J. Food Microbiol. 21, 55–68.

Barrow, P.A., 2000. Serological diagnosis of *Salmonella* by ELISA and other tests. In: Wray, C., Wray, A. (Eds.), Salmonella in Domestic Animals. CABI Publishing, New York, pp. 407–427.

Buhr, R.J., Richardson, L.J., Cason, J.A., Cox, N.A., Fairchild, B.D., 2007. Comparison of four sampling methods for the detection of *Salmonella* in broiler litter. Poult. Sci. 86, 21–25.

Caldwell, D.J., Hargis, B.M., Corrier, D.E., Vidal, L., DeLoach, J.R., 1995. Evaluation of persistence and distribution of Salmonella serotype isolation from poultry farms using drag-swab sampling. Avian Dis. 39, 617–621.

Caldwell, D.J., Hargis, B.M., Corrier, D.E., DeLoach, J.R., 1998. Frequency of isolation of Salmonella from protective foot covers worn in broiler houses as compared to drag-swab sampling. Avian Dis. 42, 381–384.

Carrique-Mas, J.J., Davies, R.H., 2008. Sampling and bacteriological detection of *Salmonella* in poultry and poultry premises: a review. Rev. Sci. Tech. 27, 665–677.

Carrique-Mas, J.J., Barnes, S., McLaren, I., Davies, R., 2009. Comparison of three plating media for the isolation of *Salmonella* from poultry environmental samples in Great Britain using ISO 6579:202 (Annex D). J. Appl. Microbiol. 107, 1976–1983.

Charlton, B.R., Walker, R.L., Kinde, B.H., Bauer, C.R., Channing-Santiago, S.E., Farver, T.B., 2005. Comparison of a *Salmonella enteritidis*-specific polymerase chain reaction assay to delayed secondary enrichment culture for the detection of *Salmonella enteritidis* in environmental drag swab samples. Avian Dis. 49, 418–422.

Chart, H., Rowe, B., Baskerville, A., Humphrey, T.J., 1990. Serological analysis of chicken flocks for antibodies to *Salmonella* Enteritidis. Vet. Rec. 127, 501–502.

Cooper, G.L., Nicholas, R.A., Bracewell, C.D., 1989. Serological and bacteriological investigations of chickens from flocks naturally infected with *Salmonella enteritidis*. Vet. Rec. 125, 567–572.

Davies, R., Breslin, M., 2001. Environmental contamination and detection of *Salmonella enterica* serovar Enteritidis in laying flocks. Vet. Rec. 149, 699–704.

Davies, R.H., Wray, C., 1996. Determination of an effective sampling regime to detect *Salmonella enteritidis* in the environment of poultry units. Vet. Microbiol. 50, 117–127.

De Rezende, E.C.L., Mallinson, E.T., Tablante, N.L., Morales, R., Park, A., Carr, L.E., Joseph, S.W., 2001. Effect of dry litter and airflow in reducing *Salmonella* and *Escherichia coli* populations in the broiler production environment. J. Appl. Poult. Res. 10, 245–251.

De Smedt, J.M., Bolderdijk, R.F., 1987. Dynamics of *Salmonella* isolation with modified semi-solid Rappaport-Vassiliadis medium. J. Food Protect. 50, 658–661.

De Smedt, J.M., Bolderdijk, R.F., Rappald, H., Lautenschlaeger, D., 1986. Rapid *Salmonella* detection in foods by motility enrichment on a modified semi-solid Rappaport-Vassiliadis medium. J. Food Protect. 49, 510–514.

Dunn, C., Martin, W.J., 1971. Comparison of media for isolation of salmonellae and shigellae from fecal specimens. Appl. Microbiol. 22, 17–22.

Dusch, H., Altwegg, M., 1995. Evaluation of five new plating media for isolation of *Salmonella* species. J. Clin. Microbiol. 33, 802–804.

Edel, W., Kampelmacher, E.H., 1973. Comparative studies on the isolations of "sub-lethally injured" salmonellae in nine European laboratories. Bull. WHO 48, 167–174.

Ellis, E.M., Williams, J.E., Mallinson, E.T., Snoeyenbos, G.H., Martin, W.J., 1976. Culture Methods for the Detection of Animal Salmonellosis and Arizonosis. The Iowa State University Press, Ames, Iowa.

European Commission, 2004. Baseline Study on the Prevalence of *Salmonella* in Laying Flocks of *Gallus gallus* in the EU: Technical Specifications. SANCO/34/2004 Rev 3.

Ewing, W.H., 1986. Edwards and Ewing's Identification of Enterobacteriaceae, fourth ed. Elsevier Science, New York.

Faddoul, G.P., Fellows, G.W., 1996. A five-year survey of the incidence of salmonellae in avian species. Avian Dis. 10, 296–304.

Fagerberg, D.J., Avens, J.S., 1976. Enrichment and plating methodology for *Salmonella* detection in food. A review. J. Milk Food Technol. 39, 628–646.

Food and Drug Administration Egg Safety Rule, July 9, 2009. Prevention of *Salmonella enteritidis* in shell eggs during production, storage, and transportation. FDA 21 CFR parts 16 and 18. Fed. Regist. 74 (130). https://www.gpo.gov/fdsys/pkg/FR-2009-07-09/pdf/E9-16119.pdf.

Food and Drug Administration, 2007. Bacteriological Analytical Manual (BAM) Chapter 5, Salmonella. May 2014 version http://www.fda.gov/Food/GuidanceRegulation/Guidance DocumentsRegulatoryInformation/Eggs/ucm228796.htm.

Food and Drug Administration, October 2008. Environmental Sampling and Detection of *Salmonella* in Poultry Houses. http://www.fda.gov/Food/FoodScienceResearch/LaboratoryMethods/ucm114716.htm.

Food and Drug Administration, December 2011. Guidance for Industry: Prevention of *Salmonella* Enteritidis in Shell Eggs During Production, Storage, and Transportation. www.fda.gov/Food/GuidanceRegulation/GuidanceDocumentsRegulatoryInformation/ucm285101.htm.

Food and Drug Administration, July 2013. Draft Guidance for Industry: Questions and Answers Regarding the Final Rule, Prevention of *Salmonella* Enteritidis in Shell Eggs During Production, Storage, and Transportation (Layers with Outdoor Access). www.fda.gov/Food/GuidanceRegulation/GuidanceDocumentsRegulatoryInformation/ucm360028.htm.

Galton, M.M., Morris, G.K., Martin, W.T., 1968. Salmonellae in Foods and Feeds. Review of Isolation Methods and Recommended Procedures. U.S. Department of Health, Education, and Welfare/Public Health Service. Communicable Disease Center, Atlanta, GA.

Gast, R.K., 2007. Serotype-specific and serotype-independent strategies for preharvest control of food-borne *Salmonella* in poultry. Avian Dis. 51, 817–818.

Gast, R.K., 2013. Paratyphoid infections. In: Swayne, D.E., Glisson, J.R., McDougald, L.R., Nolan, L.K., Suarez, D.L., Venugopal, N. (Eds.), Diseases of Poultry, thirteen ed. John Wiley & Sons, Ames, Iowa, pp. 693–706.

Goossens, H., Wauters, G., de Boeck, M., Janssens, M., Butzler, J., 1984. Semisolid selective-motility enrichment medium for isolation of salmonellae from fecal specimens. J. Clin. Microbiol. 19, 940–941.

Gradel, K.O., Andersen, J., Madsen, M., 2002. Comparisons of sampling procedures and time of sampling for the detection of *Salmonella* in Danish infected chicken flocks raised in floor systems. Acta Vet. Scand. 43, 21–30.

Grimont, P.A.D., Weill, F.-X., 2007. Antigenic Formulae of the *Salmonella* Serovars, ninth ed. WHO Collaborating Centre for Reference and Research on Salmonella.

Guth, F., 1916. Selennahrboden fur die elective Zuchtung von Typhusbacillen. Zentr. Bakteriol. 1 Abt. 77, 487–496. https://www.pasteur.fr/ip/portal/action/WebdriveActionEvent/oid/01s-000036-089.

Hajna, A.A., Damon, S.R., 1956. New enrichment and plating media for the isolation of *Salmonella* and *Shigella* organisms. Appl. Microbiol. 4, 341–345.

Hayes, J.R., Carr, L.E., Mallinson, E.T., Douglas, L.W., Joseph, S.W., 2000. Characterization of the contribution of water activity and moisture content to the population distribution of *Salmonella* spp. in commercial poultry houses. Poult. Sci. 79, 1557–1561.

Hinton, M., 1988. *Salmonella* infection in chicks following the consumption of artificially contaminated feed. Epidem. Inf. 100, 247–256.

Hoben, D.A., Aston, D.H., Peterson, A.C., 1973. Some observations on the incorporation of novobiocin into Hektoen enteric agar for improved *Salmonella* isolation. Appl. Microbiol. 26, 126–127.

Hong, Y., Liu, T., Lee, M.D., Hofacre, C.L., Maier, M., White, D.G., Ayers, S., Wang, L., Berghaus, R., Maurer, J., 2009. A rapid screen of broth enrichments for *Salmonella enterica* serovars Enteritidis, Hadar, Heidelberg, and Typhimurium by using an allelotyping multiplex PCR that targets O- and H-antigen alleles. J. Food Protect. 72, 2198–2201.

Humbert, F., 2000. In: Raes, M., Henken, A.M. (Eds.), Estimation of the Percentage of Contaminated Eggs Laid by a Naturally Contaminated *Salmonella* Enteritidis Flock RIVM Report December 2000.

Humphrey, T.J., Baskerville, A., Mawer, S.L., Rowe, B., Hopper, S., 1989. *Salmonella enteritidis* phage type 4 from the contents of intact eggs: a study involving naturally infected hens. Epidemiol. Infect. 103, 415–423.

Humphrey, T.J., Whitehead, A., Gawler, A.H.L., Henley, A., Rowe, B., 1991. Numbers of *Salmonella enteritidis* in the contents of naturally contaminated hen's eggs. Epidemiol. Infect. 106, 489–496.

International Organization for Standardization (ISO), 2002a. ISO 6579:2002. Microbiology of Food and Animal Feeding Stuffs. Horizontal Method for the Detection of *Salmonella* Spp. International Organization for Standardization, Geneva, Switzerland.

International Organization for Standardization (ISO), 2002b. ISO 6579:2002/FDAmd 1. Annex D. Detection of *Salmonella* Spp. In Animal Faeces and in Samples of the Primary Production Stage. International Organization for Standardization, Geneva, Switzerland.

Jeffries, L., 1959. Novobiocin – tetrathionate broth: a medium of improved selectivity for the isolation of salmonellae in faeces. J. Clin. Pathol. 12, 568–571.

Jones, Y.E., McLaren, I.M., Wray, C., 2000. Laboratory aspects of *Salmonella*. In: Wray, C., Wray, A. (Eds.), Salmonella in Domestic Animals. CABI Publishing, New York, pp. 393–405.

Juven, B.J., Cox, N.A., Bailey, J.S., Thompson, J.E., Charles, O.W., Shutze, J.V., 1984. Recovery of *Salmonella* from artificially contaminated poultry feeds in non-selective and selective broth media. J. Food Protect. 47, 299–302.

Kauffman, F., 1930. Die technic der Typhenbestimmung in der Typhus-paratyphus-gruppe. Zentralb. Bakteriol. Parasitenk. Infektionskr Hyg. Abt Orig 119, 152–160.

Kauffman, F., 1935. Weitereerfahrungen mit der kombinierten anreicherungsverfahren fur *Salmonella*-bacillen. Z. Hyg. Infektionskr 117, 26–32.

Kim, C.J., Nagaraja, K.V., Pomeroy, B.S., 1991. Enzyme-linked immunosorbent assay for the detection of *Salmonella* Enteritidis infections in chickens. Amer. J. Vet. Res. 52, 1069–1074.

Kinde, H., Read, D.H., Chin, R.P., Bickford, A.A., Walker, R.L., Ardans, A., Breitmeyer, R.E., Willoughby, D., Little, H.E., Kerr, D., Gardner, I.A., 1996. *Salmonella enteritidis* phage type 4 infection in a commercial layer flock in southern California: bacteriologic and epidemiologic findings. Avian Dis. 40, 665–671.

Kinde, H., Castellan, D.M., Kerr, D., Campbell, J., Breitmeyer, R., Ardans, A., 2005. Longitudinal monitoring of two commercial layer flocks and their environments for *Salmonella enterica* serovar Enteritidis and other salmonellae. Avian Dis. 49, 189–194.

Kingston, D.J., 1980. A comparison of culturing drag swabs and litter for identification of infections with *Salmonella* spp. in commercial chicken flocks. Avian Dis. 25, 513–516.

Komatsu, K.K., Restaino, L., 1981. Determination of the effectiveness of novobiocin added to two agar plating media for the isolation of *Salmonella* from fresh meat products. J. Food Saf. 3, 183–192.

Kusar, D., Pate, M., Micunovic, J., Bole-Hribovsek, V., Ocepek, M., 2010. Detection of *Salmonella* in poultry faeces by molecular means in comparison to traditional bacteriological methods. Slov. Vet. Res. 47, 45–56.

Leifson, E., 1936. New selenite enrichment media for the isolation of typhoid and paratyphoid (*Salmonella*) bacilli. Am. J. Hyg. 24, 423–432.

Liu, T., Liljebjelke, K., Bartlett, E., Hofacre, C., Sanchez, S., Maurer, J.J., 2002. Application of nested polymerase chain reaction to detection of *Salmonella* in poultry environment. J. Food Protect. 65, 1227–1232.

Lungu, B., Waltman, W.D., Berghaus, R.D., Hofacre, C.L., 2012. Comparison of a real-time PCR method with a culture method for the detection of *Salmonella enterica* serotype Enteritidis in naturally contaminated environmental samples for integrated poultry houses. J. Food Protect. 75, 743–747.

Mahe, A., Bougeard, S., Huneau-Salaun, A., Le Bouquin, S., Petetin, I., Rouxel, S., Lalande, F., Beloeil, P.A., 2008. Bayesian estimation of flock-level sensitivity of detection of *Salmonella* spp. Enteritidis and Typhimurium according to the sampling procedure in French laying-hen houses. Prev. Vet. Med. 84, 11–26.

Mallinson, E.T., Tate, C.R., Miller, R.G., Bennett, B., Russek-Cohen, E., 1989. Monitoring poultry farms for *Salmonella* by drag-swab sampling and antigen capture immunoassay. Avian Dis. 33, 684–690.

Mallinson, E.T., Miller, R.G., De Rezende, S.E., Ferris, K.E., DeGraft-Hanson, J., Joseph, S.W., 2000a. Improved plating media for the detection of *Salmonella* species with typical and atypical hydrogen sulfide production. J. Vet. Diagn. Invest. 12, 83–87.

Mallinson, E.T., DeRezende, C.E., Tablante, N.L., 2000b. A management technique to identify prime locations of *Salmonella* contamination on broiler and layer farms. J. Appl. Poult. Res. 9, 364–370.

McCarthy, N., Reen, F.J., Buckley, J.F., Frye, F.G., Boyd, E.F., Gilroy, D., 2009. Sensitive and rapid molecular detection assays for *Salmonella enterica* serovars Typhimurium and Heidelberg. J. Food Protect. 72, 2350–2357.

McCrea, B.A., Norton, R.A., Macklin, K., Hess, J.B., Bilgili, S., 2005. Recovery and genetic similarity of *Salmonella* from broiler house drag swabs versus surgical shoe covers. J. Appl. Poult. Res. 14, 694–699.

McDonough, P.L., Jacobson, R.H., Timoney, J.F., Mutalib, A., Kradel, D.C., Chang, Y.F., Shin, S.J., Lein, D.H., Trock, S., Wheeler, K., 1998. Interpretations of antibody responses to *Salmonella enterica* serotype Enteritidis gm flagellin in poultry flocks are enhanced by a kinetics-based enzyme-linked immunoassay. Clin. Diagn. Lab. Immunol. 5, 550–555.

Miller, R.G., Mallinson, E.T., 2000. Improved detection of nontyphoid and typhoid salmonellae with balanced agar formulations. J. Food Protect. 63, 1443–1446.

Miller, R.G., Tate, C.R., Mallinson, E.T., Scherrer, J.A., 1991. Xylose-lysine-tergitol 4: an improved selective agar medium for the isolation of *Salmonella*. Poult. Sci. 70, 2429–2432.

Moongkarndi, P., Rodpai, E., Kanarat, S., 2011. Evaluation of an immunochromatographic assay for rapid detection of *Salmonella enterica* serovars Typhimurium and Enteritidis. J. Vet. Diagn. Invest. 23, 797–801.

Mueller, L., 1923. Un nouveau milieu d'enrichissement pour la recherché du bacilli typhique et des paratyphiques. Comptes Rendus des Seances de la Societe de Biologie et de ses Filiales 89, 434–437.

Mutalib, A., McDonough, P., Shin, S., Patten, V., Lein, D., 1992. *Salmonella* Enteritidis in commercial layer farms in New York State; environmental survey results and significance of available monitoring tests. J. Vet. Diagn. Invest. 4, 416–418.

APHIS 91-40. National Poultry Improvement Plan (NPIP) and Auxiliary Provisions, August 1989.

National Poultry Improvement Plan (NPIP) Program Standards., August 2014a. USDA, APHIS, VS. http://www.poultryimprovement.org/documents/ProgramStandardsAugust2014.pdf.

National Poultry Improvement Plan (NPIP) and Auxiliary Provisions, August 2014b. USDA, APHIS, VS. 9 CFR Parts 145-147 and Part 56. http://www.poultryimprovement.org/documents/AuxiliaryProvisions.pdf.

Nicholas, R.A.J., Cullen, G.A., 1991. Development of an ELISA for detecting antibodies to *Salmonella* Enteritidis in chicken flocks. Vet. Rec. 128, 74–76.

Nordic Committee on Food Analysis (NMKL), 1999. *Salmonella* Detection in Foods (NMKL 71). Nordic Committee on Food Analysis. (Oslo, Norway).

North, W.R., Bartram, M.T., 1953. The efficiency of selenite broth of different compositions in the isolation of *Salmonella*. Appl. Microbiol. 1, 130–134.

OIE, 2015. Prevention, Detection and Control of *Salmonella* in Poultry, Chapter 6.5. OIE – Terrestrial Animal Health Code.

Okamura, M., Ohba, T., Kikuchi, S., Suzuki, A., Tachizaki, H., Ikedo, M., Kojima, T., Nakamura, M., 2008. Loop-mediated isothermal amplification for the rapid, sensitive, and specific detection of the O9 group of *Salmonella* in chickens. Vet. Microbiol. 132, 197–204.

Olesiuk, O.M., Carlson, V.L., Snoeyenbos, G.H., Smyser, C.F., 1969. Experimental Salmonella typhimurium infection in two chicken flocks. Avian Dis. 15, 118–124.

Opara, O.O., Carr, L.E., Russek-Cohen, E., Tate, C.R., Mallinson, E.T., Miller, R.G., Stewart, L.E., Johnston, R.W., Joseph, S.W., 1992. Correlation of water activity and other environmental conditions with repeated detection of *Salmonella* contamination on poultry farms. Avian Dis. 36, 664–671.

Park, S.H., Ricke, S.C., 2015. Development of multiplex PCR assay for simultaneous detection of *Salmonella* genus, *Salmonella* subspecies I, *Salm.* Enteritidis, *Salm.* Heidelberg, and *Salm.* Typhimurium. J. Appl. Microbiol. 118, 152–160.

Poppe, C., Johnson, R.P., Forsberg, C.M., Irwin, R.J., 1992. *Salmonella enteritidis* and other *Salmonella* in laying hens and eggs from flocks with *Salmonella* in their environment. Can. J. Vet. Res. 56, 226–232.

Pourciau, S.S., Springer, W.T., 1978. Evaluation of secondary enrichment for detecting salmonellae in bobwhite quail. Avian Dis. 22, 42–45.

Rall, V.L.M., Rall, R., Aragon, L.C., da Silva, M.G., 2005. Evaluation of three enrichment broths and five plating media for *Salmonella* detection in poultry. Braz. J. Microbiol. 36, 147–150.

Rambach, A., 1990. New plate medium for facilitated differentiation of *Salmonella* spp. from *Proteus* spp. and other enteric bacteria. Appl. Environ. Microbiol. 56, 301–303.

Rappaport, F., Konforti, N., Navon, B., 1956. A new enrichment medium for certain salmonellae. J. Clin. Pathol. 9, 261–266.

Restaino, L., Grauman, G.S., McCall, W.A., Hill, W.M., 1977. Effects of varying concentrations of novobiocin incorporated into two *Salmonella* plating media on the recovery of four Enterobacteriaceae. Appl. Environ. Microbiol. 33, 585–589.

Restaino, L., Komatsu, K.K., Syracuse, M.J., 1982. A note on novobiocin in XLD and HE agars: the optimum levels required in two commercial sources of media to improve isolation of salmonellas. J. Appl. Bacteriol. 53, 285–288.

Riemann, H., Himathongkham, S., Wiloughby, D., Tarbell, R., Breitmeyer, R., 1998. A survey for *Salmonella* by drag swabbing manure piles in California egg ranches. Avian Dis. 42, 67–71.

Rigby, C.E., Pettit, J.R., 1980. Delayed secondary enrichment for the isolation of salmonellae from broiler chickens and their environment. Appl. Environ. Microbiol. 40, 783–786.

Rolfe, D.L., Riemann, H.P., Farver, T.B., Himathongkham, S., 2000. Drag swab efficiency factors when sampling chicken manure. Avian Dis. 44, 668–675.

Rybolt, M.L., Willis, R.W., Bailey, R.H., 2005. Use of secondary enrichment for isolation of *Salmonella* from naturally contaminated samples. Poult. Sci. 84, 992–997.

Salmonella Enteritidis Pilot Project Progress Report, May 22, 1995. A Cooperative Effort of Pennsylvania Poultry Producers, Pennsylvania Poultry Federation, Egg Association of America, Pennsylvania Department of Agriculture, Pennsylvania State University, University of Pennsylvania, and USDA.

Schlosser, W.D., Henzler, D.J., Mason, J., Kradel, D., Shipman, L., Trock, S., Hurd, S.H., Hogue, A.T., Sischo, W., Ebel, E.D., 1999. The *Salmonella enterica* serovar Enteritidis pilot project. In: Saeed, A.M. (Ed.), *Salmonella enterica* Serovar Enteritidis in Humans and Animals. Iowa State University Press, pp. 353–365.

Schonenbrucher, V., Mallinson, E.T., Bulte, M., 2008. A comparison of standard cultural methods for the detection of foodborne *Salmonella* species including three new chromogenic plating media. Int. J. Food Microbiol. 123, 61–66.

Schultz, J., Jarquin, R., Ricke, S.C., Hanning, I., 2012. Optimized culturing and nucleic acid-based methods for the detection of *Salmonella enterica* in poultry environments. Poult. Sci. 91, 2761–2766.

Seo, K.H., Valentin-Bon, I.E., Brackett, R.E., Holt, P.S., 2004. Rapid, specific detection of *Salmonella* Enteritidis in pooled eggs by real-time PCR. J. Food Protect. 67, 864–869.

Skov, M.N., Carstensen, B., Tornoe, N., Madsen, M., 1999. Evaluation of sampling methods for the detection of *Salmonella* in broiler flocks. J. Appl. Microbiol. 86, 695–700.

Sommer, D., Enderlein, D., Antakli, A., Schonenbrucher, H., Slaghuis, J., Redmann, T., Lierz, M., 2012. *Salmonella* detection in poultry samples. Comparison of two commercial real-time PCR systems with culture methods for the detection of *Salmonella* spp. in environmental and fecal samples in poultry. Tierarztl. Prax. Ausg. G Grosstiere Nutztiere 40, 383–389.

Soria, M.C., Soria, M.A., Bueno, D.J., 2012. Comparison of 2 culture methods and PCR assays for *Salmonella* detection in poultry feces. Poult. Sci. 91, 616–626.

St. Louis, M.E., Morse, D.L., Potter, M.E., DeMelfi, T.M., Guzewich, J.J., Tauxe, R.V., Blake, P.A., 1988. The emergence of grade A eggs as a major source of *Salmonella enteritidis* infections: new implications for control of salmonellosis. J. Am. Med. Assoc. 259, 2103–2107.

Stokes, J.L., Osborne, W.W., 1955. A selenite brilliant green medium for the isolation of *Salmonella*. Appl. Microbiol. 3, 217–220.

Tate, C.R., Miller, R.G., Mallinson, E.T., Douglass, L.W., Johnson, R.W., 1990. The isolation of salmonellae from poultry environmental samples by several enrichment procedures using plating media with and without novobiocin. Poult. Sci. 69, 721–726.

Tate, C.R., Miller, R.G., Mallinson, E.T., 1992. Evaluation of two isolation and two nonisolation methods for detecting naturally occurring salmonellae from broiler flock environmental drag-swab samples. J. Food Protect. 55, 964–967.

Taylor, W.I., Schelhart, D., 1971. Isolation of shigellae. VIII. Comparison of gram negative broth with Rappaport's enrichment broth. Appl. Microbiol. 18, 393–395.

Tebbs, R.S., Wong, L.Y., Brzoska, P., Petrauskene, O.V., 2012. Molecular technologies for *Salmonella* detection. In: Annous, B.A., Gurther, J.B. (Eds.), Salmonella – Distribution, Adaptation, Control Measures and Molecular Technologies. InTech.

Thorns, C.J., Sojka, M.G., Chasey, D., 1990. Detection of a novel fimbrial structure on the surface of *Salmonella* Enteritidis by using a monoclonal antibody. J. Clin. Microbiol. 28, 2409–2414.

Timoney, J.F., Sikora, N., Shivaprasad, H.L., Opitz, M., 1990. Detection of antibody to *Salmonella* Enteritidis by a gm flagellin-based ELISA. Vet. Rec. 127, 168–169.

Ueda, S., Kuwabara, Y., 2009. The rapid detection of *Salmonella* from food samples by loop-mediated isothermal amplification (LAMP). Biocontrol Sci. 14, 73–76.

USDA, June 1998. *Salmonella* Enteritidis Risk Assessment. Shell Eggs and Egg Products. USDA, FSIS. http://www.fsis.usda.gov/wps/portal/frame-redirect?url=http://www.fsis.usda.gov/OPHS/risk/index.htm.

USDA, October 2005. Risk assessments of *Salmonella* enteritidis in shell eggs and *Salmonella* spp. In: Egg Products. USDA, FSIS, Office of Public Health Science. http://www.fsis.usda.gov/shared/PDF/SE_Risk_Assess_Oct2005.pdf.

USDA, 2000. *Salmonella enterica* serotype Enteritidis in table egg layers in the U.S. USDA: APHIS:VS, CEAH, National Animal Health Monitoring System, Fort Collins, CO. #N333.1000. (Layer '99).

USDA, 2014. Layers 2013 "Part II: Prevention and Control of *Salmonella* Enteritidis on Table-Egg Farms in the United States, 2013". USDA: APHIS:VS, CEAH, National Animal Health Monitoring System, Fort Collins, CO. #688.0814.

Van der Giessen, A.W., Peters, R., Berkers, P.A.T.A., Jansen, H., Notermans, S.H.W., 1991. *Salmonella* contamination of poultry flocks in the Netherlands. Vet. Q. 13, 41–46.

Van der Heijden, H.M.J.F., van der Wijngaard, J.C., 1992. *Salmonella*-ELISA for serological screening of poultry. In: Snoeyenbos, G.H. (Ed.), Proceedings of the Symposium on the Diagnosis and Control of Salmonella. United States Animal Health Association, San Diego, CA, pp. 149–156.

Van der Zee, H., Huis in't Velt, J.H.J., 2000. Methods for the rapid detection of *Salmonella*. In: Wray, C., Wray, A. (Eds.), *Salmonella* in Domestic Animals. CABI Publishing, New York, pp. 373–392.

Van der Zee, H., de Boer, E., van Netten, P., 1990. *Salmonella* isolatie met behulp van modified semi-solid Rappaport-Vassiliadis (MSRV) medium. Ware(n)-Chemicus 21, 180–189.

Van Schothorst, M., Renaud, A.M., 1983. Dynamics of *Salmonella* isolation with modified Rappaport's medium (R10). J. Appl. Bacteriol. 54, 209–215.

Van Zijderveld, F.G., Van Zijderveld-van Bemmel, A.M., Anakotta, J., 1992. Comparison of four different enzyme-linked immunosorbent assays for serological diagnosis of *Salmonella* Enteritidis infections in experimentally infected chickens. J. Clin. Microbiol. 30, 2560–2566.

Van Zijderveld, F.G., van Zijderveld-van Bemmel, A.M., Brouwers, R.A., de Vries, t.S., Landman, W.J., de Jong, W.A., 1993. Serological detection of chicken flocks naturally infected with *Salmonella* Enteritidis, using an enzyme-linked immunosorbent assay based on monoclonal antibodies against the flagellar antigen. Vet. Q. 15, 135–137.

Vassiliadis, P., Trichopoulos, D., Papadakis, G., Politi, G., 1970. *Salmonella* isolations in abattoirs in Greece. J. Hyg. 68, 601–609.

Vassiliadis, P., Pateraki, E., Papaiconomou, N., Papadakis, J.A., Trichopoulos, D., 1976. Nouveau procede d'enrichissiment de *Salmonella*. Ann. Microbiol. 127B, 195–200.

Vassiliadis, P., 1983. The Rappaport-Vassiliadis (RV) enrichment medium for the isolation of salmonellas: an overview. J. Appl. Bacteriol. 54, 69–76.

Waltman, W.D., Horne, A.M., Pirkle, C., Dickson, T.G., 1991. Use of delayed secondary enrichment for the isolation of *Salmonella* in poultry and poultry environments. Avian Dis. 35, 88–92.

Waltman, W.D., Horne, A.M., Pirkle, C., 1993. Influence of enrichment incubation time on the isolation of *Salmonella*. Avian Dis. 37, 884–887.

Waltman, W.D., Horne, A.M., Pirkle, C., 1995. Comparative analysis of media and methods for isolating *Salmonella* from poultry environmental samples. In: Proceedings of the Symposium on the Diagnosis of Salmonella Infections. US Animal Health Association, Richmond, VA, pp. 1–14.

Weinack, O.M., Smyser, C.F., Snoeyenbos, G.H., 1979. Evaluation of several methods of detecting salmonellae in groups of chickens. Avian Dis. 23, 179–193.

Wray, C., Nicholas, R.A., Thorns, C.J., Minga, U.M., 1992. Studies on the serological diagnosis of *Salmonella* Enteritidis infection in poultry. APHIS Salmonella Enteritidis control program. In: Snoeyenbos, G.H. (Ed.), Proceedings of the Symposium on the Diagnosis and Control of Salmonella. United States Animal Health Association, San Diego, CA, pp. 157–162.

CHAPTER 10

# Genetic Basis of *Salmonella* Enteritidis Pathogenesis in Chickens

**Devendra H. Shah, Jacob R. Elder, Kim L. Chiok, Narayan C. Paul**
*Washington State University, Pullman, WA, United States*

## 1. INTRODUCTION

Over the last three decades, *Salmonella* Enteritidis (SE) has remained the cause of the food-borne salmonellosis pandemic in humans. Epidemiological source-tracking studies have identified contaminated table eggs as the primary risk factor for human infection (Gould et al., 2013). Due to this risk of human infection associated with poultry products, SE also poses a significant challenge to the commercial poultry production. Consequently, experimental SE infections in young and adult chickens have been extensively used to acquire basic information about the epidemiology and pathobiology of SE in this reservoir host. Epidemiological investigations show that, unlike other non-host adapted *Salmonella* serotypes such as *S.* Typhimurium, which is isolated from variety of food animal sources, SE is most predominantly isolated from poultry, suggesting that SE has likely evolved to gain significant predilection to the poultry host. Pathobiological investigations in chickens indicate that, unlike other non-host adapted *Salmonella* serotypes, SE has evolved with the unique ability to efficiently contaminate internal contents of eggs without causing discernible illness in the infected laying hens. *S.* Enteritidis is also antigenically, phenotypically, and genetically different from other nontyphoidal *Salmonella* (NTS) serotypes including the most widely studied *S.* Typhimurium. Nevertheless, the genetic basis of pathogenesis of SE in the chicken host has only been tangentially investigated. In the following sections, we summarize the infection kinetics of SE in the chicken host and provide an overview of the current understanding of genetic factors underlying SE infection. We also discuss the important knowledge gaps that, if addressed, will improve our understanding of the complex biology of SE in young chickens and in egg laying hens.

## 2. INFECTION KINETICS OF SE IN CHICKENS

*S.* Enteritidis infection in chickens is a complex multistep process that can be broadly categorized into a few major events that include intestinal colonization, invasion, and systemic spread to internal organs such as liver and spleen. In adult laying hens,

Producing Safe Eggs. http://dx.doi.org/10.1016/B978-0-12-802582-6.00010-0

colonization of reproductive tract (RT) organs and contamination of internal contents of eggs is also an epidemiologically significant outcome of infection. Infection of the RT organs precedes contamination of internal contents of forming eggs before oviposition. Each step of the infection in this host is influenced by multiple host and pathogen-associated factors. It is important to first understand the infection kinetics of SE in chickens, and the host and pathogen factors that influence the outcomes of infection before dissecting the roles of genetic factors underlying these processes.

## 2.1 GASTROINTESTINAL TRACT COLONIZATION AND SYSTEMIC SPREAD

The primary route of infection and transmission of SE in chickens is via the feco-oral route. After ingestion, SE frequently colonizes crop (Hargis et al., 1995; Turnbull and Snoeyenbos, 1974), less frequently in proventriculus and duodenum, but most preferentially and persistently in the lower ileum, cecum, and cloaca (Turnbull and Snoeyenbos, 1974). Subsequently, SE invades intestinal epithelium and localizes in the submucosa within 4h postinfection (Berndt et al., 2007). Invasion occurs via specialized M-cells that sample the intestinal lumen as well as nonphagocytic cells. Intestinal invasion induces inflammation, which is characterized by infiltration of heterophils, macrophages, red blood cells, and other immune cells into the lamina propia and cecal luminal exudate (Macri et al., 1997; Porter and Holt, 1993; Van Immerseel et al., 2002). As inflammation progresses, SE produces virulence factors that aid its survival within macrophages and subsequent systemic spread to colonize internal organs including liver and spleen, and in laying hens this also results in colonization of the RT organs (Berchieri et al., 2001; He et al., 2010; Higgins et al., 2007). Interestingly, the infection kinetics vary markedly depending on the age or type of chickens. Porter and Holt (1993) reported that the intestinal inflammation in the adult laying hens was confined to the cecum, whereas in broiler chickens, cellular infiltrates were also reported in the lamina propia of the small intestine (Quinteiro-Filho et al., 2012). Infection of young chickens with doses as low as 100CFU at 1day posthatch can induce long-term intermittent shedding of SE for up to 24weeks postinfection (Gast and Holt, 1998; Van Immerseel, 2010). In contrast, oral infection of laying hens often results in short-term fecal shedding that may vary from 1week to up to 6weeks postinfection (Kinde et al., 2000; Shivaprasad et al., 1990). In very young birds, the internal organ colonization often results in gross pathological changes such as hepatomegaly and splenomegaly with foci of hemorrhages (Kinde et al., 2000). However, in adult chickens including laying hens, the internal organ colonization generally occurs without any overt clinical signs (Hogue et al., 1997; Prevost et al., 2006) and despite efficient colonization of chicken RT organs, egg production appears to remain unaffected or sometimes increases (Fan et al., 2014; Guard-Petter, 2001).

## 2.2 REPRODUCTIVE TRACT COLONIZATION IN LAYING HENS

Significant research has focused on the pathogenesis of RT infection in chickens because colonization of RT organs is directly associated with contamination of internal contents of eggs. Formation of different internal components of eggs occurs at different anatomical positions in the avian RT. Thus one of the key questions that have attracted the most attention is, which anatomical part(s) of chicken RT contribute significantly to the contamination of internal contents of eggs before oviposition? Some reports show that egg yolk is frequently and heavily colonized when laying hens are challenged with SE by different routes, pointing to the ovary as the primary source for egg contamination (Bichler et al., 1996; Gast et al., 2002, 2007b). Others have reported that egg albumen is most frequently and heavily colonized, suggesting colonization of oviduct structures as the primary source for contamination of egg contents (Cogan et al., 2004; Reiber and Conner, 1995; Shivaprasad et al., 1990). Systemic spread of different phage types of SE to ovaries was demonstrated in hens experimentally infected via both oral and intravenous routes (Gantois et al., 2008b; Miyamoto et al., 1997). It was suggested that SE is deposited near the basement of the highly vascularized follicular theca wall before migrating through the perivitellin layer and/or attaching and invading ovarian granulosa cells to reach the forming egg yolk (Thiagarajan et al., 1994, 1996a). Ovarian follicle maturation might also play a role in susceptibility of ovarian granulosa cells to invasion (Howard et al., 2005). In one study, invasion frequency of different phage types of SE in ovarian follicles ranged from 0.016% to 0.034% compared with 0.0003% for *Escherichia coli* (Dawoud et al., 2011). The contribution of oviduct colonization in internal contamination of the contents of freshly laid eggs has also been extensively investigated. In general, SE is isolated more frequently and in higher counts [colony forming units (CFU) per gram] from isthmus and internal contents of eggs when compared with serotypes such as Heidelberg, Virchow, and Hadar (Gantois et al., 2008b; Okamura et al., 2001). When the laying hens were infected via the intravaginal route, the oviduct (isthmus) and uterus were more frequently colonized than the ovary, a phenotype that is likely favored by the ability of SE to attach to vaginal mucosa (Miyamoto et al., 1997; Okamura et al., 2001). In contrast, intravenous infection was reported to result in more prolific colonization of the ovary and the entire oviduct (infundibulum, magnum, and isthmus), yielding more frequent recovery of SE from egg contents and developing eggs when compared with the intravaginal, intracloacal, or oral routes (Bichler et al., 1996; Kinde et al., 2000; Miyamoto et al., 1997; Petter, 1993). In addition to vaginal epithelium, SE also attaches to glandular secretions and tubular glands of isthmus in cultured primary chicken oviduct epithelial cells (COECs), pointing to the isthmus as a colonization site (De Buck et al., 2004a). In summary, it is important to note that most published studies have reported RT organ colonization in terms of either number of SE-positive tissues or number of SE-positive eggs without really establishing bacterial burden (i.e., CFU per gram) as a baseline for comparison. Several variables such as SE strain type, phage type, inoculation route, dose, breed, and age of hens represent a significant challenge for drawing meaningful interpretations from the body of published literature. Therefore there is no clear consensus regarding which parts of RT organs may contribute more significantly to egg contamination; however, the entire RT appears to be colonized by SE. The aforementioned factors (route, dose, strain type, and age of chicken) may as well significantly influence the ability of SE to colonize the avian RT. Thus it is imperative that these factors are carefully considered before designing studies to investigate RT organ infection in chickens.

## 2.3 COLONIZATION AND SURVIVAL IN THE INTERNAL CONTENTS OF EGGS

It is now widely accepted that the primary route of egg contamination is via deposition of SE from the infected RT directly into the internal contents of eggs prior to oviposition (Arnold et al., 2014; Gast and Holt, 2000; Keller et al., 1995; Shivaprasad et al., 1990). It is also known that the hen egg presents a hostile environment for efficient propagation of SE. First, the outer shell membrane and shell membrane not only act as physical barriers, but also contain chemical compounds with antibacterial properties including lysozyme (Hincke et al., 2000), ovotransferrin (Gautron et al., 2001), ovocalyxin-36 (Gautron et al., 2007), and an unknown protein extract (Gantois et al., 2009a). In addition, egg albumen, vitelline membranes, and yolk contain antimicrobial compounds such as lysozyme, ovalbumin, ovotransferrin, ovomucin, β-defensin 11, and immunoglobulins (Gantois et al., 2009a; Mageed et al., 2008; Mann, 2007; Stevens, 1991). Despite the presence of these physical and chemical barriers, certain strains of SE are efficiently able to penetrate, survive, and propagate within different egg contents. The eggshell and eggshell membrane penetration is not unique to SE because other *Salmonella* serotypes and bacterial species can also penetrate these barriers (De Reu et al., 2006). In one study, no significant differences in egg-shell penetration were found between SE and *S.* Typhimurium (Miyamoto et al., 1998). Others have compared the growth of different *Salmonella* serovars by artificially contaminating egg albumen and have reported no differences between SE ($n=8$) and *S.* Typhimurium ($n=24$) at 37°C or 42°C (Guan et al., 2006). Similarly, Messens et al. (2004) did not observe any growth advantage for SE in egg albumen when compared with other non-SE serovars including *S.* Typhimurium, Senftenberg, Stanleyville, Mbandaka, and Blockley. Interestingly, however, it was reported that *Salmonella* generally grew better in egg albumen harvested from fresh eggs than from stored eggs, presumably due to the highly alkaline environment (pH approximately 9.0) associated with storage (Messens et al., 2004). One study reported that the survival of SE (25.8%) was higher than that of Typhimurium (6.5%) or *E. coli* (1.8%) in egg albumen extracted from fresh eggs and incubated at room temperature (Clavijo et al., 2006). De Vylder et al. (2013) evaluated the survivability of 89 strains of *Salmonella* belonging to five different serogroups (B, C, D, E, and G) and 26 serotypes within fresh egg albumen incubated at 42°C. These authors reported that SE isolates generally displayed greater survivability within egg albumen compared with other serogroups, suggesting that both the age of egg albumen and temperature of incubation may have significant impact on the survival of SE in this system. More importantly, published data also show that there is wide interstrain variability in survival/growth of SE within egg albumen, suggesting that individual variation between strains is also an important factor (Clavijo et al., 2006; Shah et al., 2012a; Yim et al., 2010). Finally, it is known that SE can traverse through vitelline membrane into the egg yolk and replicate efficiently. However, this property is not unique to SE because several other serotypes including *S.* Montevideo, *S.* Infantis, *S.* Heidelberg, *S.* Typhimurium, *S.* Virchow, and *S.* Hadar have been reported to display a similar phenotype (Gantois et al., 2008b; Gast et al., 2007a; Murase et al., 2006). Unlike in egg albumen, strains of SE belonging to different phage types (4, 8, 13a, and 14b) grow equally well in egg yolks, presumably because yolk is enriched with different nutrients that can be utilized by bacteria for efficient growth and multiplication (Gast and Holt, 2001). In summary, it appears that SE may have the specialized ability to grow or survive within egg albumen; however, this trait is significantly influenced by the strain type of SE, age of egg albumen, and temperature of incubation. Therefore it is important to carefully consider these factors while designing studies that investigate genetic factors that contribute to the survivability of SE in egg albumen.

## 3. GENETIC BASIS OF SE PATHOGENESIS

To date, most of our knowledge of *Salmonella* pathogenesis in avian host has relied on the extrapolations from the research that has utilized *S.* Typhimurium as a model organism and cultured mammalian epithelial and phagocytic cells or mouse as a model host (reviewed in Garai et al., 2012; Kaiser et al., 2012; Watson and Holden, 2010). Relatively few studies have been conducted to dissect genetic basis of SE pathogenesis using chicken as a model host. Nevertheless, it is becoming increasingly evident that the genetic mechanisms underlying SE infection in chickens may have components that are distinct from the well-studied serotype *S.* Typhimurium. In the following three sections, we attempt to summarize the current understanding of the role of different genetic factors of SE in pathogenesis in the chicken host.

### 3.1 GENETIC BASIS OF GASTROINTESTINAL INFECTION IN CHICKENS

Studies focused on the SE genetic factors that contribute to the gastrointestinal (GI) infection in chickens have mostly revolved around the role of type-3 secretion system (T3SS) encoded by *Salmonella* pathogenicity islands-1 (SPI-1), SPI-2, and flagella and fimbriae factors. The role of SPI-1 and SPI-2 in pathogenesis of SE infection in chickens appears somewhat contradictory and poorly defined. Published data from our laboratory and others show that inactivation of *hilA* (SPI-1 invasion gene activator) in SE results in significantly reduced intestinal colonization (i.e., fewer bacteria recovered from the ceca) and invasion (i.e., few bacteria invade the GI tract to subsequently colonize internal organs such as liver and spleen) in orally inoculated day-old chickens (Addwebi et al., 2014; Bohez et al., 2006). In addition, suppression of *hilA* gene expression through supplementation of medium-chain fatty acids also results in reduced intestinal colonization and invasiveness of SE in orally challenged 5-day-old chicks (Van Immerseel et al., 2004). Although these data suggest that SPI-1 contributes to intestinal and internal organ colonization in young chickens, Desin et al. (2009) observed no significant differences in intestinal colonization between SPI-1 mutant and wild-type parent of SE in orally challenged day-old chickens. Moreover, the SPI-1 mutant displayed varying degrees of internal organ invasiveness, which was not always significantly different from the wild-type parent (Desin et al., 2009). In another study, when 1-week-old chickens were orally challenged with an SPI-2 mutant or a mutant with deletion of the entire SPI-1 and SPI-2, the SPI-2 mutant was impaired in colonization of the cecum, spleen, and liver early during infection (days 1–3 postinfection), but by day 4 postinfection, there were no significant differences in colonization of liver, spleen, or cecum between wild type and either of the mutants (Wisner et al., 2010). In contrast, Rychlik et al. (2009) reported that SE mutant lacking SPI-1 and SPI-2 was not impaired in its ability to colonize cecum of orally inoculated day-old chickens; however, it displayed impaired ability to colonize internal organs such as liver and spleen. In addition, the colonization of the internal organs by SPI-2 mutant of SE was impaired in 1-day-, 5-day-, and 24-week-old chickens after oral, intraperitoneal, and intravenous inoculation (Bohez et al., 2008). Finally, there are also conflicting reports on the intracellular survivability of SPI-2 mutant of SE within chicken macrophages (Bohez et al., 2008; Wisner et al., 2011). Thus more research is needed to clearly define the role of SPI-1 and SPI-2 in the pathogenesis of SE in chickens.

Besides SPI-1 and SPI-2, SE genome carries at least 13 additional SPIs that have been annotated. Of these, SPI-3, SPI-4, and SPI-5 are most extensively characterized in *S.* Typhimurium, with some studies indicating that these may contribute to pathogenesis in non-avian models (Gerlach et al., 2007; Kiss et al., 2007; Pontes et al., 2015; Wallis et al., 1999). The role of these SPIs in intestinal and internal organ colonization of SE in chickens is currently unclear. Mutation in SPI-4 impairs the intestinal colonization of *S.* Typhimurium in mouse, but not in the chicken (Kiss et al., 2007; Morgan et al., 2004, 2007). Interestingly, the competitive fitness of an SPI-4 mutant of SE in orally infected mice was less affected than a similar mutant of *S.* Typhimurium (Kiss et al., 2007). We reported that a disruption of *siiE* gene of SPI-4 in SE resulted in significantly reduced survival in chicken macrophages; however, it is currently unknown if disruption of SPI-4 would result in similar effects in vivo (Shah et al., 2012b). We have also noted that disruption of *pipA* gene of SPI-5 significantly altered the ability to colonize the GI tract and internal organs of day-old chickens (Addwebi et al., 2014). In addition, several SPI-5 genes were upregulated in the GI tract of young chicks orally infected with SE (Dhawi et al., 2011). Interestingly, Rychlik et al. (2009) reported that SPI-3, SPI-4, or SPI-5 mutants of SE were not impaired in colonization of the cecum, liver, or spleen of orally inoculated day-old chickens; however, a reduction in the ability of SE to colonize the spleen was observed when all three SPIs were deleted simultaneously. These data suggest that SPI-3, SPI-4, and SPI-5 may contribute to pathogenesis in chickens, although their exact role during the infection process still remains elusive.

Published data from our laboratory and others suggest that SE flagellum also plays a role in pathogenesis in chickens. In one study, aflagellated SE was recovered at significantly lower numbers from the ceca of orally inoculated day-old chickens when compared with the flagellated SE (Allen-Vercoe and Woodward, 1999). These authors suggested that the presence of functional flagellum, but not the motility was important for cecal colonization of SE in chickens (Allen-Vercoe et al., 1999). We reported that impaired secretion of flagellar proteins (FlgK, FljB, and FlgL) among wild-type strains of SE was associated with a low-invasive phenotype in chicken macrophages (Shah et al., 2011). We also reported that disruption of the *fljB* results in reduced invasiveness of SE in cultured chicken liver cells and reduced colonization of the small intestine in orally inoculated day-old chickens (Addwebi et al., 2014; Shah et al., 2012b). Shippy et al. (2014) reported that deletion of *flgC*, encoding the flagellum basal body protein, resulted in significantly reduced colonization of liver and spleen in orally infected 1-week-old chickens. Finally, Parker and Guard-Petter (2001) reported that disruption of *fliC* and *flhD* negatively impacted internal organ colonization when 3-week old chickens were infected subcutaneously; however, results in orally infected chickens did not show a clear negative phenotype.

The role of fimbriae in SE pathogenesis has also been suspected but not clearly defined, and the published reports are often conflicting. Early work revealed that SefA (SEF14) and FimA (SEF21) fimbriae aid in persistence of SE in the ceca of orally inoculated 30-week-old hens (Thiagarajan et al., 1996b). In contrast, Thorns et al. (1996) observed that inactivation of SEF14 resulted in reduced recovery of the mutant from liver and spleen, but did not affect cecal colonization in orally inoculated 5-day-old chickens, especially later than 15 days post-infection. Others have reported that SE strains with inactivated SEF14 (*sef*), SEF17 (*agf*), and SEF21 (*fim*) fimbriae do not show significant differences in their survival within avian macrophages or colonization/persistence in ceca, spleen, and liver of orally infected 5-day-old chickens (Rajashekara et al., 2000). Similarly, lack of SEF14 in SE did not impact intestinal colonization in orally infected 1-day-old chickens (Thorns et al., 1996). Allen-Vercoe and Woodward (1999) also reported that afimbriate mutants of SE were not impaired in colonization of the GI tract or internal organs or fecal shedding in orally infected 1-day-old chickens. We reported that disruption of *csgB* in SE impaired invasiveness in chicken liver cells, but did not significantly impact intestinal colonization and internal organ invasiveness in orally challenged 1-day-old chickens (Addwebi et al., 2014; Shah et al., 2012b). Similarly, inactivation of *stdA* in SE was reported to result in reduced colonization of cecum as well as less invasion of the internal organs in orally inoculated 7-day-old chickens (Shippy et al., 2013). Finally, Clayton et al. (2008) systematically inactivated each of the 13 annotated fimbrial operons of SE and reported that only the inactivation of *peg* operon resulted in significant reduction in the intestinal colonization in orally inoculated 18-day-old chickens. The *peg* fimbriae is interesting because it is not found in other NTS, including *S.* Typhimurium, and has been identified as a pseudogene in an avian host adapted serotype *S.* Gallinarum (Addwebi et al., 2014; Clayton et al., 2008). It would be of interest to further dissect the role of *peg* fimbriae in SE pathogenesis in chickens.

Few studies have shown that inactivation of genes affecting SE metabolism can significantly impact infection kinetics in chickens. The *aroA* (3-phosphoshikimate 1-carboxyvinyltransferase) mutant of SE,, which confers deficiency of aromatic amino acid synthesis, is significantly impaired in colonization of the cecum, spleen, and liver in orally inoculated 1- or 5-day-old chickens (Cooper et al., 1994). The *tat* operon in gram-negative bacteria transports folded enzymes across the cytoplasmic membrane to the periplasmic space. Many of the proteins predicted to be transported by this system are involved in oxidation or reduction reactions of various compounds. In one study, deletion of *tatB* in SE resulted in significantly reduced colonization of the ceca of orally inoculated 7-day-old chickens (Mickael et al., 2010). Iron metabolism genes may also play a role in the pathogenicity of SE. When SE is grown at avian body temperature, the low-pathogenic strains of SE show reduced expression of genes involved in iron metabolism, including the *suf* operon (Shah, 2014). Chickens vaccinated with purified IroN (the siderophore receptor for salmochelin) show a significant reduction in mortality when challenged intravenously with pathogenic SE strain (Kaneshige et al., 2009). Highly pathogenic strains of SE were reported to more efficiently express genes involved in protection against osmotic, oxidative, and other stresses especially when these were grown at avian body temperature (Shah et al., 2011). In addition, SE collected from ceca of young chickens showed increased expression of genes associated with ethanolamine, propanediol, sialic acid, and dicarboxylic acid metabolism (Dhawi et al., 2011). These reports suggest that SE growth in the GI tract is likely supported by diversity of metabolic substrates and that SE undergoes significant metabolic changes in the avian intestine. It is currently unknown which metabolic substrates are critical for propagation of SE in the GI tract of chickens. Finding the most critical metabolic substrates and understanding their role in SE pathogenesis in chicken may offer opportunities to find new nutritional or other alternative strategies to control SE in poultry.

There are also sporadic reports of other genetic factors that contribute to the kinetics of GI infection. For instance, the lipoprotein encoding gene, *yfgL*, was shown to affect colonization of the GI tract and spleen of orally inoculated 1-day-old chicks; specifically it was associated with lower expression of SPI-1, SPI-2, and flagellar structural proteins (Amy et al., 2004; Fardini et al., 2007). Multiple studies have reported that lipopolysaccharide (LPS) synthesis genes contribute to pathogenesis in chickens. We reported that inactivation of *rfbN* (a rhamnosyltransferase) results in a significant reduction in the colonization of the internal organs of orally challenged day-old chickens (Addwebi et al., 2014). Similarly, inactivation of *rfbM* (a mannose-1-phosphate guanylyltransferase) in SE also results in defective colonization of the GI tract and internal organs in orally challenged day-old chickens (Addwebi et al., 2014). The ribosomal maturation factor *ksgA* is under investigation in our laboratory for its contribution to pathogenicity in chickens. We have previously reported that inactivation of *ksgA* in SE results in reduced invasiveness in chicken liver cells (Shah et al., 2012b) and impaired intestinal and internal organ colonization in orally inoculated day-old chickens (Chiok et al., 2013). One study identified a few SE-specific genetic factors (*SEN1001*, *SEN1140*, *SEN1970-SEN1999*, and

*SEN4290-SEN4292*) that contribute to pathogenicity in mice, but their role remains to be tested in a chicken model (Silva et al., 2012; Vishwakarma et al., 2012). In addition, we reported a few SE-specific genes (*pegD*, *SEN1152*, *SEN1393*, and *SEN1966*) that contribute to the invasiveness of SE in human intestinal cells (Shah et al., 2012b). However, the role of several of these SE-specific genes in pathogenesis in chickens in general and laying hens in particular is currently unknown.

#### 3.1.1 Knowledge Gaps and Challenges

It is clear that there are significant gaps in our understanding of the molecular basis of pathogenesis of SE in chickens partly because of the variation in the chicken model and the strains used in different studies. Efforts are needed to standardize the chicken model and experimental approaches to study genetic basis of pathogenesis in this host. When examining the host–pathogen interaction between SE and chickens, it is important to recognize that there can be significant variation in virulence between different strains/phage types of SE (Shah et al., 2011; Shivaprasad et al., 1990; Yim et al., 2010). Much of this phenotype variation is not explained at the genotypic level (Shah, 2014; Shah et al., 2011; Yim et al., 2010). In one study, the genomes of different SE isolates were reported to display thousands of SNP differences and greater than 300 variable genes (Allard et al., 2013). In this study, 21 genes from genetic lineages representing outbreak-associated isolates showed nonsynonymous mutations and this affected least five putatively virulence-associated genes. The pseudogene content among different SE strains is also known to vary (genes *ratB*, a known virulence gene, and *mviM* are examples) (Matthews et al., 2015). Virtually no studies to date have explored the basis of this inter-strain genetic variation and its association with differential virulence of SE. It is imperative that more efforts be made to compare multiple SE strains in molecular pathogenesis studies before researchers arrive at broad conclusions about this serotype. Rapid screening of thousands or more *Salmonella* mutants by negative selection is now possible by combining random transposon mutagenesis with next-generation sequencing using transposon directed insertion sequencing (TraDIS) or TnSeq. This technology may allow simultaneous examination of multiple SE strains to identify pathogenicity factors in both young and adult chickens.

### 3.2 GENETIC BASIS OF REPRODUCTIVE TRACT PATHOGENESIS

The role of genetic factors of SE that contribute to pathogenesis of RT infection has received some attention. In general, most studies have focused on the role of intrinsic bacterial structures such as type-1 fimbriae and LPS in avian RT infection. A few high-throughput screening studies have resulted in identification of genes that may play coordinated roles in RT infection in the hen and in egg microenvironments (Gantois et al., 2008a; McKelvey et al., 2014; Raspoet et al., 2014).

Type-1 fimbriae was reported to aid in interaction of SE with epithelial cell surface or extracellular matrix of the hen RT (De Buck et al., 2003, 2004a; Li et al., 2003a; Thiagarajan et al., 1996a,b). Attachment of SE expressing SefA (SEF14), a major fimbrial protein of approximately 14 KDa (Clouthier et al., 1993), to primary chicken oviduct granulosa cells was significantly inhibited when the cells were pre-incubated with the purified fimbrial protein or with antibodies against chicken fibronectin (Thiagarajan et al., 1996a). This fimbriated strain was also able to attach to fibronectin, laminin, and collagen-IV in vitro (Thiagarajan et al., 1996a) and it was hypothesized that these interactions could aid in colonization of RT in hens. However, SEF14 fimbrial expression was not associated with ovary/oviduct colonization in orally infected hens when compared with nonfimbriated strains (Thiagarajan et al., 1996b). *S.* Enteritidis strains also display mannose-sensitive binding to avian isthmal tissues and secretions in vitro (De Buck et al., 2004a), a behavior consistent with mannose-sensitive type-1 fimbriae mediated attachment (deGraft-Hanson and Heath, 1990; Ghosh et al., 1994; van der Bosch et al., 1980). Adhesion through type-1 fimbriae was also demonstrated using defined *fimD* mutants in SE, which are unable to bind to avian ishtmal secretions in vitro (De Buck et al., 2003, 2004a). Furthermore, mutation in *fimD* yielded less frequently contaminated eggs in 19-week-old Isa-Brown-Warren hens after intravenous challenge (De Buck et al., 2004a). Another study suggested that FimA (SEF21), a major fimbrial subunit, mediates the attachment of SE to neutral glycosphingolipids, similar to glucosylceramide and ganglioside GM3, isolated from avian oviduct mucosal epithelial cells (Rajashekara et al., 2000). These molecules could represent host-cell receptors for SE type-1 fimbriae in the hen RT (Li et al., 2003a) and in chicken intestinal mucosa (Li et al., 2003b; Thiagarajan et al., 1996b). More research is needed to clearly define the role of these fimbrial proteins in RT colonization.

A role for LPS in invasion of RT has been investigated in a few studies (Coward et al., 2013; Mizumoto et al., 2005; Parker et al., 2002). Immunohistochemistry studies revealed that attachment and invasion of vaginal epithelium may, in part, rely on LPS type. One study showed that LPS type O9 (SE) was most efficient in attaching and invading avian vaginal epithelium explants when compared with LPS type O4 (*S.* Agona, *S.* Typhimurium, and *S.* Heidelberg) and LPS types O7 and O8 (*S.* Hadar, *S.* Montevideo, and *S.* Infantis) (Mizumoto et al., 2005). It was hypothesized that the differences in molecular characteristics of LPS of different serotypes may drive such phenotype. Interestingly, wild-type strains display greater diversity in LPS O-chain length and glucosylation, producing variable degrees of glucosylated low-molecular-mass (LMM) or high-molecular-mass (HMM) LPS, whereas *S.* Typhimurium mostly produce the LMM-LPS type (Parker et al., 2001). This is important because HMM-LPS is commonly found in SE isolates recovered from eggs (Parker et al., 2001). The mechanism underlying favorable interaction between HMM-LPS and eggs is not completely understood, although the length of HMM-LPS in SE provides more hydrophilicity to the outer membrane and may protect SE from antimicrobial effects of egg albumen (Guard-Petter et al., 1999). In addition, SE strains expressing HMM-LPS show discrete single nucleotide polymorphisms or SNPs, but the role of such mutations in avian RT colonization is currently unknown (Parker et al., 2001). Paradoxically, the inability to make HMM-LPS by mutation of the O-chain length determinant gene (*wzz*) in SE results in more contaminated eggs with poor egg-shell quality and heterophilic granulomas in developing eggs

(Parker et al., 2002). This study suggested that HMM-LPS is important in mitigating these pathological changes in the avian RT. It is possible that HMM-LPS confers an evolutionary advantage to SE by aiding in egg contamination without causing noticeable defects in egg-shell quality and/or the hen RT (Parker et al., 2002). More evidence supporting the importance of O-chain repeat number and length in avian RT infection was provided by the generation of mutants producing either one O-antigen attached to the LPS core, long O-chains (16–35 repeated O units), or very long O-chains (>100 repeat units) (Coward et al., 2013). In this study, SE lacking O-chains or very long O-chains was less efficient in colonizing hen ovary and oviducts, and less able to survive within egg albumen. It is likely that regulation of O-chain length impacts interaction between bacterial factors and their respective cell targets in the different microenvironments found in the hen and in the egg (Coward et al., 2013).

Few high-throughput genome-wide screening studies have pointed at contribution of other genetic factors in colonization/invasion of the RT by SE. In one study, in vivo expression technology and a promoter-trap strategy was used to identify promoters of SE genes that are induced in vivo in both hen oviduct and eggs collected from intravenously inoculated hens (Gantois et al., 2008a). Some genes involved in bacterial metabolism (*asnS* and *purA*), cell membrane integrity (*hflK* and *peg-yohN*), regulation (*lrp*), and stress-response (*uspBA* and *yrfI*) were identified (Gantois et al., 2008a). It was also reported that expression of specific stress-response and cell membrane/wall genes might aid in survival and persistence of SE in oviduct and eggs, presumably by protecting bacteria against cell membrane and DNA damage (Raspoet et al., 2011, 2014). A high-throughput screening strategy that involved microarray-based selective capture of transcribed sequences identified several genes that were overrepresented in both COECs and avian macrophages (HD-11) (McKelvey et al., 2014). Five SPI-2 genes (all belonging to the *ssa* operon) were overrepresented, suggesting a potential role for SPI-2 in survival within COEC and HD-11. Overall, published reports suggest that efficient invasion of COEC and survival within HD-11 could, in part, be attributed to SPI-2 (Bohez et al., 2008; Li et al., 2009; McKelvey et al., 2014). Another high-throughput negative selection screening of a transposon-inserted mutant library identified 81 genes that impacted SE colonization of hen oviduct loops and cultured chicken oviduct tubular gland cells (Raspoet et al., 2014). Major groups of genes included SPI-1 and SPI-2, genes involved in stress response, cell-wall and LPS structure, and the region-of-difference (ROD) genomic locus 9, 21, and 40 (Raspoet et al., 2014). RODs are interesting genomic regions because these are present in SE but absent in *S.* Typhimurium (Thomson et al., 2008). Although the exact role of several of these RODs in pathogenesis of the hen RT infection remains elusive, one study showed that deletion of ROD9 and ROD21 did not directly impact the ability of SE to colonize avian RT (Coward et al., 2012).

### 3.2.1 Knowledge Gaps and Challenges

Although several studies have focused on identifying and characterizing the role of SE genetic factors in colonization of avian RT, follow-up studies to conclusively demonstrate their role are often lacking. Both in vitro cell culture models and in vivo avian infection models have been utilized to identify and characterize genetic factors that contribute to RT infection in chickens. However, the outcome of different studies is not always comparable. For instance, COEC can be a useful in vitro model for high-throughput screening of mutants to identify genes and their possible role in avian RT infection, as well as to study physiological and immunological responses of RT. Nonetheless, critical aspects regarding the source and treatment of COEC vary among studies: primary cultures have been obtained from birds at different ages under different hormonal treatment schemes. In two separate studies, COEC derived from 7-week-old chicks repeatedly treated with estradiol resulted in successful intracellular replication of SE for up to 24 h postinfection (De Buck et al., 2004a,b), whereas COEC derived from 25- to 28-week-old mature hens resulted in failure to establish such intracellular replication (Li et al., 2009). The age, sexual maturation, and differential expression of immune mediators are suspected as potential underlying factors (Anastasiadou et al., 2013; Ebers et al., 2009; Withanage et al., 2003). This may also partially explain the differential susceptibility of ovarian follicles at different maturation stages (Wang et al., 2014). Thus, it is imperative to establish a standardized in vitro cell culture system to study interaction between SE and the avian RT, which may also serve as a model to study other relevant avian pathogens (e.g., avian coronavirus infectious bronchitis virus) with RT tissue tropism (Mork et al., 2014). In vivo studies also differ in several critical aspects that include age, breed, infection dose, route, and bacterial strain. Result outputs are sometimes reported as frequency of SE-positive organs or as bacterial burden (CFU per gram), complicating comparison among different studies. Development and validation of the laying hen model to study the role of different genetic factors in RT infection and contamination of internal content of eggs may help overcome some of the current challenges. In addition to identifying SE factors required for establishing infection in chickens, research is also needed to delineate the host factors. For instance, deciphering the metabolic requirements of SE in the chicken RT may provide some clues to factors that drive predilection of this bacterium to this organ system. In addition, how does SE interact with the chicken microbiota? what factors influence this interaction? and how does this interaction influence outcome of infection? These are some of the key questions that require additional research efforts (Ricke, 2003).

## 3.3 GENETIC BASIS OF EGG COLONIZATION AND SURVIVAL

Because the entire hen RT can be colonized by SE, contamination of internal contents of eggs could result regardless of which specific anatomic site is more heavily colonized. Consequently, there is considerable interest in understanding the genetic mechanisms underlying the ability of this serotype to colonize and survive within the internal contents of eggs, more specifically egg albumen. Early work in this field showed that nonmotile mutants of SE (Δ*fliC* and Δ*motAB*) as well as other nonmotile serovars including *S.* Gallinarum and *S.* Pullorum are impaired in their ability to propagate in egg albumen (Cogan et al., 2004). Moreover, work conducted in our laboratory revealed that wild-type strains of SE with impaired motility are also impaired in their survival within egg albumen when incubated at 25°C (Shah et al., 2012a). In addition, we reported that disruption of *fliH* and *fljB* in SE conferred motility impairment and these mutants were also impaired in their growth in egg albumen (Shah et al., 2012b). These reports collectively suggest that motility is associated with the ability of SE to survive and propagate within egg albumen. It is also speculated that flagella could be an important factor for moving *Salmonella* through albumen and toward the yolk (Baron et al., 1997). In addition to motility, curli fimbriae production appears to be important for invasion and survival within egg contents. One study showed that inactivation of *agfA* (encoding curli production) invaded yolk less frequently suggesting the importance of the role of curli in survival and persistence within the egg (Cogan et al., 2004). In addition, genes involved in DNA replication and repair have been identified in multiple studies. In one study, disruption of *yafD* (a member of exonuclease-endonuclease-phosphatase family) was reported to significantly impact the growth of SE and *S.* Typhimurium within egg albumen (Lu et al., 2003). These authors suggested that absence of *yafD* may affect the ability of bacteria to repair DNA damage and thereby provide a fitness disadvantage to SE in egg albumen. Subsequently, Clavijo et al. (2006) screened a library of 2850 transposon mutants of SE and identified a total of 32 SE mutants that showed defective survivability (≤10%) when grown in egg albumen at 37°C. These genes were broadly categorized into genes associated with cell-wall structure and functional integrity, nucleic acid and amino acid metabolism in SE; however, *yafD* gene reported earlier was not identified in this study. Interestingly, one SE-specific gene (*SEN4287*) putatively associated with restriction endonuclease system was identified by Clavijo et al. (2006). When *SEN4287* was expressed *in trans* in *S.* Typhimurium, it conferred survival advantage in egg albumen when compared with the isogenic wild-type *S.* Typhimurium; however, the survivability was not as high as the wild-type SE strain (Clavijo et al., 2006). These authors suggested that *SEN4287* may have an important role in providing fitness advantage to SE in egg albumen; however, factors other than *SEN4278* may also be required for survival in this hostile environment. We reported that disruption in genes involved in DNA recombination, replication, and repair (*SEN1152*, *SEN1393*, and *SEN1966*) and translation (*ksgA*) resulted in reduced growth in egg albumen (Shah et al., 2012b). LPS also appears to contribute to the survival of SE in egg albumen. In one study, expression of *rfbH*, an LPS biosynthesis gene, was highly upregulated during SE growth in egg albumen at room temperature (Gantois et al., 2009b). Moreover, disruption of *rfbH* resulted in impaired growth in egg albumen at both 37°C and 42°C. Additionally, we reported that disruption of *rfbN*, encoding LPS biosynthesis, resulted in reduced growth in egg albumen (Shah et al., 2012b). It is important to note that several strains of SE contain HMM-LPS, which is associated with egg contamination (Guard-Petter et al., 1999). It is possible that HMM-LPS may interfere with binding of different antimicrobial compounds including lysozyme and ovotransferrin of egg albumen with SE. Thus more research is needed to understand the specific role of the LPS in resistance to the antimicrobial compounds of egg albumen and its association with egg contamination. Collectively, the published body of literature suggests that SE likely undergoes significant metabolic adjustments in egg albumen and that certain strains of SE are likely to be more genetically fit to propagate in egg albumen. One report shows that passage of SE in the egg yolk confers a higher rate of intestinal colonization and extraintestinal organ invasion in orally inoculated mice when compared with SE passaged in laboratory media or in mouse (Moreau et al., 2016). This raises the possibility that propagation of SE within egg contents can significantly increase the pathogenicity of SE in the host; however, the underlying mechanisms are unknown. In conclusion, multiple mechanisms may be at play to regulate bacterial physiology in the egg environment and eventually aid in contamination and survival or propagation of SE within eggs. However, based on the current literature, it has been difficult to pinpoint these mechanisms unequivocally and establish cogent pathways or genetic factors. Consequently, factors or mechanisms that enable SE to colonize and survive within eggs still remain an important knowledge gap. Further research is needed to determine the genetic factors that contribute to the interaction of SE with egg contents and their impact on gene expression and virulence of SE.

## REFERENCES

Addwebi, T.M., Call, D.R., Shah, D.H., 2014. Contribution of *Salmonella* Enteritidis virulence factors to intestinal colonization and systemic dissemination in 1-day-old chickens. Poult. Sci. 93, 871–881.

Allard, M.W., Luo, Y., Strain, E., Pettengill, J., Timme, R., Wang, C., Li, C., Keys, C.E., Zheng, J., Stones, R., Wilson, M.R., Musser, S.M., Brown, E.W., 2013. On the evolutionary history, population genetics and diversity among isolates of *Salmonella* Enteritidis PFGE pattern JEGX01.0004. PLoS One 8, e55254.

Allen-Vercoe, E., Sayers, A.R., Woodward, M.J., 1999. Virulence of *Salmonella enterica* serotype Enteritidis aflagellate and afimbriate mutants in a day-old chick model. Epidemiol. Infect. 122, 395–402.

Allen-Vercoe, E., Woodward, M.J., 1999. Colonisation of the chicken caecum by afimbriate and aflagellate derivatives of *Salmonella enterica* serotype Enteritidis. Vet. Microbiol. 69, 265–275.

Amy, M., Velge, P., Senocq, D., Bottreau, E., Mompart, F., Virlogeux-Payant, I., 2004. Identification of a new *Salmonella enterica* serovar Enteritidis locus involved in cell invasion and in the colonisation of chicks. Res. Microbiol. 155, 543–552.

Anastasiadou, M., Avdi, M., Theodoridis, A., Michailidis, G., 2013. Temporal changes in the expression of avian beta-defensins in the chicken vagina during sexual maturation and *Salmonella* infection. Vet. Res. Commun. 37, 115–122.

Arnold, M.E., Martelli, F., McLaren, I., Davies, R.H., 2014. Estimation of the rate of egg contamination from *Salmonella*-infected chickens. Zoonoses Public Health 61, 18–27.

Baron, F., Gautier, M., Brule, G., 1997. Factors involved in the inhibition of growth of *Salmonella* Enteritidis in liquid egg white. J. Food Prot. 60, 1318–1323.

Berchieri Jr., A., Wigley, P., Page, K., Murphy, C.K., Barrow, P.A., 2001. Further studies on vertical transmission and persistence of *Salmonella enterica* serovar Enteritidis phage type 4 in chickens. Avian Pathol. 30, 297–310.

Berndt, A., Wilhelm, A., Jugert, C., Pieper, J., Sachse, K., Methner, U., 2007. Chicken cecum immune response to *Salmonella enterica* serovars of different levels of invasiveness. Infect. Immun. 75, 5993–6007.

Bichler, L.A., Nagaraja, K.V., Halvorson, D.A., 1996. *Salmonella enteritidis* in eggs, cloacal swab specimens, and internal organs of experimentally infected White Leghorn chickens. Am. J. Vet. Res. 57, 489–495.

Bohez, L., Ducatelle, R., Pasmans, F., Botteldoorn, N., Haesebrouck, F., Van Immerseel, F., 2006. *Salmonella enterica* serovar Enteritidis colonization of the chicken caecum requires the HilA regulatory protein. Vet. Microbiol. 116, 202–210.

Bohez, L., Gantois, I., Ducatelle, R., Pasmans, F., Dewulf, J., Haesebrouck, F., Van Immerseel, F., 2008. The *Salmonella* Pathogenicity Island 2 regulator *ssrA* promotes reproductive tract but not intestinal colonization in chickens. Vet. Microbiol. 126, 216–224.

Chiok, K.L., Addwebi, T., Guard, J., Shah, D.H., 2013. Dimethyl adenosine transferase (KsgA) deficiency in *Salmonella enterica* serovar Enteritidis confers susceptibility to high osmolarity and virulence attenuation in chickens. Appl. Environ. Microbiol. 79, 7857–7866.

Clavijo, R.I., Loui, C., Andersen, G.L., Riley, L.W., Lu, S., 2006. Identification of genes associated with survival of *Salmonella enterica* serovar Enteritidis in chicken egg albumen. Appl. Environ. Microbiol. 72(2), 1055–1064.

Clayton, D.J., Bowen, A.J., Hulme, S.D., Buckley, A.M., Deacon, V.L., Thomson, N.R., Barrow, P.A., Morgan, E., Jones, M.A., Watson, M., Stevens, M.P., 2008. Analysis of the role of 13 major fimbrial subunits in colonisation of the chicken intestines by *Salmonella enterica* serovar Enteritidis reveals a role for a novel locus. BMC Microbiol. 8, 228.

Clouthier, S.C., Muller, K.H., Doran, J.L., Collinson, S.K., Kay, W.W., 1993. Characterization of three fimbrial genes, *sefABC*, of *Salmonella enteritidis*. J. Bacteriol. 175, 2523–2533.

Cogan, T.A., Jorgensen, F., Lappin-Scott, H.M., Benson, C.E., Woodward, M.J., Humphrey, T.J., 2004. Flagella and curli fimbriae are important for the growth of *Salmonella enterica* serovars in hen eggs. Microbiology 150, 1063–1071.

Cooper, G.L., Venables, L.M., Woodward, M.J., Hormaeche, C.E., 1994. Invasiveness and persistence of *Salmonella* Enteritidis, *Salmonella typhimurium*, and a genetically defined *S. enteritidis aroA* strain in young chickens. Infect. Immun. 62, 4739–4746.

Coward, C., Sait, L., Cogan, T., Humphrey, T.J., Maskell, D.J., 2013. O-antigen repeat number in *Salmonella enterica* serovar Enteritidis is important for egg contamination, colonisation of the chicken reproductive tract and survival in egg albumen. FEMS Microbiol. Lett. 343, 169–176.

Coward, C., Sait, L., Williams, L., Humphrey, T.J., Cogan, T., Maskell, D.J., 2012. Investigation into the role of five *Salmonella enterica* serovar Enteritidis genomic islands in colonization of the chicken reproductive tract and other organs following oral challenge. FEMS Microbiol. Lett. 336, 73–78.

Dawoud, T.M., Hererra, P., Hanning, I., Kwon, Y.M., Ricke, S.C., 2011. In vitro invasion of laying hen ovarian follicles by *Salmonella* Enteritidis strains. Poult. Sci. 90, 1134–1137.

De Buck, J., Pasmans, F., Van Immerseel, F., Haesebrouck, F., Ducatelle, R., 2004a. Tubular glands of the isthmus are the predominant colonization site of *Salmonella* enteritidis in the upper oviduct of laying hens. Poult. Sci. 83, 352–358.

De Buck, J., Van Immerseel, F., Haesebrouck, F., Ducatelle, R., 2004b. Colonization of the chicken reproductive tract and egg contamination by *Salmonella*. J. Appl. Microbiol. 97, 233–245.

De Buck, J., Van Immerseel, F., Meulemans, G., Haesebrouck, F., Ducatelle, R., 2003. Adhesion of *Salmonella enterica* serotype Enteritidis isolates to chicken isthmal glandular secretions. Vet. Microbiol. 93, 223–233.

De Reu, K., Grijspeerdt, K., Messens, W., Heyndrickx, M., Uyttendaele, M., Debevere, J., Herman, L., 2006. Eggshell factors influencing eggshell penetration and whole egg contamination by different bacteria, including *Salmonella* Enteritidis. Int. J. Food Microbiol. 112, 253–260.

De Vylder, J., Raspoet, R., Dewulf, J., Haesebrouck, F., Ducatelle, R., Van Immerseel, F., 2013. *Salmonella* Enteritidis is superior in egg white survival compared with other *Salmonella* serotypes. Poult. Sci. 92, 842–845.

deGraft-Hanson, J., Heath, J.L., 1990. Effect of d-mannose on fimbriae of bacterial isolates from chicken carcasses. Poult. Sci. 69, 1582–1589.

Desin, T.S., Lam, P.K., Koch, B., Mickael, C., Berberov, E., Wisner, A.L., Townsend, H.G., Potter, A.A., Koster, W., 2009. *Salmonella enterica* serovar Enteritidis pathogenicity island 1 is not essential for but facilitates rapid systemic spread in chickens. Infect. Immun. 77, 2866–2875.

Dhawi, A.A., Elazomi, A., Jones, M.A., Lovell, M.A., Li, H., Emes, R.D., Barrow, P.A., 2011. Adaptation to the chicken intestine in *Salmonella* Enteritidis PT4 studied by transcriptional analysis. Vet. Microbiol. 153, 198–204.

Ebers, K.L., Zhang, C.Y., Zhang, M.Z., Bailey, R.H., Zhang, S., 2009. Transcriptional profiling avian beta-defensins in chicken oviduct epithelial cells before and after infection with *Salmonella enterica* serovar Enteritidis. BMC Microbiol. 9, 153.

Fan, S., Zheng, J., Duan, Z., Yang, N., Xu, G., 2014. The influences of SE infection on layers' production performance, egg quality and blood biochemical indicators. J. Anim. Sci. Biotechnol. 5, 4.

Fardini, Y., Chettab, K., Grepinet, O., Rochereau, S., Trotereau, J., Harvey, P., Amy, M., Bottreau, E., Bumstead, N., Barrow, P.A., Virlogeux-Payant, I., 2007. The YfgL lipoprotein is essential for type III secretion system expression and virulence of *Salmonella enterica* serovar Enteritidis. Infect. Immun. 75, 358–370.

Gantois, I., Ducatelle, R., Pasmans, F., Haesebrouck, F., Van Immerseel, F., 2008a. *Salmonella enterica* serovar Enteritidis genes induced during oviduct colonization and egg contamination in laying hens. Appl. Environ. Microbiol. 74, 6616–6622.

Gantois, I., Eeckhaut, V., Pasmans, F., Haesebrouck, F., Ducatelle, R., Van Immerseel, F., 2008b. A comparative study on the pathogenesis of egg contamination by different serotypes of *Salmonella*. Avian Pathol. 37, 399–406.

Gantois, I., Ducatelle, R., Pasmans, F., Haesebrouck, F., Gast, R., Humphrey, T.J., Van Immerseel, F., 2009a. Mechanisms of egg contamination by *Salmonella* Enteritidis. FEMS Microbiol. Rev. 33, 718–738.

Gantois, I., Ducatelle, R., Pasmans, F., Haesebrouck, F., Van Immerseel, F., 2009b. The *Salmonella* Enteritidis lipopolysaccharide biosynthesis gene *rfbH* is required for survival in egg albumen. Zoonoses Public Health 56, 145–149.

Garai, P., Gnanadhas, D.P., Chakravortty, D., 2012. *Salmonella enterica* serovars Typhimurium and Typhi as model organisms: revealing paradigm of host-pathogen interactions. Virulence 3, 377–388.

Gast, R.K., Guard-Petter, J., Holt, P.S., 2002. Characteristics of Salmonella enteritidis contamination in eggs after oral, aerosol, and intravenous inoculation of laying hens. Avian Dis. 46, 629–635.

Gast, R.K., Guraya, R., Guard-Bouldin, J., Holt, P.S., 2007a. In vitro penetration of egg yolks by *Salmonella* Enteritidis and *Salmonella* Heidelberg strains during thirty-six-hour ambient temperature storage. Poult. Sci. 86, 1431–1435.

Gast, R.K., Guraya, R., Guard-Bouldin, J., Holt, P.S., Moore, R.W., 2007b. Colonization of specific regions of the reproductive tract and deposition at different locations inside eggs laid by hens infected with *Salmonella* Enteritidis or *Salmonella* heidelberg. Avian Dis. 51, 40–44.

Gast, R.K., Holt, P.S., 1998. Persistence of *Salmonella enteritidis* from one day of age until maturity in experimentally infected layer chickens. Poult. Sci. 77, 1759–1762.

Gast, R.K., Holt, P.S., 2000. Deposition of phage type 4 and 13a *Salmonella* Enteritidis strains in the yolk and albumen of eggs laid by experimentally infected hens. Avian Dis. 44, 706–710.

Gast, R.K., Holt, P.S., 2001. Multiplication in egg yolk and survival in egg albumen of *Salmonella enterica* serotype Enteritidis strains of phage types 4, 8, 13a, and 14b. J. Food Prot. 64, 865–868.

Gautron, J., Hincke, M.T., Panheleux, M., Garcia-Ruiz, J.M., Boldicke, T., Nys, Y., 2001. Ovotransferrin is a matrix protein of the hen eggshell membranes and basal calcified layer. Connect Tissue Res. 42, 255–267.

Gautron, J., Murayama, E., Vignal, A., Morisson, M., McKee, M.D., Rehault, S., Labas, V., Belghazi, M., Vidal, M.L., Nys, Y., Hincke, M.T., 2007. Cloning of ovocalyxin-36, a novel chicken eggshell protein related to lipopolysaccharide-binding proteins, bactericidal permeability-increasing proteins, and plunc family proteins. J. Biol. Chem. 282, 5273–5286.

Gerlach, R.G., Jackel, D., Stecher, B., Wagner, C., Lupas, A., Hardt, W.D., Hensel, M., 2007. *Salmonella* Pathogenicity Island 4 encodes a giant non-fimbrial adhesin and the cognate type 1 secretion system. Cell Microbiol. 9, 1834–1850.

Ghosh, S., Mittal, A., Ganguly, N.K., 1994. Purification and characterization of distinct type of mannose-sensitive fimbriae from *Salmonella typhimurium*. FEMS Microbiol. Lett. 115, 229–234.

Gould, L.H., Walsh, K.A., Vieira, A.R., Herman, K., Williams, I.T., Hall, A.J., Cole, D., Centers for Disease Control and Prevention, 2013. Surveillance for foodborne disease outbreaks – United States, 1998–2008. MMWR Surveill. Summ. 62, 1–34.

Guan, J., Grenier, C., Brooks, B.W., 2006. In vitro study of *Salmonella* enteritidis and *Salmonella* typhimurium definitive type 104: survival in egg albumen and penetration through the vitelline membrane. Poult. Sci. 85, 1678–1681.

Guard-Petter, J., 2001. The chicken, the egg and *Salmonella* enteritidis. Environ. Microbiol. 3, 421–430.

Guard-Petter, J., Parker, C.T., Asokan, K., Carlson, R.W., 1999. Clinical and veterinary isolates of *Salmonella enterica* serovar Enteritidis defective in lipopolysaccharide O-chain polymerization. Appl. Environ. Microbiol. 65, 2195–2201.

Hargis, B.M., Caldwell, D.J., Brewer, R.L., Corrier, D.E., Deloach, J.R., 1995. Evaluation of the chicken crop as a source of *Salmonella* contamination for broiler carcasses. Poult. Sci. 74, 1548–1552.

He, G.Z., Tian, W.Y., Qian, N., Cheng, A.C., Deng, S.X., 2010. Quantitative studies of the distribution pattern for *Salmonella* Enteritidis in the internal organs of chicken after oral challenge by a real-time PCR. Vet. Res. Commun. 34, 669–676.

Higgins, S.E., Erf, G.F., Higgins, J.P., Henderson, S.N., Wolfenden, A.D., Gaona-Ramirez, G., Hargis, B.M., 2007. Effect of probiotic treatment in broiler chicks on intestinal macrophage numbers and phagocytosis of *Salmonella* Enteritidis by abdominal exudate cells. Poult. Sci. 86, 2315–2321.

Hincke, M.T., Gautron, J., Panheleux, M., Garcia-Ruiz, J., McKee, M.D., Nys, Y., 2000. Identification and localization of lysozyme as a component of eggshell membranes and eggshell matrix. Matrix Biol. 19, 443–453.

Hogue, A., White, P., Guard-Petter, J., Schlosser, W., Gast, R., Ebel, E., Farrar, J., Gomez, T., Madden, J., Madison, M., McNamara, A.M., Morales, R., Parham, D., Sparling, P., Sutherlin, W., Swerdlow, D., 1997. Epidemiology and control of egg-associated *Salmonella enteritidis* in the United States of America. Rev. Sci. Tech. 16, 542–553.

Howard, Z.R., Moore, R.W., Zabala-Diaz, I.B., Landers, K.L., Byrd, J.A., Kubena, L.F., Nisbet, D.J., Birkhold, S.G., Ricke, S.C., 2005. Ovarian laying hen follicular maturation and *in vitro Salmonella* internalization. Vet. Microbiol. 108, 95–100.

Kaiser, P., Diard, M., Stecher, B., Hardt, W.D., 2012. The streptomycin mouse model for *Salmonella* diarrhea: functional analysis of the microbiota, the pathogen's virulence factors, and the host's mucosal immune response. Immunol. Rev. 245, 56–83.

Kaneshige, T., Yaguchi, K., Ohgitani, T., 2009. Siderophore receptor IroN is an important protective antigen against *Salmonella* infection in chickens. Avian Dis. 53, 563–567.

Keller, L.H., Benson, C.E., Krotec, K., Eckroade, R.J., 1995. *Salmonella enteritidis* colonization of the reproductive tract and forming and freshly laid eggs of chickens. Infect. Immun. 63, 2443–2449.

Kinde, H., Shivaprasad, H.L., Daft, B.M., Read, D.H., Ardans, A., Breitmeyer, R., Rajashekara, G., Nagaraja, K.V., Gardner, I.A., 2000. Pathologic and bacteriologic findings in 27-week-old commercial laying hens experimentally infected with *Salmonella enteritidis*, phage type 4. Avian Dis. 44, 239–248.

Kiss, T., Morgan, E., Nagy, G., 2007. Contribution of SPI-4 genes to the virulence of *Salmonella enterica*. FEMS Microbiol. Lett. 275, 153–159.

Li, S., Zhang, Z., Pace, L., Lillehoj, H., Zhang, S., 2009. Functions exerted by the virulence-associated type-three secretion systems during *Salmonella enterica* serovar Enteritidis invasion into and survival within chicken oviduct epithelial cells and macrophages. Avian Pathol. 38, 97–106.

Li, W., Watarai, S., Kodama, H., 2003a. Identification of glycosphingolipid binding sites for SEF21-fimbriated *Salmonella enterica* serovar Enteritidis in chicken oviductal mucosa. Vet. Microbiol. 93, 73–78.

Li, W., Watarai, S., Kodama, H., 2003b. Identification of possible chicken intestinal mucosa receptors for SEF21-fimbriated *Salmonella enterica* serovar Enteritidis. Vet. Microbiol. 91, 215–229.

Lu, S., Killoran, P.B., Riley, L.W., 2003. Association of *Salmonella enterica* serovar Enteritidis *yafD* with resistance to chicken egg albumen. Infect. Immun. 71, 6734–6741.

Macri, N.P., Porter, R.E., Holt, P.S., 1997. The effects of induced molting on the severity of acute intestinal inflammation caused by *Salmonella enteritidis*. Avian Dis. 41, 117–124.

Mageed, A.M., Isobe, N., Yoshimura, Y., 2008. Expression of avian beta-defensins in the oviduct and effects of lipopolysaccharide on their expression in the vagina of hens. Poult. Sci. 87, 979–984.

Mann, K., 2007. The chicken egg white proteome. Proteomics 7, 3558–3568.

Matthews, T.D., Schmieder, R., Silva, G.G., Busch, J., Cassman, N., Dutilh, B.E., Green, D., Matlock, B., Heffernan, B., Olsen, G.J., Farris Hanna, L., Schifferli, D.M., Maloy, S., Dinsdale, E.A., Edwards, R.A., 2015. Genomic comparison of the closely-related *Salmonella enterica* serovars Enteritidis, Dublin and Gallinarum. PLoS One 10, e0126883.

McKelvey, J.A., Yang, M., Jiang, Y., Zhang, S., 2014. *Salmonella enterica* serovar Enteritidis antimicrobial peptide resistance genes aid in defense against chicken innate immunity, fecal shedding, and egg deposition. Infect. Immun. 82, 5185–5202.

Messens, W., Duboccage, L., Grijspeerdt, K., Heyndrickx, M., Herman, L., 2004. Growth of *Salmonella* serovars in hens' egg albumen as affected by storage prior to inoculation. Food Microbiol. 21, 25–32.

Mickael, C.S., Lam, P.K., Berberov, E.M., Allan, B., Potter, A.A., Koster, W., 2010. *Salmonella enterica* serovar Enteritidis *tatB* and *tatC* mutants are impaired in Caco-2 cell invasion in vitro and show reduced systemic spread in chickens. Infect. Immun. 78, 3493–3505.

Miyamoto, T., Baba, E., Tanaka, T., Sasai, K., Fukata, T., Arakawa, A., 1997. *Salmonella enteritidis* contamination of eggs from hens inoculated by vaginal, cloacal, and intravenous routes. Avian Dis. 41, 296–303.

Miyamoto, T., Horie, T., Baba, E., Sasai, K., Fukata, T., Arakawa, A., 1998. *Salmonella* penetration through eggshell associated with freshness of laid eggs and refrigeration. J. Food Prot. 61, 350–353.

Mizumoto, N., Sasai, K., Tani, H., Baba, E., 2005. Specific adhesion and invasion of *Salmonella* Enteritidis in the vagina of laying hens. Vet. Microbiol. 111, 99–105.

Moreau, M.R., Wijetunge, D.S., Bailey, M.L., Gongati, S.R., Goodfield, L.L., Hewage, E.M., Kennett, M.J., Fedorchuk, C., Ivanov, Y.V., Linder, J.E., Jayarao, B.M., Kariyawasam, S., 2016. Growth in egg yolk enhances *Salmonella* Enteritidis colonization and virulence in a mouse model of human colitis. PLoS One 11, e0150258.

Morgan, E., Bowen, A.J., Carnell, S.C., Wallis, T.S., Stevens, M.P., 2007. SiiE is secreted by the *Salmonella enterica* serovar Typhimurium Pathogenicity Island 4-encoded secretion system and contributes to intestinal colonization in cattle. Infect. Immun. 75, 1524–1533.

Morgan, E., Campbell, J.D., Rowe, S.C., Bispham, J., Stevens, M.P., Bowen, A.J., Barrow, P.A., Maskell, D.J., Wallis, T.S., 2004. Identification of host-specific colonization factors of *Salmonella enterica* serovar Typhimurium. Mol. Microbiol. 54, 994–1010.

Mork, A.K., Hesse, M., Abd El Rahman, S., Rautenschlein, S., Herrler, G., Winter, C., 2014. Differences in the tissue tropism to chicken oviduct epithelial cells between avian coronavirus IBV strains QX and B1648 are not related to the sialic acid binding properties of their spike proteins. Vet. Res. 45, 67.

Murase, T., Fujimoto, K., Nakayama, R., Otsuki, K., 2006. Multiplication and motility of *Salmonella enterica* serovars Enteritidis, Infantis, and Montevideo in in vitro contamination models of eggs. J. Food Prot. 69, 1012–1016.

Okamura, M., Miyamoto, T., Kamijima, Y., Tani, H., Sasai, K., Baba, E., 2001. Differences in abilities to colonize reproductive organs and to contaminate eggs in intravaginally inoculated hens and in vitro adherences to vaginal explants between *Salmonella enteritidis* and other *Salmonella* serovars. Avian Dis. 45, 962–971.

Parker, C.T., Guard-Petter, J., 2001. Contribution of flagella and invasion proteins to pathogenesis of *Salmonella enterica* serovar *enteritidis* in chicks. FEMS Microbiol. Lett. 204, 287–291.

Parker, C.T., Harmon, B., Guard-Petter, J., 2002. Mitigation of avian reproductive tract function by *Salmonella enteritidis* producing high-molecular-mass lipopolysaccharide. Environ. Microbiol. 4, 538–545.

Parker, C.T., Liebana, E., Henzler, D.J., Guard-Petter, J., 2001. Lipopolysaccharide O-chain microheterogeneity of *Salmonella* serotypes Enteritidis and Typhimurium. Environ. Microbiol. 3, 332–342.

Petter, J.G., 1993. Detection of two smooth colony phenotypes in a *Salmonella enteritidis* isolate which vary in their ability to contaminate eggs. Appl. Environ. Microbiol. 59, 2884–2890.

Pontes, M.H., Lee, E.J., Choi, J., Groisman, E.A., 2015. *Salmonella* promotes virulence by repressing cellulose production. Proc. Natl. Acad. Sci. U. S. A. 112, 5183–5188.

Porter Jr., R.E., Holt, P.S., 1993. Effect of induced molting on the severity of intestinal lesions caused by *Salmonella enteritidis* infection in chickens. Avian Dis. 37, 1009–1016.

Prevost, K., Magal, P., Beaumont, C., 2006. A model of *Salmonella* infection within industrial house hens. J. Theor. Biol. 242, 755–763.

Quinteiro-Filho, W.M., Gomes, A.V., Pinheiro, M.L., Ribeiro, A., Ferraz-de-Paula, V., Astolfi-Ferreira, C.S., Ferreira, A.J., Palermo-Neto, J., 2012. Heat stress impairs performance and induces intestinal inflammation in broiler chickens infected with *Salmonella* Enteritidis. Avian Pathol. 41, 421–427.

Rajashekara, G., Munir, S., Alexeyev, M.F., Halvorson, D.A., Wells, C.L., Nagaraja, K.V., 2000. Pathogenic role of SEF14, SEF17, and SEF21 fimbriae in *Salmonella enterica* serovar *enteritidis* infection of chickens. Appl. Environ. Microbiol. 66, 1759–1763.

Raspoet, R., Appia-Ayme, C., Shearer, N., Martel, A., Pasmans, F., Haesebrouck, F., Ducatelle, R., Thompson, A., Van Immerseel, F., 2014. Microarray-based detection of *Salmonella enterica* serovar Enteritidis genes involved in chicken reproductive tract colonization. Appl. Environ. Microbiol. 80, 7710–7716.

Raspoet, R., Gantois, I., Devloo, R., Martel, A., Haesebrouck, F., Pasmans, F., Ducatelle, R., Van Immerseel, F., 2011. *Salmonella* Enteritidis universal stress protein (*usp*) gene expression is stimulated by egg white and supports oviduct colonization and egg contamination in laying hens. Vet. Microbiol. 153, 186–190.

Reiber, M.A., Conner, D.E., 1995. Effect of mating activity on the ability of *Salmonella enteritidis* to persist in the ovary and oviduct of chickens. Avian Dis. 39, 323–327.

Ricke, S.C., 2003. The gastrointestinal tract ecology of *Salmonella* Enteritidis colonization in molting hens. Poult. Sci. 82, 1003–1007.

Rychlik, I., Karasova, D., Sebkova, A., Volf, J., Sisak, F., Havlickova, H., Kummer, V., Imre, A., Szmolka, A., Nagy, B., 2009. Virulence potential of five major pathogenicity islands (SPI-1 to SPI-5) of *Salmonella enterica* serovar Enteritidis for chickens. BMC Microbiol. 9, 268.

Shah, D.H., 2014. RNA sequencing reveals differences between the global transcriptomes of *Salmonella enterica* serovar Enteritidis strains with high and low pathogenicities. Appl. Environ. Microbiol. 80, 896–906.

Shah, D.H., Casavant, C., Hawley, Q., Addwebi, T., Call, D.R., Guard, J., 2012a. *Salmonella* Enteritidis strains from poultry exhibit differential responses to acid stress, oxidative stress, and survival in the egg albumen. Foodborne Pathog. Dis. 9, 258–264.

Shah, D.H., Zhou, X., Kim, H.Y., Call, D.R., Guard, J., 2012b. Transposon mutagenesis of *Salmonella enterica* serovar Enteritidis identifies genes that contribute to invasiveness in human and chicken cells and survival in egg albumen. Infect. Immun. 80, 4203–4215.

Shah, D.H., Zhou, X., Addwebi, T., Davis, M.A., Call, D.R., 2011. In vitro and in vivo pathogenicity of *Salmonella* Enteritidis clinical strains isolated from North America. Arch. Microbiol. 193, 811–821.

Shippy, D.C., Eakley, N.M., Mikheil, D.M., Fadl, A.A., 2013. Role of StdA in adhesion of *Salmonella enterica* serovar Enteritidis phage type 8 to host intestinal epithelial cells. Gut Pathog. 5, 43.

Shippy, D.C., Eakley, N.M., Mikheil, D.M., Fadl, A.A., 2014. Role of the flagellar basal-body protein, FlgC, in the binding of *Salmonella enterica* serovar Enteritidis to host cells. Curr. Microbiol. 68, 621–628.

Shivaprasad, H.L., Timoney, J.F., Morales, S., Lucio, B., Baker, R.C., 1990. Pathogenesis of *Salmonella enteritidis* infection in laying chickens. I. Studies on egg transmission, clinical signs, fecal shedding, and serologic responses. Avian Dis. 34, 548–557.

Silva, C.A., Blondel, C.J., Quezada, C.P., Porwollik, S., Andrews-Polymenis, H.L., Toro, C.S., Zaldivar, M., Contreras, I., McClelland, M., Santiviago, C.A., 2012. Infection of mice by *Salmonella enterica* serovar Enteritidis involves additional genes that are absent in the genome of serovar Typhimurium. Infect. Immun. 80, 839–849.

Stevens, L., 1991. Egg white proteins. Comp. Biochem. Physiol. B 100, 1–9.

Thiagarajan, D., Saeed, A.M., Asem, E.K., 1994. Mechanism of transovarian transmission of *Salmonella enteritidis* in laying hens. Poult. Sci. 73, 89–98.

Thiagarajan, D., Saeed, M., Turek, J., Asem, E., 1996a. In vitro attachment and invasion of chicken ovarian granulosa cells by *Salmonella enteritidis* phage type 8. Infect. Immun. 64, 5015–5021.

Thiagarajan, D., Thacker, H.L., Saeed, A.M., 1996b. Experimental infection of laying hens with *Salmonella enteritidis* strains that express different types of fimbriae. Poult. Sci. 75, 1365–1372.

Thomson, N.R., Clayton, D.J., Windhorst, D., Vernikos, G., Davidson, S., Churcher, C., Quail, M.A., Stevens, M., Jones, M.A., Watson, M., Barron, A., Layton, A., Pickard, D., Kingsley, R.A., Bignell, A., Clark, L., Harris, B., Ormond, D., Abdellah, Z., Brooks, K., Cherevach, I., Chillingworth, T., Woodward, J., Norberczak, H., Lord, A., Arrowsmith, C., Jagels, K., Moule, S., Mungall, K., Sanders, M., Whitehead, S., Chabalgoity, J.A., Maskell, D., Humphrey, T., Roberts, M., Barrow, P.A., Dougan, G., Parkhill, J., 2008. Comparative genome analysis of *Salmonella* Enteritidis PT4 and *Salmonella* Gallinarum 287/91 provides insights into evolutionary and host adaptation pathways. Genome Res. 18, 1624–1637.

Thorns, C.J., Turcotte, C., Gemmell, C.G., Woodward, M.J., 1996. Studies into the role of the SEF14 fimbrial antigen in the pathogenesis of *Salmonella enteritidis*. Microb. Pathog. 20, 235–246.

Turnbull, P.C., Snoeyenbos, G.H., 1974. Experimental salmonellosis in the chicken. 1. Fate and host response in alimentary canal, liver, and spleen. Avian Dis. 18, 153–177.

van der Bosch, J.F., Verboom-Sohmer, U., Postma, P., de Graaff, J., MacLaren, D.M., 1980. Mannose-sensitive and mannose-resistant adherence to human uroepithelial cells and urinary virulence of *Escherichia coli*. Infect. Immun. 29, 226–233.

Van Immerseel, F., 2010. Stress-induced survival strategies enable *Salmonella* Enteritidis to persistently colonize the chicken oviduct tissue and cope with antimicrobial factors in egg white: a hypothesis to explain a pandemic. Gut Pathog. 2, 23.

Van Immerseel, F., De Buck, J., Boyen, F., Bohez, L., Pasmans, F., Volf, J., Sevcik, M., Rychlik, I., Haesebrouck, F., Ducatelle, R., 2004. Medium-chain fatty acids decrease colonization and invasion through *hilA* suppression shortly after infection of chickens with *Salmonella enterica* serovar Enteritidis. Appl. Environ. Microbiol. 70, 3582–3587.

Van Immerseel, F., De Buck, J., De Smet, I., Mast, J., Haesebrouck, F., Ducatelle, R., 2002. Dynamics of immune cell infiltration in the caecal lamina propria of chickens after neonatal infection with a *Salmonella* Enteritidis strain. Dev. Comp. Immunol. 26, 355–364.

Vishwakarma, V., Periaswamy, B., Bhusan Pati, N., Slack, E., Hardt, W.D., Suar, M., 2012. A novel phage element of *Salmonella enterica* serovar Enteritidis P125109 contributes to accelerated type III secretion system 2-dependent early inflammation kinetics in a mouse colitis model. Infect. Immun. 80, 3236–3246.

Wallis, T.S., Wood, M., Watson, P., Paulin, S., Jones, M., Galyov, E., 1999. Sips, Sops, and SPIs but not STN influence *Salmonella* enteropathogenesis. Adv. Exp. Med. Biol. 473, 275–280.

Wang, C.L., Fan, Y.C., Chun-Hsien, T., Chiu, C.H., Tsai, H.J., Chou, C.H., 2014. *Salmonella* Enteritidis infection slows steroidogenesis and impedes cell growth in hen granulosa cells. Avian Dis. 58, 511–517.

Watson, K.G., Holden, D.W., 2010. Dynamics of growth and dissemination of *Salmonella* in vivo. Cell Microbiol. 12, 1389–1397.

Wisner, A.L., Desin, T.S., Koch, B., Lam, P.K., Berberov, E.M., Mickael, C.S., Potter, A.A., Koster, W., 2010. *Salmonella enterica* subspecies *enterica* serovar Enteritidis *Salmonella* pathogenicity island 2 type III secretion system: role in intestinal colonization of chickens and systemic spread. Microbiology 156, 2770–2781.

Wisner, A.L., Potter, A.A., Koster, W., 2011. Effect of the *Salmonella* pathogenicity island 2 type III secretion system on *Salmonella* survival in activated chicken macrophage-like HD11 cells. PLoS One 6, e29787.

Withanage, G.S., Sasai, K., Fukata, T., Miyamoto, T., Lillehoj, H.S., Baba, E., 2003. Increased lymphocyte subpopulations and macrophages in the ovaries and oviducts of laying hens infected with *Salmonella enterica* serovar Enteritidis. Avian Pathol. 32, 583–590.

Yim, L., Betancor, L., Martinez, A., Giossa, G., Bryant, C., Maskell, D., Chabalgoity, J.A., 2010. Differential phenotypic diversity among epidemic-spanning *Salmonella enterica* serovar Enteritidis isolates from humans or animals. Appl. Environ. Microbiol. 76, 6812–6820.

CHAPTER

# 11 The Relationship Between the Immune Response and Susceptibility to *Salmonella enterica* Serovar Enteritidis Infection in the Laying Hen

**Michael H. Kogut, James A. Byrd**
*USDA-ARS, College Station, TX, United States*

## 1. INTRODUCTION

A review by Wigley et al., (2014) has thoroughly detailed the basic structure and function of the avian reproductive tract and will not be discussed here. What is clear is that the avian female reproductive tract, like that of the intestinal and respiratory tracts, has a fully functional immune system made up of cells from both the innate acquired immune branches (Wigley et al., 2014). Despite this immune armament, *Salmonella enterica* serovar Enteritidis (SE) is still capable of evading detection and infecting table eggs for human consumption. This chapter will concentrate less on reviewing the response of the immune system of the reproductive tract to *Salmonella* infection, which has been reviewed (Wigley et al., 2014). Instead, we will concentrate on asking why and/or how *Salmonella* Enteritidis is able to penetrate the intestinal tract and migrate to and colonize the reproductive tract while evading detection by the host immune system. However, it is important that a thorough description of the avian immune system would be appropriate to provide the reader with a sufficient background to appreciate the infection biology of the bacteria in the avian host.

Asymptomatic carrier states are poorly understood. "Normal infections" include infection of chicks through an oral route and is characterized by a translocation through the intestinal epithelial cells followed with a splenic infection (Desmidt et al., 1997). Although asymptomatic carriers can be infected by SE and *Salmonella* Typhimurium (ST), these bacteria can survive in the gastrointestinal tract of birds for months without showing clinical signs (Barrow et al., 1987). These *Salmonella* asymptomatic carriers can be capable of having an infected gastrointestinal tract without showing clinical signs while excreting high concentrations of *Salmonella* into the environment (Aksakal et al., 2009; Perron et al., 2008). In older birds, ST

**Producing Safe Eggs. http://dx.doi.org/10.1016/B978-0-12-802582-6.00011-2**

infection leads to an asymptomatic carrier state with continuous shedding (Barrow et al., 1988; Withanage et al., 2005). These healthy carriers can be a risk to affect other birds by horizontal transmission or affect newly hatched chicks.

Hens infected with SE often shed bacteria for 1–2 weeks. But hens given large SE doses often shed low numbers or stop shedding SE within 3–4 weeks (Holt and Porter, 1993; Saeed, 1999; Seo et al., 2000). Feed also plays a role in the shedding of SE. In a study by Holt and Porter (Holt and Porter, 1993), 59-week-old hens when challenged with 7.0 $\log_{10}$ SE had a 71% shedding rate. At 28 days post infection, 20% of the hens were shedding SE. At 30 days post infection, feed and water were removed and caused an increase in SE shedding to up to 65%. These hens may have had subclinical infections and became chronic shedders and will continue to lay contaminated eggs (Shivaprasad, 2000; Velge et al., 2005). *S.* Enteritidis colonization depends on the infecting strain, age of host, host immune status, and host species (Barrow et al., 1987; Morgan et al., 2004; Shivaprasad, 2000).

## 2. OVERVIEW OF THE AVIAN IMMUNE SYSTEM

The immune system of vertebrates is a multifaceted network of molecules and cells that coordinate responses against infectious agents, toxins, and danger signals while maintaining tolerance to self-antigens. The mechanisms of the host immune response of vertebrates are classically separated into two interdependent branches: (1) the "hard-wired" responses encoded by genes in the host's germline and that recognize molecular patterns shared by multiple microorganisms/dangers that are not in the host, the innate immune response, and (2) responses encoded by gene components that somatically rearrange to assemble antigen (Ag)-binding molecules, with specificity for individual, unique foreign structures, the acquired/adaptive immune response (Table 11.1).

### 2.1 INNATE IMMUNITY

The host immune response to pathogens in the earliest stages of infection is a critical determinant of disease resistance and susceptibility. These early responses, the innate host defenses, are dedicated to the containment of the pathogens holding infections to a level that can be resolved by the ensuing development of acquired immune mechanisms. The first "layer" of the innate immune system includes physical (epithelial lining and mucous layers of intestinal and reproductive tracts, skin), soluble [proteins and bioactive small molecules that are present in biological fluids (complement, β-defensins, cathelicidins) or released by cells when activated (cytokines, prostaglandins, leukotrienes, enzymes)] (Janeway and Medzhitov, 2002). The second "layer" of the innate immune system is a rapidly induced, phylogenitically conserved response of all multicellular organisms (Medzhitov and Janeway, 1997). Innate immunity depends on a collection of germ-line-encoded pattern recognition receptors (PRRs) for detection of microbe-associated molecular patterns (PAMPs) on or

**Table 11.1** Characteristics of the Different Arms of Avian Immune Response

| Characteristics | Innate (Constitutive) | Acquired (Inducible) |
|---|---|---|
| Receptors | MAMPs, DAMPs | Specific T and B cell antigen receptors (TCR/BCR) |
| Receptor features | Germline encoded, invariant, nonclonal | Random somatic gene rearrangement, clonal |
| Specificity of response | Nonspecific | Specific (Ag) |
| Response time | Rapid: immediate | Slow: days/weeks |
| | All multicellular | Only in vertebrates |
| Memory | Trained immunity? | Memory |
| Cellular components | Monocytes, macrophages, polymorphonuclear cells, natural killer, dendritic cells, epithelial cells | T and B lymphocytes |
| Humoral components | Antimicrobial peptides, complement, C-reactive protein, mannose-binding protein | Immunoglobulins |

in major groups of microbes (Janeway and Medzhitov, 2002) or damage-associated molecular patterns and includes uric acid, ATP, DNA fragments, and mitochondrial contents (Hanson et al., 2011). Microbial- or pathogen-associated molecular patterns (MAMPs or PAMPs) are critical for pathogen replication and/or survival and are unique to large groups of microorganisms and not host cells, thus providing the host with an efficient, nonself means of detecting invading pathogens. Recognition of MAMPs induces various extracellular activation cascades and intracellular signaling pathways, leading to the inflammatory response, recruitment of phagocytic cells for clearance of the pathogens, and mobilization of professional antigen-presenting cells. PRRs are present in two separate compartments of the host: cell membranes and cell cytoplasm. PRRs on cell membranes have an assortment of functional activities including promotion of phagocytosis, presentation of MAMPs to other PRRs, and the initiation of major intracellular signaling pathways. Recognition of MAMPs by PRRs, either alone or in heterodimerization with other PRRs [Toll-like receptors (TLRs); nucleotide-binding oligomerization domain proteins (NLR); retinoic-acid inducible gene-I (RLRs); C-type lectins], induces intracellular signals responsible for the activation of genes that encode for proinflammatory cytokines, antiapoptotic factors, and antimicrobial peptides (AMPs) (Carpenter and O'Neill, 2007; Underhill, 2007; Takeuchi and Akira, 2010). At least 11 different TLRs have been identified in the chicken (Keestra et al., 2013). All TLRs have an extracellular sensing leucine-rich repeat (LRR) domain, a transmembrane domain, and a highly conserved cytoplasmic Toll- and interleukin (IL)-1 receptor (TIR) (Trinchieri and Sher, 2007). PAMPs recognized by TLRs include three general categories of ligands: proteins, nucleic acids, and lipid-based elements derived from a wide range of organisms such as bacteria, viruses, parasites, and fungi. Two additional families of innate receptors,

RLRs and NLRs, have been described and join the TLRs as key pathogen sensors. In mammals, the RLR family contains retinoic acid-inducible gene 1 (RIG-1), melanoma differentiation-associated protein 5 (MDA5), and laboratory of genetics and physiology 2 as members, but chickens lack RIG-1 (Magor et al., 2013). The lack of the RIG-1 family undoubtedly accounts for the susceptibility of chickens to single-stranded RNA viruses such as influenza A and Newcastle disease virus. However, the MDA5 family can, at least partially, detect avian influenza to generate an interferon response (Karpala et al., 2011). NLRs are a group of intracellular microbial sensors that sense microbial products or the products of damaged cells, such as ATP and uric acid. NLRs are composed of a central nucleotide-binding domain and C-terminal LRRs (Maekawa et al., 2011). NLRs have been classified into four subfamilies, in mammals, with members such as NOD1, NOD2, NACHT, NALPs, and IPAF (Kawai and Akira, 2009). However, to date, there have been four NLRs found on the chicken genome [NOD1 (Tao et al., 2015, NLRP3 (Ye et al., 2015, NLRC5 (Lian et al., 2012; Ciraci et al., 2010)], and only two have been functionally described (Ciraci et al., 2010; Ye et al., 2015; Liu et al., 2015).

Although a certain degree of redundancy exists between signals induced by various PRRs, in general, no single PRR is likely to be the sole mediator of activation of the innate immune response. Therefore, a variety of pathogens, each containing different MAMPs, can interact with a certain combination of PRRs on or in a host cell. The variety of PRR complexes trigger specific intracellular signal transduction pathways that will induce specific gene expression profiles, particularly cytokine/chemokine expression, best suited for the invasive pathogen (Chang et al., 2006; Akira et al., 2006; Gowan et al., 2007; Medzhitov, 2007; Geijtenboek and Gringhuis, 2009; Kawai and Akira, 2009; Cao, 2016). Hence a combination of TLR and non-TLR PRRs will be triggered during infections (Kawai and Akira, 2011).

Ligation of MAMPs to TLRs activate intracellular signaling cascades that induce nuclear factor (NF)-κβ-dependent genes and the synthesis of multiple inflammatory mediators that include soluble proteins and bioactive molecules that are either constitutively present in biological fluids such as defensins or are released from cells as they are activated including cytokines, chemokines, and lipid inflammatory mediators (prostaglandins, leukotrienes) (Kaiser et al., 2005; Scott and Owens, 2008; Kogut et al., 2012; Guo et al., 2013; Cuperus et al., 2013; Goosens et al., 2013).

## 2.2 ACQUIRED IMMUNITY

The initiation of the acquired immune response is based on recognition of an antigen by a specific receptor expressed on the surface of T or B lymphocytes (Erf, 2004). The antigen specificity of the receptors is encoded by genes that are assembled by somatic rearrangement of a number of gene components to form intact T-cell receptor and immunoglobulin B-cell receptor genes. The assembly of antigen receptors from a collection of a few hundred germline genes permits the formation of millions of different antigen receptors each with potentially unique specificity for a different antigen.

The main effector mechanisms of adaptive immunity include the production of antibodies by B cells, the killing of infected host cells by cytotoxic T cells, and various helper T-cell-mediated actions. In particular, activation of adaptive immunity results in the production of memory B cells and T cells, which can provide life-long specific protection against subsequent infections with a pathogen bearing the same antigens. Like mammals, chickens have both humoral and cell-mediated branches of the acquired immune system (Wigley, 2013). The bursa of Fabricius is a unique organ of birds that is essential for B lymphocyte development and humoral immunity (Glick, 1991). Embryonic stem cells migrate to the bursa and undergo proliferation that persists for several weeks after hatch. These precursor B cells have already rearranged their immunoglobulin genes before entering the bursa; unlike mammals, the chicken has a very limited number of variable genes using a process called gene conversion to create antibody diversity. In gene conversion, the variable heavy and light chains are replaced with upstream pseudogenes (Benatar et al., 1992). The cell-mediated component of the acquired response includes $\alpha\beta$ and $\gamma\delta$ T cells and natural killer cells (Myers and Schat, 1990; Trout and Lillehoj, 1996; Straub et al., 2013; Guo et al., 2013). Both CD4 and CD8 T-cell subsets are present with Th1 CD4+ cells having similar function as found in mammals (Gobel et al., 2003; Guo et al., 2013). Furthermore, Th2 cells have been shown to function as well (Kaiser et al., 2005; Degen et al., 2005).

## 3. IMMUNITY IN THE HEN REPRODUCTIVE TRACT

Chicken reproductive tissues are not considered immunological, but immune cells and innate and acquired immune mechanisms have been shown to be functional throughout the tract (Subedi et al., 2007; Woods et al., 2009; Michailidis et al., 2012; Wigley et al., 2014; Yoshimura et al., 2014).

### 3.1 INNATE IMMUNITY

There is a large and growing body of literature citing the presence of complex immune responses, especially the innate immune response, within the female reproductive system. The female reproductive tract is lined by an epithelial cell layer that produces a protective mucous barrier and itself provides a tight barrier (Yoshimura et al., 2014). Furthermore, multiple cells in the organs of the reproductive tract, including various epithelial and macrophage-like cells in ovary, oviduct, isthmus, and ovarian follicles, express a number of TLRs (Woods et al., 2009; Michailidis et al., 2010, 2011), which can respond to MAMPs by an increased mRNA expression of a number of cytokines (Yoshimura et al., 2006; Sundaresan et al., 2007, 2008; Nii et al., 2011; Abdelsalam et al., 2012; Abdel Mageed et al., 2011; Sonoda et al., 2013), chemokines (Sundaresan et al., 2007, 2008), and lipid inflammatory mediators (Yoshimura et al., 2014). However, the largest body of literature has detailed the expression of a number of AMPs (summarized by Yoshimura, 2015). A total of 14 avian β-defensin (AvBD, characterized by conserved cysteine-rich domains) genes have been identified in the chicken (Yoshimura, 2015) with 11 AvBDs having been found to be expressed in the various cell types of the ovary and oviduct of laying hens (Abdel Mageed et al., 2008; Michailidis et al., 2012) with differential expression influenced by the bird sexual maturity, age, and breed (Yoshimura et al., 2006; Ebers et al., 2009; Michailidis et al., 2012). Interestingly, seven AvBDs have also been found in the egg, suggesting the presence of protective bioactive molecules in ovo (Mann, 2007; Mann and Mann, 2008; Mine and D'Silva, 2008). Lastly, a separate AMP, chicken liver expressed AMP-2 (LEAP-2; characterized by disulfide-bond, cysteine-rich motifs) has been constitutively expressed in the chicken ovary (Michailidis et al., 2010).

### 3.2 ACQUIRED IMMUNITY

B and T lymphocytes are found scattered throughout the hen reproductive tract (Wigley et al., 2014; Yoshimura et al., 2014). IgM+, IgY+, and IgA + B cells have been found beneath the epithelial layer of the oviduct (Lebacq-Verheyden et al., 1974; Withanage et al., 1997). Furthermore, both CD4+ and CD8+ T cells, and γδ T cells are found in the ovaries, vagina, and oviduct as single cells and in small lymphoid aggregates (Withanage et al., 1997; Johnston et al., 2012). Although dendritic cells, per se, have not been identified in the reproductive tract, major histocompatibility complex class II+ macrophages that do function as antigen presenting cells have been found in the mucosal epithelium and lamina propria layers of the oviduct as well (Zheng and Yoshimura, 1999). Therefore, the appropriate cell types are available in the hen reproductive tract for a functional acquired response. Proinflammatory cytokines mRNA expression in the reproductive tract has been reported (Sundaresan et al., 2007, 2008); the scattered distribution of T lymphocytes makes it difficult to measure the relative expression of Th1 and Th2 cytokine mRNA (Johnston et al., 2012). However, these authors were able to detect IL-4 and IL-6 expression (Johnston et al., 2012).

## 4. IMMUNE RESPONSE OF THE LAYING HEN REPRODUCTIVE TRACT TO *SALMONELLA* INFECTION

The immune response of the avian reproductive tract to *Salmonella* infection has been comprehensively reviewed (Wigley et al., 2014). Thus, we will only briefly summarize the ability of the immune system of the hen reproductive tract to respond to colonization by *Salmonella*. The innate immune response appears to be characterized by the upregulation of TLR4, -5, and -15, which recognize lipopolysaccharide (LPS), flagellin, and bacterial proteases, respectively (Keestra et al., 2013), in response to SE infection (Ozoe et al., 2009; Michailidis et al., 2011; Yoshimura, 2015). Recognition of these MAMPs activates the TLR signaling pathways that resulted in the upregulation of five avian defensin genes (AvβD4, 5, 7, 11, and 12) in the ovary (Michailidis et al.,

2012) and six AvβDs in the vagina (AvβD5, 7, 10, 11, 12, 14; Anastasiadou et al., 2013) of sexually mature hens. Furthermore, TLR stimulation also results in the increased expression of the proinflammatory cytokines IL-1β and IL-6 and chemokines CxCLi2 and LPS-induced tumor necrosis factor-α (Yoshimura et al., 2006; Sundarersan et al., 2007, 2008; Abdel Mageed et al., 2008, 2011). Together, these findings suggest a functional innate response to *Salmonella* infection manifested by TLR recognition of bacterial MAMPs and the production of AMPs.

*Salmonella* infection in the reproductive tract of mature hens also evokes both humoral and cell-mediated acquired immune response. *S.* Enteritidis-specific antibodies were found locally (oviducts) and systemically (serum) in hens after infection (Withange et al., 1998). Furthermore, infection with SE induced the local secretion of IgA, IgY, and IgM in the oviducts (Withange et al., 1999). This *S.* Enteritidis-specific antibody production correlated with a reduction in bacterial load in the oviducts. An increase in T lymphocyte numbers was found in the oviducts and ovaries of infected hens 1 week post infection, which was followed by an increase in B lymphocytes at 14 days post infection (Withange et al., 1998). The number of the CD4+ T helper and CD8+ T cytotoxic cells also increased in the reproductive tracts of *S.* Enteritidis-infected hens during the first 14 days post infection (Withange et al., 2003). An increase in macrophage numbers in the reproductive tract of mature hens at the same time as the increase in T cells suggests the clonal expansion of bacteria-specific acquired memory response.

So the main question that needs to be asked is with such a complex arsenal of weapons within the innate and acquired immune systems designed to identify, limit the spread, and remove bacterial pathogens such as *Salmonella*, how does this pathogen avoid detection by the immune system and persist systemically in a carrier state in the reproductive tract of hens throughout sexual development? We will concentrate the rest of this chapter on attempting to answer this question.

## 5. HOST AND BACTERIAL FACTORS THAT CONTRIBUTE TO ASYMPTOMATIC *S.* ENTERITIDIS COLONIZATION OF REPRODUCTIVE TRACT

Although colonization of the reproductive tract can occur via an ascending infection from the shared opening of the reproductive and intestinal tracts, the primary route of infection is the systemic spread of the bacteria from the intestinal tract (Gast et al., 2003; Gantois et al., 2009; Wigley et al., 2014). In mammals, *Salmonella* have two different routes to systemic infections. First, the bacteria can travel directly through the lymphatic system directly to systemic organs (Worley et al., 2006). However, chickens possess neither lymph nodes nor a lymphatic system (Kaiser et al., 2005). The second route is directly through the bloodstream carried by phagocytic cells (macrophages and/or dendritic cells) (Vasquez-Torres et al., 1999). In poultry, it is via macrophages that *Salmonella* are carried to the reproductive tract (Gast et al., 2007; Bernt et al., 2007; Gantois et al., 2008; Chappell et al., 2009; He et al., 2012).

*Salmonella* possess a set of important factors genetically determined in two type III secretion systems (TTSS-1 andTTSS-2) encoded on *Salmonella* Pathogenicity Islands (SPI), particularly SPI-1 and SPI-2. The SPI-1 is essential for colonization in the gut (Dieye et al., 2009), whereas both SPI-1 and SPI-2 are required for colonization of systemic organs (Dieye et al., 2009; Rychlik et al., 2009). Therefore, a successful SE infection of the avian reproductive tract depends initially on the outcome of the bacteria's encounter with the macrophage followed by the ability to persistently colonize the ovary/oviduct/vagina despite the presence of a functional immune surveillance system. *Salmonella*'s ability to survive, colonize, and persist depends on a series of bacterial and host factors that leads to the asymptomatic infection of the reproductive tract.

## 6. BACTERIAL FACTORS

### 6.1 *SALMONELLA*–MACROPHAGE INTERACTIONS: EVASION OF PHAGOCYTE DEFENSE MECHANISMS

The first "phase" of a *Salmonella enterica* infection in the reproductive tract of maturing hens begins with the invasion of and penetration through the epithelium lining the intestinal tract (Chappell et al., 2009; Wigley et al., 2014) where the bacteria encounter resident macrophages lining the basolateral side of the epithelium (Chappell et al., 2009; He et al., 2012; Brauckmann et al., 2015). These macrophages can serve as both host cells and transport cells for the bacteria, so their ability to survive and replicate in the immune cells is of tantamount importance for persistence in extraintestinal organs (Chappell et al., 2009; Wigley et al., 2014). Survival, growth, and persistence in macrophages is dependent on the SPI-2 type III secretion system (Chappell et al., 2009; Setta et al., 2012; He et al., 2012, 2013). In mammalian models, a second virulence system, the PhoP/Q two-component regulatory system, has also been shown to be a factor in *Salmonella* survival in macrophages (Thompson et al., 2011; Lathrop et al., 2015). However, there are no reports of the PhoP/Q system involvement in *Salmonella*-avian macrophage interactions. The SPI-2 virulence system is induced in *Salmonella* by intracellular signals as the bacteria are growing in the acidified, modified phagosome [*Salmonella*-containing vacuole (SCV)], which prevent phagolysosomal fusion and thus avoid exposure to lysosomal antimicrobial contents (Vasquez-Torres et al., 2000; Haraga et al., 2008; Steele-Mortimer, 2008). Furthermore, as yet unidentified specific SPI-2 effector proteins have been shown to protect *Salmonella* from the macrophage reactive oxygen intermediate, NADPH oxidase, and reactive nitrogen species, nitric oxide (NO) (Vazquez-Torres et al., 2000; Das et al., 2009; Aussel et al., 2011; Henard and Vazquez-Torres, 2011). Specifically, *Salmonella* downregulates NO production and inducible nitric oxide synthase induction in interferon-γ-activated macrophages in an SPI-2-dependent manner (Das et al., 2009). Moreover, an SPI-2 effector protein blocks correct colocalization of NADPH oxidase vesicles with SCV (Vasquez-Torres et al., 2000; van der Heijden et al., 2015).

An emerging host defense mechanism against *Salmonella* is the process of autophagy. Autophagy, under normal homeostatic condition, is an evolutionarily conserved cellular response to remove defective proteins and organelles, but has been shown to be involved in the capture and removal of intracellular bacteria (Levine, 2005; Cemma and Brumell, 2012; Deretic et al., 2013). The autophagic response involves development of autophagosomes that engulf cytosolic components or bacteria that then fuse with lysosomes for degradation (Levine, 2005). *Salmonella* have been shown to interact with and are contained by autophagy systems in both phagocytic and nonphagocytic host cells (Birmingham et al., 2006). In this process, during the formation of the SCV, cytoplasmic aggregates form that are ubiquitinated by host ligases that enable the aggregates to be recognized by the autophagy pathway (Mesquita et al., 2012; Narayanan and Edelmann, 2014). Evidence has shown that *Salmonella* inhibits its antibacterial autophagy through SPI-2-dependent effector proteins that target two different posttranslational protein modification pathways. Posttranslational modifications, such as ubiquitination and phosphorylation, play vital roles in bacterial evasion of phagocytic cell killing (Narayanan and Edelmann, 2014). With the first, the intracellular bacteria release the SPI-2 effector protein, SseL, which deubiquitinates the cytoplasmic aggregates, which inhibits the ubiquitin-driven autophagy process in the macrophage (Mesquita et al., 2012). With the second mechanism, unknown SPI-2 factor(s) specifically phosphorylate the nonreceptor tyrosine kinase, focal adhesion kinase, which in turn phosphorylates Akt, an upstream regulator of the serine/threonine kinase, mammalian target of rapamycin (mTOR). Activated Akt then phosphorylates mTOR, which as a principle regulator of autophagy (Powell et al., 2012), suppresses the autophagic process (Owen et al., 2014). mTOR plays a vital role in cell growth and metabolism by sensing environmental cues, including when nutrients are in abundance and when immune cells are in metabolically demanding situations such as stimulation with growth factors, nutrient availability and immune regulatory signals (Laplante and Sabatini, 2012; Cobbald, 2013).

Once survivability inside the macrophage is assured, *Salmonella* have been shown to influence the motility of the infected macrophage, thereby exploiting the macrophage as a Trojan horse to spread from the intestine to internal organs (Worley et al., 2006). Amazingly, the SPI-2 effector protein SseI appears to play a dual role in affecting macrophage motility both early during intestinal infection and later during the colonization of the internal organs (Worley et al., 2006; McLaughlin et al., 2009). Following the resolution of an inflammatory response in the intestine, normal CD18+ cells (macrophages and dendritic cells) can reenter the bloodstream by traversing the basement endothelium in a process called reverse transmigration (Thornborough and Worley, 2012). Infected CD18+ cells do not normally reverse transmigrate, thus balancing resolving inflammation with inhibiting the spread of microbes. However, *Salmonella* actively exploits the reverse transmigration process by secreting SseI (also known as SrfH) that binds to the host protein, TRIP6, to stimulate reverse transmigration to enhance dissemination away from the intestine to internal organs (Worley et al., 2006; Thornborough and Worley, 2012). TRIP6 is an adaptor protein that regulates cellular motility (Yi et al.,

2009; Lai et al., 2005). This is an extraordinary mechanism wherein "an intracellular pathogen overcomes host defenses designed to immobilize infected host cells" (Worley et al., 2006). Remarkably, once the pathogen is in the internal organs, SseI then plays a paradoxical role in maintaining a chronic infection (McLaughlin et al., 2009). Here, SseI inhibits phagocyte mobility by interacting with a different host regulator of cell migration, IQGAP1 (McLaughlin et al., 2009). Furthermore, the authors also reported that an SseI-dependent decrease of macrophage migration was also associated with a reduction in CD4+ T-cell numbers in the spleens of infected animals. Previous reports have demonstrated reduced T-cell activation due to an SPI-2-dependent suppression of antigen presentation (Cheminay et al., 2005; Tobar et al., 2006; Bueno et al., 2007). Therefore, the authors hypothesize that the data are suggestive that SseI indirectly controls CD4+ T-cell numbers by inhibiting migration of CD18+ cells and limiting their ability to effectively prime naive T cells. Taken together, these results suggest reduced capacity of the host to clear *Salmonella* from extraintestinal sites of infection consequently leading to asymptomatic long-term infection.

## 6.2 MACROPHAGE POLARIZATION: A *SALMONELLA* METABOLIC SURVIVAL NICHE AND IMMUNE ESCAPE MECHANISM?

Upon recruitment into tissues, mononuclear phagocytes respond to local environmental signals (pro- or antiinflammatory cytokines, microbial products, dead and/or damaged cells, tissue metabolism, activated lymphocytes) by changing their physiology to acquire distinct functional phenotypes (Gordon and Martinez, 2010; Biswas and Mantovani, 2010), specifically, the so-called classically activated M1 macrophages and the "alternatively activated" M2 macrophages (Gordon, 2003; Biswas and Mantovani, 2010; Sica and Mantovani, 2011). The terminology is based on the Th1- and Th2-derived immune responses (Mills et al., 2000). Although the Th1/Th2 paradigm has been defined in chickens (Guo et al., 2013; Chausse et al., 2014), there is no direct evidence that chicken macrophages can polarize into the M1/M2 phenotypes (Wigley et al., 2014). However, *Salmonella* do appear to prefer the M2 phenotype macrophage for long-term persistent infections of both murine and human macrophages (Eisele et al., 2013; Lathrop et al., 2015).

We and others have demonstrated the development of a Th2, antiinflammatory response in the cecum of chickens that begins at least 4 days after an initial infection with *Salmonella* and continues for weeks (Chausse et al., 2014; Kogut et al., 2015, 2016). Moreover, we have noted a significant increase in T regulatory cells in the cecum that corresponds to this shift from a proinflammatory to an antiinflammatory environment (Shanmugasundaram et al., 2015). Lastly, we also found alterations in the metabolic signatures of the cecum of the *Salmonella*-infected animals that are linked to an M2 phenotype, albeit, in a tissue and not macrophages. However, macrophage polarization is linked with dramatic alterations in multiple metabolic pathways (Shapiro et al., 2011; Biswas and Mantovani, 2012). Specifically, lipid oxidation metabolism mediated by the peroxisome proliferator-activated receptors (PPAR) γ/δ pathways within the M2 macrophage provide an advantageous niche for a number of intracellular microbial pathogens including *S.* Typhimurium in mice (Eisele et al., 2013), *Mycobacterium tuberculosis* (Almeida et al., 2012), *Brucella abortus* (Xavier et al., 2013), *Listeria monocytogenes* (Abdullah et al., 2012), *Francisella tularensis* (Shirey et al., 2008), *Leishmania* (Chan et al., 2012), and *Toxoplasma gondii* (Jensen et al., 2011). We submit that a persistent, carrier-state *Salmonella* infection in the chicken cecum induces a number of environmental cues that can potentially alter the polarization of infiltrating mononuclear phagocytes from an M1 state early infection to a preferential M2 state. The M2 macrophages "represent a unique niche for long-term intracellular bacterial survival" (Eisele et al., 2013) as well as an excellent mechanism for: (1) evading the host immune response, (2) promoting bacterial replication, and (3) dissemination throughout the reticuloendothelial system (RES) to internal organs. Therefore, the susceptibility of a hen to *Salmonella* colonization of the reproductive tract and successful egg contamination may depend on the ability of the bacteria to encounter and infect the M2-type macrophage subset as opposed to the proinflammatory M1-type macrophage. Additional experiments are required to confirm this hypothesis.

## 6.3 *SALMONELLA* EVASION OF HOST DEFENSE MECHANISMS IN THE REPRODUCTIVE TRACT

Precisely how *Salmonella* colonizes individual internal organs is still not well understood. However, *Salmonella*-infected macrophages are disseminated through the RES; ending up in internal organs with large numbers of macrophages such as the spleen and liver, thus providing a ready source of potential host cells residing in these tissues. Although not normally recognized as a component of the RES, the reproductive tract, as discussed in the previous section, contains a large number of functional macrophages and lymphocytes (Wigley et al., 2014). Once the *Salmonella*-infected macrophage reaches an organ of the reproductive tract (or any other internal organ), the bacteria can induce an SPI-2-dependent delayed apoptosis of the host macrophages (Guiney, 2005; McGhie et al., 2009). Consequently, either apoptotic cells containing bacteria can be phagocytized by additional macrophages or free bacteria are able to infect neighboring nonphagocytic cells within the reproductive tract (Guiney, 2005; Ruby et al., 2011).

### *6.3.1 Subversion of Antimicrobial Peptides*

*Salmonella* are able to subvert the AMP killing activities of the reproductive innate immune system by a number of physical and genetic means (McKelvey et al., 2014; Matmouros and Miller, 2015). For example, the bacteria are able to cloak their presence by remodeling their envelope thus increasing the hydrophobicity resulting in decreased binding of AMPs (Lee et al., 2004; Herrera et al., 2010; Kato et al., 2012). In addition, *Salmonella* possess an outer membrane protease that can target and degrade cationic AMPs (Guina et al., 2000). *S.* Enteritidis possesses two antimicrobial resistance genes, *virK* and *ybjX*, on its genome that confer bacterial resistance to polymixin B and avian β-defensins (McKelvey et al., 2014). Both genes are part of the PhoP/PhoQ regulon and are involved in modulation of the outer membrane of the bacteria that results in resistance to AMPs.

### 6.3.2 Modification of PAMPs/Evasion of PRRs/Subvert PRR Signaling

LPS comprises the major portion of the Gram-negative bacterial cell wall (Raetz and Whitfield, 2002). LPS consists of three components: the lipid A domain, a core oligosaccharide, and a variable number of repeat units of a polysaccharide O antigen. In response to host signals, *Salmonella* is capable of modifying the lipid A portion of LPS by activating the PhoP/PhoQ and PmrA-PmrB regulons (Ernst et al., 2001; Raetz and Whitfield, 2002; Kawasaki et al., 2005). Activation of these two-component regulatory systems results in the production of bacterial enzymes that palmitoylate, hydroxylate, deacylate, and attach aminoarabinose to lipid A (Ernst et al., 2001). These *Salmonella*-induced modifications increase resistance to AMPs and alter host recognition of LPS by TLR4 resulting in altering host cell signaling that mediate the innate immune response (Kawasaki et al., 2004a,b; Lee et al., 2004). Moreover, it has been found that the length of the O antigen, i.e., the number of repeating units, is under genetic control of the bacteria. *S.* Enteritidis appears to be able to increase the length of the O antigen component of the LPS molecule to increase colonization of the reproductive tract and increase bacterial survival in the egg during its formation (Coward et al., 2013).

*Salmonella* flagellin is required by the bacterium for motility (Stecher et al., 2004), but it is also a target of two components of the mammalian innate immune system: (1) TLR5, which detects extracellular flagellin, and (2) Naip5-Naip6/NirC4/caspase-1, which detects cytosolic flagellin (Gewirtz et al., 2001; Hayashi et al., 2001; Miao et al., 2006; Kofoed and Vance, 2011). However, as of this writing, there are no reports in the literature that poultry possess cytosolic PRR for the detection of flagellin. Flagellin production is tightly regulated and as such expression can be altered by host environmental cues. *Salmonella* actively inhibits flagellin expression during the establishment of systemic infections as a means of avoiding immune detection (Stecher et al., 2004; Cummings et al., 2005; Lai et al., 2013; Kilroy et al., 2016).

Another mechanism by which *Salmonella* can evade detection by PRRs is to directly antagonize signaling components. For example, once a TLR is activated by a PAMP an interaction between the TIR domains that are present on both the TLR and on adaptor proteins (MyD88 or TIRAP) activate specific signal transduction pathways in the host to generate a protective innate immune response (Patterson and Werling, 2013). The importance of this interaction is obvious with the report that *Salmonella* possess a gene (*tipA*) that mimics the TIR domain of TLR and their adapter proteins that compete with the endogenous TIR domains and thus prevents downstream TLR4 signaling (Newman et al., 2006). Specifically, *Salmonella* TipA protein modulates NF-κB activation and IL-1β production (Newman et al., 2006).

Members of the cytosolic Nod-like receptor protein family (NLR) direct the assembly of multiprotein complexes termed inflammasomes in response to detection of microbial products in the cytosol or disruption of cellular membranes by microbial virulence factors (Shin and Brodsky, 2015; Storek and Monack, 2015). Inflammasome assembly induces activation of caspase-1-dependent cleavage and secretion of IL-1 family cytokines and a caspase-1-dependent proinflammatory cell death (pyroptosis). Inflammasome activation plays a major role in host defense against a variety of pathogens, but a number of viral and bacterial pathogens have been found to interfere with inflammasome activation (Shin and Brodsky, 2015). For example, it has been shown that pathogen-derived metabolites can be recognized by NLR resulting in the activation of inflammasome-mediated immunity (reviewed in Shin and Brodsky, 2015). Specifically, the *Salmonella*-derived tricarboxylic acid cycle (TCA) cycle metabolite citrate is recognized by the NLRP3 inflammasome resulting in extraintestinal clearance of a systemic infection (Wynosky-Dolfi et al., 2014). However, screening an *S.* Typhimurium transposon library, it was found that three *Salmonella* genes that code for TCA cycle enzymes active during intracellular infection (*acnB*, aconitase; *aceA*, isocitrate lyase; *icdA*, isocitrate dehydrogenase) were found to modulate inflammasome activation through the metabolism of citrate (Wynosky-Dolfi et al., 2014).

### 6.3.3 Targeting Host Signaling Cascades

*Salmonella* use effector proteins to divert, inhibit, and otherwise influence host cell signaling pathways to the advantage of the bacteria obstructing immune signaling pathways such as the transcription factor NF-κB and the mitogen-activated protein kinase (MAPK) signaling cascade (Collier-Hyams et al., 2002; Haraga and Miller, 2002; La Negrate et al., 2008; Mazurkieicz et al., 2008; Wu et al., 2012; Li et al., 2013). Both NF-κB and MAPK activation lead to transcription of proinflammatory cytokines and antimicrobial molecules genes.

### 6.3.4 Target NF-κB

AvrA is an effector protein that functions as an immunological brake inhibiting the activation of NF-κB by stabilizing two inhibitors of NF-κB pathway, IκBα and β-catenin, which prevent release of the NF-κB for translocation to the nucleus thereby inhibiting the inflammatory responses (Collier-Hyams et al., 2002; Ye et al., 2007). Similarly, the effector protein *Salmonella* secreted factor L (SseL), a deubiquitinase that suppresses NF-κB activation by removing ubiquitin from IκBα preventing NF-κB translocation (La Negrate et al., 2008). Another effector protein, SspH1, localizes to the host cell nucleus and inhibits NF-κB–dependent gene expression (Haraga and Miller, 2002).

### 6.3.5 Target MAPK

AvrA has also shown to possess acetyltransferase activity that targets upstream kinases of the c-Jun-NH2-terminal kinase (JNK) pathway, thereby suppressing apoptotic removal of the bacterial intracellular niche and avoiding acquired immune mechanisms (Wu et al., 2012). SpvC inhibits inflammation by the dephosphorylation of the extracellular signal-regulated kinase (ERK) signaling pathway (Mazurkieicz et al., 2008).

# 7. HOST FACTORS

## 7.1 RESISTANCE VERSUS TOLERANCE

We have thus far focused this review on: (1) mechanisms of host immune resistance to control *Salmonella* infection and (2) mechanisms the bacteria have developed to evade, suppress, and avoid immune detection to colonize the reproductive tract of developing hens. However, the fact remains that *Salmonella* can be carried by the hen with virtually no ill effects on the host. This asymptomatic condition suggests a different host defense strategy against infection, a process known as tolerance. Resistance is a defense strategy to detect infection and eliminate the invading microbe, whereas tolerance is the ability of the host to limit the damage caused by both the pathogen and the host immune response (immunopathology; Ayres and Schneider, 2012). Although a relatively new immunological concept, tolerance as a host defense strategy has been ignored in veterinary infectious disease studies (Schneider and Ayres, 2008). However, in studying the interactions between the host response and *Salmonella* infections in the reproductive tract of poultry, disease tolerance appears to play a major role because of the asymptomatic nature of infection. Therefore, the chicken and bacteria appear to have evolved a relationship that minimizes both the normal host response and the normal bacterial virulence. However, this tolerant state is "detrimental to food safety" in humans (Calenge and Beaumont, 2012).

It is important to point out that infection tolerance is not immune tolerance, which is defined as "unresponsiveness of the immune system to substances or tissue that have the capacity to elicit an immune response" (Suzuki et al., 2010). As reviewed earlier, the chicken does respond immunologically to *Salmonella* infection, so other mechanisms are involved in the induction of disease tolerance. With this understanding, disease tolerance then depends on the host regulation, through physiological means of prevention, reduction, and avoidance of the pathological damage triggered by an infection (Schneider and Ayres, 2008). Therefore, "healthy carriers that remain asymptomatic despite being infected are likely to have a high level of tolerance to the pathogen" (Medzitov et al., 2012) such as found in persistent *Salmonella* infections in poultry.

Since a pathogen and the induced immunopathology can theoretically affect any physiological system, disease tolerance would involve a number of processes that will reduce host susceptibility to damage. Therefore, any physiological mechanism that typically maintains homeostasis and functional integrity of host tissues could contribute to disease tolerance. Mechanistically, limiting tissue damage is regulated by a number of evolutionarily conserved stress and/or damage responses. These responses confer tissue damage control, by providing cellular adaptation to environmental changes (Hayes and Dinkova-Kostova, 2014). For example, stress responses maintain cellular functions by activating metabolic processes in response to local alterations in oxygen tension (hypoxia), redox status (oxidative stress), osmolarity, and metabolite concentrations (ADP/ATP, glucose). All are essential mechanisms of cell and tissue homeostasis (Soames and Ribeiro, 2015). Damage responses attempt to preserve cellular functions while minimizing damage to macromolecules (DNA, lipids, proteins) and/or organelles (mitochondria, Golgi, endoplasmic reticulum) (Medzitov et al., 2012; Soames and Ribeiro, 2015).

Unlike immune responses that have measureable outputs to evaluate effectiveness, disease tolerance lacks clear-cut outputs (Schneider and Ayres, 2008). However, measurement of local cell metabolic processes and function, redox status, concentrations of metabolites, and organelle function of parenchymal cells and tissues (host's cells/tissues that do not have a direct impact on pathogens) would be beneficial in evaluating stress and damage responses. For example, during the first 3 weeks after infection of day-old broilers with *S.* Typhimurium, we have observed key metabolic changes that affected fatty acid and glucose metabolism through the 5'-adenosine monophosphate-activated protein kinase (AMPK) and the insulin/mTOR signaling pathway in the skeletal muscle were also altered (Arsenault et al., 2013). In addition, we found dramatic phenotypical alterations in the cecal tissue of *Salmonella*-infected chickens from the early response (4–48 h), which is proinflammatory, fueled by glycolysis and mTOR-mediated protein synthesis to the later phase (4–5 d) where the local environment has undergone an immune-metabolic reprogramming to an antiinflammatory state driven by AMPK-directed oxidative phosphorylation (Kogut et al., 2016). Therefore, metabolism appears to provide a potential measurement that characterizes a state of infection tolerance.

## 8. CONCLUSIONS AND FUTURE DIRECTION

*Salmonella* exercise a number of mechanisms to surmount host defenses. The bacterium has evolved means to mimic, subvert, exploit, and/or antagonize host defense. Future research should be directed at a better understanding of the means by which *Salmonella* invade the intestinal mucosa and migrate to extraintestinal organs. A fundamental strategy for the treatment of *Salmonella* in poultry is to alter host immune responses to enhance the clearance of infectious agents and prevent or reduce tissue damage due to inflammation. Unlike conventional antibiotics that are designed to target a pathogen, modulating the immune system exerts their protective effects by acting on the host. Vaccines are still the definitive immune-based prophylactic strategy for human and veterinary infectious agents. Yet, modern veterinary vaccines have multiple inadequacies including lack of cross-protection against several different strains of a pathogen, multiantigenicity of pathogens, slow development of protective immunity (days versus hours), lack of adequate adjuvants, and a need for site-specific immune responses that require further advances in infection immunobiology to address these challenges and improve efficacy.

*Salmonella* Enteritidis and Typhimurium are capable of survival and replication in some subsets of macrophages in mammals. Multiple factors, of both the host and pathogen, profoundly affect the outcome of infection. Studies to determine whether *Salmonella* uses similar strategies in poultry are needed to better understand the interactions critical to poultry infections.

Technical breakthroughs have highlighted the host–microbe interactome to explore the immune system and the role for microorganisms in the development and function of the immune system. Although these interaction studies have provided a window on the pathogenesis of the disease, there is a distinct lack of information on the effect of these interactions in the overall host physiology. The immune response and nutrient metabolism are two fundamental biological systems indispensable to maintaining and preserving life. Each of these systems is capable of modulating the activity of the other to ensure that the host animal is capable of coordinating the appropriate responses under any conditions. Thus, metabolic systems are integrated with pathogen-sensing and immune responses, and these pathways are evolutionarily conserved. Yet we know very little about the effect of infections on host metabolism. Several important networks sense and manage nutrients and integrate with immune and inflammatory pathways to influence the physiological and pathological metabolic states.

Other prospective areas of research for defining the relationship between the immune response and susceptibility to SE infection in the laying hen include:

- a better understanding of the microenvironment of the reproductive tract that allows for the asymptomatic carrier state focusing on immunity. How does this environment compare to that of the intestine and other extraintestinal sites of infection (liver and spleen)? Are the immune modifications (mimicry, subversion, exploitation, and/or antagonize) induced by the bacteria site specific?
- determine whether constituents of the intestinal microbiome can manipulate host physiology to promote extraintestinal tolerance to infection and/or inflammation.
- study *Salmonella*–host interactions that regulate metabolism and immune pathways in *Salmonella*-infected macrophages. For example, are there mechanisms that allow SE to persist in avian macrophages while other serovars induce cell death.
- determine whether organ-specific T cells are present in the reproductive tract and other extraintestinal organs (liver, spleen) and whether these T cells regulate *Salmonella* persistence.

## REFERENCES

Abdelsalam, M., Isobe, N., Yoshimura, Y., 2012. Effects of lipopolysaccharide and interleukins on the expression of avian β-defensins in hen ovarian follicular tissue. Poult. Sci. 91, 2877–2884.

Adel Mageed, A.M., Isobe, N., Yoshimura, Y., 2008. Expression of avian beta defensins in the oviduct and effects of lipopolysaccharide on their expression in the vagina of hens. Poult. Sci. 87, 979–984.

Abdel Mageed, A.M., Isobe, N., Yoshimura, Y., 2011. Changes in the density of immunoreactive avian β-defensin-3 and -11 in the hen uterus in response to lipopolysaccharide inoculation. J. Poult. Sci. 48, 73–77.

Abdullah, Z., Geiger, S., Nino-Castro, A., Böttcher, J.P., Muraliv, E., Gaidt, M., Schildberg, F.A., Riethausen, K., Flossdorf, J., Krebs, W., Chakraborty, T., Kurts, C., Schultze, J.L., Knolle, P.A., Klotz, L., 2012. Lack of PPARγ in myeloid cells confers resistance to *Listeria monocytogenes* infection. PLoS One 7, e37349.

Akira, S., Uematsu, S., Takeuchi, O., 2006. Pathogen recognition and innate immunity. Cell 124, 783–801.

Aksakal, A., Boynukara, B., Solmaz, H., Ilhan, Z., Kutlu, I.a., Gulhan, T., 2009. Occurrence and antibiotic susceptibility of *Salmonella* serotypes in apparently healthy slaughtered sheep in Van, Turkey. J. Ani. Vet. Adv. 8, 1455–1460.

Almeida, P.E., Carneiro, A.B., Silva, A.R., Bozza, P.T., 2012. PPARγ expression and function in mycobacterial infection: roles in lipid metabolism, immunity, and bacterial killing. PPAR Res. 2012, 383829.

Anastasiadou, M., Avdi, M., Theodoridis, A., Michailidis, G., 2013. Temporal changes in the expression of avian β-defensins in the chicken vagina during sexual maturation and *Salmonella* infection. Vet. Res. Comm. 37, 115–122.

Arsenault, R.J., Napper, S., Kogut, M.H., 2013. *Salmonella enterica* Typhimurium infection causes metabolic changes in chicken muscle involving AMPK, fatty acid and insulin/mTOR signaling. Vet. Res. 44, 35–50.

Aussel, L., Zhao, W., Hébrard, M., Guilhon, A.-A., Viala, J.P.M., Henri, S., Chasson, L., Gorvel, J.-P., Barras, F., Méresse, S., 2011. *Salmonella* detoxifying enzymes are sufficient to cope with the host oxidative burst. Mol. Microbiol. 80, 628–640.

Ayres, J.S., Schneider, D.S., 2012. Tolerance of infections. Annu. Rev. Immunol. 30, 271–294.

Barrow, P.A., Huggins, M.B., Lovell, M.A., Simpson, J.M., 1987. Observations on the pathogenesis of experimental *Salmonella typhimurium* infection in chickens. Res. Vet. Sci. 42, 194–199.

Barrow, P.A., Simpson, J., Lovell, M., 1988. Intestinal colonisation in the chicken by food-poisoning *salmonella* serotypes; microbial characteristics associated with faecal excretion. Avian Pathol. 17, 571–588.

Benatar, T., Tkalec, L., Ratcliffe, M.J.H., 1992. Stochastic rearrangement of immunoglobulin variable region genes in chicken B cell development. Proc. Natl. Acad. Sci. 89, 7615–7619.

Berndt, A., Wilhelm, A., Jugert, C., Pieper, J., Sachse, K., Methner, U., 2007. Chicken cecum immune response to *Salmonella enterica* serovars of different levels of invasiveness. Infect. Immun. 75, 5993–6007.

Birmingham, C.L., Smith, A.C., Bakowski, M.A., Yoshimori, T., Brumel, J.H., 2006. Autophagy controls *Salmonella* infection in response to damage to the *Salmonella*-containing vacuole. J. Biol. Chem. 281, 11374–11383.

Biswas, S.K., Mantovani, A.K., 2010. Macrophage plasticity and interaction with lymphocyte subsets: cancer as a paradigm. Nat. Immunol. 11, 889–896.

Biswas, S.K., Mantovani, A.K., 2012. Orchestration of metabolism by macrophages. Cell. Metab. 15, 432–437.

Braukmann, M., Methner, U., Berndt, A., 2015. Immune reaction and survivability of *Salmonella* Typhimurium and *Salmonella* Infantis after infection of primary avian macrophages. PloS One 10, e0122540.

Bueno, S.M., González, P.A., Schwebach, J.R., Kalergis, A.M., 2007. T cell immunity evasion by virulent *Salmonella enterica*. Immunol. Lett. 111, 14–20.

Calenge, F., Beaumont, C., 2012. Toward integrative genomics study of genetic resistance to *Salmonella* and *Campylobacter* intestinal colonization in fowl. Front. Genet. 3 Article 261.

Carpenter, S., O'Neill, L.A., 2007. How important are toll-like receptors for antimicrobial responses? Cell. Microbiol. 9, 1891–1901.

Cao, X., 2016. Self-regulation and cross-regulation of pattern-recognition receptor signaling in health and disease. Nat. Rev. Immunol. 18, 35–50.

Cemma, M., Brumell, J.H., 2012. Interactions of pathogenic bacteria with autophagy systems. Curr. Biol. 22, R50–R546.

Chan, M.M., Adapala, N., Chen, C., 2012. Peroxisome proliferator-activated receptor-γ-mediated polarization of macrophages in *Leishmania* infection. PPAR Res. 2012, 796235.

Cheng, T.-F., Brzostek, S., Ando, O., Van Scoy, S., Kumar, K.P., Reich, N.C., 2006. Differential activation of IFN regulatory factor (IRF)-3 and IRF-5 transcription factors during viral infection. J. Immunol. 176, 7462–7470.

Chappell, L., Kaiser, P., Barrow, P., Jones, M.A., Johnston, C., Wigley, P., 2009. The immunobiology of avian systemic salmonellosis. Vet. Immunol. Immunopathol. 128, 53–59.

Chaussé, A.-M., Grépinet, O., Bottreau, E., Robert, V., Hennequet-Antier, C., Lalmanach, A.-C., Lecardonnel, J., Beaumont, C., Velge, P., 2014. Susceptibility to *Salmonella* carrier-state: a possible Th2 response in susceptible chicks. Vet. Immunol. Immunopathol. 159, 16–28.

Cheminay, C., Möhlenbrink, A., Hensel, M., 2005. Intracellular *Salmonella* inhibit antigen presentation by dendritic cells. J. Immunol. 174, 2892–2899.

Ciraci, C., Tuggle, C.K., Wannemuehle, M.J., Nettleton, D., Lamont, S.J., 2010. Unique genome-wide transcriptome profiles of chicken macrophages exposed to *Salmonella*-derived endotoxin. BMC Genomics 9, 256.

Cobbald, S.P., 2013. The mTOR pathway and integrating immune regulation. Immunology 140, 391–398.

Collier-Hyams, L.S., Zeng, H., Sun, J., Tomlinson, A.D., Bao, Z.Q., Chen, H., Madara, J.L., Orth, K., Neish, A.S., 2002. Cutting edge: *Salmonella* AvrA effector inhibits the key proinflammatory, anti-apoptotic NF-κB pathway. J. Immunol. 169, 2846–2850.

Coward, C., Sait, L., Cogan, T., Humphrey, T.J., Maskell, D.J., 2013. O-antigen repeat number in *Salmonella enterica* serovar Enteritidis is important for egg contamination, colonization of the chicken reproductive tract and survival in egg albumen. FEMS Microbiol. Lett. 343, 169–176.

Cummings, L.A., Rassoulian Barrett, S.L., Wilkerson, W.D., Fellnerova, I., Cookson, B.T., 2005. FliC-specific $CD4^+$ T cell responses are restricted by bacterial regulation of antigen expression. J. Immunol. 174, 7929–7938.

Cuperus, T., Coorens, M., van Dijk, A., Haagsman, H.P., 2013. Avian host defense peptides. Dev. Comp. Immunol. 41, 352–369.

Das, P., Lahiri, A., Lahiri, A., Charavortty, D., 2009. Novel role of the nitrite transporter NirC in *Salmonella* pathogenesis: SPI2-dependent suppression of inducible nitric oxide synthase in activated macrophages. Microbiology 155, 2476–2489.

Degen, W.G., Daal, N., Rothwell, L., Kaiser, P., Schijins, V.E., 2005. Th1/Th2 polarization by viral and helminth infections in birds. Vet. Microbiol. 105, 163–167.

Deretic, V., Saitoh, T., Akira, S., 2013. Autophagy in infection, inflammation and immunity. Nat. Rev. Immunol. 13, 722–737.

Desmidt, Ducatelle, Haesebrouck, 1997. Pathogenesis of *Salmonella enteritidis* phage type four after experimental infection of young chickens. Vet. Microbiol. 56, 99–109.

Dieye, Y., Ameiss, K., Mellata, M., Curtiss III, R., 2009. The *Salmonella* pathogenicity island (SPI) 1 contributes more than SPI2 to the colonization of the chicken by *Salmonella enterica* serovar Typhimurium. BMC Microbiol. 9, 3.

Ebers, K.L., Zhang, C.Y., Zhang, M.Z., Bailey, R.H., Zhang, S., 2009. Transcriptional profiling avian beta-defensins in chicken oviduct epithelial cells before and after infection with *Salmonella enterica* serovar Enteritidis. BMC Microbiol. 9, 153.

Eisele, N.A., Ruby, T., Jacobson, A., Manzanillo, P.S., Cox, J.S., Lam, L., Mukundan, L., Chawla, A., Monack, D.M., 2013. *Salmonella* require the fatty acid regulator PPARδ for the establishment of a metabolic environment essential for long-term persistence. Cell Host Microb. 14, 171–182.

Erf, G., 2004. Cell mediated immunity in poultry. Poult. Sci. 83, 580–590.

Ernst, R.K., Guina, T., Miller, S.I., 2001. *Salmonella typhimurium* outer membrane remodeling: role in resistance to host innate immunity. Microbes Infect. 3, 1327–1334.

Gantois, I., Eeckhaut, V., Pasmans, F., Haesebrouck, F., Ducatelle, R., Van Immerseel, F., 2008. A comparative study on the pathogenesis of egg contamination by different serotypes of *Salmonella*. Avian Pathol. 37, 379–406.

Gantois, I., Ducatelle, R., Pasmans, F., Haesbrouck, F., Gast, R., Humphrey, T.J., Van Immerseel, F., 2009. Mechanisms of egg contamination by *Salmonella* Enteritidis. FEMS Microbiol. Rev. 33, 718–738.

Gast, R.K., Guard-Petter, J., Holt, P., 2003. Effect of prior serial in vivo passage on the frequency of *Salmonella* Enteritidis in eggs from experimentally infected laying hens. Avian Dis. 47, 633–639.

Gast, R.K., Guraya, R., Guard-Bouldin, J., Holt, P.S., Moore, R.W., 2007. Colonization of specific regions of the reproductive tract and deposition at different locations inside eggs laid by hens infected with *Salmonella* Enteritidis or *Salmonella* Heidelberg. Avian Dis. 51, 40–44.

Geijtenbeek, T.B., Gringhuis, S.I., 2009. Signaling through C-type lectin receptors: Shaping immune responses. Nat. Rev. Immunol. 9, 465–479.

Gewirtz, A.T., Navas, T.A., Lyons, S., Godowski, P.J., Madara, J.A., 2001. Cutting edge: bacterial flagellin activates basolaterally expressed TLR5 to induce epithelial proinflammatory gene expression. J. Immunol. 167, 1882–1885.

Glick, B., 1991. Historical perspective: the bursa of Fabricius and its influence on B-cell development, past and present. Vet. Immunol. Immunopathol. 30, 3–12.

Göbel, T.W., Schneider, K., Schaerer, B., Mejri, I., Puehler, F., Weigend, S., Staeheli, P., Kaspers, B., 2003. IL-18 stimulates the proliferation and IFN-gamma release of CD4+ T cells in the chicken: conservation of a Th1-like system in a nonmammalian species. J. Immunol. 171, 1809–1815.

Goosens, K.E., Ward, A.C., Lowenthal, J.W., Bean, A.G.D., 2013. Chicken interferons, their receptors and interferon-stimulated genes. Dev. Comp. Immunol. 41, 370–376.

Gordon, S., Martinez, F.O., 2010. Alternative activation of macrophages: mechanism and functions. Immunity 32, 593–604.

Gordon, S., 2003. Alternative activation of macrophages. Nat. Rev. Immunol. 3, 23–35.

Gowen, A.B., Wong, M.-H., Jung, K.-H., Sanders, A.B., Mitchell, W.M., Alexopoulou, L., Flavell, R.A., Sidwell, R.W., 2007. TLR3 Is essential for the induction of protective immunity against Punta Toro virus infection by the double-stranded RNA (dsRNA), Poly(I: $C_{12}$U), but not Poly(I: C): differential recognition of synthetic dsRNA molecules. J. Immunol. 178, 5200–5208.

Guina, T., Yi, E.C., Wang, H., Hackett, M., Miller, S.I., 2000. A PhoP-regulated outer membrane protease of *Salmonella enterica* serovar Typhimurium promotes resistance to alpha-helical antimicrobial peptides. J. Bacteriol. 182, 4077–4086.

Guiney, D.G., 2005. The role of host cell death in *Salmonella* infections. Curr. Top. Microbiol. Immunol 289, 131–150.

Guo, P., Thomas, J.D., Bruce, M.P., Hinton, T.M., Bean, A.G.D., Lowenthal, J.W., 2013. The chicken Th1 response: potential therapeutic applications of ChIFN-$\gamma$. Dev. Comp. Immunol. 41, 389–396.

Hansen, J.D., Vojtech, L.N., Laing, K.J., 2011. Sensing disease and danger: a survey of vertebrate PRRs and their origins. Dev. Comp. Immunol. 35, 386–397.

Haraga, A.I., Miller, S.I., 2002. A *Salmonella enterica* serovar Typhimurium translocated leucine-rich repeat effector protein inhibits NF-$\kappa$B-dependent gene expression. Infect. Immun. 71, 4052–4058.

Haraga, A.I., Ohlson, M.B., Miller, S.I., 2008. Salmonellae interplay with host cells. Nat. Rev. Microbiol. 6, 53–66.

Hayashi, F., Smith, K.D., Ozinsky, A., Hawn, T.R., Yi, E.C., Goodlett, D.R., Eng, J.K., Akira, S., Underhill, D.M., Aderem, A., 2001. The innate immune response to bacterial flagellin is mediated by Toll-like receptor 5. Nature 410, 1099–1103.

Hayes, J.D., Dinkova-Kostova, A.T., 2014. The Nrf2 regulatory network provides an interface between redox and intermediary metabolism. Trends Biochem. Sci. 39, 199–218.

He, H., Genovese, K.J., Kogut, M.H., 2012. A comparative study on invasion, survival, modulation of oxidative burst and nitric oxide responses of macrophages (HD11), and systemic infection in chickens by prevalent poultry *Salmonella* serovars. Foodborne Pathog. Dis. 9, 1104–1110.

He, H., Genovese, K.J., Swaggerty, C.L., Nisbet, D.J., Kogut, M.H., 2013. Nitric oxide as a biomarker of intracellular *Salmonella* viability and identification of the bacteriostatic activity of protein kinase A inhibitor h-89. PLos One 8, e58873.

Herrera, C.M., Hankins, J.V., Trent, M.S., 2010. Activation of PmrA inhibits LpxT-dependent phosphorylation of lipid A promoting resistance to antimicrobial peptides. Mol. Microbiol. 76, 1444–1460.

Henard, C.A., Vazquez-Torres, A., 2011. Nitric oxide and *Salmonella* pathogenesis. Front. Microbiol. 2 Art. 84.

Holt, P.S., Porter, R.E., 1993. Effect of induced molting on the recurrence of a previous *Salmonella enteritidis* infection. Poult. Sci. 72, 2069–2078.

Janeway, C.A., Medzhitov, R., 2002. Innate immune recognition. Annu. Rev. Immunol. 20, 197–216.

Jensen, K.D.C., Wang, Y., Tait Wojno, E.D., Shastri, A.J., Hu, K., Cornel, L., Boedec, E., Ong, Y.-C., Chien, Y.-H., Hunter, C.A., Boothroyd, J.C., Saeij, J.P.J., 2011. Toxoplasma polymorphic effectors determine macrophage polarization and intestinal inflammation. Cell Host Microbe 9, 472–483.

Johnston, C.E., Hartley, C., Salisbury, A.M., Wigley, P., 2012. Immunological changes at point-of-lay increase susceptibility to *Salmonella enterica* serovar Enteritidis infection in vaccinated chickens. PLoS One 7, e48195.

Kaiser, P., Poh, T.Y., Rothwell, L., Avery, S., Balu, S., Pathania, U.S., Hughes, S., Goodchild, M., Morrell, S., Watson, M., Bumstead, N., Kaufman, J., Young, J.R., 2005. A genomic analysis of chicken cytokines and chemokines. J. Interferon. Cytokine Res. 25, 467–484.

Karpala, A., Stewart, C., McKay, J., Lowenthal, J.W., Bean, A.G.D., 2011. Characterization of chicken MDA5 activity: regulation of IFN-β in the absence of RIG-I functionality. J. Immunol. 186, 5397–5405.

Kato, A., Chen, H.D., Latifi, T., Groisman, E.A., 2012. Reciprocal control between a bacterium's regulatory system and the modification status of its lipopolysaccharide. Mol. Cell 47, 897–908.

Kawai, T., Akira, A., 2009. The roles of TLRs, RLRs and NLRs in pathogen recognition. Int. Immunol. 21, 317–337.

Kawai, T., Akira, A., 2011. Toll-like receptors and their crosstalk with other innate receptors in infection and immunity. Immunity 34, 637–650.

Kawasaki, K., Ernst, R.K., Miller, S.I., 2004a. 3-O-Deacylation of Lipid A by PagL, a PhoP/PhoQ-regulated deacylase of *Salmonella typhimurium*, modulates signaling through Toll-like receptor 4. J. Biol. Chem. 279, 20044–28848.

Kawasaki, K., Ernst, R.K., Miller, S.I., 2004b. Deacylation and palmitoylation of lipid A by salmonellae outer membrane enzymes modulate host signaling through Toll-like receptor 4. J. Endotoxin Res. 10, 439–444.

Kawasaki, K., Ernst, R.K., Miller, S.I., 2005. Inhibition of *Salmonella enterica* serovar Typhimurium lipopolysaccharide deacylation by aminoarabinose membrane modification. J. Bacteriol. 187, 2448–2457.

Keestra, A.M., de Zoete, M.R., Bouwman, L.I., Vaezirad, M.M., van Patten, J.P.M., 2013. Unique features of chicken Toll-like receptors. Dev. Comp. Immunol. 41, 316–323.

Kilroy, S., Raspoet, R., Martel, A., Bosseler, L., Appia-Ayme, C., Thompson, A., Haesebrouck, F., Ducatelle, R., Van Immerseel, F., 2016. *Salmonella* Enteritidis persists in the laying hen oviduct by suppressing flagellin-induced inflammation. Vet. Res.

Kofoed, E.M., Vance, R.E., 2011. Innate immune recognition of bacterial ligands by NAIPs determines inflammasome specificity. Nature 477, 592–595.

Kogut, M.H., Chiang, H.I., Swaggerty, C.L., Pevzner, I.Y., Zhou, H., 2012. Gene expression analysis of the toll-like receptor pathways in heterophils from genetic chicken lines that differ in their susceptibility to *Salmonella enteritidis*. Front. Genet. 3 Art. 121.

Kogut, M.H., Genovese, K.J., Arsenault, R., 2015. A role for non-canonical Wnt-β-catenin and TGF-β signaling pathways in the induction of tolerance during the establishment of a *Salmonella enterica* serovar Enteritidis persistent cecal infection in chickens. Front. Vet. Infect. Dis. 2 Art. 33.

Kogut, M.H., Genovese, K.J., He, H., Arsenault, R.J., 2016. AMPK and mTOR: sensors and regulators of immunometabolic changes during *Salmonella* infection in the chicken. Poult. Sci. 95, 345–353.

Lebacq-Verheyden, A.M., Vaerman, J.P., Heremans, J.F., 1974. Quantification and distribution of chicken immunoglobulins IgA, IgM, and IgG in serum and secretions. Immunology 27, 683–692.

Lai, Y.-U., Chen, C.-S., Lin, W.-C., Lin, F.T., 2005. c-Src-mediated phosphorylation of TRIP6 regulates its function in lysophosphatidic acid-induced cell migration. Mol. Cell. Biol. 25, 5859–5868.

Lai, M.A., Quarles, E.K., Lopez-Yglesias, A.H., Zhao, X., Hajjar, A.M., Smith, K.D., 2013. Innate immune detection of flagellin positively and negatively regulates *Salmonella* infection. PLoS One 8, e72047.

Le Negrate, G., Faustin, B., Welsh, K., Loeffler, M., Krajewska, M., Hasegawa, P., Mukherje, S., 2008. *Salmonella* secreted factor L deubiquitinase of *Salmonella typhimurium* inhibits NF-κB, suppresses IκBα ubiquitination and modulates innate immune responses. J. Immunol. 180, 5045–5056.

Lathrop, S.K., Binder, K.A., Starr, T., Cooper, K.G., Chong, A., Carmody, A.B., Steele-Mortimer, O., 2015. Replication of *Salmonella enterica* serovar Typhimurium in human monocyte-derived macrophages. Infect. Immun. 83, 2661–2667.

Lee, H., Hsu, F.-F., Turk, J., Groisman, E.A., 2004. The PmrA-regulated *pmrC* gene mediates phosphoethanolamine modification of Lipid A and polymyxin resistance in *Salmonella enterica*. J. Bacteriol. 186, 4124–4133.

Levine, B., 2005. Eating oneself and uninvited guests. Autophagy-related pathways in cellular defense. Cell 120, 159–162.

Li, S., Zhang, L., Yao, Q., Li, L., Dong, N., Rong, J., Gao, W., Ding, X., Sun, L., Chen, X., Chen, S., Shao, F., 2013. Pathogen blocks host death receptor signaling by arginine GlcNAcylation of death domains. Nature 501, 242–246.

Lian, L., Ciraci, C., Chang, G., Hu, J., Lamont, S.J., 2012. NLRC5 knockdown in chicken macrophages alters response to LPS and poly (I: C) stimulation. BMC Vet. Res. 8, 23.

Liu, X., Ma, T., Wang, H., Sheng, Z., Dou, X., Wang, K., Li, Z., Pan, Z., Chang, G., Chen, G., 2015. Up-regulation of NLRC5 and NF-κB signaling pathway in carrier chickens challenged with *Salmonella* enterica Serovar Pullorum at different persistence periods. Indian J. Biochem. Biophys. 52, 132–139.

Laplante, M., Sabatini, D.M., 2012. mTOR signaling in growth control and disease. Cell 149, 274–293.

Maekawa, T., Kufer, T.A., Schulze-Lefert, P., 2011. NLR functions in plant and animal immune systems: so far and yet so close. Nat. Immunol. 12, 817–825.

Magor, K.E., Navarro, D.M., Barber, M.R.W., Petkau, K., Fleming-Canepa, X., Blyth, G.A.D., Blaine, A.H., 2013. Defense genes: missing from the flight division. Dev. Comp. Immunol. 41, 377–388.

Mann, K., 2007. The chicken egg white proteome. Proteomics 7, 3558–3568.

Mann, K., Mann, M., 2008. The chicken egg yolk plasma and granule proteomes. Proteomics 8, 178–191.

Matamouros, S., Miller, S.I., 2015. *S.* Typhimurium strategies to resist killing by cationic antimicrobial peptides. Biochim. Biophys. Acta 1848, 3021–3025.

Mazurkiewicz, P., Thomas, J., Thompson, J.A., Liu, M., Arbibe, L., Sansonetti, P., Holden, D.W., 2008. SpvC is a *Salmonella* effector with phosphothreonine lyase activity on host mitogen-activated protein kinases. Mol. Microbiol. 67, 1371–1383.

McGhie, E.J., Brawn, L.C., Hume, P.J., Humphreys, D., Koronakis, V., 2009. Salmonella takes control: effector-driven manipulation of the host. Curr. Opin. Microbiol. 12, 117–124.

McKelvey, J.A., Yang, M., Jiang, Y., Zhang, S., 2014. *Salmonella enterica* serovar Enteritidis antimicrobial peptide resistance genes aid in defense against chicken innate immunity, fecal shedding, and egg deposition. Infect. Immun. 82, 5185–5202.

McLaughlin, L.M., Govoni, G.R., Gerke, C., Gopinath, S., Peng, K., Laidlaw, G., Chien, Y.H., Jeong, H.W., Li, Z., Brown, M.D., Sacks, D.B., Monack, D., 2009. The *Salmonella* SPI2 effector SseI mediates long-term systemic infection by modulating host cell migration. PLoS Pathog. 5, e1000671.

Medzhitov, R., 2007. Recognition of microorganisms and activation of the immune response. Nature 449, 819–826.

Medzhitov, R., Janeway Jr., C.A., 1997. Innate immunity: the virtues of a nonclonal system of recognition. Cell 91, 295–298.

Medzhitov, R., Schneider, D.S., Soares, M.P., 2012. Disease tolerance as a defense strategy. Science 335, 936–941.

Mesquita, F.S., Thomas, M., Sachee, M., Santos, A.J., Figueira, R., Holden, D.W., 2012. The *Salmonella* deubiquitinase SseL inhibits selective autophagy of cytosolic aggregates. PLoS Pathog. 8, e1002743.

Miao, E.A., Alpuche-Aranda, C.M., Dors, M., Clark, A.E., Bader, M.W., Miller, S.I., Aderem, A., 2006. Cytoplasmic flagellin activates caspase-1 and secretion of interleukin-1β via Ipaf. Nat. Immunol. 7, 569–575.

Michailidis, G., Theodoridis, A., Avdi, M., 2010. Transcriptional profiling of Toll-like receptors in chicken embryos and in the ovary during sexual maturation and in response to *Salmonella enteritidis* infection. Anim. Reprod. Sci. 122, 294–302.

Michailidis, G., Theodoridis, A., Avdi, M., 2011. Effects of sexual maturation and *Salmonella* infection on the expression of Toll-like receptors in the chicken vagina. Anim. Reprod. Sci. 123, 234–241.

Michailidis, G., Avdi, M., Argirou, 2012. Transcriptional profiling of antimicrobial peptides avian β-defensins in the chicken ovary during sexual maturation and in response to *Salmonella enteritidis* infection. Res. Vet. Sci. 92, 60–65.

Mills, C.D., Kincaid, K., Alt, J.M., Heilman, M.J., Hill, A.M., 2000. Cellular immunology and immune regulation M-1/M-2 macrophages and the Th1/Th2 paradigm. J. Immunol. 164, 6166–6173.

Mine, Y., D'Silva, I., 2008. Bioactive Components in Egg White. Egg Bioscience and Biotechnology. John Wiley & Sons, Inc, pp. 148–150.

Morgan, E., Campbell, J.D., Rowe, S.C., Bispham, J., Stevens, M.P., Bowen, A.J., Barrow, P.A., Maskell, D.J., Wallis, T.S., 2004. Identification of host-specific colonization factors of *Salmonella enterica* serovar Typhimurium. Mol. Microbiol. 54, 994–1010.

Myers, T.J., Schat, K.A., 1990. Natural killer cell activity of chicken intraepithelial leukocytes against rotavirus-infected target cells. Vet. Immunol. Immunopathol. 26, 157–170.

Narayanan, L.A., Edelmann, M.J., 2014. Ubiquitination as an efficient molecular strategy employed in *Salmonella* infection. Front. Immunol. 5 Art. 558.

Newman, R.M., Salunkhe, P., Godzik, A., Reed, J.C., 2006. Identification and characterization of a novel bacterial virulence factor that shares homology with mammalian Toll/interleukin-1 receptor family proteins. Infect. Immun. 74, 594–601.

Nii, T., Sonoda, Y., Isobe, N., Yoshimura, Y., 2011. Effects of lipopolysaccharide on the expression of proinflammatory cytokines and chemokines and the subsequent recruitment of immunocompetent cells in the oviduct of laying and molting hens. Poult. Sci. 90, 2332–2341.

Owen, K.A., Meyer, C.B., Bouton, A.H., Casanova, J.E., 2014. Activation of focal adhesion kinase by *Salmonella* suppresses autophagy via an Akt/mTOR signaling pathway and promotes survival in macrophages. PLos Pathog. 10, e1004159.

Ozoe, A., isobe, N., Yoshimura, Y., 2009. Expression of Toll-like receptors and TLR4 response to lipopolysaccharide in hen oviduct. Vet. Immunol. Immunopathol. 127, 259–268.

Patterson, N.J., Werling, D., 2013. To con protection: TIR-domain containing proteins (Tcp) and innate immune evasion. Vet. Immunol. Immunopathol. 155, 147–154.

Perron, G.G., Quessy, S., Bell, G., 2008. A reservoir of drug-resistant pathogenic bacteria in asymptomatic hosts. PLoS One 3, e3749.

Powell, J.D., Pollizzi, K.N., Heikamp, E.M., Horton, M.R., 2012. Regulation of immune responses by mTOR. Annu. Rev. Immunol. 30, 39–68.

Raetz, C.R.H., Whitfield, C., 2002. Lipopolysaccharide endotoxins. Annu. Rev. Biochem 71, 635–700.

Ruby, T., McLaughlin, L., Gopinath, S., Monack, D.M., 2011. *Salmonella's* long-term relationship with its host. FEMS Microbiol. Rev. 36, 600–615.

Rychlik, I., Karasova, D., Sebkova, A., Volf, J., sisak, F., Havlickova, H., Kummer, V., Imre, A., Szmolka, A., Nagy, B., 2009. Virulence potential of five major pathogenicity islands (SPI-1 to SPI-5) of *Salmonella enterica* serovar Enteritidis for chickens. BMC Microbiol. 9, 268.

Saeed, A.M., 1999. Experimental infection of four strains of commercial laying hens with *Salmonella enterica* Enteritidis phage type 8. In: Saeed, A.M., G, R.K., Potter, M.E., Wall, P.G. (Eds.), *Salmonella enterica* serovar Enteritidis in Humans and Animals: Epidemiology, Pathogenesis, and Control. Iowa State University Press, Ames, Iowa.

Schneider, D.S., Ayres, J.S., 2008. Two ways to survive infection: what resistance and tolerance can teach us about treating infectious diseases. Nat. Rev. Immunol. 8, 889–895.

Scott, T., Owens, M.D., 2008. Thrombocytes respond to lipopolysaccharide through Toll-like receptor-4, and MAP kinase and NF-κB pathways leading to expression of interleukin-6 and cyclooxygenase-2 with production of prostaglandin E2. Mol. Immunol. 45, 1001–1008.

Seo, K.H., Holt, P.S., Gast, R.K., Hofacre, C.L., 2000. Combined effect of antibiotic and competitive exclusion treatment on *Salmonella enteritidis* fecal shedding in molted laying hens. J. Food Prot. 63, 545–548.

Setta, A.M., Barrow, P.A., Kaiser, P., Jones, M.A., 2012. Early immune dynamics following infection with *Salmonella enterica* serovars Enteritidis, Infantis, Pullorum, and Gallinarum: cytokine and chemokine gene expression profile and cellular changes in chicken cecal tonsils. Comp. Immunol. Microbiol. Infect. Dis. 35, 397–410.

Shanmugasundaram, R., Kogut, M.H., Arsenault, R.L., Swaggerty, C.L., Cole, K.Y., Reddish, M.J., Selvaraj, R.K., 2015. Effect of *Salmonella* infection on cecal tonsil regulatory T cell properties in chickens. Poult. Sci. 94, 1828–1835.

Shapiro, H., Lukaty, A., Ariel, A., 2011. Macrophages, meta-inflammation, and immunometabolism. Sci. World J. 11, 2509–2529.

Shirey, K.A., Cole, L.E., Keegan, A.D., Vogel, S.N., 2008. *Francisella tularensis* live vaccine strain induces macrophage alternative activation as a survival mechanism. J. Immunol. 181, 4159–4167.

Shin, S., Brodsky, I.E., 2015. The inflammasome: Learning from bacterial evasion strategies. Sem. Immunol. 27, 102–110.

Shivaprasad, H.L., 2000. Fowl typhoid and pullorum disease. Rev. Sci. Tech. 19, 405–424.

Sica, A., Mantovani, A., 2011. Macrophage plasticity and polarization: *in vivo veritas*. J. Clin. Invest 122, 787–795.

Soares, M.P., Ribeiro, A.M., 2015. Nrf2 as a master regulator of tissue damage control and disease tolerance to infection. Biochem. Soc. Trans. 43, 663–668.

Sonoda, Y., Abdel Mageed, A.M., Isobe, N., Yoshimura, Y., 2013. Induction of avian β-defensins by CpG oligodeoxynucleotides and proinflammatory cytokines in hen vaginal cells in vitro. Reproduction 145, 621–631.

Stecher, B., Hapfelmeier, S., Müller, C., Kremer, M., Stallmach, T., Hardt, W.-D., 2004. Flagella and chemotaxis are required for efficient induction of *Salmonella enterica* serovar Typhimurium colitis in streptomycin-pretreated mice. Infect. Immun. 72, 4138–4150.

Steele-Mortimer, O., 2008. The *Salmonella*-containing vacuole—moving with the times. Curr. Opin. Microbiol. 11, 38–45.

Storek, K.M., Monack, D.M., 2015. Bacterial recognition pathways that lead to inflammasome activation. Immunol. Rev. 265, 112–129.

Straub, C., Neulen, M.-L., Sperling, B., Windau, K., Zechman, M., Jansen, C.A., Viertlboeck, B.C., Gobel, T.W., 2013. Chicken NK cell receptors. Dev. Comp. Immunol. 41, 324–333.

Subedi, K., Isobe, N., Nishibori, M., Yoshimura, Y., 2007. Changes in the expression of Toll-like receptor mRNAs during follicular growth and in response to lipopolysaccharide in ovarian follicles of laying hens. J. Reprod. Dev. 53, 1227–1235.

Sundaresan, N.R., Anish, D., Sastry, K.V., Saxena, V.K., Mohan, J., Ahmed, K.A., 2007. Cytokines in reproductive remodeling of molting White Leghorn hens. J. Reprod. Immunol. 73, 39–50.

Sundaresan, N.R., Anish, D., Sastry, K.V., Saxena, V.K., Nagarajan, K., Subramani, J., Leo, M.D., Sit, N., Mohan, J., Saxena, M., Ahmed, K.A., 2008. High doses of dietary zinc induce cytokines, chemokines, and apoptosis in reproductive tissues during regression. Cell Tissue Res. 332, 543–554.

Suzuki, J., Ricordi, C., Chen, Z., 2010. Immune tolerance induction by integrating innate and adaptive immune regulators. Cell Transplant. 19, 253–268.

Takeuchi, O., Akira, S., 2010. Pattern recognition receptors and inflammation. Cell 140, 805–820.

Tao, Z.Y., Zhu, C.H., Shi, Z.H., Song, C., Xu, W.J., Song, W.T., Zou, J.M., Qin, A.T., 2015. Molecular characterization, expression, and functional analysis of NOD1 in Qingyun partridge chicken. Genet. Mol. Res. 14, 2691–2701.

Thompson, J.A., Liu, M., Helaine, S., Holden, D.W., 2011. Contribution of the PhoP/Q regulon to survival and relocation of *Salmonella enterica* serovar Typhimurium in macrophages. Microbiology 157, 2084–2093.

Thornbrough, J.M., Worley, M.J., 2012. A naturally occurring single nucleotide polymorphism in the *Salmonella* SPI-2 type III effector srfH/sseI controls early extraintestinal dissemination. PLos One 7, e45245.

Tobar, J.A., Carreño, L.J., Bueno, S.M., González, P.A., Mora, J.E., Quezada, S.A., Kalergis, A.M., 2006. Virulent *Salmonella enterica* serovar Typhimurium evades adaptive immunity by preventing dendritic cells from activating T cells. Infect. Immun. 74, 6438–6448.

Trinchieri, G., Sher, A., 2007. Cooperation of Toll-like receptor signals in innate immune defense. Nat. Rev. Immunol. 7, 179–190.

Trout, J.M., Lillehoj, H.S., 1996. T lymphocyte roles during *Eimeria acervulina* and *Eimeria tenella* infections. Vet. Immunol. Immunopathol. 53, 163–172.

Underhill, D.M., 2007. Collaboration between the innate immune receptors dectin-1, TLRs, and Nods. Immunol. Rev. 219, 75–87.

Van der Heijden, J., Bosman, E.S., Reynolds, L.A., Finlay, B.B., 2015. Direct measurement of oxidative and nitrosative stress dynamics in *Salmonella* inside macrophages. Proc. Natl. Acad. Sci. 112, 560–565.

Vázquez-Torres, A., Jones-Carson, J., Bäumler, A.J., Falkow, S., Valdivia, R., Brown, W., Le, M., Berggren, R., Parks, W.T., Fang, F.C., 1999. Extraintestinal dissemination of *Salmonella* by CD18-expressing phagocytes. Nature 401, 804–809.

Vazquez-Torres, A., Xu, Y., Jones-Carson, J., Holden, D.W., Lucia, S.M., Dinauer, M.C., Mastroeni, P., Fang, F.C., 2000. *Salmonella* pathogenicity island 2-dependent evasion of the phagocyte NADPH oxidase. Science 287, 1655–1658.

Velge, P., Cloeckaert, A., Barrow, P., 2005. Emergence of *Salmonella* epidemics: the problems related to *Salmonella enterica* serotype Enteritidis and multiple antibiotic resistance in other major serotypes. Vet. Res. 36, 267–288.

Wigley, P., 2013. Immunity to bacterial infections in the chicken. Dev. Comp. Immunol. 41, 413–417.

Wigley, P., Barrow, P. and Schat, K.A., 2014. The avian reproductive immune system. In: Schat, K.A., Kaspers, B., Kaiser, P. (Eds.), Avian Immunology, second ed. Elsevier Science, Amsterdam, pp. 265–274.

Withanage, G.S.K., Baba, E., Sasai, K., Fukata, T., Kuwamura, M., Miyamoto, T., Arakawa, A., 1997. Localization and enumeration of T and B lymphocytes in the reproductive tract of laying hens. Poult. Sci. 76, 671–676.

Withanage, G.S.K., Sasai, K., Fukata, T., Miyamoto, T., Baba, E., Lillehoj, H.S., 1998. T lymphocytes, B lymphocytes, and macrophages in the ovaries and oviducts of laying hens experimentally infected with *Salmonella enteritidis*. Vet. Immunol. Immunopathol. 66, 173–184.

Withanage, G.S.K., Sasai, K., Fukata, T., Miyamoto, T., Baba, E., 1999. Secretion of *Salmonella*-specific antibodies in the oviducts of hens experimentally infected with *Salmonella enteritidis*. Vet. Immunol. Immunopathol. 67, 185–193.

Withanage, G.S., Sasai, K., Fukata, T., Miyamoto, T., Lillehoj, H.S., Baba, E., 2003. Increased lymphocyte subpopulations and macrophages in the ovaries and oviducts of laying hens infected with *Salmonella enterica* serovar Enteritidis. Avian Pathol. 32, 383–390.

Withanage, G.S., Wigley, P., Kaiser, P., Mastroeni, P., Brooks, H., Powers, C., Beal, R., Barrow, P., Maskell, D., McConnell, I., 2005. Cytokine and chemokine responses associated with clearance of a primary *Salmonella enterica* serovar Typhimurium infection in the chicken and in protective immunity to rechallenge. Infect. Immun. 73, 5173–5182.

Woods, D.C., Schorey, J.S., Johnson, A.L., 2009. Toll-like receptor signaling in hen ovarian granulosa cells is dependent on stage of follicle maturation. Reproduction 137, 987–996.

Worley, M.J., Nieman, G.S., geddes, K., Heffron, F., 2006. *Salmonella typhimurium* disseminates within its host by manipulating the motility of infected cells. Proc. Natl. Acad. Sci. 103, 17915–17920.

Wu, H., Jones, R.M., Neish, A.S., 2012. The Salmonella effector AvrA mediates bacterial intracellular survival during infection in vivo. Cell. Microbiol. 14, 23–39.

Wynoski-dolfi, M.A., Snyder, A.G., Philip, N.H., Doonan, P.J., Poffenberger, M.C., Avizonis, D., Zwack, E.E., Riblett, A.M., Hu, B., Strowig, T., Flavell, R.A., Jones, R.G., Freedman, B.D., Brodsky, I.E., 2014. Oxidative metabolism enables *Salmonella* evasion of the NLRP3 inflammasome. J. Exp. Med. 211, 653–668.

Xavier, M.N., Winter, M.G., Spees, A.M., den Hartigh, A.D., Nguyen, K., Roux, C.M., Silva, T.M.A., Atluri, V.L., Kerrinnes, T., Keestra, A.M., Monack, D.M., Luciw, P.A., Eigenheer, R.A., Bäumler, A.J., Santos, R.L., Tsolis, R.M., 2013. PPARγ-mediated increase in glucose availability sustains chronic *Brucella abortus* infection in alternatively activated macrophages. Cell Host Microbe 14, 159–170.

Ye, I., Meng, Y., Zheng, K., Liu, J., Wang, Q., Tao, P., Jia, K., Liao, M., Ying, Z., 2015. Tissue specific expression pattern and histological distribution of NLRP3 in Chinese Yellow chickens. Vet. Res. Comm. 39, 171–177.

Ye, Z., Petrof, E.O., Boone, D., Claud, E.C., Sun, J., 2007. *Salmonella* effector AvrA regulation of colonic epithelial cell inflammation by deubiquitination. Am. J. Pathol. 171, 880–892.

Yi, J., Kloeker, S., Jensen, C.C., Bockholt, S., Honda, H., Hirai, H., Beckerle, M.C., 2009. Members of the zyxin family of LIM proteins interact with members of the $p130^{Cas}$ family of signal transducers. J. Biol. Chem. 277, 9580–9589.

Yoshimura, Y., 2015. Avian β-defensins expression for the innate immune system in hen reproductive organs. Poult. Sci. 94, 804–809.

Yoshimura, Y., Obashi, H., Subedi, K., Nishbori, M., Isobe, N., 2006. Effects of age, egg-laying activity, and *Salmonella* inoculation on the expansion of gallinacin mRNA in the vagina of the hen oviduct. J. Reprod. Dev. 52, 211–218.

Yoshimura, Y., Abdel Megeed, A.M., Nii, T., Ariyadi, B., Isobe, N., 2014. Innate immune functions in hen reproductive organs. Avian Biol. Res. 7, 39–47.

Zheng, W.M., Yoshimura, Y., 1999. Localization of macrophages in the chicken oviduct: effects of age and gonadal steroids. Poult. Sci. 78, 1014–1018.

CHAPTER

# *Salmonella* Heidelberg in Layer Hens and Egg Production: Incidence and Potential Issues

12

**Pravin R. Kaldhone[1], Steven L. Foley[2], Steven C. Ricke[1]**

*[1]University of Arkansas, Fayetteville, AR, United States; [2]U.S. Food and Drug Administration, Jefferson, AR, United States*

## 1. INTRODUCTION

Pathogenic bacteria cause an estimated 3.6 million cases of food-borne illnesses per year in the United States, approximately one-third (1.04 million cases) of which are caused by the nontyphoidal *Salmonella* (Scallan et al., 2011). Food-borne illnesses caused by organisms such as *Salmonella* associated with food animals have continued to remain a prominent concern over the past several decades. Several factors have contributed to this continued prevalence of food-borne illnesses. For example, centralization of food production, food processing, and distribution system increases the possibility for larger outbreaks (Rose et al., 2002; Bhatt and Zhang, 2013). In addition, increased and divergent sources of foods and food ingredients such as eggs add to the uncertainty (Ricke et al., 2013a). Likewise, changes in consumer dining habits, food preferences, and increased consumption of raw foods such as vegetables are likely contributors as well (Hanning et al., 2009).

Among the bacterial food-borne pathogens *Salmonella*-related infections annually account for 35% of hospitalizations and 65% of deaths in the United States (Scallan et al., 2011). Human salmonellosis is dominated by the broad host range serovars (Foley et al., 2011, 2013; Ricke et al., 2013b). Similarly, several of the most commonly detected serovars in the US poultry industry are able to colonize multiple host species (Foley et al., 2011, 2013). Not all serovars behave the same way in their respective host and some serovars exhibit a broader host range than others. Invasive diseases associated with these serovars can arise when gastrointestinal organisms are able to undergo extraintestinal spreading leading to bacteremia and focal (localized to one organ or system) infection causing systemic manifestations (Jones et al., 2008; Suez et al., 2013). For example, invasive nontyphoidal *Salmonella* have surfaced as a leading cause of bloodstream infections in sub-Saharan Africa in juveniles and adults (Feasey et al., 2012).

**Producing Safe Eggs. http://dx.doi.org/10.1016/B978-0-12-802582-6.00012-4**

Poultry continues to be one of the primary reservoirs of *Salmonella* among animals (Edwards, 1958; Foley et al., 2008, 2011, 2013; Finstad et al., 2012; Howard et al., 2012). It has been estimated that 90% of the cases of salmonellosis in the United States originated from chickens, eggs, or egg products (Chittick et al., 2006; Foley et al., 2008). In the United States, 3.9% of the whole chicken, 18% of the ground chicken, 1.6% of the ground beef, 15% of the ground turkey, and 2.3% of the turkey samples tested positive for *Salmonella* during 2013 (CDC, 2013). Among the known human disease-causing *Salmonella* serovars, a limited number are considered to be significant causes of food-borne infection. *Salmonella enterica* serovar Heidelberg, *S.* Kentucky, *S.* Senftenberg, *S.* Enteritidis, *S.* Typhimurium, and *S.* Hadar have all been identified as prominent serovars in samples isolated from chickens over the past 20 years (Foley et al., 2008, 2011, 2013). In more recent times, *S. enterica* serovar Heidelberg has emerged as a leading food-borne disease-causing serovar. Among the documented outbreaks, *S.* Heidelberg has been associated with those originating from eggs and egg production suggesting that it can occupy this specific niche. Given this concern, the overall objective of this review is to discuss what is currently known about *S.* Heidelberg as a food-borne pathogen in general and future perspectives for understanding its incidence in poultry and eggs and potential issues associated with antibiotic resistance.

## 2. EMERGENCE OF *SALMONELLA* HEIDELBERG AS A FOOD-BORNE PATHOGEN

*Salmonella* Heidelberg was initially discovered in 1933 in Heidelberg, Germany (Habbs, 1933). In 1954, it was isolated for the first time in the United States (Smyser et al., 1965). More recently, the serovar has consistently remained among the top 10 most common etiologic agents for nontyphoidal *Salmonella* infection (Harris et al., 1990; Stanley et al., 1992; Threlfall et al., 1992; Foley and Lynne, 2008; Donado-Godoy et al., 2015). Specifically in the United States, *S.* Heidelberg is typically among the top five most common serovars causing human salmonellosis (CDC, 2011) and has been responsible for several outbreaks in the United States and Canada (Dutil et al., 2010; Hoffman et al., 2013). For example, a multistate outbreak that occurred in 2013 and 2014 was traced back to contaminated chicken that managed to sicken 634 people in 29 different states (CDC, 2014).

According to the US Department of Agriculture (USDA) National Veterinary Service Laboratory data from 1968 to 2010, 71% of the *S.* Heidelberg isolates collected over that time period originated from poultry-related sources (CDC, 2013). Among isolates at slaughter, poultry accounted for 86% of the food animal isolates, which is noteworthy, given the high percentage of the isolates that originates from one class of hosts. Thus the serovar likely has evolved to survive in the chicken gastrointestinal environment. Genetic adaptation to environmental conditions associated with the host could be one reason for this predominance as has been indicated for other serovars (Johnson et al., 2010). The environments where birds are raised can influence the patterns of predominant organisms in their gut ecosystem (Nordentoft et al., 2011). The avian gut provides a diverse polymicrobial environment that could potentially provide selective pressure to alter the genetic composition of *S.* Heidelberg in such a manner to better adapt to the poultry environment (Han et al., 2012). In addition, antimicrobial exposure may impact the populations of organisms present in an environment, especially since several *S.* Heidelberg strains have been reported to be multidrug resistant (Lynne et al., 2009). If these strains outcompete other bacteria under selective pressure, it may help explain the relative prominence of this pathogen along the food production continuum, where *S.* Heidelberg isolates displaying antimicrobial resistance have been recovered at many different steps (Lynne et al., 2009). A detailed examination of the incidence and distribution in poultry and egg production is warranted to develop a more in-depth understanding regarding potential explanations for the prevalence of this serovar.

## 3. EPIDEMIOLOGY OF EGG-ASSOCIATED *SALMONELLA* OUTBREAKS

A substantial proportion of the increase in poultry-associated salmonellosis is no doubt due to an increase in poultry and egg market sales. Since 1910, per capita consumption of poultry products in the United States has increased 6.5-fold (Buzby and Farah, 2006). Egg consumption in the United States has also reached a 30-year high in the last few years, with the per-person consumption increasing by 12 eggs over the past 4 years (Clarke, 2015). To meet this demand, the USDA reported that 242 million cases of shell eggs (15 dozen eggs per case) were produced in the United States during 2014 by 305 million shell egg laying hens (Egg facts, 2015). In 1 month alone, 5.814 millions of cases of eggshells were broken for egg products in the United States (Ibarburu, 2015). Global egg production also grew from 35.2 million tons to 62.6 million tons in the last few years (Windhorst, 2009). Regional imbalances in egg production and demand have led to substantial growth in egg transportation globally. In North America, the United States is a major exporter and Canada and Mexico are importers (Windhorst, 2009). South and Southeast Asian countries such as Malaysia and Thailand are now exporters, whereas the U.A.E. and Kazakhstan are leading importers in Western and Central Asian countries (Windhorst, 2009). In the European Union, the ban on cage rearing for birds has led to an increase in the import of eggs from neighboring non–European Union countries such as Belarus (Windhorst, 2009). These increased levels of imports have increased the odds of food-borne diseases caused by eggs, because the corresponding effects of transit time adds to the likelihood of a break in the cold chain and opportunities for cross-contamination (Carrasco et al., 2012).

Egg consumption has been identified as a risk factor for *S.* Heidelberg infection (Hennessy et al., 2004). Hennessy et al. (2004) estimated that approximately 37% of *S.* Heidelberg population-attributable infections originate from consuming eggs prepared outside the home. In addition, eggs can be consumed in many different forms, including either eggs alone or through products containing eggs that may be raw or lightly cooked, such as Caesar salad dressing, homemade ice cream, hollandaise sauce, and fresh pasta dishes (Ricke et al., 2013a). This diverse use of eggs in a wide range of food products makes it a critical potential vehicle for *S.* Heidelberg transmission and also a challenge to pinpoint specific sources of infections.

In 2010, the nation's largest egg recall was due to *S. enterica* serovar Enteritidis contamination on egg layer farms in Iowa (Flynn, 2012). When one of the farms associated with the 2010 outbreak was inspected in 2012, investigators isolated *S.* Heidelberg from the poultry houses (Hoffman et al., 2013). They were confirmed initially using pulsed-field gel electrophoresis (PFGE) followed by whole genome sequencing (WGS) to differentiate isolates with the same PFGE pattern (Hoffman et al., 2013). The four isolates related to this outbreak were linked to the JF6X01.0022 XbaI and JF6A26.0001 BlnI patterns.

In another instance, outbreak strains isolated from clinical patients exhibited a specific PFGE pattern (MMWR, 2013) that matched the PFGE pattern of a strain isolated from the suspected slaughterhouse examined during the traceback investigation with similar PFGE patterns being subsequently differentiated by WGS (Evans et al., 2014). Two isolates from this outbreak exhibited a multidrug-resistant phenotype. The patients, from whom these two isolates were collected, were under 1 year of age. The variable antimicrobial resistance profile of these isolates and patients' immature immune status would have made outpatient treatment difficult, thus the infants required hospitalization for their illness. It is probable that these clinical isolates could have acquired their antimicrobial resistance via the poultry house environment (MMWR, 2013). Due to the occurrence of *S.* Heidelberg in poultry houses, poultry could be considered as a primary source of the pathogen, thus making poultry-associated *S.* Heidelberg a potential public health concern (FDA, 2012).

## 4. ECOLOGY OF *SALMONELLA* HEIDELBERG COLONIZATION AND INVASION IN POULTRY

In general, *S.* Heidelberg possesses potential extensive colonization capacity for poultry associated with its ability to attach and invade the host intestinal epithelial cells. Some *S.* Heidelberg strains have been identified that contain mobile genetic elements such as the incompatibility group (Inc.) FIB plasmids. These plasmids often contain genes for iron acquisition, toxin (colicin) production, serum survival, and antimicrobial resistance (Han et al., 2012). The presence of these plasmids and their aerobactin operon and *sit* iron transport systems likely play a role in allowing *S.* Heidelberg to successfully colonize the epithelial lining of poultry.

Chicken macrophages are an important component in the defense against bacterial invasion (He et al., 2012). Newly hatched chickens possess a naïve immune system and rely on transferred maternal immunity for their defense against infection. The macrophage response in these young birds is attenuated (He et al., 2012). Therefore young birds are particularly susceptible to *S.* Heidelberg infection due to the pathogen's ability to weaken the macrophage response coupled with the weaker host immune system. These factors make *S.* Heidelberg an important player in terms of invasive infection in newly hatched chicks.

Macrophage survival is a key for boosting virulence leading to invasive *Salmonella* infections. Gokulan et al. (2013) examined the infection of J774 mouse macrophages by *S.* Heidelberg and concluded that *S.* Heidelberg was able to enter and survive in this macrophage cell line with differing abilities (Gokulan et al., 2013; Agnihothram et al., 2015). Those strains that survived the best contained a plasmid-encoded type 4 secretion system (T4SS), which likely diminished host immune response that resulted in increased uptake and survival of *S.* Heidelberg in the macrophages after 24h of incubation (Gokulan et al., 2013). He et al. (2012) conducted a study examining the interaction of *S.* Heidelberg and HD-11 chicken macrophage cells, which are MC29 virus-transformed chicken macrophage cells (Beug et al., 1979). An effective, robust oxidative burst was considered an indicator of effective defense functions of the cell line; however, the study demonstrated that HD-11's phorbol myristate acetate–stimulated oxidative burst decreased after infection by *S.* Heidelberg (He et al., 2012). Thus this ability of *S.* Heidelberg to limit macrophage function could be critical to its extraintestinal survival.

If *S.* Heidelberg infects the reproductive tract of egg laying hens, eggs produced have the potential to be contaminated as occurs with other serovars particularly *S.* Enteritidis. *Salmonella* Enteritidis can contaminate eggs in two distinct ways, either by external penetration of the eggshell or internally via transovarian infection (Gantois et al., 2009; Howard et al., 2012; Martelli and Davies, 2012). The external eggshell penetration route includes transmission from the feces of colonized birds to the egg surface followed by penetration to the interior of eggs and growth during the storage (Cockburn and Vemon, 1956). *Salmonella* Heidelberg would appear to be a candidate for external egg contamination as it has been isolated from layer feces in commercial layer houses (Li et al., 2007). Gast et al. (2007b) studied in vitro egg contamination by *S.* Heidelberg and *S.* Enteritidis under ambient temperature. They observed *S.* Enteritidis exhibits a remarkably greater rate of eggshell penetration and yolk multiplication as compared to *S.* Heidelberg. They also reported a significantly lower rate of penetration and multiplication at incubation temperatures between 20°C and 30°C for *S.* Heidelberg.

Although *S.* Heidelberg appears to be less capable of penetration, prevention measures focused on temperature control at the poultry farm and during processing and transportation may still be important for controlling *S.* Heidelberg. Although little direct evidence has been established that *S.* Heidelberg possesses characteristics that allow it to be prevalent in eggs, it has been reported to grow in Brain Heart Infusion broth at 19°C and 37°C, with only slight variation when compared with other *Salmonella* serovars including *S.* Enteritidis (Juneja et al., 2003). When McConnell and Schaffner (2014) incubated *S.* Heidelberg as part of a cocktail of *Salmonella* serovars in raw ground beef, they validated the recommended US Food and Drug Administration (FDA) guidelines for the length of time that food can be kept out of temperature control if the food product starts at 5°C and does not exceed 21°C (McConnell and

Schaffner, 2014). Although it remains to be determined if similar criteria would be applicable for *S.* Heidelberg in eggs, there may be potential for a relatively high prevalence of *S.* Heidelberg in eggs if the opportunity for initial contamination arises and sufficient temperature abuse occurs to allow substantial growth. Certainly, improper transport and a break in the cold chain could enhance growth of *S.* Heidelberg in contaminated eggs (Schoeni et al., 1995). Although pasteurizing egg whites appears to cause a greater than eight log reduction of *S.* Heidelberg (Muriana, 1997), cooking may not be able to always eliminate the organism as several *Salmonella* serovars are capable of surviving simulated domestic conditions for various forms of cooking eggs (Humphrey et al., 1989).

Among *Salmonella* serovars, *S.* Heidelberg, *S.* Enteritidis, and *S.* Typhimurium are able to colonize the reproductive tract of layer hens with *S.* Enteritidis exhibiting tissue tropism for the reproductive tract (Gast et al., 2004, 2005, 2007a, 2011; Gantois et al., 2008). Gast et al. (2011) observed the same rate of isolation for *S.* Heidelberg and *S.* Enteritidis in ovaries and oviducts of chicken. This indicates both serovars might possess similar capabilities to colonize the reproductive tract and also implies that factors other than colonization of the bird's main reproductive tract play an important role in the contamination of egg. *Salmonella* Enteritidis has been shown to produce high-molecular-weight lipopolysaccharides and be able to grow to high cell densities (Guard-Petter, 1998; Parker et al., 2001; Gast et al., 2011). These characteristics could have a role in colonization of bacteria to the epithelium of the gastrointestinal tract and could be the reason for the greater ability of *S.* Enteritidis to colonize and invade gastrointestinal tract than other serovars such as *S.* Heidelberg (Gast et al., 2011). *Salmonella* Enteritidis uses diverse types of fimbriae such as SEF 14, SEF 17, and SEF 21 to attach to the host luminal lining. It also possesses long polar and plasmid-mediated fimbriae (Foley et al., 2008), whereas *S.* Heidelberg expresses fimbriae such as FliA, FliB, and FliC. *Salmonella* Heidelberg has been isolated from ovaries of naturally infected chickens (Barnhart et al., 1991), which may provide an opportunity for transovarian contamination of eggs. The egg-contamination ability could be attributed to expression of potential virulence factors such as the outer membrane proteins, fimbriae and flagella. Environmental factors including temperature and pH might affect the expression of these virulence factors as well (McDermid et al., 1996; Morales et al., 2007; Gast et al., 2011) and impact the ability of *Salmonella* to infect eggs. Clearly, more research needs to be done to elucidate whether *S.* Heidelberg possesses specific mechanisms associated with colonization of the layer hen reproductive tract.

## 5. VIRULENCE AND PATHOGENESIS OF *SALMONELLA* HEIDELBERG

Although the *S.* Heidelberg association with laying hens and eggs remains to be fully explored, its pathogenesis in humans is assumed to be fairly typical of other food-borne disease-causing *Salmonella* serovars. *Salmonella* infections in humans can lead to gastrointestinal illness, which is characterized by nausea, vomiting, abdominal pain, and diarrhea that begins 12–36h following consumption of the contaminated food. The severity of the symptoms depends on various factors including the level of virulence gene expression of the organism and the host immune status (Robertson et al., 2003).

The type of diarrhea caused by *S.* Heidelberg and other pathogenic serovars is inflammatory diarrhea, which is the result of the interaction of bacterial enterotoxin and host epithelium (Foley et al., 2011, 2013). Following ingestion, *Salmonella* adheres to the intestinal epithelium with the help of flagella and fimbriae (Van Asten and Van Dijk, 2005; Foley et al., 2013). Conserved and host-specific factors expressed by *Salmonella* helps the organism to colonize the host gastrointestinal epithelium (Stevens et al., 2009; Foley et al., 2013). The pathogen crosses the intestinal epithelial barrier with the aid of the *Salmonella* pathogenicity island (SPI) 1-encoded type 3 secretion system (T3SS), which is a molecular transporter that facilitates the transfer of toxins and effector proteins such as InvJ, SpaO, PrgI/J, SipA/B/C/D, SptP, AvrA, SopA/B/D/E/E2, SlrP, and SspH1 from the cytoplasm of the bacteria into the host cells (Galán and Wolf-Watz, 2006; Schlumberger and Hardt, 2006; Foley et al., 2013). Thus the T3SS promotes cellular uptake and invasion. Some of the effector proteins such as SopA/B/D/E2 and SipA activate the host signal transduction cascade (Hopkins and Threlfall, 2004; Foley et al., 2013), which leads to induction of membrane ruffling at the contact site where *Salmonella* interacts with the host cell (Al-Mousawi et al., 2010). Membrane ruffling leads to engulfment of the bacterium and the formation of *Salmonella*-containing vacuoles within the host cells.

The virulence phenotype displayed by the pathogen is largely determined by the virulence factors that the organism carries. In addition to the SPI-1-coded T3SS, there is a second T3SS coded by SPI-2 that plays an important role in the virulence of serovars such as *S.* Enteritidis (Hensel et al., 1998; Rosselin et al., 2011; Foley et al., 2013; Ricke et al., 2013b). The SPI-2-coded T3SS is involved in postinvasion changes in the intracellular environment (Malik-Kale et al., 2011). Each of the T3SSs forms complex systems that deliver at least 40 distinct virulence effectors into the host cells to facilitate invasion, survival, and replication within host cells (Malik-Kale et al., 2011). These virulence effectors are responsible for various functions including decreasing the activating and trafficking of free oxygen radicals and inhibiting phagocyte maturation. Free oxygen radicals, such as nitrous oxide, are one of the macrophages' primary defense tools against microbial pathogens (Rosselin et al., 2011). Defective macrophages can be responsible for intracellular survival and proliferation of bacterial pathogens (Withanage et al., 2005).

*Salmonella* Heidelberg, along with other invasive nontyphoidal serovars, possess additional genetic elements that can facilitate invasive infections. In addition to SPI-1 and 2, *Salmonella* can carry several additional SPIs, including 3–6, 9, 13, and 14, which are important for *Salmonella* virulence (Suez et al., 2013). For example, SPI-6 encodes genes such as *invasin*, *pagN*, *CS54*, and *sinH*, which contribute to *Salmonella*'s invasive phenotype. Fimbriae gene clusters, such as *bcf*, *csg*, *stb*, *sth*, and *sti* also aid infection of the host by forming filamentous structure on the cell surface that assists colonization in chicken gastrointestinal epithelium (Foley et al., 2013). *Salmonella* Heidelberg, like other *Salmonella* serovars is able to penetrate the intestinal epithelium, spread from one epithelial cell to another and eventually enter into the macrophages and dendritic cells (Wallis et al., 1986; Richter-Dahlfors et al., 1987; Jones et al., 1994; Rescigno et al., 2001; Salcedo et al., 2001; Meyerholz et al., 2002; Geddes et al., 2007; Malik-Kale et al., 2011).

## 6. ANTIMICROBIAL RESISTANCE IN *SALMONELLA* HEIDELBERG

The other phenotype characteristic that *S.* Heidelberg shares with several *Salmonella* serovars is antimicrobial resistance. Several strains of *S.* Heidelberg have been shown to cause invasive disease, which often requires antimicrobial therapy for treatment (Suez et al., 2013). Consequently, antimicrobial resistance is a major health concern due to potential clinical treatment failure. Analysis of resistance trends has shown that *S.* Heidelberg isolates collected in recent years are more likely to be more resistant to clinically important antimicrobial agents than they were historically (Folster et al., 2012). These findings may be due, at least in part, to selective pressure from continued use of antimicrobial agents in animal feeds and veterinary and human medicine (Crump et al., 2011). For example, the numbers of cephalosporin-resistant *S.* Heidelberg infections occurring in chickens and humans exhibited a distinctive trend in Quebec, Canada, during the past few years. From 2004 to 2007 the number of resistant infections was decreasing; however, from 2007 to 2011 the trend reversed and subsequently the numbers of resistant infections increased. This trend has been suggested to correlate with the reintroduction of the use of ceftiofur in hatcheries in Quebec that began in late 2006 after a period of disuse (Otto et al., 2014).

Overall, antimicrobial resistance increases the cost of illness by increasing the number of cases, severity, and duration of illness (Rabsch et al., 2001; Foley et al., 2008). Resistance also leads to clinical treatment failure if the administered therapy is ineffective and the health care provider is forced to apply the next line of therapy. This regimen lengthens the time of recovery and heightens the odds of the patient developing bacteremia, septicemia, and organ system failure. Once organ system failure ensues, it can cause irreversible damage to the body, potentially leading to death. In this way, increased resistance can lead to chronic sequelae and increased mortality (Barza and Travers, 2002).

Human antimicrobial use is also a risk factor for salmonellosis. Antimicrobial treatment can disrupt the normal microbiome of the host. The microbiome serves an important function by occupying the epithelial surface and preventing colonization by new organisms (Rashid et al., 2015). This colonization resistance can be hampered by antimicrobial therapy (Barza and Travers, 2002; Molbak, 2005). As noted earlier, antimicrobial therapy can disrupt colonization resistance in the host. If there are *Salmonella* present or subsequent ingestion of organisms that are resistant to the drug used for previous therapy, it increases the likelihood of infection and proliferation increasing the severity of the food-borne illness (Koningstein et al., 2010).

As in Canada, *Salmonella* strains isolated in the United States have shown extended spectrum cephalosporin (ESC) resistance (Taylor et al., 2015). This phenomenon is due in large part to the ability of *Salmonella* to synthesize AmpC-like β-lactamase (Philippon et al., 2002). This enzyme is coded by the $bla_{CMY}$ genes, which are often located on plasmids, including those of incompatibility groups (Inc.) A/C and IncI1. Both IncI1 and IncA/C plasmids have been identified in *S.* Heidelberg isolated from poultry (Han et al., 2012). The IncI1 plasmids carrying $bla_{CMY}$ gene have been observed to acquire kanamycin resistance along with cephalosporins (Folster et al., 2011). IncI1 plasmids carrying $bla_{CMY}$ genes typically belong to sequence type 12 of the plasmid multilocus sequencing typing (pMLST) analysis scheme (Jolley and Maiden, 2010). This sequence-based technique relies on genes contained on the plasmid. If genes carrying antimicrobial resistance are present on the plasmid, then there are enhanced odds that resistance could have been acquired through horizontal genetic transfer (Kaldhone et al., 2008; Krauland et al., 2010; Cain and Hall, 2012). Similar findings from pMLST have been reported for *S.* Kentucky isolates originating from poultry (Fricke et al., 2009) and *Salmonella* and *Escherichia coli* from environmental, animal, and human sources in Canada (Mataseje et al., 2010). The increased dissemination of ESC resistance in the North American continent is likely due to the transmission of the plasmid-encoded $bla_{CMY}$ genes. The fact that ESC strains display different PFGE patterns indicates the ability of the plasmids to incorporate into a variety of genetic backgrounds across multiple serovars and even species (Folster et al., 2012).

Ceftiofur resistance has been reported to be frequent among *S.* Heidelberg isolates from chicken (9%) and human sources (33%) in Quebec, Canada (CIPARS, 2011). A rotational administration of antibiotics was implemented to mitigate increasing antimicrobial resistance. It was subsequently observed that ceftiofur resistance decreased from 70 cases per 100,000 in 2004 to 29 cases per 100,000 in 2007 (CIPARS, 2011). Changes in antimicrobial agent use in veterinarian and agricultural practices might act as a risk reduction strategy to decrease antimicrobial resistance while treating food-borne illnesses.

## 7. ISOLATION, IDENTIFICATION, AND DETECTION

There are numerous methods for the isolation, identification, and detection of *Salmonella* (Ricke et al., 2013b; Park et al., 2014). Several conventional isolation and identification methods are described by the US FDA, European Committee for Standardization, and International Organization of Standardization and are culture based. These classical techniques follow a standard sequence beginning with nonselective preenrichments, followed by selective enrichments, isolation on selective agar media, and finally biochemical screening with triple sugar and lysine iron agars. Serological testing using poly-O and poly-H antisera is used as a step to identify the specific serotype of the isolate based on the Kaufmann–White scheme. In the case of *S.* Heidelberg, Maurischat et al. (2015) listed the serological profile (serotype) as 4,[5],12:r:1,2 when using it as one of the serovars for a real-time multiplex polymerase chain reaction (PCR) assay developed to differentiate *S.* Enteritidis and *S.* Typhimurium. Numerous broths and plating media have been employed for *Salmonella* growth and some of them have experienced significant modifications during the course of time to increase their efficacy. The media used may have an impact on the efficacy of isolating different *Salmonella* serovars. This difference could be a reason for the variability in isolation rates and prevalence of certain *Salmonella* serovars in different locales (Richardson et al., 2011).

PCR approaches represent a more recent molecular-based methodology used to detect and identify food-borne pathogens including *Salmonella*. The technique uses enzymatic amplification of specific DNA sequences in an isolate (Gharieb et al., 2015). Over time several different variations of PCR have been used for *Salmonella* identification; these include multiplex PCR, SYBR Green based real-time (RT) PCR (Roche Diagnostics, Indianapolis, IN), and the BAX System (DuPont, Wilmington, DE) (Park et al., 2014; Gunel et al., 2015). These molecular methods are more rapid and reproducible, yet a positive result is only considered presumptive and needs to be confirmed by a standard method that leads to isolating an organism.

Although molecular methods are more rapid than culture identification, their speed is reduced by the fact that many of these rapid methods require that the samples undergo a culture-enrichment step that can take several hours before analysis. Issues with including an enrichment step arise when considering that certain enrichment protocols may favor the odds of detection of certain serotypes over others (Gorski, 2012). For example, *S.* Heidelberg spent media has been shown to limit the growth of certain *S.* Typhimurium isolates (Rivera Calo et al., 2015). Although the mechanism is not clear, specific metabolites produced by *S.* Heidelberg could play a role in this inhibition. In addition, several different techniques have been used across laboratories (Singer et al., 2009). In a study comparing common culturing methods for *Salmonella* it was found that each of the five methods compared resulted in a different *Salmonella* prevalence from swine fecal samples (Love and Rostagno, 2008). However, only the combination of results from two or more methods agreed most closely with the known level of positives (Love and Rostagno, 2008).

Advanced molecular typing techniques can be used to identify the effects of horizontal gene transfer among bacterial strains and thus used for traceback identification of pathogen sources. The techniques often employed include PFGE, clustered regulatory interspaced short palindromic repeats (CRISPR)–multiple variable locus sequence typing (MVLST) analysis (Young et al., 2012) and more recently WGS. The CRISPR are unique genetic elements that are made up of short sequences called spacers and conserved direct repeats (Haft et al., 2005). Analysis of CRISPR loci has been used to differentiate clinical isolates of *Salmonella* (Fabre et al., 2012). *Salmonella*-associated MVLST is a sequencing-based typing method that relies on the comparison of sequences of two virulence genes, *fimH1* and *sseL* (Liu et al., 2011). Molecular subtyping methods can be used in concert, for example, the combined CRISPR-MVLST and PFGE analysis has been shown to possess more discriminatory power than individual methods for *S.* Heidelberg isolates

(Shariat et al., 2013). This means that if isolates have similar PFGE profiles, then they can be differentiated from each other using CRISPR-MVLST. Among these techniques, PFGE has proved to have utility for identifying horizontal gene transfer among *S.* Heidelberg strains from turkey-associated sources (Kaldhone et al., 2008).

Additional typing methods include multiple loci variable number tandem repeat analysis (MLVA), which is a PCR-based method that relies on differences in the number of tandem repeats that are observed at multiple loci known to have strings of repetitive sequence in the bacterial genome (Broschat et al., 2010). Multiple amplification of prophage locus typing (MAPLT) represents another sequence-based technique that depends on loci located in integrated prophage sequences (Ross and Heuzenroeder, 2005). When MLVA and MAPLT were combined, they were found to be better able to distinguish among *S.* Heidelberg isolates of phage type (PT) 1, compared with the respective individual methods (Young et al., 2012). These approaches have been used to distinguish between human and nonhuman-associated isolates among PT1 isolates (Demczuk et al., 2003).

With increasing ease and availability of sequencing, WGS has been promoted as the ultimate tool for the investigation of food-borne pathogens such as *S.* Heidelberg (Hoffman et al., 2013, 2014). In a retrospective study of a recent *Salmonella* outbreak, Hoffmann et al. (2014) used WGS to examine the genetic relatedness of *S.* Heidelberg isolates associated with the 2011 multistate outbreak. The sequencing confirmed the presence of multiple antimicrobial resistance genes and likely enhanced virulence genes associated with T4SS. Single-nucleotide polymorphism (SNP) analysis based on WGS data has proved to be helpful for in-depth differentiation of isolates; in one study, 284 significant SNPs were found in 44 *S.* Heidelberg isolates that exhibited nearly identical PFGE patterns (Hoffmann et al., 2014). Similarly, Bekal et al. (2016) used a high-quality core genome single-nucleotide variant (hqSNV) to discriminate among the more prevalent and highly clonal *S.* Heidelberg isolates. More than 59 hqSNVs were measured among 46 *S.* Heidelberg isolates from three different outbreaks in Quebec that possessed the same PFGE and PT patterns. The ability to use SNP analyses to discriminate highly clonal isolates demonstrates that the WGS-based approach could be a superior typing tool while working with events where the isolates were previously considered identical with conventional subtyping methods.

## 8. FUTURE ISSUES—EVOLUTION OF *SALMONELLA* HEIDELBERG

*Salmonella* serotypes vary in their host specificities, for example, *S.* Gallinarum and *S.* Dublin are very host-specific serovars, whereas *S.* Typhimurium, *S.* Enteritidis, and *S.* Heidelberg represent examples of broader host range serovars (Baümler et al., 1998). Host range is dictated by several factors including genome plasticity and interaction with the host and its immune system (Foley and Lynne, 2008). The acquisition of new genes that allow for the attachment or colonization of new host environments can facilitate expansion of a respective microorganism's host range (Methner et al., 2011). These new genes can be obtained through horizontal gene transfer by a variety of vehicles including phages, plasmids, and transposons (Foley et al., 2013). Particularly troublesome is the continued isolation of strains resistant to multiple antimicrobial agents that makes them more problematic for treatment (Hennessy et al., 2004; Foley et al., 2008). Mutations in virulence-related genes and presence of pseudogenes indicate that alterations of its virulence profile may also be occurring (Chiu et al., 2005). *Salmonella* Heidelberg appears to be continually evolving, likely in response to exposure to different external pressures (Beltran et al., 1988; Suez et al., 2013). These changing phenotypic characteristics of *S.* Heidelberg are brought about by alterations in genomic composition of strains through acquisition of new genes or mutation of existing gene contents (Foley et al., 2013) allowing bacteria to adapt to external stress and to alter their genetic content (Onchman and Moran, 2001; Maurelli, 2007).

In general, there are approximately $10^{-10}$ mutations per base pair in bacteria (Bars et al., 2012). Some of the bacteria express mutations higher than this frequency and are referred to as "mutators" (Bars et al., 2012). *Salmonella* Heidelberg strain, SHB182, is one example of such a mutator. The SHB182 strain has been associated with the bovine intestinal microbiome (Le Gall et al., 2009). Mutations that this strain has accumulated are believed to facilitate increased adaption to the changing bovine intestinal lining. To study this phenomenon, Bars et al. (2012) created a 12-base pair-deletion in a methyl mismatch repair system for SHB182. This led to enhanced adherence to epithelial cells through increased expression of *fliC* and decreased expression of *fliA* and *fliB*. Allelic differences in *fliC* gene among *S.* Heidelberg, *S.* Typhimurium, and *S.* Muenchen are the result of recombination (Milkman and Stoltzfus, 1988; Smith et al., 1990). These findings from studies with SHB182 indicate that at least some strains of *S.* Heidelberg are able to undergo genetic changes in response to environmental stress factors that are reflected in altered pathogenic phenotypes.

Thus a historical analysis of *S.* Heidelberg strains may explain the genetic adaptations that members of the serovar have undergone to survive in their respective environments (Kivisaar, 2003). For example, in a study exploring *E. coli*, a mutation phenotype enabled organisms to adapt rapidly to the mouse gut environment (Giraud et al., 2001). *Escherichia coli* can acquire and accumulate mutations to adapt rapidly to its environment. Genomic plasticity enables organisms to alter their genomic content (Liu et al., 2007), and in *S.* Heidelberg's case to potentially increase its host range to infect a broad range of species. Therefore a high-resolution genomic map will be useful for identifying the correlation between parent strains and newer strains. Genomic rearrangements could be either in the form of an insertion or a deletion. Addition or deletion of genes helps that organism to divert their resources toward more critical functions such as survival in a stressful environment. Consequently, *S.* Heidelberg with its ability of genomic plasticity is able to survive across a diverse host range.

## 9. CONCLUSIONS

Eggs have been an important source of protein in human diets and as such there has been increasing demand for egg production on a global scale, which makes identifying ways to improve egg safety imperative. *Salmonella* Heidelberg has become one of the more common organisms isolated along the poultry and egg production, processing and consuming continuum (Foley et al., 2011). The continued isolation of strains of *S.* Heidelberg that are resistant to multiple antimicrobial agents adds to the importance of studying ways to mitigate the risk of this organism in egg production.

To survive in the chicken intestinal tract, *S.* Heidelberg likely has displayed genomic plasticity to adapt to the host environment, either through the acquisition of required genes or deletion of unnecessary genes. Mobile genetic elements such as plasmids make this genetic information exchange possible. For example, the acquisition of IncFIB plasmids carrying iron transport and toxin production genes is an example of genetic adaptability of some *S.* Heidelberg strains (Han et al., 2012). Other factors such as the expression of certain fimbriae may enhance transovarian spread and perhaps indicate a phenotypic adaptation of *S.* Heidelberg to gain an ecological advantage in the avian environment. Increased virulence coupled with antimicrobial resistance makes *S.* Heidelberg a challenge for egg safety. Several different antimicrobial resistance genes have been identified on different plasmids detected in *S.* Heidelberg (Folster et al., 2011), which can facilitate the horizontal spread of antimicrobial resistance among bacteria and makes resistance difficult to manage in *Salmonella*.

Technical advances dictate the methods for *Salmonella* detection. Conventionally used culture methods are still considered as the gold standard; however, variability in outcome based on the media used and a longer duration to obtain results are drawbacks of culture methods (Richardson et al., 2011). PCR-based techniques are rapid and replicable, but often require culture confirmation to verify the results (Gunel et al., 2015). Advanced molecular techniques such as PFGE, MVLST, and CRISPR analyses, especially in concert with one another, are important to understanding the molecular epidemiology of disease transmission (Liu et al., 2011; Young et al., 2012; Shariat et al., 2013). The rise of WGS-based methods provide valuable tools to gain detailed information on the genetics and natural history of *S.* Heidelberg strains (Bekal et al., 2016), which may provide useful data for identifying tactics to intervene and decrease bacterial contamination. *Salmonella* Heidelberg has become an important food-borne pathogen in eggs as well other food products, so there is critical need to understand the genetic mechanisms this organism uses to adapt to the avian environment and cause human disease, to provide better strategies to intervene and improve food safety.

## DISCLAIMER

The views presented in this manuscript do not necessarily reflect those of the US Food and Drug Administration.

## REFERENCES


Agnihothram, S.S., Basco, M.D.S., Mullis, L., Foley, S.L., Hart, M.E., Sung, K., Azevedo, M.P., 2015. Infection of murine macrophages by *Salmonella enterica* serovar Heidelberg blocks murine norovirus infectivity and virus-induced apoptosis. PLoS One 10, e0144911. http://dx.doi.org/10.1371/journal.pone.0144911.

Al-Mousawi, A., Eissa, A., Abu-Zant, F., Drobiova, H., Al-Saif, I., Al-Saleh, E., 2010. Correlation between cluster analyses of *Salmonella* strains isolated from diarrhetic patients in Kuwait and biofilm formation. WIT Trans. Ecol. Environ. Environ. Toxicol. 132, 67–74.

Barnhart, H.M., Dreesen, D.W., Bastein, R., Pancorbo, O.C., 1991. Prevalence of *Salmonella* Enteritidis and other serovars in ovaries of layer hens at time of slaughter. J. Food Prot. 54, 488–491.

Bars, H.E., Gall-David, S.L., Renoux, V.M., Bonnaure-Mallet, M., Jolivet-Gougeon, A., Bousarghin, L., 2012. Impact of a mutator phenotype on motility and cell adherence in *Salmonella* Heidelberg. Vet. Microbiol. 159, 99–106.

Barza, M., Travers, K., 2002. Excess infections due to antimicrobial resistance: the "Attributable Fraction". Clin. Infect. Dis. 34, S126–S130.

Baümler, A.J., Tsolis, R.M., Ficht, T.A., Adams, L.G., 1998. Evolution of host adaptation in *Salmonella enterica*. Infect. Immun. 66, 4579–4587.

Bekal, S., Berry, C., Reimer, A.R., Van Domselaar, G., Beaudry, G., Fournier, E., Doualla-Bell, F., Levac, E., Gaulin, C., Ramsay, D., Huot, C., Walker, M., Sieffert, C., Tremblay, C., 2016. Usefulness of high-quality core genome single-nucleotide variant analysis for subtyping the highly clonal and the most prevalent *Salmonella* Heidelberg clone in the context of outbreak investigations. J. Clin. Microbiol. 54, 289–295.

Beltran, P., Musser, J.M., Helmuth, R., Farmer III, J.J., Frerichs, W.M., Wachsmuth, I.K., Ferris, K., McWhorter, A.C., Wells, J.G., Cravioto, A., Selander, R.K., 1988. Toward a population genetic analysis of *Salmonella*: genetic diversity and relationships among strains of serotypes *S. choleraesuis*, *S. derby*, *S. dublin*, *S. enteritidis*, *S. heidelberg*, *S. infantis*, *S. newport*, and *S. typhimurium*. Proc. Natl. Acad. Sci. U. S. A. 85, 7753–7757.

Beug, H., Von-Kirchbach, A., Doderlein, G., Conscience, J.F., Graf, T., 1979. Chicken hematopoietic cells transformed by seven strains of detective avian leukemia viruses display three distinct phenotypes of differentiation. Cell 18, 375–390.

Bhatt, T., Zhang, J., 2013. Food product tracing technology capabilities and interoperability. J. Food Sci. 78 (Suppl. 2), B28–B33.

Broschat, S.L., Call, D.R., Davis, M.A., Meng, D., Lockwood, S., Ahmed, R., Besser, T.E., 2010. Improved identification of epidemiologically related strains of *Salmonella enterica* by use of a fusion algorithm based on pulsed-field gel electrophoresis and multiple-locus variable-number tandem-repeat analysis. J. Clin. Microbiol. 48, 4072–4082.

Buzby, J.C., Farah, H.A., 2006. Chicken Consumption Continues Longrun Rise, vol. 4. Amber Waves, p. 5.

Cain, A.K., Hall, R.M., 2012. Evolution of a multiple antibiotic resistance region in IncHI1 plasmids: reshaping resistance regions in situ. J. Antimicrob. Chemother. 67, 2848–2853.

Carrasco, E., Morales-Rueda, A., Garcia-Gimeno, R.M., 2012. Cross-contamination and recontamination by *Salmonella* in foods: a review. Food Res. Int. 45, 545–556.

Center for Disease Control and Prevention, 2011. *Salmonella* surveillance. In: Annual Summary, 2009 Atlanta, GA.

Center for Disease Control and Prevention, 2013. An Atlas of *Salmonella* in the United States, 1968–2011.

Center for Disease Control and Prevention, 2014. http://www.cdc.gov/Salmonella/heidelberg-10-13access/.

Chittick, P., Sulka, A., Tauxe, R., Fry, A., 2006. A summary of national reports of foodborne outbreaks of *Salmonella* Heidelberg infections in the United States: clues for disease prevention. J. Food Prot. 69, 1150–1153.

Chiu, C.H., Tang, P., Chu, C., Hu, S., Bao, Q., Yu, J., Chou, Y.Y., Wang, H.S., 2005. The genome sequence of *Salmonella enterica* serovar Choleraesuis, a highly invasive and resistant zoonotic pathogen. Nucleic Acids Res. 33, 1690–1698.

Canadian integrated program for antimicrobial resistance surveillance (CIPARS), 2011. Antimicrobial Resistance Short Report. Public Health Agency of Canada, Guelph, Ontario. http://www.phac-aspc.gc.ca/cipars-picra/pubs-eng.php.

Clarke, P., 2015. Bacon helped US lead the way in boosting egg sales. Poult. World 170, 18.

Cockburn, W., Vemon, E., 1956. Food Poisoning in England and Wales, 1956. Public Health Lab. Serv. Rep. Sect. II 233–241.

Crump, J.A., Medalla, F.M., Joyce, K.W., Krueger, A.L., Hoekstra, R.M., Whichard, J.M., Barzilay, E.J., 2011. Antimicrobial resistance among invasive nontyphoidal *Salmonella enterica* isolates in the United States: National Antimicrobial Resistance Monitoring system, 1996 to 2007. Emerging infections program NARMS working group. Antimicrob. Agents Chemother. 55, 1148–1154.

Demczuk, W., Soule, G., Clark, C., Ackermann, H.-W., Easy, R., Khakria, R., Rodgers, F., Ahmed, R., 2003. Phage-based typing scheme for *Salmonella enterica* serovar Heidelberg, a causative agent of food poisoning in Canada. J. Clin. Microbiol. 41, 4279–4284.

Donado-Godoy, P., Byrne, B.A., Hume, M., Leon, N., Perez-Gutierrez, E., Flores, M.J.V., Clavijio, V., Holguin, A., Romer-Zunniga, J.J., Castellanos, R., Tafur, M., Smith, W.A., 2015. Molecular characterization of *Salmonella* Paratyphi B dT+ and *Salmonella* Heidelberg from poultry and retail chicken meat in Colombia by pulsed-field gel electrophoresis. J. Food Prot. 78, 802–807.

Dutil, L., Irwin, R., Finley, R., Ng, L.K., Avery, B., 2010. Ceftiofur resistance in *Salmonella enterica* serovar Heidelberg from chicken meat and humans, Canada. Emerging Infect. Dis. 16, 48–54. http://dx.doi.org/10.3201/eid 1601.090729.

Edwards, P., 1958. Salmonellosis: observation on incidence and control. Ann. N. Y. Acad. Sci. 70, 598–613.

Egg facts, 2015. The Egg Business. http://www.aeb.org/farmers-and-marketers/industry-overview.

Evans, P.S., Luo, Y., Muruvanda, T., Ayers, S., Hiatt, B., Hoffman, M., Zhao, S., Allard, M.W., Brown, E., 2014. Complete genome sequences of *Salmonella enterica* serovar Heidelberg strains associated with a multistate food-borne illness investigation. Genome Announc. 2, e01154–13.

Fabre, L., Zhang, J., Guigon, G., LeHello, S., Guibert, V., Accou-Demartin, M., De Romans, S., Lim, C., Roux, C., Passet, V., Diancourt, L., Guibourdenche, M., Issenhuth-Jeanjean, S., Achtman, M., Brisse, S., Sola, C., Weill, F.-X., 2012. CRISPR typing and subtyping for improved laboratory surveillance of *Salmonella* infections. PLoS One 7, e36995.

Feasey, N.A., Dougan, G., Kingsley, R.A., Heyderman, R.S., Gordan, M.A., 2012. Invasive non-typhoidal *Salmonella* disease: an emerging and neglected tropical disease in Africa. Lancet 379, 2489–2499.

Finstad, S., O'Bryan, C.A., Marcy, J.A., Crandall, P.G., Ricke, S.C., 2012. *Salmonella* and broiler processing in the United States: relationship to foodborne salmonellosis. Food Res. Int. 45, 789–794.

Flynn, D., 2012. *Salmonella* Heidelberg Found in Iowa Poultry Houses. Food Safety News. http://www.foodsafetynews.com/2012/09/Salmonella-heidelberg-found-in-iowa-poultry-houses/#.VZQIpKbcG50.

Foley, S.L., Lynne, A.M., 2008. Food-animal associated *Salmonella* challenges: pathogenicity and antimicrobial resistance. J. Anim. Sci. 86, E173–E187.

Foley, S.L., Lynne, A.M., Nayak, R., 2008. *Salmonella* challenges: prevalence in swine and poultry and potential pathogenicity of such isolates. J. Anim. Sci. 86, E149–E162.

Foley, S.L., Nayak, R., Hanning, I.B., Johnson, T.J., Han, J., Ricke, S.C., 2011. Population dynamics of *Salmonella enterica* serotypes in commercial egg and poultry production. Appl. Environ. Microbiol. 77, 4273–4279.

Foley, S.L., Johnson, T.J., Ricke, S.C., Nayak, R., Danzeisen, J., 2013. *Salmonella* pathogenicity and host adaptation in chicken-associated serovars. Microbiol. Mol. Biol. Rev. 77, 582–607.

Folster, J.P., Pecic, G., McCullough, A., Rickert, R., Whichard, J.M., 2011. Characterization of bla(CMY)-encoding plasmids among *Salmonella* isolates in the United States in 2007. Foodborne Pathog. Dis. 8, 1289–1294.

Folster, J.P., Pecic, G., Singh, A., Duval, B., Rickert, R., Ayers, S., Abbott, J., McGlinchey, B., Bauer-Turpin, J., Haro, J., Hise, K., Zhao, S., Fedorka-Cray, P.J., Whichard, J., McDermott, P.F., 2012. Characterization of extended-spectrum cephalosporin-resistant *Salmonella enterica* serovar Heidelberg isolated from food animals, retail meat, and humans in the United States 2009. Foodborne Pathog. Dis. 9, 638–645.

Food and Drug Administration, 2012. Inspections, Compliance, Enforcement and Criminal Investigations (Warning letters) http://www.fda.gov/ICECI/EnforcementActions/WarningLetters/2012/ucm315877.htm.

Fricke, W.F., McDermott, P.F., Mammel, M.K., Zhao, S., Johnson, T.J., Rasko, D.A., Fedorka-Cray, P.J., Pedroso, A., Whichard, J.M., Leclerc, J.E., White, D.G., Cebula, T.A., Ravel, J., 2009. Antimicrobial resistance-conferring plasmids from avian pathogenic *Escherichia coli* strains in *Salmonella enterica* serovar Kentucky isolates from poultry. Appl. Environ. Microbiol. 75, 5963–5971.

Galán, J.E., Wolf-Watz, H., 2006. Protein delivery into eukaryotic cells by type III secretion machines. Nature 444, 567–573.

Gantois, I., Eeckhaut, V., Pasmans, F., Haesebrouck, F., Ducatelle, R., Van Immerseel, F., 2008. A comparative study on pathogenesis of egg contamination by different serotypes of *Salmonella*. Avian Pathol. 37, 399–406.

Gantois, I., Ducatelle, R., Pasmans, F., Haesebrouck, F., Gast, R., Humphrey, T.J., Van Immerseel, F., 2009. Mechanisms of egg contamination by *Salmonella* Enteritidis. FEMS Micobiol. Rev. 33, 718–738.

Gast, R.K., Guard-Bouldin, J., Holt, P.S., 2004. Colonization of reproductive organs and internal contamination of eggs after experimental infection of laying hens with *Salmonella heidelberg* and *Salmonella enteritidis*. Avian Dis. 48, 863–869.

Gast, R.K., Guard-Bouldin, J., Holt, P.S., 2005. The relationship between the duration of fecal shedding and the production of contaminated eggs by laying hens infected with strains of *Salmonella* Enteritidis and *Salmonella* Heidelberg. Avian Dis. 49, 382–386.

Gast, R.K., Guraya, R., Guard-Bouldin, J., Holt, P.S., Moore, R.W., 2007a. Colonization of specific regions of the reproductive tract and deposition at different locations inside eggs by hens infected with *Salmonella* Enteritidis or *Salmonella* Heidelberg. Avian Dis. 51, 40–44.

Gast, R.K., Guraya, R., Guard-Bouldin, J., Holt, P.S., 2007b. In vitro penetration of egg yolks by *Salmonella* Enteritidis and *Salmonella* Heidelberg strains during thirty-six-hour ambient temperature storage. Poult. Sci. 86, 1431–1435.

Gast, R.K., Guraya, R., Guard, J., Holt, P.S., 2011. The relationship between the numbers of *Salmonella* Enteritidis, *Salmonella* Heidelberg, or *Salmonella* Hadar colonizing reproductive tissues of experimentally infected laying hens and deposition inside eggs. Avian Dis. 55, 243–247.

Geddes, K., Cruz, F., Heffron, F., 2007. Analysis of cells targeted by *Salmonella* type III secretion in vivo. PLoS Pathog. 3, e196. http://dx.doi.org/10.1371/journal.ppat.0030196.

Gharieb, R.M., Tartor, Y.H., Khedr, M.H., 2015. Non-typhoidal *Salmonella* in poultry meat and diarrhoeic patients: prevalence, antibiogram, virulotyping, molecular detection and sequencing of class I integrons in multidrug resistant strains. Gut Pathog. 7. http://dx.doi.org/10.1186/s13099-015-0081-1.

Giraud, A., Matic, I., Tenaillon, O., Clara, A., Radman, M., Fons, M., Taddei, F., 2001. Costs and benefits of high mutation rates: adaptive evolution of bacteria in the mouse gut. Science 291, 2606–2608.

Gokulan, K., Khare, S., Rooney, A.W., Han, J., Lynne, A.M., Foley, S.L., 2013. Impact of plasmids, including those encodingVirB4/D4 type IV secretion systems, on *Salmonella enterica* serovar Heidelberg virulence in macrophages and epithelial cells. PLoS One 8, e77866. http://dx.doi.org/10.1371/journal.pone.0077866.

Gorski, L., 2012. Selective enrichment media bias the types of *Salmonella enterica* strains isolated from mixed strain cultures and complex enrichment broths. PLoS One 7, e34722. http://dx.doi.org/10.1371/journal.pone.0034722.

Guard-Petter, J., 1998. Variants of smooth *Salmonella enterica* serovar Enteritidis that grow to higher cell density than the wild type are more virulent. Appl. Environ. Microbiol. 64, 2166–2172.

Gunel, E., Killic, G.P., Bulut, E., Durul, B., Acar, S., Alpas, H., Soyer, Y., 2015. *Salmonella* surveillance on fresh produce in retail in Turkey. Int. J. Food Microbiol. 199, 72–77.

Habbs, V.H., 1933. About a new type of bacteria form the paratyphoid enteritis group. J. Bacteriol. 130, 367–374.

Haft, D.H., Selengut, J., Mongodin, E.F., Nelson, K.E., 2005. A guild of 45 CRISPR associated (Cas) protein families and multiple CRISPR/Cas subtypes exists in prokaryotic genomes. PLoS Computat. Biol. 1, e60.

Han, J., Lynne, A.M., David, D.E., Tang, H., Xu, J., Nayak, R., Kaldhone, P., Logue, C.M., Foley, S.L., 2012. DNA sequence analysis of plasmids from multidrug resistant *Salmonella enterica* serotype Heidelberg isolates. PLoS One 7, e51160.

Hanning, I.B., Nutt, J.D., Ricke, S.C., 2009. Salmonellosis outbreaks in the United States due to fresh produce: sources and potential intervention measures. Foodborne Pathog. Dis. 6, 635–648.

Harris, A., Cherubin, C., Biek, R., Edwards, L.C., 1990. Frequency of *Salmonella* Typhimurium the year after a massive outbreak. Diagn. Microbiol. Infect. Dis. 13, 25–30.

He, H., Genovese, K.J., Swaggerty, C.L., Nisbet, D.J., Kogut, M.H., 2012. A comparative study of invasion, survival, modulation of oxidative burst, and nitric oxide responses of macrophages (HD11), and systemic infection in chickens by prevalent poultry *Salmonella* serovars. Foodborne Pathog. Dis. 9, 1104–1110.

Hennessy, T., Cheng, L., Kassenborg, H., Ahuja, S., Mohle-Boetani, J., Marcus, R., Shiferaw, B., Angulo, F., 2004. Egg consumption is the principal risk factor for sporadic *Salmonella* serotype Heidelberg infections: a case-control study in FoodNet sites. Clin. Infect. Dis. 38, S237–S243.

Hensel, M., Shea, J.E., Waterman, S.R., Mundy, R., Nikolaus, T., Banks, G., Vazquez-Torres, A., Gleeson, C., Fang, F.C., Holden, D.W., 1998. Genes encoding putative effector proteins of the type III secretion system of *Salmonella* pathogenicity island 2 are required for bacterial virulence and proliferation in macrophages. Mol. Microbiol. 30, 163–174.

Hoffman, M., Luo, Y., Lafon, P.C., Timme, R., Allard, M.W., 2013. Genome sequences of *Salmonella enterica* serovar Heidelberg isolates isolated in the United States from a multistate outbreak of human *Salmonella* infections. Genome Announc. 1, e00004–12 Pubmed:23405335.

Hoffmann, M., Zhao, S., Pettengill, J., Luo, Y., Monday, S.R., Abbott, J., Ayers, S.L., Cinar, H.N., Muruvanda, T., Li, C., Allard, M.W., Whichard, J., Meng, J., Brown, E.W., McDermott, P.F., 2014. Comparative genomic analysis and virulence differences in closely related *Salmonella enterica* serotype Heidelberg isolates from humans, retail meats, and animals. Genome Biol. Evol. 6, 1046–1068.

Hopkins, K.L., Threlfall, E.J., 2004. Frequency and polymorphism of *sopE* in isolates of *Salmonella enterica* belonging to the ten most prevalent serotypes in England and Wales. J. Med. Microbiol. 53, 539–543.

Howard, Z.R., O'Bryan, C.A., Crandall, P.G., Ricke, S.C., 2012. *Salmonella* Enteritidis in shell eggs: current issues and prospects for control. Food Res. Int. 45, 755–764.

Humphrey, T.J., Greenwood, M., Gilbert, R.J., Rowe, B., Chapman, P.A., 1989. The survival of salmonellas in shell eggs cooked under simulated domestic conditions. Epidemiol. Infect. 102, 35–45.

Ibarburu, M., 2015. U. S. Flock Trends and Projections. The Egg Industry Center Market Reports and Industry Analysis. American Egg Board. October 7, 2015.

Johnson, T.J., Thorsness, J.L., Anderson, C.P., Lynne, A.M., Foley, S.L., 2010. Horizontal gene transfer of a ColV plasmid has resulted in a dominant avian clonal type of *Salmonella enterica* serovar Kentucky. PLoS One 5, e15524.

Jolley, K.A., Maiden, M.C., 2010. BIGSdb: scalable analysis of bacterial genome variation at the population level. BMC Bioinformatics 11, 595.

Jones, B.D., Ghori, N., Falkow, S., 1994. *Salmonella typhimurium* initiates murine infection by penetrating and destroying the specialized epithelial M cells of the Peyer's Patches. J. Exp. Med. 180, 15–23.

Jones, T.F., Ingram, L.A., Cieslak, P.R., Vugia, D.J., Dangelo, M.T., Hurs, S., Medus, C., Cronquirst, A., Angulo, F.J., 2008. Salmonellosis outcomes differ substantially by serotype. J. Infect. Dis. 198, 109–114.

Juneja, V.K., Marks, H.M., Huang, L., 2003. Growth and heat resistance kinetic variation among isolates of *Salmonella* and its application to risk assessment. Risk Anal. 23, 199–213.

Kaldhone, P.R., Nayak, R., Lynn, A.M., David, D.E., McDermott, P.F., Logue, C.M., Foley, S.L., 2008. Characterization of *Salmonella enterica* serovar Heidelberg from turkey-associated sources. Appl. Environ. Microbiol. 74, 5038–5046.

Kivisaar, M., 2003. Stationary phase mutagenesis: mechanisms that accelerates adaptations of microbial populations under environmental stress. Environ. Microbiol. 5, 814–827.

Koningstein, M., Simonsen, J., Helms, M., Molbak, K., 2010. The interaction between prior antimicrobial drug exposure and resistance in human *Salmonella* infections. J. Antimicrob. Chemother. 65, 1819–1825.

Krauland, M., Harrison, L., Paterson, D., Marsh, J., 2010. Novel integron gene cassette arrays identified in a global collection of multi-drug resistant non-typhoidal *Salmonella enterica*. Curr. Microbiol. 60, 217–223.

Le Gall, S., Desbordes, L., Gracieux, P., Saffroy, S., Bousarghin, L., Bonnaure-Mallet, M., Jolivet-Gougeon, A., 2009. Distribution of mutation frequencies among *Salmonella enterica* isolates from animal and human sources and genetic characterization of *Salmonella* Heidelberg hypermutator. Vet. Microbiol. 137, 306–312.

Li, X., Payne, J.B., Santos, F.B., Levine, J.F., Anderson, K.E., Sheldon, B.W., 2007. *Salmonella* populations and prevalence in layer feces from commercial high-rise houses and characterization of the *Salmonella* isolates by serotyping, antibiotic resistance analysis, and pulsed field gel electrophoresis. Poult. Sci. 86, 591–597.

Liu, W.Q., Liu, G.R., Li, J.Q., Xu, G.M., Danni, Q., He, X.Y., Juan, D., Zhang, H.N., Randal, J.N., Liu, S.L., 2007. Diverse genome structures of *Salmonella* paratyphi C. BMC Genomics 8, 1–10.

Liu, F., Barrangou, R., Gerner-Smidt, P., Ribot, E.M., Knabel, S.J., Dudley, E.G., 2011. Novel virulence gene and clustered regularly interspaced short palindromic repeat (CRISPR) multilocus sequence typing of the major serovars of *Salmonella enterica* subsp. *enterica*. Appl. Environ. Microbiol. 77, 1946–1956.

Love, B.C., Rostagno, M.H., 2008. Comparison of five culture methods for *Salmonella* isolation from swine fecal samples of known infection status. J. Vet. Diagn. Invest. 20, 620–624.

Lynne, A.M., Kaldhone, P., David, D., White, D.G., Foley, S.L., 2009. Characterization of antimicrobial resistance in *Salmonella enterica* serotype Heidelberg isolated from food animals. Foodborne Pathog. Dis. 6, 206–215.

Malik-Kale, P., Jolly, C.E., Lanthrop, S., Winfree, S., Luterbach, C., Steele-Mortimer, O., 2011. *Salmonella* – at home in the host cell. Front. Microbiol. 2, 125.

Martelli, F., Davies, R.H., 2012. *Salmonella* serovars isolated from table eggs: an overview. Food Res. Int. 45, 745–754.

Mataseje, L.F., Baudry, P.J., Zhanel, G.G., Morck, D.W., Read, R.R., Louie, M., Mulvey, M.R., 2010. Comparison of CMY-2 plasmids isolated from human, animal and environmental *Escherichia coli* and *Salmonella* spp. from Canada. Diagn. Microbiol. Infect. Dis. 67, 387–391.

Maurelli, A.T., 2007. Black holes, antiviral genes, and gene inactivation in the evolution of bacterial pathogens. FEMS Microbiol. Lett. 267, 1–8.

Maurischat, S., Baumann, B., Martin, A., Malorny, B., 2015. Rapid detection and specific differentiation of *Salmonella enterica* subsp. *enterica* Enteritidis, Typhimurium and its monophasic variant 4,[5],12:i:-by real-time multiplex PCR. Int. J. Food Microbiol. 193, 8–14.

McConnell, J.A., Schaffner, D.W., 2014. Validation of mathematical models for *Salmonella* growth in raw ground beef under dynamic temperature conditions representing loss of refrigeration. J. Food Prot. 77, 1110–1115.

McDermid, A.S., McKee, A.S., Dowsett, A.B., Marsh, P.D., 1996. The effect of environmental pH on the physiology and surface structures of *Salmonella* serotype enteritidis phage type 4. J. Med. Microbiol. 45, 452–458.

Methner, U., Haase, A., Berndt, A., Martin, G., Nagy, B., Barrow, P.A., 2011. Exploitation of intestinal colonization-inhibition between *Salmonella* organisms for live vaccines in poultry: potential and limitations. Zoonoses Public Health 58, 540–548.

Meyerholz, D.K., Stabel, T.J., Ackermann, M.R., Carlson, S.A., Jones, B.D., Pohlenz, J., 2002. Early epithelial invasion by *Salmonella enterica* serovar Typhimuium DT104 in swine ileum. Vet. Pathol. 39, 712–720.

Milkman, R., Stoltzfus, A., 1988. Molecular evolution of the *Escherichia coli* chromosome II clonal segments. Genetics 120, 359–366.

Morbidity and Mortality Weekly Report (MMWR), 2013. Outbreak of *Salmonella* Heidelberg Infections Linked to a Single Poultry Producer – 13 States, 2012–2013, vol. 62, pp. 553–556.

Molbak, K., 2005. Human health consequences of antimicrobial drug-resistant *Salmonella* and other foodborne pathogens. Clin. Infect. Dis. 41, 1613–1620.

Morales, C.A., Musgrove, M., Humphrey, T.J., Cates, C., Gast, R., Guard-Bouldin, J., 2007. Pathotyping of *Salmonella enterica* by analysis of single-nucleotide polymorphism in *cyaA* and flanking 23S ribosomal sequences. Environ. Microbiol. 9, 1047–1059.

Muriana, P.M., 1997. Effect of pH and hydrogen peroxide on heat inactivation of *Salmonella* and *Listeria* in egg white. Food Microbiol. 14, 11–19.

Nordentoft, S., Molbak, L., Bjerrum, L., De-Vylder, J., Immerseel, F., Pedersen, K., 2011. The influence of the cage system and colonisation of *Salmonella* Enteritidis on the microbial gut flora of laying hens studied by T-RFLP and 454 pyrosequencing. BMC Microbiol. 11, 187.

Onchman, H., Moran, N.A., 2001. Genes lost and genes found: evolution of bacterial pathogenesis and symbiosis. Science 292, 1096–1099.

Otto, S.J.G., Carson, C.A., Finley, R.L., Thomas, M.K., Reid-Smith, R.J., McEwen, S.A., 2014. Estimating the number of human cases of ceftiofur-resistant *Salmonella enterica* serovar Heidelberg in Quebec and Ontario, Canada. Clin. Infect. Dis. 59, 1281–1290.

Park, S.H., Aydin, M., Khatiwara, A., Dolan, M.C., Gilmore, D.F., Bouldin, J.L., Ahn, S., Ricke, S.C., 2014. Current and emerging technologies for rapid detection and characterization of *Salmonella* in poultry and poultry products. Food Microbiol. 38, 250–262.

Parker, C.T., Liebana, E., Henzler, D.J., Guard-Petter, J., 2001. Lipopolysaccharide O-chain microheterogeneity of *Salmonella* serotypes Enteritidis and Typhimurium. Environ. Microbiol. 3, 332–342.

Philippon, A., Arlet, G., Jacoby, G.A., 2002. Plasmid-determined AmpC-type beta-lactamases. Antimicrob. Agents Chemother. 46, 1–11.

Rabsch, W., Tschäpe, H., Bäumler, A.J., 2001. Non-typhoidal salmonellosis: emerging problems. Microbes Infect. 3, 237–247.

Rashid, M.-U., Rosenborg, S., Panagiotidis, G., Löfdal, K.S., Weintraub, A., Nord, C.E., 2015. Ecological effects of ceftaroline-avibactam on the normal human intestinal microbiota. Antimicrob. Agents Chemother. 59, 4504–4509.

Rescigno, M., Urbano, M., Valzasina, B., Francolini, M., Rotta, G., Bonascio, R., Granucci, F., Kraehenbuhl, J.P., Riccicardi-Castagnoli, P., 2001. Dendritic cells express tight junction proteins and penetrate gut epithelial monolayers to sample bacteria. Nat. Immunol. 2, 361–367.

Richardson, E.J., Limaye, B., Inamdar, H., Datta, A., Manjiri, A.S., Pullinger, G.D., Thomson, N.R., Joshi, R.R., Watson, M., Stevens, M.P., 2011. Genome sequences of *Salmonella enterica* serovar Typhimurium, Choleraesuis, Dublin, and Gallinarum strains of well-defined virulence in food-poisoning animals. J. Bacteriol. 193, 3162–3163.

Richter-Dahlfors, A., Buchan, A.M., Finlay, B.B., 1987. Murine salmonellosis studied by confocal microscopy: *Salmonella* Typhimurium resides intracellularly inside macrophages and exerts a cytotoxic effect on phagocytes in vivo. J. Exp. Med. 186, 569–580.

Ricke, S.C., Jones, D.R., Gast, R.K., 2013a. Egg and egg products. Chapter 46. In: Doores, S., Salfinger, Y., Tortorello, M.L. (Eds.), Compendium of Methods for the Microbiological Examinations of Foods, fifth ed. American Public Health Association, Washington, D.C., pp. 1–11.

Ricke, S.C., Koo, O., Foley, S., Nayak, R., 2013b. *Salmonella*. Chapter 7. In: Labbé, R.G., García, S. (Eds.), Guide to Foodborne Pathogens, second ed. John Wiley & Sons, Ltd., pp. 112–136.

Rivera Calo, J., Baker, C.A., Park, S.H., Ricke, S.C., 2015. Specificity of *Salmonella* Typhimurium strain (ATCC 14028) growth responses to *Salmonella* serovar-generated spent media. J. Environ. Sci. Health B 50, 423–429.

Robertson, J., McKenzie, J., Duncan, N., Vercoe, M., Woodward, E., Flint, M., Grant, G., 2003. Lack of flagella disadvantages *Salmonella enterica* serovar Enteritidis during the early stages of infection in the rat. J. Med. Microbiol. 52, 91–99.

Rose, B.E., Hill, W.E., Umholtz, R., Ransom, G.M., James, W.O., 2002. Testing for *Salmonella* in raw meat and poultry products collected at federally inspected establishments in the United States 1998 through 2000. J. Food Prot. 65, 937–947.

Ross, I.L., Heuzenroeder, M.W., 2005. Discrimination within phenotypically closely related definitive types of *Salmonella enterica* serovar Typhimurium by the multiple amplification of phage locus typing technique. J. Clin. Microbiol. 43, 1604–1611.

Rosselin, M., Abed, N., Virloguex-Payant, I., Bottreau, E., Sizaret, P.Y., Velge, P., Wiedemann, A., 2011. Heterogeneity of type III secretion system (T3SS)-1-independent entry mechanisms used by *Salmonella* Enteritidis to invade different cell types. Microbiology 157, 839–847.

Salcedo, S.P., Noursadeghi, M., Cohen, J., Holden, D.W., 2001. Intracellular replication of *Salmonella* Typhimurium strains in specific subsets of splenic macrophages in vivo. Cell Microbiol. 3, 587–597.

Scallan, E., Hoekstra, R.M., Angulo, F.J., Tauxe, R.V., Widdowson, M.A., Roy, S.L., Jones, J.L., Griffin, P.M., 2011. Foodborne illness acquired in the United States – major pathogens. Emerging Infect. Dis. 17, 7–15.

Schlumberger, M.C., Hardt, W.D., 2006. *Salmonella* type III secretion effectors: pulling the host cell's strings. Curr. Opin. Microbiol. 9, 46–54.

Schoeni, J.L., Glass, K.A., McDermott, J.L., Wong, A.C.L., 1995. Growth and penetration of *Salmonella enteritidis*, *Salmonella heidelberg* and *Salmonella typhimurium* in eggs. Int. J. Food Microbiol. 24, 385–396.

Shariat, N., Sandt, C., DiMarizo, M., Barrangou, R., Dudley, E., 2013. CRISPR-MVLST subtyping of *Salmonella enterica* subsp. *enterica* serovars Typhimurium and Heidelberg and application in identifying outbreak isolates. Biomed. Cent. Microbiol. 13, 1–31.

Singer, R.S., Mayer, A.E., Hanson, T.E., Isaacson, R.E., 2009. Do microbial interactions and cultivation media decrease the accuracy of *Salmonella* detection systems and outbreak investigations? J. Food Prot. 72, 707–713.

Smith, N.S., Beltran, P., Selander, R.K., 1990. Recombination of *Salmonella* phase 1 flagellin genes generates new serovars. J. Bacteriol. 172 (5), 2209–2216.

Smyser, C.F., Adinarayanan, N., Roekel, V.H., Snoeyenbos, G.H., 1965. Field and laboratory observations on *Salmonella* Heidelberg infection in three chicken breeding flocks. In: The 37$^{th}$ Annual Meeting of the Northeastern Conference on Avian Diseases.

Stanley, J., Burnens, A., Powell, N., Chowdry, N., Jones, C., 1992. The insertion sequence IS200 fingerprints chromosomal genotypes and epidemiological relationships in *Salmonella* Heidelberg. J. Gen. Microbiol. 138, 2329–2336.

Stevens, M.P., Humphrey, T.J., Maskell, D.J., 2009. Molecular insights into farm animal and zoonotic *Salmonella* infections. Philos. Trans. R. Soc. Lond. B Biol. Sci. 364, 2709–2723.

Suez, J., Portola, S., Dagan, A., Marcel, A., Shirr, Y.I., Desai, P.T., Gammon, V., McClelland, M., Rehab, G., Gal-Moor, O., 2013. Virulence gene profiling and pathogenicity characterization of non-typhoid *Salmonella* accounted for invasive disease in humans. PLoS One 8, e58449.

Taylor, A.L., Murphree, R., Ingram, L.A., Garman, K., Soloman, D., Coffey, E., Walker, D., Rogers, M., Marder, E., Bottomley, M., Woron, A., Thomas, L., Roberts, S., Hardin, H., Arjmandi, P., Green, A., Simmons, L., Cornell, A., Dunn, J., 2015. Multidrug-resistant *Salmonella* Heidelberg associated with mechanically separated chicken at a correctional facility. Foodborne Pathog. Dis. 12, 950–952.

Threlfall, E.J., Hall, M.L.M., Ward, L.R., Rowe, B., 1992. Plasmid profiles demonstrate that an upsurge of *Salmonella* Beria in humans in England and Wales is associated with imported poultry meat. Eur. J. Epidemiol. 8, 27–33.

Van Asten, A.J., Van Dijk, J.E., 2005. Distribution of "classic" virulence factors among *Salmonella* spp. FEMS Immunol. Med. Microbiol. 44, 251–259.

Wallis, T.S., Starkey, W.G., Stephen, J., Haddon, S.J., Osborne, M.P., Candy, D.C., 1986. The nature and role of mucosal damage in relation to *Salmonella* Typhimurium-induced fluid secretion in the rabbit ileum. J. Med. Microbiol. 22, 39–49.

Windhorst, H.W., 2009. Recent patterns of egg production and trade: a status report on regional basis. World's Poult. Sci. J. 65, 685–708.

Withanage, G.S., Mastroeni, P., Brooks, H.J., Maskell, D.J., McConnell, I., 2005. Oxidative and nitrosative responses of the chicken macrophage cell line MQ-NCSU to experimental *Salmonella* infection. Br. Poult. Sci. 46, 261–267.

Young, C.-C., Ross, I.L., Heuzenroeder, M.W., 2012. A new methodology for differentiation and typing of closely related *Salmonella enterica* serovar Heidelberg isolates. Curr. Microbiol. 65, 481–487.

SECTION

# 3

# Development of Specific Interventions for *Salmonella* in Laying Hens and Table Eggs: Present and Future Prospects

CHAPTER

# Preharvest Measures to Improve the Safety of Eggs

13

**Wolfgang De Cort, Richard Ducatelle, Filip Van Immerseel**
*Ghent University, Gent, Belgium*

## 1. INTRODUCTION: *SALMONELLA* AND EGGS

Of the more than 2500 *Salmonella* serotypes, only a small fraction is common in poultry flocks (Gast, 2007; EFSA, 2015). These few serotypes, however, often cause disease in humans, indicating that there is an epidemiologically important connection between poultry products and human *Salmonella* infections (Gast, 2007; Callaway et al., 2008). Humans often become infected with *Salmonella* after consumption of contaminated poultry products. Thus minimizing the contamination of poultry products with *Salmonella* should reduce the number of *Salmonella* infections in humans. Human illness caused by the consumption of *Salmonella*-contaminated poultry products results in considerable costs for public health care. It was estimated that the total combined cost resulting from food-borne *Salmonella* infections in humans accumulates to $2.7 billion annually (US Department of Agriculture and Economic Research Service, 2010).

In 2013 in Europe, *Salmonella* was most frequently detected in poultry meat and less often in pork or beef. *Salmonella* could be found most often in fresh turkey meat, with 5.4% of samples positive for *Salmonella*, followed by fresh broiler, pig, and finally bovine meat (EFSA, 2015). *Salmonella* was rarely found in table eggs, with only 0.03% of single samples or 0.5% of batch samples being positive for *Salmonella* (EFSA, 2015). Despite this low prevalence, eggs and egg products remain the most important source of human food-borne *Salmonella* outbreaks because these food products are consumed in very large quantities (Kimura et al., 2004; Gillespie et al., 2005; Zhang et al., 2006; Greig and Ravel, 2009). In addition, these egg products are often consumed raw or undercooked and thus present a greater threat for human health than food products that are cooked before consumption (EFSA, 2015).

To maintain this low prevalence of *Salmonella* in eggs and egg products or even further reduce it, different control measures can be applied during different stages of the egg production process. In addition to measures at the postharvest level (e.g., cooling of eggs), preharvest control measures can efficiently be implied to reduce *Salmonella* contamination of eggs. The goal of these preharvest measures is to minimize opportunities for the introduction, persistence, and transmission of flock

Producing Safe Eggs. http://dx.doi.org/10.1016/B978-0-12-802582-6.00013-6

infections with *Salmonella* and other human pathogens (Gast, 2007). Risk assessment studies have shown that interventions at multiple steps in the farm-to-table process and thus also implementing preharvest control measures, constitute the most effective approach to achieve a decrease in the number of *Salmonella*-contaminated eggs that are being produced (Hope et al., 2002; Gast, 2007).

The remainder of this chapter will discuss several control measures that can be applied preharvest to lower the infection pressure in a flock, to reduce gut colonization, and to decrease the susceptibility of laying hens to *Salmonella* infections, as well as their limitations.

## 2. BIOSECURITY MEASURES

Laying hen flocks can become infected with *Salmonella* from a myriad of sources. Often *Salmonella* is acquired through horizontal transfer rather than vertical transmission (van de Giessen et al., 1994). It has been shown that a history of previous *Salmonella* infections in the poultry house, absence of cleaning and disinfection, presence of rodents, induced molting, larger flock size (>30,000 hens), multiple flocks of different ages on the farm, cage housing, in-line egg processing, rearing pullets on the floor, pests with access to feed before movement to the feed trough, visitors allowed in the layer houses, and trucks near farms and air inlets, all are major risk factors associated with *Salmonella* contamination of the laying hen premises (Denagamage et al., 2015). High level of manure contamination, middle and late phase of production, high degree of egg-handling equipment contamination, a flock size of over 30,000 hens, and egg production rate over 96% are the risk factors most strongly associated with *Salmonella* contamination of shell eggs (Denagamage et al., 2015). Consequently, biosecurity, flock management, and sanitation are of major importance when trying to reduce the *Salmonella* prevalence in laying hens and ultimately, in eggs.

Biosecurity is defined as a collection of rules and procedures that minimize exposure of a susceptible population to an infectious (biological) agent, such as *Salmonella* (Wenzel and Nusbaum, 2007; Cox and Pavic, 2009). In practice, this means that all activities on the farm should be planned and carried out in such a way that hygienic risks are minimized, that a continuous monitoring plan should provide up-to-date information on the flock's *Salmonella* status, and that a sanitation and decontamination plan should allow immediate action in the case that detection of *Salmonella* occurs (Ducatelle and Van Immerseel, 2011). These hygienic and biosecurity measures should be integrated in the general management plan of the layer farm to minimize the chance to expose the laying hens to *Salmonella* (Ducatelle and Van Immerseel, 2011).

It is known that *Salmonella* can survive for lengthy durations in litter, even after removal of the flock (Davies and Breslin, 2003b). Therefore thorough cleaning and disinfection of the layer house between production cycles will strongly reduce the risk of animal exposure to *Salmonella* or other enteric pathogens (Gast, 2007). It has been shown that *Salmonella* Enteritidis infections in laying hen flocks is mainly a problem of persistent contamination of laying houses and associated wildlife vectors, underlining the importance of thorough cleaning (Wales et al., 2006). Survival of *Salmonella* in the litter and the feed is correlated with the moisture levels in the poultry houses, and higher *Salmonella* numbers have been reported at higher water-activity levels (Turnbull and Snoeyenbos, 1973; Eriksson et al., 2001). Similarly, *Salmonella* contamination levels are higher in areas of the poultry house where there is a reduced air flow (Gast, 2007). In addition, the pH of the litter plays a role in the survival of *Salmonella* as well (Turnbull and Snoeyenbos, 1973). A higher pH, caused naturally over time by dissolved ammonia or the addition of lime, can lower *Salmonella* numbers in the litter (Bennett et al., 2003). Stressful housing or flock management conditions can promote or exacerbate infections with pathogens as well (Gast, 2007). As an example, when chickens are infected after water deprivation, they shed *Salmonella* Typhimurium in their feces for a longer period of time (Brownell et al., 1969). Also molting induced by feed withdrawal will render the chickens far more susceptible for *Salmonella* infections and will result in an increase in intestinal colonization, fecal shedding, invasion of internal organs, and transmission to other birds (Porter and Holt, 1993; Holt et al., 1994, 1995; Holt, 1995). Alternative molting methods, such a feeding a diet based on alfalfa, can induce molting in laying hens with less stress and thus avoid the feed deprivation–associated gut microenvironment that causes increased *Salmonella* susceptibility (Ricke et al., 2013).

Laying hens can thus also become infected with *Salmonella* through the exposure to contaminated wildlife vectors, such as rodents, insects, and other birds. In addition, most salmonellae have a broad host range and thus a large number of potential reservoirs that may serve as sources for introduction of *Salmonella* in the laying house (Gast, 2007). Therefore contact with potential carriers such as wild birds, rodents, and insects, as well as humans in the laying house, should be avoided (Cox and Pavic, 2009). Insects such as flies, litter beetles, and cockroaches can carry *Salmonella* internally or externally, whereas a single mouse dropping can contain up to $10^5$ *S.* Enteritidis cells (Henzler and Opitz, 1992; Gast, 2007). Finally, it has been shown that *Salmonella* isolates found in wildlife vectors show close similarity to those found in the associated flock, underlining the epidemiological connection between both (Liebana et al., 2003).

Contaminated feed and drinking water can be important sources of horizontally acquired *Salmonella* infections in laying hens. Therefore feed and drinking water offered to the animals should be free of *Salmonella.* Feed infected with even low levels of *Salmonella* can be the cause of persistent *Salmonella* infections in young chickens (Hinton, 1988). A connection between the serotypes found in feed and the ones found in poultry has been suggested (Veldman et al., 1995). The feed can become infected by incorporating contaminated feed ingredients in the feed itself, or through the incorporation in the feed of *Salmonella*-contaminated dust, present at the manufacturing facility (Jones, 2011). In addition, *Salmonella* is able to survive up to 2 years in feed (Davies and Wray, 1996). It is possible to reduce the numbers of *Salmonella* in the feed by avoiding the use of contaminated ingredients, reducing the introduction of dust in the feed, and treating the feed to kill *Salmonella* present

(Jones, 2011). Killing *Salmonella* present in the feed can be done by pelleting the feed, adding antimicrobial chemicals, or both, which might act synergistically (Tabib et al., 1984; Wales et al., 2010; Jones, 2011).

Drinking water provided to the laying hens should be free of enteropathogens and of potable quality. In practice, on-farm water is frequently drawn from natural sources and should be treated with chemicals or filtered before being presented to the animals (Cox and Pavic, 2009; Wales et al., 2010). In addition, the drinking systems used in poultry production are often susceptible to biofilm formation, which makes sanitation and regular cleaning essential (Cox and Pavic, 2009).

In addition to horizontal transmission, *Salmonella* can also spread by vertical transmission. The hatching process generates large amounts of dust and airborne fluff, which, in the case of contaminated eggs, can contribute to the spread of *Salmonella* to other eggs and hatched chicks (Cox et al., 1990; Mitchell et al., 2002; Cox and Pavic, 2009). To avoid this, numerous methods are employed to disinfect eggs, such as disinfection with ultraviolet light, ozone, or chemicals (Davies and Breslin, 2003a; Rodriguez-Romo and Yousef, 2005; Cox and Pavic, 2009). Finally, it has been reported that *Salmonella* infection can spread by airborne routes, which is possibly mediated through contaminated dust (Gast et al., 1998).

## 3. GASTROINTESTINAL COLONIZATION CONTROL

Eggs can become contaminated with *S.* Enteritidis by penetration through the egg shell from contaminated feces during or after oviposition (Barrow and Lovell, 1991; Humphrey et al., 1991). Egg contamination before oviposition originates from the infection of the hen's reproductive organs (Shivaprasad et al., 1990; De Buck et al., 2004; Gantois et al., 2006b). Reducing gastrointestinal colonization by *Salmonella* will reduce the fecal *Salmonella* load, and thus the chance of egg contamination by fecal matter. Similarly, limiting gastrointestinal colonization will also reduce the spread of *Salmonella* to internal and reproductive organs, and thus lower the chance of egg contamination before oviposition. As such, gastrointestinal colonization control can be used to reduce the contamination of eggs by *Salmonella*.

### 3.1 ORGANIC ACIDS AS FEED OR DRINKING WATER ADDITIVES TO CONTROL *SALMONELLA* IN POULTRY

Organic acids were originally added to feed and drinking water to decontaminate them and thus prevent uptake of *Salmonella* by the animals. However, it became clear that these organic acids could lower intestinal colonization and fecal shedding of *Salmonella* as well when they were administered to chickens. Organic acids possess the ability to cross the bacterial cell membrane and dissociate in the cytoplasm, resulting in anion accumulation, which is a major cytotoxic event for bacteria (Van Immerseel and Atterbury, 2013). In addition, at much lower (physiological) concentrations organic acids such as butyrate (10 mmol/L) have been shown to decrease expression of the *hilA* gene, a major regulator for expression of the *Salmonella* pathogenicity island 1 (SPI-1), which plays an important role in epithelial invasion, resulting in reduced invasive capabilities (Van Immerseel et al., 2004a; Boyen et al., 2008). It has also been shown that even very low concentrations of butyrate affect SPI-1 gene expression, while not having effects on metabolic gene expression (Gantois et al., 2006a).

Early studies in which formic acid and propionic acid were added to poultry feed (at high concentrations up to 1%) showed significant decreases of *Salmonella* in the feed, and were associated with lower cecal colonization when animals were given treated feed when compared with nontreated feed (Hinton and Linton, 1988; Iba and Berchieri, 1995). Antibacterial properties of organic acids, however, depend on temperature and water activity, and as such, it is difficult to obtain sufficient decontamination of properly stored dry feed using organic acids (Van Immerseel and Atterbury, 2013). Because of this, it is also believed that organic acids are more optimally added to the drinking water rather than the feed, as this would result in a much higher activity of the acids (Van Immerseel and Atterbury, 2013). In addition, it has been shown that short-chain fatty acids present in the feed can no longer be detected further down the gastrointestinal tract, presumably because they are absorbed by the intestinal mucosa (Hume et al., 1993; Thompson and Hinton, 1997). As such, the administered organic acids do not reach the ceca, which are important sites for invasion of the intestinal epithelium by *Salmonella* (Barrow et al., 1994; Hu and Guo, 2007). Therefore to optimize the efficacy of the acid formulations, novel technologies have been developed to release the acids further down the gastrointestinal tract. One possibility to achieve this is by incorporating the organic acids into a carrier, which prevents absorption of the acids in the upper gastrointestinal tract and allows them to be released further down (Van Immerseel et al., 2006; Van Immerseel and Atterbury, 2013). It has been shown that a coated butyrate feed additive decreased fecal shedding and cecal colonization of *Salmonella* in broilers, whereas a powder formulation of butyrate had no effect (Van Immerseel et al., 2005a). A study in which broiler chickens were given feed supplemented with microbeads containing formic, acetic, propionic, or butyric acid showed that butyrate and propionate decreased colonization of internal organs and ceca, whereas acetic acid increased colonization (Van Immerseel et al., 2004c). Although these studies show the potential of coated organic acids as a control measure to reduce *Salmonella* colonization in poultry, it also illustrates that the choice of organic acid is crucial to achieve this reduction when releasing acids in the intestine. This is less important for drinking water acidification where pH effects are more influential, although sensory issues can play a role there. For instance, butyrate cannot be used to decontaminate drinking water because of the pronounced odor.

Alternative strategies are currently being explored to increase the organic acid concentration in the intestinal tract. One of these possible strategies is the supplementation of poultry feed with certain compounds that can be converted to butyrate by the intestinal microbiota, or by the administration of butyrate-producing bacteria (Van Immerseel and Atterbury, 2013).

## 3.2 PREBIOTICS

Prebiotics are nondigestible feed ingredients that beneficially affect the host by selectively stimulating the growth and/or activity of one or more bacterial species already present in the gastrointestinal tract of the host (Gibson et al., 2005; Van Immerseel et al., 2009). Prebiotics are oligo- and polysaccharides such as lactose, lactulose, fructooligosaccharides, galactooligosaccharides and arabinoxylooligosaccharides, which confer beneficial effects on poultry health by indirect means, as the fermentable prebiotics provide a substrate for metabolism and growth of the normal intestinal microbiota (Eeckhaut et al., 2008; Van Immerseel et al., 2009; Vandeplas et al., 2010). They thus indirectly inhibit pathogen colonization by competitive exclusion (CE) (see later discussion) and may stimulate production of antibacterial metabolites such as lactic acid, fatty acids, or bacteriocins (Vandeplas et al., 2010). As an example, xylooligosaccharides were shown to induce butyrate production because of stimulation of butyrate-producing Clostridial cluster XIVa bacteria in the ceca of broilers (De Maesschalck et al., 2015). It is also possible that prebiotics (such as mannanoligosaccharides) can bind to pathogens in the intestinal lumen, and therefore block adhesion of these pathogens to epithelial cells (Spring et al., 2000; Vandeplas et al., 2010). However, studies on the ability of prebiotics to control colonization by enteropathogens such as *Salmonella* have produced inconsistent results (Rehman et al., 2009; Vandeplas et al., 2010). Lactose, for instance, was shown to reduce *Salmonella* organ invasion when added to the feed (Tellez et al., 1993), but when administered through the drinking water, it failed to reduce *Salmonella* colonization in crop and ceca (Barnhart et al., 1999). Therefore, to achieve more unambiguous results, it might actually be advisable to simply directly administer beneficial microorganisms to poultry.

## 3.3 PROBIOTICS AND COMPETITIVE EXCLUSION PRODUCTS

It is possible to induce changes in the chicken intestinal microbiota composition by administering probiotics to feed or drinking water (Netherwood et al., 1999). Administration of probiotics can also stimulate the immune system, elicit cross-talk with other beneficial bacteria, and induce the production of host enzymes, which in turn may lead to a beneficial nutritional and growth-promoting effect (Jin et al., 1996; Patterson and Burkholder, 2003; Vandeplas et al., 2010). Other potential mechanisms by which probiotics can exclude enteric pathogens such as *Salmonella* include competition for receptor sites, competition for nutrients, and the production of antimicrobial metabolites (such as bacteriocins, fatty acids, and hydrogen peroxide) (Patterson and Burkholder, 2003; Van Immerseel et al., 2009; Vandeplas et al., 2010). Microorganisms used as probiotics in animal feed are mainly strains belonging to the genera *Lactobacillus*, *Enterococcus*, *Pediococcus*, and *Bacillus*, whereas yeasts such as *Saccharomyces* have also been used (Simon et al., 2001; Vandeplas et al., 2010). It has been shown that certain probiotic strains are able to reduce *Salmonella* colonization (Jin et al., 1996; Line et al., 1998; Audisio et al., 1999; Van Coillie et al., 2007), although the observed effects are often of minor magnitude (Van Immerseel et al., 2009).

Newly hatched chicks are highly susceptible to infection with *Salmonella* and can become infected even when they are exposed to only low doses of the bacterium (Van Immerseel et al., 2004b). However, young chicks quickly become more resistant to infection due to the development of a normal gut microbiota and their immune system (Coloe et al., 1984; Vainio and Imhof, 1995; Bar-Shira et al., 2003; Bar-Shira and Friedman, 2006; Crhanova et al., 2011). Normally these chickens acquire their gut microbiota from the poultry house environment. However, it is also possible to administer bacterial cultures to newly hatched chicks to decrease their susceptibility to *Salmonella* infection (Rantala and Nurmi, 1973; Borie et al., 2009). This concept is referred to as CE, and implies the introduction of healthy, nonpathogenic, and mature microbial cultures into the intestinal tract of an animal creating an environment in which the healthy microbiota outcompete pathogens like *Salmonella*, reducing the chance of pathogen colonization (Kerr et al., 2013). As such, this concept can be exploited in practice to protect poultry against *Salmonella* infection. Furthermore, CE is also applied in human medicine to treat *Clostridium difficile* infections in humans (Aas et al., 2003; Brandt and Reddy, 2011). It is believed that orally administered CE cultures offer protection against pathogen colonization through competition for attachment sites in the gastrointestinal tract, and inhibit bacterial growth indirectly by lowering the intestinal pH and the production of volatile fatty acids (Snoeyenbos et al., 1979; Doyle and Erickson, 2006).

There are two different types of CE preparations, namely, undefined and defined preparations. Undefined preparations are derived from intestinal contents of mature birds, and when administered to young birds, these have been shown to protect young chickens against *Salmonella* infections by reducing intestinal colonization and internal organ invasion (Bolder et al., 1992; Nuotio et al., 1992; Revolledo et al., 2009; Kerr et al., 2013). In addition, it has been shown that administration of a CE preparation to egg-type pullets before transfer into a *Salmonella*-contaminated laying house resulted in a significantly lower frequency of isolating *Salmonella* from fecal and environmental samples (Davies and Breslin, 2003c). Although these preparations are often able to confer strong protection against *Salmonella* infections, their application in practice does raise some safety concerns. Because these preparations are undefined, they might contain potential pathogens and their application might thus bring about the spread of these pathogens. These concerns can, however, be overcome by thoroughly testing undefined CE preparations for the presence of these potential pathogens. Defined CE treatments have been evaluated for their capabilities to control *Salmonella* infections as well. However, when mixtures are applied that consist of only a few different bacterial species or strains, much like simple probiotics, only limited protection is observed. Increasingly complex defined mixtures seem to be better at protecting against *Salmonella* infections (Stavric et al., 1985; Corrier et al., 1994). This was confirmed in a study in which administration of culture consisting of a mixture of 11 lactic acid bacteria to young chickens protected better against *Salmonella* challenge than a combination of three *Lactobacillus* isolates (Higgins et al., 2010). Similarly, a study in which a mixture of 29 defined bacterial strains was administered to chicks resulted in strong protection against *S.* Typhimurium challenge (Corrier et al., 1995). Finally, application of these defined preparations should raise fewer safety concerns and performance of these preparations should be more consistent when compared with undefined preparations (Gast, 2011).

## 4. VACCINATION

Vaccination of poultry against *Salmonella* infection has been investigated extensively and has already been successfully applied in numerous locations around the world (Gast, 2011). For example, a vaccination program for laying hens was successfully implemented in the United Kingdom, where it resulted in the observation of significantly fewer human *S.* Enteritidis infections (Cogan and Humphrey, 2003; Gast, 2007). In addition, implementation of a vaccination program in commercial laying flocks in Japan decreased the isolation of *S.* Enteritidis from liquid egg samples (Toyota-Hanatani et al., 2009a), which illustrates the positive effects associated with the vaccination of laying hens. The goal of vaccination is to reduce the consequences of *Salmonella* infection in poultry. It will lower susceptibility of the birds to infection, reduce horizontal transmission of the infection within a flock, reduce the pathogen load in the poultry house, reduce horizontal transmission of the infection to progeny, and lower the frequency of egg contamination (Gast, 2011). In most cases, vaccination with either live or killed vaccines reduces fecal shedding, organ invasion, and egg contamination after experimental challenge (Gast et al., 1992, 1993). In addition, it has been observed that vaccinated chickens pass along some degree of immunity to their progeny (Hassan and Curtiss, 1996; Avila et al., 2006). No vaccine is, however, able to consistently prevent infection completely, especially against high challenge doses of salmonellae (Gast et al., 1992; Roland et al., 2004). Both killed and live vaccines have already been used to immunize commercial flocks, and both can provide significant protection against *Salmonella* (Van Immerseel et al., 2005b; Gast, 2007). Despite this, it has been previously reported that insufficient rodent control or general sanitation problems in the laying house can have a negative impact on vaccine performance (Davies and Breslin, 2003a). In addition, animal stress, caused by excess heat, or food and water deprivation, can compromise vaccine performance as well (Nakamura et al., 1994).

### 4.1 KILLED *SALMONELLA* VACCINES

Killed *Salmonella* vaccines, or bacterins, have already been implemented successfully in *Salmonella* control programs in the past, and contributed strongly to the decline of *S.* Enteritidis prevalence in laying hen flocks where implemented in control programs. Currently, there are several inactivated vaccine preparations commercially available for several economically and epidemiologically important serovars, such as *S.* Enteritidis. Use of these killed vaccines is associated with a reduced *Salmonella* load in feces, internal tissues, and eggs when they are challenged orally, and a lower mortality, lesions, and clinical signs when they are challenged intravenously or intramuscularly (Timms et al., 1994; Clifton-Hadley et al., 2002; Woodward et al., 2002; Van Immerseel et al., 2005b; Gast, 2007, 2011). In addition, administration of killed vaccines to laying hens undergoing induced molting reduces the increased fecal shedding of *S.* Enteritidis associated with the molting (Nakamura et al., 2004). Another experimental study showed that fecal and environmental samples of a laying flock were negative for the presence of *Salmonella* after bacterin administration, even when the vaccinated hens were transferred into previously contaminated facilities, further illustrating the protective effect of these killed vaccines (Davies and Breslin, 2003a). These killed vaccines are quite versatile as well, as autogenous bacterins can be developed quickly and made available when there are problems with a specific *Salmonella* serotype or certain variants, or are associated with a certain location (Gast, 2011). Another advantage of killed vaccines is that there is no safety concern about the introduction of live vaccine strains in the food chain through their administration to food producing animals (Van Immerseel et al., 2005b; Desin et al., 2013).

A limitation of killed vaccines is their inability to effectively elicit a protective cell-mediated immune response, possibly because important bacterial antigens are lost during the bacterial inactivation procedures (Muotiala et al., 1989; Barrow, 2007; Gast, 2011). In addition, killed vaccines may be less able to elicit an immune response to antigenically unrelated *Salmonella* serovars, and can thus potentially offer only limited cross-protection (Van Immerseel et al., 2005b). This can, however, be counteracted to a certain extent by developing multivalent bacterins comprising a mixture of strains or serotypes, to obtain a broader spectrum of protection. This was shown when fecal shedding and egg contamination for both *S.* Enteritidis and Typhimurium were reduced in laying hens vaccinated with a bacterin after challenge with either serotype (Okamura et al., 2007). In addition, treatment with a trivalent killed vaccine resulted in a reduced fecal and cecal recovery of *Salmonella* in chickens after heterologous challenge with *S.* Enteritidis, Typhimurium, or Infantis (Deguchi et al., 2009), further illustrating that it is possible to obtain protection against multiple serotypes using a multivalent bacterin. Another limitation of killed vaccines is that, although they might be able to elicit a strong humoral immune response, high antibody titers do not necessarily correspond with a higher degree of protection against infection (Mizumoto et al., 2006). There are also concerns about local pathological consequences following parenteral injection of toxic bacterial cell constituents such as lipopolysaccharides, or oil-emulsion adjuvants (Gast, 2011). Finally, administration of killed vaccines is done by injecting each bird individually, which is labor intensive and thus brings about high labor costs (Gast, 2011).

## 4.2 LIVE *SALMONELLA* VACCINES

Live attenuated *Salmonella* vaccines are *Salmonella* strains containing mutations or deletions in genes that are essential for metabolism, virulence, or survival in the host, which should result in an avirulent strain that is still able to induce a protective immune response (Van Immerseel et al., 2005b; Desin et al., 2013). Different *aroA* mutants of *S.* Enteritidis have already been developed, resulting in vaccine strains that are no longer able to synthesize certain aromatic compounds required for in vivo growth (Cooper et al., 1994b). Such vaccine strains are able to induce a long-lasting reduction of fecal shedding, horizontal transmission, organ invasion, and egg contamination in chickens experimentally infected with *S.* Enteritidis (Cooper et al., 1993, 1994a, 1996). A *S.* Typhimurium Δ*cya*-Δ*crp* mutant strain, unable to produce functional adenylate cyclase and cyclic adenosine monophosphate receptor proteins, was able to induce a strong protection against intestinal and internal organ colonization by a wild-type challenge strain (Hassan and Curtiss, 1994). A Δ*guaBfliC S.* Enteritidis double-deletion mutant, which is nonflagellated and deficient for guanine synthesis, was able to reduce internal organ colonization after challenge, while still allowing for serological differentiation between infected birds and vaccinated birds (Adriaensen et al., 2007). Other *Salmonella* mutants that are able to protect against *S.* Enteritidis infection have been developed as well, such as a temperature-sensitive mutant strain (Cerquetti and Gherardi, 2000), and a mutant strain obtained after repeated passage through chicken heterophils (Kramer, 1998).

It has been shown that certain live *Salmonella* vaccines are able to induce cross-protection between different serotypes. A *Salmonella* Gallinarum 9R (SG9R) strain, able to protect poultry against *Salmonella* Gallinarum infection, offers protection against *S.* Enteritidis infection as well (Feberwee et al., 2001). In addition, it has been shown that an avirulent *S.* Typhimurium vaccine strain reduced colonization, organ invasion, and egg contamination by *S.* Enteritidis (Hassan and Curtiss, 1997). Other live vaccines, however, are often not able to induce cross-protection against other serotypes, and as such, findings concerning this aspect of live vaccines are somewhat inconsistent (Gast, 2007). One such example is that an attenuated *S.* Enteritidis *aroA* mutant strain was not cross-protective against *S.* Typhimurium (Cooper et al., 1993).

Live *Salmonella* vaccines have several advantages over inactivated vaccines, as they stimulate both cell-mediated and humoral immune responses and usually express a wider array of antigens in vivo (Babu et al., 2004; Van Immerseel et al., 2005b). As such, live *Salmonella* vaccines are often considered to offer better protection against *Salmonella* infection than killed vaccines. Moreover, they can easily be administered orally (Gast, 2007; Desin et al., 2013). A major disadvantage of live vaccine strains is that these might persist in the chickens as well as in the environment, possibly resulting in the introduction of the vaccine in the food chain with the potential risk of posing a threat to human health (Tan et al., 1997; Barbezange et al., 2000b). On the other hand, some persistence can be desirable, as horizontal spread of the live vaccine would result in the protection of birds that were not originally immunized (Desin et al., 2013). Other disadvantages of live vaccines are possible interference with *Salmonella* testing procedures and the possibility of reversion to virulence (Barbezange et al., 2000a; Adriaensen et al., 2007). This reversion to virulence has already been observed in the field (Van Immerseel et al., 2013), yet can be avoided by using vaccine strains in which a genetic modification is introduced by deleting the target genes instead of mutating them. In addition, using sensitive culturing methods, certain live vaccine strains could also be detected in vaccinated hens long after administration, which is undesirable as well (Tan et al., 1997). Several challenges are associated with administering live vaccines and vaccines in general to poultry, such as the high costs of vaccine administration, a too late induction of immunity, and difficulties with delivering the vaccine in a uniform way to the poultry flock (Desin et al., 2013). However, because live vaccines can easily be administered orally, different methods to administer these on a large scale have been developed. Applications through the drinking water or by spray are most widely used for *Salmonella* vaccines, allowing for a relatively uniform, easy, and economical administration.

### 4.3 SUBUNIT VACCINES

Subunit vaccines, consisting of or displaying defined antigens, can be used to protect poultry against *Salmonella* infections as well. It has been shown that subunit vaccines consisting of *Salmonella* outer membrane proteins administered in combination with adjuvants or incorporated into immunostimulating complexes are able to offer protection against *S.* Enteritidis infection (Charles et al., 1994; Khan et al., 2003). Immunization of laying hens with purified *S.* Enteritidis fimbria resulted in a reduced reproductive organ invasion and egg contamination by *S.* Enteritidis (De Buck et al., 2005). Cecal colonization by *S.* Enteritidis was reduced after administration of a flagellar subunit vaccine (Toyota-Hanatani et al., 2009b). And although subunit vaccines are currently infrequently applied in practice, they thus represent another viable option to protect poultry against *Salmonella* infections. Therefore, there are some that believe that the next generation of poultry vaccines will consist of subunit vaccines delivering antigens from possibly multiple pathogens using viral or DNA vectors (Kaiser, 2010).

## 5. ALTERNATIVE PREHARVEST STRATEGIES TO CONTROL *SALMONELLA* IN POULTRY

### 5.1 BACTERIOPHAGES

Another possible, yet more experimental alternative to combat *Salmonella* is bacteriophage therapy. Bacteriophages are natural predators of their bacterial hosts and can cause lysis of bacterial cells as part of their life cycle. Although the potential of bacteriophages for pathogen control was already being investigated before the discovery of antibiotics, interest in bacteriophages has been rekindled due to widespread microbial antibiotic resistance following the overzealous usage of antibiotics in both human and veterinary medicine (Sulakvelidze et al., 2001). Indeed, it has been shown that certain bacteriophages are able to reduce carriage of *Salmonella* in live birds. One study showed that administration of bacteriophages to chickens in the drinking water or via coarse spray before experimental infection resulted in a significant reduction of *Salmonella* colonization (Borie et al., 2008). Another study showed that different bacteriophages can be used to reduce colonization by *S.* Enteritidis, Typhimurium, and Hadar (Atterbury et al., 2007). However, other studies showed no or minimal reduction in *Salmonella* colonization after bacteriophage therapy, illustrating the importance of the choice of bacteriophage, method of administration, and titer administered (Berchieri et al., 1991; Sklar and Joerger, 2001).

A major advantage of bacteriophages is their flexibility, as different phages targeting different *Salmonella* serovars can be combined in a single treatment, offering protection against multiple serotypes (Van Immerseel and Atterbury, 2013; Grant et al., 2016). A disadvantage is that bacteriophages are able to mediate horizontal transfer of DNA between bacteria through transduction, which might result in the undesirable transfer of genes associated with pathogenicity, survival, and virulence (Ho and Slauch, 2001; Van Immerseel and Atterbury, 2013). In addition, bacteriophage therapy can strongly reduce *Salmonella* colonization, but complete elimination of the pathogen will be difficult to achieve (Van Immerseel and Atterbury, 2013).

## 5.2 BREEDING

Another option to decrease the susceptibility of poultry for *Salmonella* infections is the targeted breeding of *Salmonella*-resistant chicken lines. It is clear that some chicken lines are more resistant to *Salmonella* infections than others (Guillot et al., 1995; Berchieri et al., 2001). As such, it is believed that the genetic traits that confer this higher resistance to the chickens can be selected for through breeding, resulting in chicken lines more resistant to *Salmonella* infections (Beaumont et al., 2008).

The observed variation between chicken lines in *Salmonella* susceptibility has been attributed to both the innate and the adaptive immune system. It has been shown that resistant chicken lines possess macrophages that are able to clear *Salmonella* infection more quickly than those of more susceptible lines, and that this clearance is accompanied by a strong respiratory burst (Wigley et al., 2002). In addition, macrophages from resistant chicken lines also express proinflammatory cytokines and chemokines more quickly and to a higher degree after challenge with *Salmonella* (Wigley et al., 2006). Beal et al. showed that resistance to *S.* Typhimurium infection correlates with an increased T-cell response (Beal et al., 2005). In addition, it has been shown that certain chicken lines expressing different levels of β-defensins are not equally susceptible to *Salmonella* infection, suggesting a role for β-defensins and thus highlighting their potential importance as a genetic trait (Derache et al., 2009). This was confirmed in another study, in which chicken lines that are susceptible to *Salmonella* were shown to express higher levels of chemokines, cytokines, antimicrobial mediators, and defensin genes (Sadeyen et al., 2006).

Genetic selection is an attractive control option for *Salmonella*, as this would increase resistance of the chickens without bringing about the additional costs associated with administration of additives, vaccination, or other control measures. However, at the moment, little progress in breeding *Salmonella*-resistant laying lines is being made. Conversely, it has even been reported that commercial layer lines, selected for reproduction and egg production, display lower cytokine gene expression levels after *S.* Enteritidis infection when compared with a native chicken line, suggesting that the currently used chicken lines are more susceptible to *Salmonella* as a consequence of selective breeding for other traits than resistance to *Salmonella* (Redmond et al., 2009). In addition, if a newly bred *Salmonella*-resistant chicken line would be less productive than the current laying lines, the use of this new line would be limited. Consequently, developing a new chicken that is resistant to *Salmonella* infection while maintaining its productive capabilities will be an ambitious approach to control *Salmonella* in poultry.

## 6. FUTURE TRENDS—ZERO *SALMONELLA* PREVALENCE CONCEPT?

At the moment, control efforts for *Salmonella* mainly focus on quality assurance programs that embody risk reduction measures throughout the entire production cycle. An important objective within these programs is reducing the susceptibility of laying hens to *Salmonella* infections, as it is not always possible to prevent introduction of a pathogen in a flock. In addition, it is believed that an integrated approach in which protective measures are applied at multiple steps in the production process is the most effective approach to reduce the number of *Salmonella*-contaminated eggs (Hope et al., 2002). Therefore both pre- and postharvest measures contribute strongly to the reduction of *Salmonella*, and even limited improvements in the efficacy of these protective treatments can enhance the overall success of control programs (Gast, 2011).

However, despite these control measures, it is still possible to find *Salmonella*-positive eggs that are introduced in the food chain (EFSA, 2015). In Europe, legislation dictates that no *Salmonella* is to be found in 5 × 25 g of eggs at retail, otherwise these products are to be eliminated (EU legislation 2073/2005). Although this is a good strategy to monitor for highly positive egg batches, introduction of *Salmonella*-contaminated eggs into the food chain cannot be excluded. However, achieving truly zero *Salmonella*-positive eggs will be difficult due to the fact that *Salmonella* is ubiquitous in the environment, and currently, no control measures ensure full protection of hens against the pathogen. In addition, the cost associated with the measures needed to ensure the production of only *Salmonella*-free eggs would be tremendously high. As such, the acceptable level of protection currently enforced by European legislation, which minimizes the number of *Salmonella*-positive eggs, is both a technologically and economically better alternative to strive for.

## REFERENCES

Aas, J., Gessert, C.E., Bakken, J.S., 2003. Recurrent *Clostridium difficile* colitis: case series involving 18 patients treated with donor stool administered via a nasogastric tube. Clin. Infect. Dis. 36, 580–585.

Adriaensen, C., De Greve, H., Tian, J.Q., De Craeye, S., Gubbels, E., Eeckhaut, V., Van Immerseel, F., Ducatelle, R., Kumar, M., Hernalsteens, J.P., 2007. A live *Salmonella enterica* serovar Enteritidis vaccine allows serological differentiation between vaccinated and infected animals. Infect. Immun. 75, 2461–2468.

Atterbury, R.J., Van Bergen, M.A., Ortiz, F., Lovell, M.A., Harris, J.A., De Boer, A., Wagenaar, J.A., Allen, V.M., Barrow, P.A., 2007. Bacteriophage therapy to reduce *Salmonella* colonization of broiler chickens. Appl. Environ. Microbiol. 73, 4543–4549.

Audisio, M.C., Oliver, G., Apella, M.C., 1999. Antagonistic effect of *Enterococcus faecium* J96 against human and poultry pathogenic *Salmonella* spp. J. Food Prot. 62, 751–755.

Avila, L.A., Nascimento, V.P., Salle, C.T., Moraes, H.L., 2006. Effects of probiotics and maternal vaccination on *Salmonella* Enteritidis infection in broiler chicks. Avian Dis. 50, 608–612.

Babu, U., Dalloul, R.A., Okamura, M., Lillehoj, H.S., Xie, H., Raybourne, R.B., Gaines, D., Heckert, R.A., 2004. *Salmonella* Enteritidis clearance and immune responses in chickens following *Salmonella* vaccination and challenge. Vet. Immunol. Immunopathol. 101, 251–257.

Barbezange, C., Ermel, G., Ragimbeau, C., Humbert, F., Salvat, G., 2000a. Some safety aspects of *Salmonella* vaccines for poultry: in vivo study of the genetic stability of three *Salmonella* Typhimurium live vaccines. FEMS Microbiol. Lett. 192, 101–106.

Barbezange, C., Humbert, F., Rose, V., Lalande, F., Salvat, G., 2000b. Some safety aspects of *Salmonella* vaccines for poultry: distribution and persistence of three *Salmonella* Typhimurium live vaccines. Avian Dis. 44, 968–976.

Barnhart, E.T., Caldwell, D.J., Crouch, M.C., Byrd, J.A., Corrier, D.E., Hargis, B.M., 1999. Effect of lactose administration in drinking water prior to and during feed withdrawal on *Salmonella* recovery from broiler crops and ceca. Poult. Sci. 78, 211–214.

Barrow, P.A., Lovell, M.A., 1991. Experimental infection of egg-laying hens with *Salmonella* Enteritidis phage type 4. Avian Pathol. 20, 335–348.

Barrow, P.A., Huggins, M.B., Lovell, M.A., 1994. Host specificity of *Salmonella* infection in chickens and mice is expressed in vivo primarily at the level of the reticuloendothelial system. Infect. Immun. 62, 4602–4610.

Barrow, P.A., 2007. *Salmonella* infections: immune and non-immune protection with vaccines. Avian Pathol. 36, 1–13.

Bar-Shira, E., Friedman, A., 2006. Development and adaptations of innate immunity in the gastrointestinal tract of the newly hatched chick. Dev. Comp. Immunol. 30, 930–941.

Bar-Shira, E., Sklan, D., Friedman, A., 2003. Establishment of immune competence in the avian GALT during the immediate post-hatch period. Dev. Comp. Immunol. 27, 147–157.

Beal, R.K., Powers, C., Wigley, P., Barrow, P.A., Kaiser, P., Smith, A.L., 2005. A strong antigen-specific T-cell response is associated with age and genetically dependent resistance to avian enteric salmonellosis. Infect. Immun. 73, 7509–7516.

Beaumont, C., Lecerf, F., Protais, J., Calenge, F., Prevost, K., Lalmanach, A.C., Chapuis, H., Pitel, F., Burlot, T., Sellier, N., Fravalo, P., Vignal, A., Velge, P., 2008. An integrated approach of genetic resistance to *Salmonella* carrier state in fowls: from genetics to genomics and modelling. Dev. Biol. (Basel) 132, 353–357.

Bennett, D.D., Higgins, S.E., Moore, R.W., Beltran, R., Caldwell, D.J., Byrd, J.A., Hargis, B.M., 2003. Effects of lime on *Salmonella* Enteritidis survival in vitro. J. Appl. Poult. Res. 12, 65–68.

Berchieri Jr., A., Lovell, M.A., Barrow, P.A., 1991. The activity in the chicken alimentary tract of bacteriophages lytic for *Salmonella typhimurium*. Res. Microbiol. 142, 541–549.

Berchieri Jr., A., Wigley, P., Page, K., Murphy, C.K., Barrow, P.A., 2001. Further studies on vertical transmission and persistence of *Salmonella enterica* serovar Enteritidis phage type 4 in chickens. Avian Pathol. 30, 297–310.

Bolder, N.M., van Lith, L.A., Putirulan, F.F., Jacobs-Reitsma, W.F., Mulder, R.W., 1992. Prevention of colonization by *Salmonella* Enteritidis PT4 in broiler chickens. Int. J. Food Microbiol. 15, 313–317.

Borie, C., Albala, I., Sanchez, P., Sanchez, M.L., Ramirez, S., Navarro, C., Morales, M.A., Retamales, A.J., Robeson, J., 2008. Bacteriophage treatment reduces *Salmonella* colonization of infected chickens. Avian Dis. 52, 64–67.

Borie, C., Sanchez, M.L., Navarro, C., Ramirez, S., Morales, M.A., Retamales, J., Robeson, J., 2009. Aerosol spray treatment with bacteriophages and competitive exclusion reduces *Salmonella* enteritidis infection in chickens. Avian Dis. 53, 250–254.

Boyen, F., Haesebrouck, F., Vanparys, A., Volf, J., Mahu, M., Van Immerseel, F., Rychlik, I., Dewulf, J., Ducatelle, R., Pasmans, F., 2008. Coated fatty acids alter virulence properties of *Salmonella* Typhimurium and decrease intestinal colonization of pigs. Vet. Microbiol. 132, 319–327.

Brandt, L.J., Reddy, S.S., 2011. Fecal microbiota transplantation for recurrent *Clostridium difficile* infection. J. Clin. Gastroenterol. 45 (Suppl.), S159–S167.

Brownell, J.R., Sadler, W.W., Fanelli, M.J., 1969. Factors influencing the intestinal infection of chickens with Salmonella typhimurium. Avian Dis. 13, 804–816.

Callaway, T.R., Edrington, T.S., Anderson, R.C., Byrd, J.A., Nisbet, D.J., 2008. Gastrointestinal microbial ecology and the safety of our food supply as related to *Salmonella*. J. Anim. Sci. 86, E163–E172.

Cerquetti, M.C., Gherardi, M.M., 2000. Orally administered attenuated *Salmonella* Enteritidis reduces chicken cecal carriage of virulent *Salmonella* challenge organisms. Vet. Microbiol. 76, 185–192.

Charles, S.D., Hussain, I., Choi, C.U., Nagaraja, K.V., Sivanandan, V., 1994. Adjuvanted subunit vaccines for the control of *Salmonella* Enteritidis infection in turkeys. Am. J. Vet. Res. 55, 636–642.

Clifton-Hadley, F.A., Breslin, M., Venables, L.M., Sprigings, K.A., Cooles, S.W., Houghton, S., Woodward, M.J., 2002. A laboratory study of an inactivated bivalent iron restricted *Salmonella enterica* serovars Enteritidis and Typhimurium dual vaccine against Typhimurium challenge in chickens. Vet. Microbiol. 89, 167–179.

Cogan, T.A., Humphrey, T.J., 2003. The rise and fall of *Salmonella* Enteritidis in the UK. J. Appl. Microbiol. 94 (Suppl.), 114S–119S.

Coloe, P.J., Bagust, T.J., Ireland, L., 1984. Development of the normal gastrointestinal microflora of specific pathogen-free chickens. J. Hyg. (Lond.) 92, 79–87.

Cooper, G.L., Venables, L.M., Nicholas, R.A., Cullen, G.A., Hormaeche, C.E., 1993. Further studies of the application of live *Salmonella enteritidis aroA* vaccines in chickens. Vet. Rec. 133, 31–36.

Cooper, G.L., Venables, L.M., Woodward, M.J., Hormaeche, C.E., 1994a. Vaccination of chickens with strain CVL30, a genetically defined *Salmonella enteritidis aroA* live oral vaccine candidate. Infect. Immun. 62, 4747–4754.

Cooper, G.L., Venables, L.M., Woodward, M.J., Hormaeche, C.E., 1994b. Invasiveness and persistence of *Salmonella enteritidis*, *Salmonella typhimurium*, and a genetically defined *Salmonella enteritidis aroA* strain in young chickens. Infect. Immun. 62, 4739–4746.

Cooper, G.L., Venables, L.M., Lever, M.S., 1996. Airborne challenge of chickens vaccinated orally with the genetically-defined *Salmonella* Enteritidis *aroA* strain CVL30. Vet. Rec. 139, 447–448.

Corrier, D.E., Nisbet, D.J., Scanlan, C.M., Tellez, G., Hargis, B.M., Deloach, J.R., 1994. Inhibition of *Salmonella* Enteritidis cecal and organ colonization in Leghorn chicks by a defined culture of cecal bacteria and dietary lactose. J. Food Prot. 57, 377–381.

Corrier, D.E., Nisbet, D.J., Scanlan, C.M., Hollister, A.G., Deloach, J.R., 1995. Control of *Salmonella typhimurium* colonization in broiler chicks with a continuous-flow characterized mixed culture of cecal bacteria. Poult. Sci. 74, 916–924.

Cox, J.M., Pavic, A., 2009. Advances in enteropathogen control in poultry production. J. Appl. Microbiol. 108, 745–755.

Cox, N.A., Bailey, J.S., Mauldin, J.M., Blankenship, L.C., 1990. Presence and impact of *Salmonella* contamination in commercial broiler hatcheries. Poult. Sci. 69, 1606–1609.

Crhanova, M., Hradecka, H., Faldynova, M., Matulova, M., Havlickova, H., Sisak, F., Rychlik, I., 2011. Immune response of chicken gut to natural colonization by gut microflora and to *Salmonella enterica* serovar Enteritidis infection. Infect. Immun. 79, 2755–2763.

Davies, R., Breslin, M., 2003a. Effects of vaccination and other preventive methods for *Salmonella* Enteritidis on commercial laying chicken farms. Vet. Rec. 153, 673–677.

Davies, R.H., Breslin, M., 2003b. Persistence of *Salmonella* Enteritidis phage type 4 in the environment and arthropod vectors on an empty free-range chicken farm. Environ. Microbiol. 5, 79–84.

Davies, R.H., Breslin, M.F., 2003c. Observations on the distribution and persistence of *Salmonella enterica* serovar Enteritidis phage type 29 on a cage layer farm before and after the use of competitive exclusion treatment. Br. Poult. Sci. 44, 551–557.

Davies, R.H., Wray, C., 1996. Persistence of *Salmonella* Enteritidis in poultry units and poultry food. Br. Poult. Sci. 37, 589–596.

De Buck, J., Van Immerseel, F., Haesebrouck, F., Ducatelle, R., 2004. Colonization of the chicken reproductive tract and egg contamination by *Salmonella*. J. Appl. Microbiol. 97, 233–245.

De Buck, J., Van Immerseel, F., Haesebrouck, F., Ducatelle, R., 2005. Protection of laying hens against *Salmonella* Enteritidis by immunization with type 1 fimbriae. Vet. Microbiol. 105, 93–101.

De Maesschalck, C., Eeckhaut, V., Maertens, L., De Lange, L., Marchal, L., Nezer, C., De Baere, S., Croubels, S., Daube, G., Dewulf, J., Haesebrouck, F., Ducatelle, R., Taminau, B., Van Immerseel, F., 2015. Effects of xylo-oligosaccharides on broiler chicken performance and microbiota. Appl. Environ. Microbiol. 81, 5880–5888.

Deguchi, K., Yokoyama, E., Honda, T., Mizuno, K., 2009. Efficacy of a novel trivalent inactivated vaccine against the shedding of *Salmonella* in a chicken challenge model. Avian Dis. 53, 281–286.

Denagamage, T., Jayarao, B., Patterson, P., Wallner-Pendleton, E., Kariyawasam, S., 2015. Risk factors associated with *Salmonella* in laying hen farms: systematic review of observational studies. Avian Dis. 59, 291–302.

Derache, C., Esnault, E., Bonsergent, C., Le Vern, Y., Quere, P., Lalmanach, A.C., 2009. Differential modulation of beta-defensin gene expression by *Salmonella* Enteritidis in intestinal epithelial cells from resistant and susceptible chicken inbred lines. Dev. Comp. Immunol. 33, 959–966.

Desin, T.S., Koster, W., Potter, A.A., 2013. *Salmonella* vaccines in poultry: past, present and future. Expert Rev. Vaccines 12, 87–96.

Doyle, M.P., Erickson, M.C., 2006. Reducing the carriage of foodborne pathogens in livestock and poultry. Poult. Sci. 85, 960–973.

Ducatelle, R., Van Immerseel, F., 2011. Management and sanitation procedures to control *Salmonella* in laying hen flocks. In: Van Immerseel, F., Nys, Y., Bain, M. (Eds.), Improving the Safety and Quality of Eggs and Egg Products, vol. 2. Woodhead Publishing Limited, Cambridge, pp. 146–162.

Eeckhaut, V., Van Immerseel, F., Dewulf, J., Pasmans, F., Haesebrouck, F., Ducatelle, R., Courtin, C.M., Delcour, J.A., Broekaert, W.F., 2008. Arabinoxylooligosaccharides from wheat bran inhibit *Salmonella* colonization in broiler chickens. Poult. Sci. 87, 2329–2334.

EFSA, 2015. The European Union Summary Report on trends and sources of zoonoses, zoonotic agents and food-borne outbreaks in 2013. EFSA J.

Eriksson, C.L., De Rezende, C.L.E., Mallinson, E.T., Tablante, N.L., Morales, R., Park, A., Carr, L.E., Joseph, S.W., 2001. Effect of dry litter and airflow in reducing *Salmonella* and *Escherichia coli* populations in the broiler production environment. J. Appl. Poult. Res. 10, 245–251.

Feberwee, A., de Vries, T.S., Hartman, E.G., de Wit, J.J., Elbers, A.R., de Jong, W.A., 2001. Vaccination against *Salmonella* Enteritidis in Dutch commercial layer flocks with a vaccine based on a live *Salmonella* gallinarum 9R strain: evaluation of efficacy, safety, and performance of serologic *Salmonella* tests. Avian Dis. 45, 83–91.

Gantois, I., Ducatelle, R., Pasmans, F., Haesebrouck, F., Hautefort, I., Thompson, A., Hinton, J.C., Van Immerseel, F., 2006a. Butyrate specifically down-regulates *Salmonella* pathogenicity island 1 gene expression. Appl. Environ. Microbiol. 72, 946–949.

Gantois, I., Ducatelle, R., Timbermont, L., Boyen, F., Bohez, L., Haesebrouck, F., Pasmans, F., Van Immerseel, F., 2006b. Oral immunisation of laying hens with the live vaccine strains of TAD *Salmonella* vac E and TAD *Salmonella* vac T reduces internal egg contamination with *Salmonella* Enteritidis. Vaccine 24, 6250–6255.

Gast, R.K., Stone, H.D., Holt, P.S., Beard, C.W., 1992. Evaluation of the efficacy of an oil-emulsion bacterin for protecting chickens against Salmonella enteritidis. Avian Dis. 36, 992–999.

Gast, R.K., Stone, H.D., Holt, P.S., 1993. Evaluation of the efficacy of oil-emulsion bacterins for reducing fecal shedding of Salmonella enteritidis by laying hens. Avian Dis. 37, 1085–1091.

Gast, R.K., Mitchell, B.W., Holt, P.S., 1998. Airborne transmission of Salmonella enteritidis infection between groups of chicks in controlled-environment isolation cabinets. Avian Dis. 42, 315–320.

Gast, R.K., 2007. Serotype-specific and serotype-independent strategies for preharvest control of food-borne *Salmonella* in poultry. Avian Dis. 51, 817–828.

Gast, R.K., 2011. Pre-harvest measures to control *Salmonella* in laying hens. In: Van Immerseel, F., Nys, Y., Bain, M. (Eds.), Improving the Safety and Quality of Eggs and Egg Products, vol. 2. Woodhead Publishing Limited, Cambridge, pp. 120–145.

Gibson, G.R., McCartney, A.L., Rastall, R.A., 2005. Prebiotics and resistance to gastrointestinal infections. Br. J. Nutr. 93 (Suppl. 1), S31–S34.

Gillespie, I.A., O'Brien, S.J., Adak, G.K., Ward, L.R., Smith, H.R., 2005. Foodborne general outbreaks of *Salmonella* Enteritidis phage type 4 infection, England and Wales, 1992–2002: where are the risks? Epidemiol. Infect. 133, 795–801.

Grant, A., Hashem, F., Parveen, S., 2016. *Salmonella* and *Campylobacter*: antimicrobial resistance and bacteriophage control in poultry. Food Microbiol. 53, 104–109.

Greig, J.D., Ravel, A., 2009. Analysis of foodborne outbreak data reported internationally for source attribution. Int. J. Food Microbiol. 130, 77–87.

Guillot, J.F., Beaumont, C., Bellatif, F., Mouline, C., Lantier, F., Colin, P., Protais, J., 1995. Comparison of resistance of various poultry lines to infection by *Salmonella enteritidis*. Vet. Res. 26, 81–86.

Hassan, J.O., Curtiss 3rd, R., 1994. Development and evaluation of an experimental vaccination program using a live avirulent *Salmonella* typhimurium strain to protect immunized chickens against challenge with homologous and heterologous *Salmonella* serotypes. Infect. Immun. 62, 5519–5527.

Hassan, J.O., Curtiss 3rd, R., 1996. Effect of vaccination of hens with an avirulent strain of *Salmonella* typhimurium on immunity of progeny challenged with wild-type *Salmonella* strains. Infect. Immun. 64, 938–944.

Hassan, J.O., Curtiss 3rd, R., 1997. Efficacy of a live avirulent *Salmonella* typhimurium vaccine in preventing colonization and invasion of laying hens by *Salmonella* typhimurium and *Salmonella* enteritidis. Avian Dis. 41, 783–791.

Henzler, D.J., Opitz, H.M., 1992. The role of mice in the epizootiology of Salmonella enteritidis infection on chicken layer farms. Avian Dis. 36, 625–631.

Higgins, J.P., Higgins, S.E., Wolfenden, A.D., Henderson, S.N., Torres-Rodriguez, A., Vicente, J.L., Hargis, B.M., Tellez, G., 2010. Effect of lactic acid bacteria probiotic culture treatment timing on *Salmonella* Enteritidis in neonatal broilers. Poult. Sci. 89, 243–247.

Hinton, M., Linton, A.H., 1988. Control of *Salmonella* infections in broiler chickens by the acid treatment of their feed. Vet. Rec. 123, 416–421.

Hinton, M., 1988. *Salmonella* infection in chicks following the consumption of artificially contaminated feed. Epidemiol. Infect. 100, 247–256.

Ho, T.D., Slauch, J.M., 2001. OmpC is the receptor for Gifsy-1 and Gifsy-2 bacteriophages of *Salmonella*. J. Bacteriol. 183, 1495–1498.

Holt, P.S., Buhr, R.J., Cunningham, D.L., Porter Jr., R.E., 1994. Effect of two different molting procedures on a *Salmonella enteritidis* infection. Poult. Sci. 73, 1267–1275.

Holt, P.S., Macri, N.P., Porter Jr., R.E., 1995. Microbiological analysis of the early Salmonella enteritidis infection in molted and unmolted hens. Avian Dis. 39, 55–63.

Holt, P.S., 1995. Horizontal transmission of Salmonella enteritidis in molted and unmolted laying chickens. Avian Dis. 39, 239–249.

Hope, B.K., Baker, R., Edel, E.D., Hogue, A.T., Schlosser, W.D., Whiting, R., McDowell, R.M., Morales, R.A., 2002. An overview of the *Salmonella* Enteritidis risk assessment for shell eggs and egg products. Risk Anal. 22, 203–218.

Hu, Z.H., Guo, Y.M., 2007. Effects of dietary sodium butyrate supplementation on the intestinal morphological structure, absorptive function and gut flora in chickens. Anim. Feed Sci. Technol. 132, 240–249.

Hume, M.E., Corrier, D.E., Ambrus, S., Hinton Jr., A., DeLoach, J.R., 1993. Effectiveness of dietary propionic acid in controlling *Salmonella* typhimurium colonization in broiler chicks. Avian Dis. 37, 1051–1056.

Humphrey, T.J., Chart, H., Baskerville, A., Rowe, B., 1991. The influence of age on the response of SPF hens to infection with *Salmonella enteritidis* PT4. Epidemiol. Infect. 106, 33–43.

Iba, A.M., Berchieri Jr., A., 1995. Studies on the use of a formic acid-propionic acid mixture (Bio-add$^{TM}$) to control experimental *Salmonella* infection in broiler chickens. Avian Pathol. 24, 303–311.

Jin, L.Z., Ho, Y.W., Ali, M.A., Abdullah, N., Jalaludin, S., 1996. Effect of adherent *Lactobacillus* spp. on in vitro adherence of salmonellae to the intestinal epithelial cells of chicken. J. Appl. Bacteriol. 81, 201–206.

Jones, F.T., 2011. A review of practical *Salmonella* control measures in animal feed. J. Appl. Poult. Res. 20, 102–113.

Kaiser, P., 2010. Advances in avian immunology – prospects for disease control: a review. Avian Pathol. 39, 309–324.

Kerr, A.K., Farrar, A.M., Waddell, L.A., Wilkins, W., Wilhelm, B.J., Bucher, O., Wills, R.W., Bailey, R.H., Varga, C., McEwen, S.A., Rajic, A., 2013. A systematic review-meta-analysis and meta-regression on the effect of selected competitive exclusion products on *Salmonella* spp. prevalence and concentration in broiler chickens. Prev. Vet. Med. 111, 112–125.

Khan, M.I., Fadl, A.A., Venkitanarayanan, K.S., 2003. Reducing colonization of *Salmonella* Enteritidis in chicken by targeting outer membrane proteins. J. Appl. Microbiol. 95, 142–145.

Kimura, A.C., Reddy, V., Marcus, R., Cieslak, P.R., Mohle-Boetani, J.C., Kassenborg, H.D., Segler, S.D., Hardnett, F.P., Barrett, T., Swerdlow, D.L., 2004. Chicken consumption is a newly identified risk factor for sporadic *Salmonella enterica* serotype Enteritidis infections in the United States: a case-control study in FoodNet sites. Clin. Infect. Dis. 38 (Suppl. 3), S244–S252.

Kramer, T.T., 1998. Effects of heterophil adaptation on *Salmonella* enteritidis fecal shedding and egg contamination. Avian Dis. 42, 6–13.

Liebana, E., Garcia-Migura, L., Clouting, C., Clifton-Hadley, F.A., Breslin, M., Davies, R.H., 2003. Molecular fingerprinting evidence of the contribution of wildlife vectors in the maintenance of *Salmonella* Enteritidis infection in layer farms. J. Appl. Microbiol. 94, 1024–1029.

Line, J.E., Bailey, J.S., Cox, N.A., Stern, N.J., Tompkins, T., 1998. Effect of yeast-supplemented feed on *Salmonella* and *Campylobacter* populations in broilers. Poult. Sci. 77, 405–410.

Mitchell, B.W., Buhr, R.J., Berrang, M.E., Bailey, J.S., Cox, N.A., 2002. Reducing airborne pathogens, dust and *Salmonella* transmission in experimental hatching cabinets using an electrostatic space charge system. Poult. Sci. 81, 49–55.

Mizumoto, N., Toyota-Hanatani, Y., Sasai, K., Tani, H., Ekawa, T., Ohta, H., Baba, E., 2006. Survey of Japanese layer farms for *Salmonella* Enteritidis with vaccination- and infection-specific antigens for egg yolk antibodies. J. Food Prot. 69, 17–21.

Muotiala, A., Hovi, M., Makela, P.H., 1989. Protective immunity in mouse salmonellosis: comparison of smooth and rough live and killed vaccines. Microb. Pathog. 6, 51–60.

Nakamura, M., Nagamine, N., Takahashi, T., Suzuki, S., Sato, S., 1994. Evaluation of the efficacy of a bacterin against Salmonella enteritidis infection and the effect of stress after vaccination. Avian Dis. 38, 717–724.

Nakamura, M., Nagata, T., Okamura, S., Takehara, K., Holt, P.S., 2004. The effect of killed *Salmonella* enteritidis vaccine prior to induced molting on the shedding of *Salmonella* Enteritidis in laying hens. Avian Dis. 48, 183–188.

Netherwood, T., Gilbert, H.J., Parker, D.S., O'Donnell, A.G., 1999. Probiotics shown to change bacterial community structure in the avian gastrointestinal tract. Appl. Environ. Microbiol. 65, 5134–5138.

Nuotio, L., Schneitz, C., Halonen, U., Nurmi, E., 1992. Use of competitive exclusion to protect newly-hatched chicks against intestinal colonisation and invasion by *Salmonella enteritidis* PT4. Br. Poult. Sci. 33, 775–779.

Okamura, M., Tachizaki, H., Kubo, T., Kikuchi, S., Suzuki, A., Takehara, K., Nakamura, M., 2007. Comparative evaluation of a bivalent killed *Salmonella* vaccine to prevent egg contamination with *Salmonella enterica* serovars Enteritidis, Typhimurium, and Gallinarum biovar Pullorum, using 4 different challenge models. Vaccine 25, 4837–4844.

Patterson, J.A., Burkholder, K.M., 2003. Application of prebiotics and probiotics in poultry production. Poult. Sci. 82, 627–631.

Porter Jr., R.E., Holt, P.S., 1993. Effect of induced molting on the severity of intestinal lesions caused by Salmonella enteritidis infection in chickens. Avian Dis. 37, 1009–1016.

Rantala, M., Nurmi, E., 1973. Prevention of the growth of *Salmonella infantis* in chicks by the flora of the alimentary tract of chickens. Br. Poult. Sci. 14, 627–630.

Redmond, S.B., Chuammitri, P., Andreasen, C.B., Palic, D., Lamont, S.J., 2009. Chicken heterophils from commercially selected and non-selected genetic lines express cytokines differently after in vitro exposure to *Salmonella* Enteritidis. Vet. Immunol. Immunopathol. 132, 129–134.

Rehman, H., Vahjen, W., Kohl-Parisini, A., Ijaz, A., Zentek, J., 2009. Influence of fermentable carbohydrates on the intestinal bacteria and enteropathogens in broilers. World's Poult. Sci. J. 65, 75–89.

Revolledo, L., Ferreira, C.S., Ferreira, A.J., 2009. Prevention of *Salmonella* Typhimurium colonization and organ invasion by combination treatment in broiler chicks. Poult. Sci. 88, 734–743.

Ricke, S.C., Dunkley, C.S., Durant, J.A., 2013. A review on development of novel strategies for controlling *Salmonella* Enteritidis colonization in laying hens: fiber-based molt diets. Poult. Sci. 92, 502–525.

Rodriguez-Romo, L.A., Yousef, A.E., 2005. Inactivation of *Salmonella enterica* serovar Enteritidis on shell eggs by ozone and UV radiation. J. Food Prot. 68, 711–717.

Roland, K., Tinge, S., Warner, E., Sizemore, D., 2004. Comparison of different attenuation strategies in development of a *Salmonella* Hadar vaccine. Avian Dis. 48, 445–452.

Sadeyen, J.R., Trotereau, J., Protais, J., Beaumont, C., Sellier, N., Salvat, G., Velge, P., Lalmanach, A.C., 2006. *Salmonella* carrier-state in hens: study of host resistance by a gene expression approach. Microbes Infect. 8, 1308–1314.

Shivaprasad, H.L., Timoney, J.F., Morales, S., Lucio, B., Baker, R.C., 1990. Pathogenesis of Salmonella enteritidis infection in laying chickens. I. Studies on egg transmission, clinical signs, fecal shedding, and serologic responses. Avian Dis. 34, 548–557.

Simon, O., Jadamus, A., Vahjen, W., 2001. Probiotic feed additives – effectiveness and expected modes of action. J. Anim. Feed Sci. 10, 51–67.

Sklar, I.B., Joerger, R.D., 2001. Attempts to utilize bacteriophage to combat *Salmonella enterica* serovar Enteritidis infection in chickens. J. Food Saf. 21, 15–29.

Snoeyenbos, G.H., Weinack, O.M., Smyser, C.F., 1979. Further studies on competitive exclusion for controlling salmonellae in chickens. Avian Dis. 23, 904–914.

Spring, P., Wenk, C., Dawson, K.A., Newman, K.E., 2000. The effects of dietary mannaoligosaccharides on cecal parameters and the concentrations of enteric bacteria in the ceca of *Salmonella*-challenged broiler chicks. Poult. Sci. 79, 205–211.

Stavric, S., Gleeson, T.M., Blanchfield, B., Pivnick, H., 1985. Competitive exclusion of *Salmonella* from newly hatched chicks by mixtures of pure bacterial cultures isolated from fecal and cecal contents of adult birds. J. Food Prot. 48, 778.

Sulakvelidze, A., Alavidze, Z., Morris Jr., J.G., 2001. Bacteriophage therapy. Antimicrob. Agents Chemother. 45, 649–659.

Tabib, Z., Jones, F.T., Hamilton, P.B., 1984. Effect of pelleting of poultry feed on the activity of molds and mold inhibitors. Poult. Sci. 63, 70–75.

Tan, S., Gyles, C.L., Wilkie, B.N., 1997. Evaluation of an *aroA* mutant *Salmonella typhimurium* vaccine in chickens using modified semisolid Rappaport Vassiliadis medium to monitor faecal shedding. Vet. Microbiol. 54, 247–254.

Tellez, G., Dean, C.E., Corrier, D.E., DeLoach, J.R., Jaeger, L., Hargis, B.M., 1993. Effect of dietary lactose on cecal morphology, pH, organic acids, and *Salmonella enteritidis* organ invasion in Leghorn chicks. Poult. Sci. 72, 636–642.

Thompson, J.L., Hinton, M., 1997. Antibacterial activity of formic and propionic acids in the diet of hens on *Salmonellas* in the crop. Br. Poult. Sci. 38, 59–65.

Timms, L.M., Marshall, R.N., Breslin, M.F., 1994. Laboratory and field trial assessment of protection given by a *Salmonella enteritidis* PT4 inactivated, adjuvant vaccine. Br. Vet. J. 150, 93–102.

Toyota-Hanatani, Y., Ekawa, T., Ohta, H., Igimi, S., Hara-Kudo, Y., Sasai, K., Baba, E., 2009a. Public health assessment of *Salmonella enterica* serovar Enteritidis inactivated-vaccine treatment in layer flocks. Appl. Environ. Microbiol. 75, 1005–1010.

Toyota-Hanatani, Y., Kyoumoto, Y., Baba, E., Ekawa, T., Ohta, H., Tani, H., Sasai, K., 2009b. Importance of subunit vaccine antigen of major FliC antigenic site of *Salmonella* Enteritidis II: a challenge trial. Vaccine 27, 1680–1684.

Turnbull, P.C., Snoeyenbos, G.H., 1973. Roles of ammonia, water activity, and pH in salmonellacidal effect of long-used poultry litter. Avian Dis. 17, 72–86.

U.S. Department of Agriculture and Economic Research Service, 2010. Foodborne Illness Cost Calculator. Retrieved January, 2016, from: http://www.foodsafetynews.com/2010/06/foodborne-illness-cost-calculator-updated/#.VqoI5E_EP-k.

Vainio, O., Imhof, B.A., 1995. The immunology and developmental biology of the chicken. Immunol. Today 16, 365–370.

Van Coillie, E., Goris, J., Cleenwerck, I., Grijspeerdt, K., Botteldoorn, N., Van Immerseel, F., De Buck, J., Vancanneyt, M., Swings, J., Herman, L., Heyndrickx, M., 2007. Identification of lactobacilli isolated from the cloaca and vagina of laying hens and characterization for potential use as probiotics to control *Salmonella* Enteritidis. J. Appl. Microbiol. 102, 1095–1106.

van de Giessen, A.W., Ament, A.J., Notermans, S.H., 1994. Intervention strategies for *Salmonella enteritidis* in poultry flocks: a basic approach. Int. J. Food Microbiol. 21, 145–154.

Van Immerseel, F., Atterbury, R.J., 2013. Other approaches to infection control. In: Barrow, P.A., Methner, U. (Eds.), *Salmonella* in Domestic Animals, vol. 2. CAB International, Oxfordshire, pp. 518–535.

Van Immerseel, F., De Buck, J., De Smet, I., Pasmans, F., Haesebrouck, F., Ducatelle, R., 2004a. Interactions of butyric acid- and acetic acid-treated *Salmonella* with chicken primary cecal epithelial cells in vitro. Avian Dis. 48, 384–391.

Van Immerseel, F., De Buck, J., Pasmans, F., Bohez, L., Boyen, F., Haesebrouck, F., Ducatelle, R., 2004b. Intermittent long-term shedding and induction of carrier birds after infection of chickens early posthatch with a low or high dose of *Salmonella* Enteritidis. Poult. Sci. 83, 1911–1916.

Van Immerseel, F., Fievez, V., de Buck, J., Pasmans, F., Martel, A., Haesebrouck, F., Ducatelle, R., 2004c. Microencapsulated short-chain fatty acids in feed modify colonization and invasion early after infection with *Salmonella* enteritidis in young chickens. Poult. Sci. 83, 69–74.

Van Immerseel, F., Boyen, F., Gantois, I., Timbermont, L., Bohez, L., Pasmans, F., Haesebrouck, F., Ducatelle, R., 2005a. Supplementation of coated butyric acid in the feed reduces colonization and shedding of *Salmonella* in poultry. Poult. Sci. 84, 1851–1856.

Van Immerseel, F., Methner, U., Rychlik, I., Nagy, B., Velge, P., Martin, G., Foster, N., Ducatelle, R., Barrow, P.A., 2005b. Vaccination and early protection against non-host-specific *Salmonella* serotypes in poultry: exploitation of innate immunity and microbial activity. Epidemiol. Infect. 133, 959–978.

Van Immerseel, F., Russell, J.B., Flythe, M.D., Gantois, I., Timbermont, L., Pasmans, F., Haesebrouck, F., Ducatelle, R., 2006. The use of organic acids to combat *Salmonella* in poultry: a mechanistic explanation of the efficacy. Avian Pathol. 35, 182–188.

Van Immerseel, F., De Zutter, L., Houf, K., Pasmans, F., Haesebrouck, F., Ducatelle, R., 2009. Strategies to control *Salmonella* in the broiler production chain. World's Poult. Sci. J. 65, 367–391.

Van Immerseel, F., Studholme, D.J., Eeckhaut, V., Heyndrickx, M., Dewulf, J., Dewaele, I., Van Hoorebeke, S., Haesebrouck, F., Van Meirhaeghe, H., Ducatelle, R., Paszkiewicz, K., Titball, R.W., 2013. *Salmonella* Gallinarum field isolates from laying hens are related to the vaccine strain SG9R. Vaccine 31, 4940–4945.

Vandeplas, S., Dubois Dauphin, R., Beckers, Y., Thonart, P., Thewis, A., 2010. *Salmonella* in chicken: current and developing strategies to reduce contamination at farm level. J. Food Prot. 73, 774–785.

Veldman, A., Vahl, H.A., Borggreve, G.J., Fuller, D.C., 1995. A survey of the incidence of *Salmonella* species and *Enterobacteriaceae* in poultry feeds and feed components. Vet. Rec. 136, 169–172.

Wales, A., Breslin, M., Davies, R., 2006. Assessment of cleaning and disinfection in *Salmonella*-contaminated poultry layer houses using qualitative and semi-quantitative culture techniques. Vet. Microbiol. 116, 283–293.

Wales, A.D., Allen, V.M., Davies, R.H., 2010. Chemical treatment of animal feed and water for the control of *Salmonella*. Foodborne Pathog. Dis. 7, 3–15.

Wenzel, J.G., Nusbaum, K.E., 2007. Veterinary expertise in biosecurity and biological risk assessment. J. Am. Vet. Med. Assoc. 230, 1476–1480.

Wigley, P., Hulme, S.D., Bumstead, N., Barrow, P.A., 2002. In vivo and in vitro studies of genetic resistance to systemic salmonellosis in the chicken encoded by the SAL1 locus. Microbes Infect. 4, 1111–1120.

Wigley, P., Hulme, S., Rothwell, L., Bumstead, N., Kaiser, P., Barrow, P., 2006. Macrophages isolated from chickens genetically resistant or susceptible to systemic salmonellosis show magnitudinal and temporal differential expression of cytokines and chemokines following *Salmonella enterica* challenge. Infect. Immun. 74, 1425–1430.

Woodward, M.J., Gettinby, G., Breslin, M.F., Corkish, J.D., Houghton, S., 2002. The efficacy of Salenvac, a *Salmonella enterica* subsp. *enterica* serotype Enteritidis iron-restricted bacterin vaccine, in laying chickens. Avian Pathol. 31, 383–392.

Zhang, L., Yan, Z., Ryser, E.T., 2006. Comparison of the reveal test, the U.S. Food and Drug Administration culture method, and selective media for recovery of *Salmonella* Enteritidis from commercial egg layer flock environments. J. Food Prot. 69, 2766–2769.

CHAPTER

# Current and Future Perspectives on Development of *Salmonella* Vaccine Technologies

14

**Sandra Aehle[1], Roy Curtiss III[2]**

[1]*Elanco Animal Health, Inc., Pacific, MO, United States;* [2]*University of Florida, Gainesville, FL, United States*

## 1. INTRODUCTION

There are over 2500 serotypes of *Salmonella enterica*. Among them, *Salmonella* Enteritidis (SE), *Salmonella* Heidelberg (SH), and *Salmonella* Typhimurium (ST) are commonly associated with chickens and to various extents with other food animals and human infections. Over the past two decades, the Centers for Disease Control and Prevention (CDC) and US Department of Agriculture (USDA) annual surveillance summaries consistently show SE as the most common serotype implicated in human illness in the United States (CDC, 2014a,b; USDA, 2015) and most commonly associated with chickens and eggs (CDC, 2012, 2014a,b; Elson et al., 2005; Kuehn, 2010; Hennessy et al., 2004; Roy et al., 2002; USDA, 2015). SH is also found in most major food animal species, eggs, and retail meat samples and is among the top five most common serotypes associated with human disease (Chittick et al., 2006; Han et al., 2011; Hennessy et al., 2004, 1996).

## 2. CHALLENGES TO *SALMONELLA* CONTROL

Traditionally, antibiotics have been used to treat bacterial infections and for growth promotion in food animals. However, these applications of antibiotics contributed to increasing rates of antibiotic resistance (Bailar and Travers, 2002; CDC, 2013), resulting in contamination of flocks and food products by antibiotic-resistant *Campylobacter*, *Salmonella*, *Enterococcus*, and *Escherichia coli* and thereby increasing risks of difficult-to-treat human infections (Cohen and Tauxe, 1986). Public concerns over the spread of antibiotic resistance in zoonotic bacterial pathogens, which poses a threat to the effectiveness of existing antibiotic therapy in both clinical medical and

**Producing Safe Eggs. http://dx.doi.org/10.1016/B978-0-12-802582-6.00014-8**

veterinary practices (Batz et al., 2005; CDC, 1997; Hoffmann et al., 2012; Schleifer et al., 1984; Williams and Benson, 1978), led the European Union, in 1999, to ban use of most antibiotics for growth promotion to preserve the effectiveness of important human drugs (Casewell et al., 2003). In 2004, the US FDA banned enrofloxacin in food animals on the grounds that its use contributed to fluoroquinolone resistance in human pathogens (FDA, 2005). More recently, FDA and the food animal industry have agreed to cease use of growth-promoting antimicrobials. However, there are concerns that reductions in antibiotic use in animal production may lead to an increase in food-borne pathogens and incidence of other bacterial infections such as *Clostridium perfringens*, the etiologic cause of necrotic enteritis (Mot et al., 2014).

## 3. *SALMONELLA* CONTROL

It is necessary to develop other effective ways to control *Salmonella* in poultry flocks besides relying entirely on management practices to reduce *Salmonella* infections. The most widely used control measures against *Salmonella* are preventive biosecurity and vaccination with live and inactivated *Salmonella* vaccines. Vaccines are a key component in preharvest *Salmonella* control programs. Since contaminated food and especially poultry products are the major source of human *Salmonella* infection, vaccination of chickens is an important strategy to reduce the levels of *Salmonella* in poultry flocks, which should ultimately lower the risk of *Salmonella* zoonosis. The European Food Safety Authority baseline survey into the prevalence of *Salmonella* in large-scale laying hen holdings reported that vaccination reduces the probability of SE infection by 88% (EFSA, 2007).

## 4. VACCINATION TO REDUCE *SALMONELLA* ENTERITIDIS INFECTION IN LAYERS

The ultimate goal of vaccination against a disease is to develop protective immunity so that when the host encounters the pathogen, it is "remembered" by the immune system and the infection is either prevented or the disease is significantly reduced. Live attenuated and inactivated whole-cell or subunit vaccines against *Salmonella* spp. have been used to reduce *Salmonella* prevalence in chickens, which in turn reduces the risk for food-borne disease transmission (Berghaus et al., 2011; Khan et al., 2003; Van Immerseel et al., 2005). The cellular immune response, induced by live vaccines enhances phagocytes and T cells, is necessary for control and clearance of *Salmonella* from blood and tissues. Inactivated vaccines, although able to boost the phagocyte response by way of production of antibodies, do not stimulate T-cell immunity and thus is unable to prevent bacterial growth in infected organs or bacteremia and are unable to control the spread of *Salmonella* (Babu et al., 2003, 2004; Coward et al., 2014). However, the long-term immunity stimulated by vaccination with both live and inactivated *Salmonella* vaccines increases resistance against infection by reducing the duration, severity, and shedding of *Salmonella* (Berghaus et al., 2011; Dórea et al., 2010; FDA, 2009; UEP, 2016). Live and inactivated *Salmonella* vaccines are therefore an integral part of a *Salmonella* control program when used in conjunction with good sanitation, biosecurity, and management practices.

## 5. LIVE ATTENUATED *SALMONELLA* VACCINES

Live attenuated *Salmonella* vaccines are a weakened form of the bacteria that stimulate an intensified immune response in poultry that reduces or eliminates the pathogen. Two commonly used methods to attenuate ST or SE bacteria involve either chemical mutagenesis or modification of the bacterial genome by classical or recombinant methods. Live attenuated vaccines have the advantage of ease of mass application to young chicks and pullets through coarse spray or in drinking water, delivering a bolus of antigens to the host, and can induce both antibody and cell-mediated immune responses (Atterbury et al., 2010; Babu et al., 2003, 2004; Holt et al., 2003; Linde et al., 1997). These live vaccines have used diverse attenuation strategies for strains of ST (Curtiss and Kelly, 1987; Hassan and Curtiss, 1994; Dórea et al., 2010; Gantois et al., 2006; Kelly et al., 1992; Linde et al., 1997) and SE (Cerquetti and Gherardi, 2000; Cooper et al., 1994; De Cort et al., 2015a; Gantois et al., 2006; Linde et al., 1997; Methner et al., 2011; Nandre et al., 2012; Springer et al., 2011). Although chicks are immunologically immature, administering attenuated *Salmonella* bacteria orally to newly hatched chickens results in extensive colonization of the intestinal tract and persistence in internal effector lymphoid tissues (Cooper et al., 1994; Hassan and Curtiss, 1990, 1994, 1996; Van Immerseel et al., 2002). The application of live attenuated *Salmonella* vaccines provides the 1-day-old chick with direct resistance against colonization by *Salmonella* and prevents establishment of a persistent infection. Table 14.1 shows a list of the currently available live *Salmonella* vaccines approved for commercial use in poultry.

Methner et al. (1999) demonstrated the advantage of administration of a *Salmonella* vaccine and a competitive exclusion (CE) culture used in combination to produce a better protective effect than either single prophylactic application alone. The best protective effect was observed when the *Salmonella* vaccine strain was administered before or simultaneously with the CE culture because the combination ensured an adequate persistence of the vaccine strain as a prerequisite for the expression of the colonization inhibition effect of the CE product and induction of a strong immune response. This CE effect was also demonstrated by Barrow et al. (1987) when *Salmonella* strains were observed to competitively exclude each other in the intestinal tract of chicks, suggesting that live *Salmonella* vaccines colonizing attachment sites in the intestinal tract may preclude later colonization by wild-type *Salmonella* sp. Van Immerseel et al. (2002) also demonstrated that a rapid onset of resistance to intestinal colonization by other *Salmonella* strains was induced when newly hatched chicks were inoculated with an attenuated live *Salmonella* vaccine. The cecal lamina propria was shown to infiltrate with heterophils as early as 12h

**Table 14.1** Registered Live Attenuated *Salmonella* Vaccines for Commercial Use in Poultry

| Product | Company | Bacteria | Claims |
|---|---|---|---|
| Cevac SG9R<br>Chickens | Ceva[a] | *Salmonella* Gallinarum | For the active immunization of healthy chickens from 4 weeks of age against infections due to *Salmonella* Gallinarum, *S.* Pullorum |
| AviPro Megan Vac 1<br>For young growing chickens | Elanco Animal Health[b] | *Salmonella* Typhimurium Δ*cya* Δ*crp* (Registered in the United States, Canada, South Africa, New Zealand, and Dominican Republic) | For the reduction of *S.* Heidelberg<br>*S.* Enteritidis<br>*S.* Typhimurium infections |
| AviPro Megan Egg<br>For pullets, layers, breeders, and turkeys | Elanco Animal Health | *Salmonella* Typhimurium Δ*cya* Δ*crp* (Registered in the United States, Canada) | For the reduction of *S.* Enteritidis infection of organs, oviduct, and ovaries in pullet chickens and *S.* Typhimurium infection of turkeys |
| AviPro SalVac T<br>Layers and breeders | Elanco Animal Health | *Salmonella* Typhimurium<br>Nal/Rif/Rtt metabolic drift mutant | Active immunization against *Salmonella* Typhimurium infections |
| AviPro SalVac E<br>Layers and breeders | Elanco Animal Health | *Salmonella* Enteritidis<br>Nal/Rif/Rtt metabolic drift mutant | Active immunization against *Salmonella* Enteritidis infections |
| AviPro *Salmonella* Duo<br>Layers and breeders | Elanco Animal Health | *Salmonella* Typhimurium and *S.* Enteritidis<br>Nal/Rif/Rtt metabolic drift mutant | Active immunization against *Salmonella* Typhimurium and *S.* Enteritidis infections |
| Salmovac SE<br>Chickens, layers, and breeders | IDT Biologika Animal Health[c] | *Salmonella* Enteritidis<br>Adenine and histidine auxotroph | Active immunization against *Salmonella* Typhimurium and *S.* Enteritidis infections |
| – | Merck Animal Health[d] | – | – |
| Gallivac SE<br>Chickens, layers, breeders | Merial[e] | *Salmonella* Enteritidis<br>Adenine and histidine auxotroph | For the reduction of colonization, persistence, and invasion of the intestinal tract by *Salmonella* Enteritidis and *S.* Typhimurium |
| Poulvac ST<br>For young growing chickens | Zoetis[f] | *Salmonella* Typhimurium Δ*aroA* *serC* | Reduction of *S.* Heidelberg, *S.* Enteritidis, *S.* Typhimurium infections |

[a] *http://www.ceva.us/.*
[b] *https://www.elanco.us/products-services/poultry/vaccines/.*
[c] *https://www.idt-animal-health.com/veterinarian/poultry/products-poultry/zoosaloral-h/.*
[d] *http://www.merck-animal-health-usa.com/species/poultry/index.aspx.*
[e] *http://www.merial.co.uk/Avian/Pages/Layer.aspx.*
[f] *https://www.zoetisus.com/species/poultry_new.aspx.*

postvaccination. Infiltration of macrophages and T-lymphocytes followed from 20 h and B-lymphocytes from 24 h postvaccination. When challenged with a wild-type ST strain, vaccinated animals had a lower number of the wild-type ST bacteria in their organs and cecal contents in the first days after challenge compared with nonvaccinated animals. Van Immerseel et al. (2005) suggested that the early protective effect of vaccination may be a result of interactions with the host, either by competition for intestinal adhesion sites or through inducing innate immunity or both. This colonization inhibition effect provided by early application of a live *Salmonella* vaccine was observed by McReynolds et al. (2007) when chicks were protected as early as 48 h posthatch from colonization by wild-type *Salmonella*. Additional information on the benefits of CE products provided through direct-fed microbial products to reduce *Salmonella* colonization is available in the detailed review by Lee et al. (2010).

## 6. INACTIVATED *SALMONELLA* VACCINES

Live *Salmonella* vaccines generally confer better protection than inactivated ones by providing local, cell-mediated, and humoral immunities, whereas inactivated *Salmonella* vaccines stimulate long-term specific immunity and extend protection through the egg-laying period. Inactivated *Salmonella* whole-cell vaccines have the advantage of being prepared rapidly from specific serotypes identified on the premises as prevalently attributing to product contamination. Inactivated vaccine preparations have been shown to provide strong protection to hens and progeny against targeted species of *Salmonella* (Bailey et al., 2007; Gast et al., 1992; Paiva et al., 2009; Tran et al., 2010). Inactivation of *Salmonella* spp. is achieved by either formalin or acetone treatment, by heating or by electron-beam irradiation (Jesudhasan et al., 2015). An adjuvant is an ingredient of the inactivated vaccine product that enhances the immune response to the bacterial antigen. Water-in-oil and water-in-oil-in-water emulsions are commonly used in the preparation of bacterin poultry vaccines (Vazquez, 2014). Table 14.2 shows a list of inactivated ND-IB-*Salmonella* combination, mono-, and polyvalent, vaccines approved for use in commercial poultry.

Because administering injectable vaccines to pullets is labor intensive, using polyvalent vaccines is cost-effective and reduces multiple handling of birds. Vaccine manufacturers often combine other antigens, such as Newcastle disease virus and infectious bronchitis virus, with *Salmonella* Enteritidis whole-cell bacterin in one vaccine presentation to provide a wider spectrum of protection (see Table 14.2). Pavic et al. (2010) showed that a trivalent autogenous vaccine [serogroups B (Typhimurium), C (Mbandaka) and E (Orion)] significantly reduced serogroups B and C cecal colonization after challenge infections, induced higher serological and maternal antibody titers, and protected progeny of vaccinated hens from cecal colonization after a challenge with wild-type ST. Inactivated commercial SE vaccines decreased fecal shedding of SE in molted hens (Nakamura et al., 2004).

Although there are currently no subunit *Salmonella* vaccines commercially available in the United States, some studies have shown promise in their use to reduce SE

**Table 14.2** Registered Inactivated Mono- and Polyvalent Vaccines for Commercial Use in Breeders and Layers

| Product | Company | Bacterial Strain(s) | Viral Strain(s) |
|---|---|---|---|
| Layermune SE | Ceva[a] | *Salmonella* Enteritidis | – |
| Layermune 3 | Ceva | *Salmonella* Enteritidis | Newcastle disease virus (LaSota), infectious bronchitis (Mass, Mass Holland) |
| Cevac Salmune TEK | Ceva | *Salmonella* Typhimurium, *S.* Enteritidis, *S.* Kentucky | – |
| Cevac Corymune 4 | Ceva | *Avibacterium paragallinarum* serotypes A, B, and C, and *Salmonella* Enteritidis | – |
| Cevac Corymune 7 | Ceva | *Avibacterium paragallinarum* serotypes A, B, and C, and *Salmonella* Enteritidis | Newcastle disease virus (LaSota), infectious bronchitis (Mass), and B8/78 strain of the egg drop syndrome (EDS) virus |
| AviPro 109 SE4 | Elanco Animal Health[b] | *Salmonella* Enteritidis | – |
| AviPro 329 ND-IB2-SE4 | Elanco Animal Health | *Salmonella* Enteritidis | Newcastle disease virus (LaSota), infectious bronchitis (Mass DG, Ark) |
| Avisan Secure Layers and breeders | Hipra[c] | *Salmonella* Enteritidis *S. Typhimurium* | – |
| Bron-Newcavac-SE | Merck Animal Health[d] | *Salmonella* Enteritidis | Newcastle disease virus (LaSota), infectious bronchitis (Mass) |
| Garasol | Merck Animal Health | *Escherichia coli, Salmonella* Typhimurium, *Pseudomonas aeruginosa* | – |
| Nobilis Salenvac T | Merck Animal Health | *Salmonella* Typhimurium Iron-regulated proteins subunit vaccine | – |
| Nobilis Salenvac E | Merck Animal Health | *Salmonella* Enteritidis | – |
| SE Guard | Merck Animal Health | *Salmonella* Enteritidis | – |
| Gallimune SE+ST | Merial[e] | *Salmonella* Enteritidis *Salmonella* Typhimurium | – |
| Poulvac SE | Zoetis[f] | *Salmonella* Enteritidis | – |
| Poulvac SE-ND-IB | Zoetis | *Salmonella* Enteritidis | Newcastle disease virus (LaSota), infectious bronchitis (Mass) |

[a]*http://www.ceva.us/.*
[b]*https://www.elanco.us/products-services/poultry/vaccines/.*
[c]*https://www.hipra.com/.*
[d]*http://www.merck-animal-health-usa.com/species/poultry/index.aspx.*
[e]*http://www.merial.co.uk/Avian/Pages/Layer.aspx.*
[f]*https://www.zoetisus.com/species/poultry_new.aspx.*

infections in layers. De Buck et al. (2005) demonstrated that immunization of layers with a subunit vaccine comprising type 1 fimbriae from SE reduced colonization of the reproductive organs and reduced the number of contaminated eggs. An SE FliC polypeptide (SEp 9) was found to be as efficacious in reducing a wild-type SE challenge strain in the ceca as a commercial whole-cell inactivated vaccine (Toyota-Hanatani et al., 2009). Khan et al. (2003) reported that SE outer membrane proteins 75.6 and 82.3 kDa were effective in reducing colonization of SE on intestinal mucosa and ceca in chickens. Whether used alone or in combination with other antigens, inactivated and subunit vaccines have been shown to provide significant protection to egg-laying hens against *Salmonella* and poultry diseases of economic importance.

## 7. *SALMONELLA* VACCINATION PROGRAMS

Immunization with commercial live and inactivated *Salmonella* vaccines is widely used in the United States and has been shown to reduce *Salmonella* infections in broiler breeders (Berghaus et al., 2011; Dórea et al., 2010) and in molted hens (Holt et al., 2003; Jesudhasan et al., 2015; Nakamura et al., 2004). When used as a component in a comprehensive *Salmonella* control program, attenuated live and inactivated vaccines stimulate optimal immunity and provide a first-line of defense that directly increases the level of resistance to *Salmonella* infections in layers. Vaccination schemes using a combination of live and inactivated *Salmonella* vaccines have been shown to be effective in reducing egg contamination and shedding by SE and wild-type *Salmonella* spp. in poultry and the environment (Arnold et al., 2014; Atterbury et al., 2009; Berghaus et al., 2011; Davies and Breslin, 2003; Feberwee et al., 2001; Young et al., 2007). By reducing the amount of *Salmonella* in layers and breeders through vaccination with live and inactivated vaccines and implementing risk reduction practices, the amount of *Salmonella* entering the food supply is also reduced.

## 8. RECENT IMPROVEMENTS IN ENHANCING SAFETY, IMMUNOGENICITY, AND EFFICACY OF LIVE *SALMONELLA* VACCINES TO CONTROL *SALMONELLA* INFECTIONS IN POULTRY

**Choice of strains to attenuate.** The *Salmonella* strain selected to be the parent of live vaccine strains needs to be highly invasive for the target animal host to effectively colonize internal lymphoid tissues to high levels and persist for at least 2 weeks to be able to induce maximal mucosal, systemic, and cellular immunities (Curtiss et al., 2010; McSorley, 2014). In general, the strain will have the lowest $LD_{50}$ (median lethal dose; 50%) of all tested strains of a given serotype or species and attenuated derivatives of the selected strain should be able to induce protective immunity against challenge with all available strains of the same serotype or species. In terms of *Salmonella* it is critically important for an attenuated vaccine to confer protective immunity against all (or at least most) serotypes infecting poultry. Years ago, our group (Curtiss et al., 1991) chose ST UK-1 as the parent of vaccine strains. At the time, this strain had maximal ability to cause lethal infections in horses, cattle, mice, and chickens with an $LD_{50}$ of 3000 colony forming units (CFU) for day-of-hatch chicks (Curtiss et al., 1991; Hassan and Curtiss, 1990). The strain was much more virulent than any strain of SE tested as well as all other tested strains of ST. It was subsequently shown that isogenic attenuated vaccine derivatives of UK-1 could induce protection against all other wild-type strains of ST, whereas the derivatives of SL1344, while inducing protective immunity against most ST strains, could not induce complete protection against wild-type UK-1 (Zhang et al., 1997). Genome sequencing demonstrated that UK-1 possessed additional virulence genes not present in other sequenced ST strains (Luo et al., 2011). In comparative studies of ST strains for the ability to destroy tumors, the UK-1 derivative was far more effective than derivatives of the widely used ATCC 14028 ST strain (Felgner et al., 2016).

## 9. ACHIEVING THE BALANCE BETWEEN ATTENUATION AND IMMUNOGENICITY

Traditional means of attenuating bacteria have invariably reduced immunogenicity by reducing the ability of the vaccine to access and persist in lymphoid tissues. This is true by using the continuous passage means to select attenuated strains as worked out by Louis Pasteur over 150 years ago (Schwartz, 2009) or the more recent approach used for generating live vaccines by introducing specific mutations and especially multiple deletion mutations unable to revert (Cerquetti and Gherardi, 2000; Cooper et al., 1994; Curtiss and Kelly, 1987; De Cort et al., 2015a; Dórea et al., 2010; Gantois et al., 2006; Hassan and Curtiss, 1994; Kelly et al., 1992; Linde et al., 1997; Methner et al., 2011; Nandre et al., 2012; Springer et al., 2011). To address problems of reduced immunogenicity associated with attenuation, improved live *Salmonella* vaccines have been designed to display a regulated delayed attenuation in vivo phenotype (Curtiss et al., 2007, 2009, 2010). This approach enables the vaccine strain to display all the attributes of the wild-type invasive parent at the time of administration to animals to withstand host-mediated defense mechanisms and effectively invade and colonize effector internal lymphoid tissues before displaying the attenuated phenotype to be unable to cause harm or reduce growth performance. This has been achieved by using three different strategies.

The first strategy is to turn off the ability to synthesize lipopolysaccharide (LPS) O-antigen (and/or parts of the LPS core) in vivo, which can be achieved using mutants unable to synthesize phosphomannose isomerase that synthesize LPS O-antigen when grown in the presence of mannose (Collins et al., 1991) and fail to do so in vivo since mannose is not present (Curtiss et al., 2007). Such vaccine strains gradually lose O-antigen as a function of cell division in the immunized animal host and become susceptible to complement-mediated cytotoxicity and are also more readily phagocytized and killed by macrophages. This regulated delayed attenuation phenotype is also achievable by use of *galE* mutants (Clarke and Gyles, 1986; Germanier and Furer, 1971), which cease to synthesize LPS O-antigen and the LPS outer core in vivo due to the absence of galactose.

The second means of achieving regulated delayed attenuation is to replace promoters for genes needed to display virulence with regulatable promoters that will be turned on for vaccine strain growth and initial colonization of the gastrointestinal tract and lymphoid tissues and then be turned off in vivo to preclude disease symptoms (Curtiss et al., 2009). In these cases, the promoters for the virulence genes are replaced with sugar-regulated promoters that are only expressed when strains are grown in the presence of sugars such as arabinose or rhamnose. Since these sugars are not present in vivo, the products of these virulence genes decrease by at least half at every vaccine cell division in vivo (Curtiss et al., 2009; Li et al., 2009; Shi et al., 2010).

The third means of attenuation employs vaccine strains exhibiting regulated delayed lysis in vivo as the ultimate means of attenuation with complete biological containment and no shedding of survivors (Kong et al., 2008). This strategy uses arabinose-regulated synthesis of muramic acid and diaminopimelic acid, two unique essential constituents of the rigid peptidoglycan layer of the cell wall. Vaccine strains exhibiting such regulated delayed lysis in vivo have been shown to induce better immune responses and higher levels of protective immunity to challenge than similar vaccine strains not undergoing regulated delayed lysis (Ameiss et al., 2010; Ashraf et al., 2011; Jiang et al., 2015; Juárez-Rodríguez et al., 2012).

## 10. INDUCTION OF CROSS-PROTECTIVE IMMUNITY AGAINST DIVERSE *SALMONELLA* SEROTYPES AND CLOSELY RELATED ENTERIC BACTERIA

There are some 2500 serotypes of *Salmonella enterica* and, insofar as is known, all have the same LPS core structure (Jansson et al., 1981; Lüderitz et al., 1971; Rick, 1987) such that antibodies to the LPS core of either ST or SE react with the LPS core of all other serotypes. Only *Salmonella* Arizonae has a slightly different LPS core chemistry (Olsthoorn et al., 1998). Thus attenuated *Salmonella* vaccine strains with an inability to synthesize LPS O-antigen due to a mutation of the *pmi* gene encoding phosphomannose isomerase (Markovitz et al., 1967) cease to synthesize the O-antigen in vivo since free mannose is unavailable (Curtiss et al., 2007). Similarly, attenuated *Salmonella* vaccines with *galE* mutations also fail to synthesize LPS O-antigen and the outer part of the LPS core (Raetz, 1996) in vivo due to the unavailability of free galactose. In this regard, synthesis of GDP-mannose and UDP-galactose, which cease to be synthesized in vivo due to the *pmi* and *galE* mutations, respectively, are required precursors for LPS synthesis. Growing such vaccine strains with mannose and galactose yields strains with complete LPS that are maximally invasive to colonize internal effector lymphoid tissues, but gradually lose LPS components in vivo to expose the LPS core and surface outer membrane proteins to surveillance by the immune system. These strains become sensitive to complement-mediated toxicity (Reeves, 1995; Rowley, 1968) and increased phagocytosis (Fields et al., 1986), which is the basis for their attenuation. Thus vaccine strains with *pmi* and *galE* mutations induce antibodies directed against the LPS core and thus confer some level of cross-protective immunity to other *Salmonella* serotypes (Curtiss et al., 2007).

It has been learned from multiple studies that iron-regulated outer membrane proteins (IROMPs) can induce protective immune responses against various enteric bacterial pathogens (Bolin and Jensen, 1987). Thus a bacterin derived from growing virulent *Salmonella* strains under iron starvation conditions that leads to overproduction of IROMPs, induces better protective immunity to heterologous *Salmonella* serotypes than a bacterin derived from *Salmonella* grown with iron excess (Clifton-Hadley et al., 2002). Unfortunately, this bacterin is licensed in Europe and is not available in the United States (see Nobilis Salenvac T in Table 14.2). In terms of live vaccines, the regulated delayed cessation in the synthesis of the Fur protein (Curtiss et al., 2007), which controls expression of multiple genes for iron acquisition including those for IROMPs (Earhart, 1996), causes an upregulation of IROMP synthesis in vivo to induce protective immunity (Curtiss et al., 2009). In fact, the antibodies induced react with IROMPs from all *Salmonella* serotypes tested (Curtiss et al., 2007). It would thus be surmised that live *Salmonella* vaccines that included regulated delayed attenuation features that would expose the common LPS core and cause synthesis of IROMPs would induce better cross-protective immunity against diverse *Salmonella* serotypes than would a traditionally attenuated live *Salmonella* vaccine.

## 11. SIMULTANEOUS CONTROL OF *SALMONELLA* INFECTION AND PERSISTENCE BY CONTROL OF OTHER AVIAN COLONIZERS AND PATHOGENS

Attenuated *Salmonella* strains have been widely used experimentally as vectors for delivery of protective antigens specified by DNA sequences encoding such antigens derived from other pathogens (Galen and Curtiss, 2014). Recombinant attenuated *Salmonella* vaccines (RASVs) make use of plasmid DNA vectors complementing a *Salmonella* chromosomal mutation so as to avoid using antibiotic-resistance genes. The uses of plasmid–host combinations that establish a balanced lethal situation are ideal in that loss of the plasmid vector leads to death of the RASV. Mutations that confer a requirement for a peptidoglycan-specific amino acid such as diaminopimelic acid (Nakayama et al., 1988) or D-alanine (Xin et al., 2012) have been particularly useful. In terms of reducing *Salmonella* persistence in poultry, the development and use of RASVs delivering *C. perfringens* antigens to reduce necrotic enteritis (Kulkarni et al., 2008; Jiang et al., 2014, 2015; Zekarias et al., 2008), avian pathogenic *E. coli* antigens to reduce colisepticemia (Roland et al., 1999, 2004), *Campylobacter jejuni* antigens to reduce intestinal colonization (Buckley et al., 2010; Laniewski et al., 2014; Layton et al., 2011; Theoret et al., 2012), and *Eimeria* antigens to reduce coccidiosis (Konjufca et al., 2006, 2008) would all be beneficial.

## 12. IMPROVED VACCINATION PRACTICES FOR ULTIMATE CONTROL OF PERSISTENCE OF *SALMONELLA* AND ENTERIC PATHOGENS IN POULTRY TO ENHANCE POULTRY HEALTH, REDUCE ANTIBIOTIC USAGE, AND ENSURE FOOD SAFETY

Live attenuated *Salmonella* vaccines as well as RASVs designed to protect against other bacteria and parasite infections of poultry can be administered by coarse spray to newly hatched chicks in the hatchery (Atterbury et al., 2010; De Cort et al., 2015b; Kelly-Aehle, 2004) with second or third immunizations administered in the drinking water or by coarse spray from 10 days of age or older. Since mixtures of two RASVs each delivering a different protective antigen have been found to induce protective immunity equivalent to that induced by immunization with a single RASV delivering both antigens (Jiang et al., 2015; Xin et al., 2012), it can be surmised that mixtures of live vaccines will be efficacious in controlling infections and thereby reduce use of antibiotics. Booster immunizations with RASVs 2–4 weeks before administering inactivated bacterin vaccines at 12–17 weeks of age will be appropriate for pullets to maintain the level of protective immunity through the lay period, and administering RASVs by coarse spray or drinking water 1 or 2 weeks before molting (Holt et al., 2003) will be important to boost immunity during a second period of egg production. Since vaccination of breeding hens leads to transfer of maternal immunity to chicks and since such maternal immunity appears to enhance success of immunizing chicks with the same vaccine (Hassan and Curtiss, 1996), it would be logical to immunize breeders, as well as chicks and pullets destined to be egg layers, as an important intervention to reduce *Salmonella* prevalence vertically from top-down and to reduce colonization in progeny, and most importantly, transmission through the food chain to humans.

## 13. SUMMARY

Live attenuated and killed bacterin vaccines, used alone but often in combination to lessen presence of important *Salmonella enterica* serotypes in egg-producing layer hens, are currently available. These vaccines are safe and have significantly reduced the presence of *Salmonella* in and on eggs and have thus enhanced food safety. New improved live vaccines are being developed that induce protective immunity to other pathogens that cause economic losses to poultry producers in addition to inducing immunity against *Salmonella* carriage and persistence. It can be expected that more widespread use of existing and newly developed vaccines for breeders as well as for production layer hens will significantly reduce problems associated with *Salmonella* and other bacterial pathogens. Increased reliance on using such vaccines should also reduce use of antibiotics during poultry production and thus decrease selection pressure for proliferation of antibiotic-resistant commensal and pathogenic bacterial strains.

## REFERENCES


Ameiss, K., Ashraf, S., Kong, W., Pekosz, A., Wu, W.H., Milich, D., Billaud, J.N., Curtiss III, R., 2010. Delivery of woodchuck hepatitis virus-like particle presented influenza M2e by recombinant attenuated *Salmonella* displaying a delayed lysis phenotype. Vaccine 28, 6704–6713.

Arnold, M.E., Gosling, R.J., La Ragione, R.M., Davies, R.H., Martelli, F., 2014. Estimation of the impact of vaccination on faecal shedding and organ and egg contamination for *Salmonella* Enteritidis, *Salmonella* Typhimurium and monophasic *Salmonella* Typhimurium. Avian Pathol. 43, 155–163.

Ashraf, S., Kong, W., Wang, S., Yang, J., Curtiss III, R., 2011. Protective cellular responses elicited by vaccination with influenza nucleoprotein delivered by a live recombinant attenuated *Salmonella* vaccine. Vaccine 29, 3990–4002.

Atterbury, R.J., Carrique-Mas, J.J., Davies, R.H., Allen, R.M., 2009. *Salmonella* colonization of laying hens following vaccination with killed and live attenuated commercial *Salmonella* vaccines. Vet. Rec. 165, 493–496.

Atterbury, R.J., Davies, R.H., Carrique-Mas, J.J., Morris, V., Harrison, D., Tucker, V., Allen, V.M., July 31, 2010. Effect of delivery method on the efficacy of *Salmonella* vaccination in chickens. Vet. Rec. 167 (5), 161–164. http://dx.doi.org/10.1136/vr.b4884.

Babu, U., Dalloul, R.A., Okamura, M., Lillehoj, H.S., Xie, H., Raybourne, R.B., Gaines, D., Heckert, R.H., 2004. *Salmonella enteritidis* clearance and immune responses in chickens following *Salmonella* vaccination and challenge. Vet. Immunol. Immunopathol. 101, 251–257.

Babu, U., Scott, M., Myers, M.J., Okamura, M., Gaines, D., Yancy, H.F., Lillehoj, H., Heckert, R.H., Raybourne, R.B., 2003. Effects of live attenuated and killed *Salmonella* vaccine on T-lymphocyte mediated immunity in laying hens. Vet. Immunol. Immunopathol. 91, 39–44.

Bailar 3rd, J.C., Travers, K., 2002. Review of assessments of the human health risk associated with the use of antimicrobial agents in agriculture. Clin. Infect. Dis. 34 (Suppl. 3), S135–S143.

Bailey, J.S., Rolon, A., Hofacre, C.L., Holt, P.S., Wilson, J.L., Cosby, D.E., Richardson, L.J., Cox, N.A., 2007. Intestinal humoral immune response and resistance of *Salmonella* challenge of progeny from breeders vaccinated with killed antigen. Int. J. Poult. Sci. 6, 417–423.

Barrow, P.A., Tucker, J.F., Simpson, J.M., 1987. Inhibition of colonization of the chicken alimentary tract with *Salmonella typhimurium* by gram-negative facultatively anaerobic bacteria. Epidemiol. Infect. 98, 311–322.

Batz, M.B., Doyle, M.P., Morris Jr., G., Painter, J., Singh, R., Tauxe, R.V., Taylor, M.R., Lo Fo Wong, D.M., Food Attribution Working Group, 2005. Attributing illness to food. Emerg. Infect. Dis. 11, 993–999.

Berghaus, R.D., Thayer, S.G., Maurer, J.J., Hofacre, C.L., 2011. Effect of vaccinating breeder chickens with a killed *Salmonella* vaccine on *Salmonella* prevalences and loads in breeder and broiler chicken flocks. J. Food Prot. 74, 727–734.

Bolin, C.A., Jensen, A.E., 1987. Passive immunization with antibodies against iron-regulated outer membrane proteins protects turkeys from *Escherichia coli* septicemia. Infect. Immun. 55 (5), 1239–1242.

Buckley, A.M., Wang, J., Hudson, D.L., Grant, A.J., Jones, M.A., Maskell, D.J., Stevens, M.P., 2010. Evaluation of live-attenuated *Salmonella* vaccines expressing *Campylobacter* antigens for control of *C. jejuni* in poultry. Vaccine 28 (4), 1094–1105.

Casewell, M., Friis, C., Marco, E., McMullin, P., Phillips, I., 2003. The European ban on growth-promoting antibiotics and emerging consequences for human and animal health. J. Antimicrob. Chemother. 52, 159–161.

CDC, 2013. Centers for Disease Control and Prevention. Antibiotic Resistance Threats in the United States, 2013. http://www.cdc.gov/drugresistance/threat-report-2013/pdf/ar-threats-2013-508.pdf.

CDC, 2012. Foodborne Diseases Active Surveillance Network (FoodNet): FoodNet Surveillance Report for 2011 (Final Report) U.S. Department of Health and Human Services, Centers for Disease Control, Atlanta, Georgia. http://www.cdc.gov/foodnet/PDFs/2012_annual_report_508c.pdf.

CDC, 2014a. National *Salmonella* Surveillance Annual Report, 2012. U.S. Department of Health and Human Services, Centers for Disease Control, Atlanta, Georgia. http://www.cdc.gov/ncezid/dfwed/pdfs/salmonella-annual-report-2012-508c.pdf.

CDC, 1997. *Salmonella* serotype Montevideo infections associated with chicks–Idaho, Washington, and Oregon, spring 1995 and 1996. MMWR 46, 237–239.

CDC, 2014b. Surveillance for Foodborne Disease Outbreaks, United States, 2012. U.S. Department of Health and Human Services, CDC, Atlanta, Georgia. http://www.cdc.gov/ncezid/dfwed/pdfs/salmonella-annual-report-2012-508c.pdf.

Cerquetti, M.C., Gherardi, M.M., 2000. Vaccination of chickens with a temperature-sensitive mutant of *Salmonella enteritidis*. Vaccine 18, 1140–1145.

Chittick, P., Sulka, A., Tauxe, R.V., Fry, A.M., 2006. A summary of national reports of foodborne outbreaks of *Salmonella* Heidelberg infections in the United States: clues for disease prevention. J. Food Prot. 69 (5), 1150–1153.

Clarke, R.C., Gyles, C.L., 1986. Galactose epimeraseless mutants of *Salmonella typhimurium* as live vaccines for calves. Can. J. Vet. Res. 50 (2), 165–173.

Clifton-Hadley, F.A., Breslin, M., Venables, L.M., Sprigings, K.A., Cooles, S.W., Houghton, S., Woodward, M.J., 2002. A laboratory study of an inactivated bivalent iron restricted *Salmonella enterica* serovars Enteritidis and Typhimurium dual vaccine against Typhimurium challenge in chickens. Vet. Microbiol. 89 (2–3), 167–179.

Cohen, M.L., Tauxe, R.V., 1986. Drug-resistant *Salmonella* in the United States: an epidemiologic perspective. Science 234, 964–969.

Collins, L.V., Attridge, S., Hackett, J., 1991. Mutations at *rfc* or *pmi* attenuate *Salmonella typhimurium* virulence for mice. Infect. Immun. 59, 1079–1085.

Cooper, G.L., Venables, L.M., Woodward, M.J., Hormaeche, C.E., 1994. Vaccination of chickens with strain CVL30, a genetically defined *Salmonella enteritidis aroA* live oral vaccine candidate. Infect. Immun. 62, 4747–4754.

Coward, C., Restif, O., Dybowski, R., Grant, A.J., Maskell, D.J., Mastroeni, P., 2014. The effects of vaccination and immunity on bacterial infection dynamics in vivo. PLoS Pathog. 10 (9), e1004359. http://dx.doi.org/10.1371/journal.ppat.1004359.

Curtiss III, R., Kelly, S.M., 1987. *Salmonella typhimurium* deletion mutants lacking adenylate cyclase and cyclic AMP receptor protein are avirulent and immunogenic. Infect. Immun. 55, 3035–3043.

Curtiss III, R., Porter, S.B., Munson, M., Tinge, S.A., Hassan, J.O., Gentry-Weeks, C., Kelly, S.M., 1991. Nonrecombinant and Recombinant Avirulent Salmonella Live Vaccines for Poultry. In: Blankenship, L.C. (Eds.), Colonization control of human bacterial enteropathogens in poultry. Academic Press, Inc., San Diego, CA, pp. 169–198.

Curtiss III, R., Wanda, S.-Y., Gunn, B.M., Zhang, X., Tinge, S.A., Ananthnarayan, V., Mo, H., Wang, S., Kong, W., 2009. *Salmonella enterica* serovar Typhimurium strains with regulated delayed attenuation in vivo. Infect. Immun. 77, 1071–1082.

Curtiss III, R., Xin, W., Li, Y., Kong, W., Wanda, S.-Y., Gunn, B., Wang, S., 2010. New technologies in using recombinant attenuated *Salmonella* vaccine vectors. Crit. Rev. Immunol. 30 (3), 255–270.

Curtiss III, R., Zhang, X., Wanda, S.-Y., Kang, H.Y., Konjufca, V., Li, Y., Gunn, B., Wang, S., Scarpellini, G., Lee, I.S., 2007. Induction of host immune responses using *Salmonella*-vectored vaccines. In: Brogden, K.A., Minion, F.C., Cornick, N., Stanton, T.B., Zhang, Q., Nolan, L.K., Wannemuehler, M.J. (Eds.), Virulence Mechanisms of Bacterial Pathogens, fourth ed. ASM Press, Washington, D.C., pp. 297–313.

Davies, R., Breslin, M., 2003. Effects of vaccination and other preventive methods for *Salmonella* Enteritidis on commercial laying chicken farms. Vet. Rec. 153, 673–677.

De Buck, J., Van Immerseel, F., Haesebrocuk, F., Ducatelle, R., 2005. Protection of laying hens against *Salmonella* Enteritidis by immunization with type 1 fimbriae. Vet. Microbiol. 105, 93–101.

De Cort, W., Mot, D., Haesebrouck, F., Ducatelle, R., Van Immerseel, F., 2015a. A colonization-inhibition culture consisting of *Salmonella* Enteritidis and Typhimurium Δ*hilAssrAfliG* strains protects against infection by strains of both serotypes in broilers. Vaccine 32 (36), 4633–4638.

De Cort, W., Haesebrouck, F., Ducatelle, R., van Immerseel, F., 2015b. Administration of a *Salmonella* Enteritidis Δ*hilAssrAfliG* strain by coarse spray to newly hatched broilers reduces colonization and shedding of a *Salmonella* Enteritidis challenge strain. Poult. Sci. 94 (1), 131–135.

Dórea, F.C., Cole, D.J., Hofacre, C., Zamperini, K., Mathis, D., Doyle, M.P., Lee, M.D., Maurer, J.J., 2010. Effect of *Salmonella* vaccination of chicken breeders on reducing carcass contamination of broiler chickens in integrated poultry operations. Appl. Environ. Microbiol. 76, 7820–7825.

Earhart, C.F., 1996. Uptake and metabolism of iron and molybdenum. In: Neidhardt, F.C., Curtiss III, R., Ingraham, J.L., Lin, E.C.C., Low, K.B., Magasanik, B., Reznikoff, W.S., Riley, M., Schaechter, M., Umbarger, H.E. (Eds.), *Escherichia coli* and *Salmonella*: Cellular and Molecular Biology, second ed. ASM Press, Washington, DC, pp. 1075–1090.

EFSA, 2007. Report of the task force on zoonoses data collection on the analysis of the baseline survey on the prevalence of *Salmonella* in broiler flocks of *Gallus gallus*, in the EU, 2005–2006. EFSA J. 98, 1–85.

Elson, R., Little, C.L., Mitchell, R.T., 2005. *Salmonella* and raw shell eggs: results of a cross-sectional study of contamination rates and egg safety practices in the United Kingdom catering sector in 2003. J. Food Prot. 68, 256–264.

Feberwee, A., deVries, T.S., Hartman, E.G., deWit, J.J., Elbers, A.R.W., deJong, W.A., 2001. Vaccination against *Salmonella enteritidis* in Dutch commercial layer flocks with a vaccine based on a live *Salmonella gallinarum* 9R strain: evaluation of efficacy, safety, and performance of serologic *Salmonella* tests. Avian Dis. 45, 83–91.

Felgner, S., Kocijancic, D., Frahm, M., Curtiss III, R., Erhardt, M., Weiss, S., 2016. Optimizing *Salmonella enterica* serovar Typhimurium for bacteria-mediated tumor therapy. Gut Microbes 7 (2), 171–177. Epub ahead of print: http://www.ncbi.nlm.nih.gov/pubmed/26939530?dopt=Abstract.

Fields, P.I., Swanson, R.V., Haidaris, C.G., Heffron, F., 1986. Mutants of *Salmonella typhimurium* that cannot survive within the macrophage are avirulent. Proc. Natl. Acad. Sci. U. S. A. 83 (14), 5189–5193.

Food and Drug Administration, U.S., 2005. FDA Announces Final Decision About Veterinary Medicine, pp. P05–P48 Docket No. 2000N-1571.

Food and Drug Administration, U.S., 2009. Prevention of *Salmonella Enteritidis* in Shell Eggs during Production, Storage, and Transportation (Final Rule. Title 21 CFR Parts 16 and 118).

Galen, J.E., Curtiss 3rd, R., 2014. The delicate balance in genetically engineering live vaccines. Vaccine 32, 4376–4385.

Gantois, I., Ducatelle, R., Timbermont, L., Boyen, F., Bohez, L., Haesebrouk, F., Pasmans, F., Van Immerseel, F., 2006. Oral immunisation of laying hens with the live vaccine strains of TAD *Salmonella* vac® E and TAD *Salmonella* vac® T reduces internal egg contamination with *Salmonella* Enteritidis. Vaccine 24, 6250–6255.

Gast, R.K., Stone, H.D., Holt, P.S., Beard, C.W., 1992. Evaluation of the efficacy of an oil-emulsion bacterin for protecting chickens against Salmonella Enteritidis. Avian Dis. 36 (4), 992–999.

Germanier, R., Furer, E., 1971. Immunity in experimental salmonellosis. II. Basis for the avirulence and protective capacity of *galE* mutants of *Salmonella typhimurium*. Infect. Immun. 4 (6), 663–673.

Han, J., David, D.E., Deck, J., Lynne, A.M., Kaldhone, P., Nayak, R., Stefanova, R., Foley, S.L., 2011. Comparison of *Salmonella enterica* serovar Heidelberg isolates from human patients with those from animal and food sources. J. Clin. Microbiol. 49, 1130–1133.

Hassan, J.O., Curtiss III, R., 1990. Control of colonization by virulent *Salmonella typhimurium* by oral immunization of chickens with avirulent Δ*cya* Δ*crp S. typhimurium*. Res. Microbiol. 141, 839–850.

Hassan, J.O., Curtiss III, R., 1994. Development and evaluation of oral vaccination program using live avirulent *Salmonella typhimurium* to protect vaccinated chickens against challenge with homologous and heterologous *Salmonella* serotypes. Infect. Immun. 62, 5519–5527.

Hassan, J.O., Curtiss III, R., 1996. Effect of vaccination of hens with an avirulent strain of *Salmonella typhimurium* on immunity of progeny challenged with wild-type *Salmonella* strains. Infect. Immun. 64, 938–944.

Hennessy, T.W., Cheng, L.H., Kassenborg, H., Ahuja, S.D., Mohle-Boetani, J., Marcus, R., Shiferaw, B., Angulo, F.J., Emerging Infections Program FoodNet Working Group, 2004. Egg consumption is the principal risk factor for sporadic *Salmonella* serotype Heidelberg infections: a case-control study in FoodNet sites. Clin. Infect. Dis. 38 (Suppl. 3), S237–S243.

Hennessy, T.W., Hedberg, C.W., Slutsker, L., White, K.E., Besser-Wiek, J.M., Moen, M.E., Feldman, J., Coleman, W.W., Edmonson, L.M., MacDonald, K.L., Osterholm, M.T., 1996. A national outbreak of *Salmonella enteritidis* infections from ice cream. The Investigation Team. N. Engl. J. Med. 334, 1281–1286.

Hoffmann, S., Batz, M.B., Morris Jr., J.G., 2012. Annual cost of illness and quality-adjusted life year losses in the United States due to 14 foodborne pathogens. J. Food Prot. 75, 1292–1302.

Holt, P.S., Gast, R.K., Kelly-Aehle, S., 2003. Use of a live attenuated Salmonella typhimurium vaccine to protect hens against Salmonella enteritidis infection while undergoing molt. Avian Dis. 47, 656–661.

Jansson, P.E., Lindberg, A.A., Lindberg, B., Wollin, R., 1981. Structural studies on the hexose region of the core in lipopolysaccharides from *Enterobacteriaceae*. Eur. J. Biochem. 115 (3), 571–577.

Jesudhasan, P.R., McReynolds, J.L., Byrd, A.J., He, H., Genovese, K.J., Droleskey, R., Swaggerty, C.L., Kogut, M.H., Duke, S., Nisbet, D.J., Praveen, C., Pillai, S.D., 2015. Electron-beam-inactivated vaccine against *Salmonella* Enteritidis colonization in molting hens. Avian Dis. 59, 165–170.

Jiang, Y., Kong, Q., Roland, K.L., Curtiss III, R., 2014. Membrane vesicles of *Clostridium perfringens* type A strains induce innate and adaptive immunity. Int. J. Med. Microbiol. 304, 431–443.

Jiang, Y., Mo, H., Willingham, C., Wang, S., Park, J., Kong, W., Roland, K.L., Curtiss III, R., 2015. Protection against necrotic enteritis in broiler chickens by regulated delayed lysis *Salmonella* vaccines. Avian Dis. 59, 475–485.

Juárez-Rodríguez, M.D., Arteaga-Cortés, L.T., Kader, R., Curtiss III, R., Clark-Curtiss, J.E., 2012. Live attenuated *Salmonella* vaccines against *Mycobacterium tuberculosis* with antigen delivery via the type III secretion system. Infect. Immun. 80, 798–814.

Kelly, S.M., Bosecker, B.A., Curtiss III, R., 1992. Characterization and protective properties of attenuated mutants of *Salmonella choleraesuis*. Infect. Immun. 60, 4881–4890.

Kelly-Aehle, S.M., 2004. Method of Vaccination of Newly Hatched Chicks (US Patent # 6866847).

Khan, M.I., Fadl, A.A., Venkitanarayanan, K.S., 2003. Reducing colonization of *Salmonella* Enteritidis in chickens by targeting outer membrane proteins. J. App. Microbiol. 95, 142–145.

Kong, W., Wanda, S.-Y., Zhang, X., Bollen, W., Tinge, S.A., Roland, K.L., Curtiss III, R., 2008. Regulated programmed lysis of recombinant *Salmonella* in host tissues to release protective antigens and confer biological containment. Proc. Natl. Acad. Sci. U. S. A. 105, 9361–9366.

Konjufca, V., Jenkins, M., Wang, S., Juarez-Rodriguez, M.D., Curtiss III, R., 2008. Immunogenicity of recombinant attenuated *Salmonella enterica* serovar Typhimurium vaccine strains carrying a gene that encodes *Eimeria tenella* antigen SO7. Infect. Immun. 76, 5745–5753.

Konjufca, V., Wanda, S.-Y., Jenkins, M.C., Curtiss III, R., 2006. A recombinant attenuated *Salmonella enterica* serovar Typhimurium vaccine encoding *Eimeria acervulina* antigen offers protection against *E. acervulina* challenge. Infect. Immun. 74, 6785–6796.

Kuehn, B.M., 2010. *Salmonella* cases traced to egg producers: findings trigger recall of more than 500 million eggs. JAMA 304, 1316.

Kulkarni, R.R., Parreira, V.R., Sharif, S., Prescott, J.F., 2008. Oral immunization of broiler chickens against necrotic enteritis with an attenuated *Salmonella* vaccine vector expressing *Clostridium perfringens* antigens. Vaccine 26, 4194–4203.

Laniewski, P., Kuczkowski, M., Chrzastek, K., Wozniak, A., Wyszynska, A., Wieliczko, A., Jagusztyn-Krynicka, E.K., 2014. Evaluation of the immunogenicity of *Campylobacter jejuni* CjaA protein delivered by *Salmoenlla enterica* sv. Typhimurium strain with regulated delayed attenuation in chickens. World J. Microbiol. Biotechnol. 30, 281–292.

Layton, S.L., Morgan, M.J., Cole, K., Kwon, Y.M., Donoghue, D.J., Hargis, B.M., Pumford, N.R., 2011. Evaluation of *Salmonella*-vectored *Campylobacter* peptide epitopes for reduction of *Campylobacter jejuni* in broiler chickens. Clin. Vaccine Immunol. 18 (3), 449–454.

Lee, K., Lillehoj, H.S., Siragusa, G.R., 2010. Direct-fed microbials and their impact on the intestinal microflora and immune system of chickens. J. Poult. Sci. 47, 106–114.

Li, Y., Wang, S., Scarpellini, G., Gunn, B., Xin, W., Wanda, S.-Y., Roland, K.L., Curtiss III, R., 2009. Evaluation of new generation *Salmonella enterica* serovar Typhimurium vaccines with regulated delayed attenuation to induce immune responses against PspA. Proc. Natl. Acad. Sci. U. S. A. 106, 593–598.

Linde, K.L., Hahn, I., Vielitz, E., 1997. Development of Live Salmonella Vaccines Optimally Attenuated for Chickens, vol. 20. Lohmann Information, pp. 23–31.

Lüderitz, O., Westphal, O., Staub, A.M., Nikaido, H., 1971. Isolation and chemical and immunological characterization of bacterial lipopolysaccharides. In: Weinbaum, G., Kadis, S., Ajl, S. (Eds.), Microbial Toxins, vol. 4. Academic Press, New York.

Luo, Y., Kong, Q., Yang, J., Golden, G., Wanda, S.-Y., Jensen, R.V., Ernst, P.B., Curtiss III, R., 2011. Complete genome sequence of the universal killer, *Salmonella enterica* serovar Typhimurium UK-1 (ATCC 68169). J. Bacteriol. 193, 4035–4036.

Markovitz, A., Sydiskis, R.J., Lieberman, M.M., 1967. Genetic and biochemical studies on mannose-negative mutants that are deficient in phosphomannose isomerase in *Escherichia coli* K-12. J. Bacteriol. 94 (5), 1492–1496.

McReynolds, J.L., Moore, R.W., McElroy, A.P., Hargis, B.M., Caldwell, D.J., 2007. Evaluation of a competitive exclusion culture and Megan Vac 1 on *Salmonella typhimurium* colonization in neonatal broiler chickens. J. Appl. Poult. Res. 16, 456–463.

McSorley, S.J., 2014. Immunity to intestinal pathogens: lessons learned from *Salmonella*. Immunol. Rev. 260 (1), 168–182.

Methner, U., Barrow, P.A., Berndt, A., Rychlik, I., 2011. *Salmonella* Enteritidis with double deletion in *phoPfliC* – a potential live *Salmonella* vaccine candidate with novel characteristics for use in chickens. Vaccine 29, 3248–3253.

Methner, U., Barrow, P.A., Berndt, A., Steinbach, G., 1999. Combination of vaccination and competitive exclusion to prevent *Salmonella* colonization in chickens: experimental studies. Int. J. Food Microbiol. 49, 35–42.

Mot, D., Timbermont, L., Haesebrouck, F., Ducatelle, R., Van Immerseel, F., 2014. Progress and problems in vaccination against necrotic enteritis in broiler chickens. Avian Pathol. 43 (4), 290–300.

Nakamura, M., Nagata, T., Okamura, S., Takehara, K., Holt, P.S., 2004. The effect of killed *Salmonella enteritidis* vaccine prior to induced molting on the shedding of *S. enteritidis* in laying hens. Avian Dis. 48, 183–188.

Nakayama, K., Kelly, S.M., Curtiss III, R., 1988. Construction of an $Asd^+$ expression-cloning vector: Stable maintenance and high level expression of cloned genes in a *Salmonella* vaccine strain. Nat. Biotech. 6, 693–697.

Nandre, R.M., Matsuda, K., Chaudhari, A.A., Kim, B., Lee, J.H., 2012. A genetically engineered derivative of *Salmonella* Enteritidis as a novel live vaccine candidate for salmonellosis in chickens. Res. Vet. Sci. 93, 596–603.

Olsthoorn, M.M., Petersen, B.O., Schlecht, S., Haverkamp, J., Bock, K., Thomas-Oates, J.E., Holst, O., 1998. Identification of a novel core type in *Salmonella* lipopolysaccharide. Complete structural analysis of the core region of the lipopolysaccharide from *Salmonella enterica* sv. Arizonae O62. J. Biol. Chem. 273 (7), 3817–3829.

Paiva, J.B., Penha Filho, R.A.C., Arguello, Y.M.S., Silva, M.D., Gardin, Y., Resende, F., Berchieri Jr., A., Sesti, L., 2009. Efficacy of several *Salmonella* vaccination programs against experimental challenge with *Salmonella* Gallinarum in commercial brown layer and broiler breeder hens. Braz. J. Poult. Sci. 1, 65–72.

Pavic, A., Groves, P.J., Cox, J.M., 2010. Utilization of a novel autologous killed tri-vaccine (serogroups B [Typhimurium], C [Mbandaka] and E [Orion]) for *Salmonella* control in commercial poultry breeders. Avian Pathol. 39 (1), 31–39.

Raetz, C.R.H., 1996. Bacterial lipopolysaccharides: a remarkable family of bioactive macroamphiphiles. In: Neidhardt, F.C., Curtiss III, R., Ingraham, J.L., Lin, E.C.C., Low, K.B., Magasanik, B., Reznikoff, W.S., Riley, M., Schaechter, M., Umbarger, H.E. (Eds.), *Escherichia coli* and *Salmonella*: Cellular and Molecular Biology, second ed. ASM Press, Washington, DC, pp. 1035–1063.

Reeves, P., 1995. Role of O-antigen variation in the immune response. Trends Microbiol. 3 (10), 381–386.

Rick, P.D., 1987. Lipopolysaccharide biosynthesis. In: Neidhardt, F.C., Ingraham, J.L., Low, K.B., Magasanik, B., Schaechter, M., Umbarger, H.E. (Eds.), *Escherichia coli* and *Salmonella*: Cellular and Molecular Biology, first ed. ASM Press, Washington, D.C., pp. 648–662.

Roland, K., Curtiss III, R., Sizemore, D., 1999. Construction and evaluation of a Δ*cya* Δ*crp* *Salmonella typhimurium* strain expressing avian pathogenic *Escherichia coli* O78 LPS as a vaccine to prevent air sacculitis in chickens. Avian Dis. 43, 429–441.

Roland, D., Karaca, K., Sizemore, D., 2004. Expression of *Escherichia coli* antigens in *Salmonella typhimurium* as a vaccine to prevent air sacculitis in chickens. Avian Dis. 48, 595–605.

Rowley, D., 1968. Sensitivity of rough gram-negative bacteria to the bactericidal action of serum. J. Bacteriol. 95 (5), 1647–1650.

Roy, P., Dhillon, A.S., Lauerman, L.H., Schaberg, D.M., Bandli, D., Johnson, S., 2002. Results of *Salmonella* isolation from poultry products, poultry, poultry environment, and other characteristics. Avian Dis. 46, 17–24.

Schleifer, J.H., Juven, B.J., Beard, C.W., Cox, N.A., 1984. The susceptibility of chicks to Salmonella montevideo in artificially contaminated poultry feed. Avian Dis. 28, 497–503.

Schwartz, M., 2009. Dr. Jekyll and Mr. Hyde: a short history of anthrax. Mol. Aspects Med. 30, 347–355.

Shi, H., Wang, S., Roland, K.L., Gunn, B.M., Curtiss III, R., 2010. Immunogenicity of a live recombinant *Salmonella enterica* serovar Typhimurium vaccine expressing *pspA* in neonates and infant mice born from naïve and immunized mothers. Clin. Vaccine Immunol. 17, 363–371.

Springer, S., Lindner, T., Ahrens, M., Woitow, G., Prandini, F., Selbitz, H.J., 2011. Duration of immunity induced in chickens by an attenuated live *Salmonella enteritidis* vaccine and an inactivated *Salmonella enteritidis/typhimurium* vaccine. Berl. Munch. Tierarztl. Wochenschr. 124, 89–93.

Theoret, J.R., Cooper, K.K., Zekarias, B., Roland, K.L., Law, B.F., Curtiss III, R., Joens, L.A., 2012. The *Campylobacter jejuni* Dps homologue is important for in vitro biofilm formation and cecal colonization of poultry and may serve as a protective antigen for vaccination. Clin. Vaccine Immunol. 19, 1426–1431.

Toyota-Hanatani, Y., Kyoumoto, Y., Baba, E., Ekawa, T., Ohta, H., Tani, H., Sasai, K., 2009. Importance of subunit vaccine antigen of major Fli C antigenic site of *Salmonella* Enteritidis II: a challenge trial. Vaccine 27, 1680–1684.

Tran, Q.L., Quessy, S., Letellier, A., Derosiers, A., Boulianne, M., 2010. Immune response following vaccination against *Salmonella* Enteritidis using 2 commercial bacterins in laying hens. Can. J. Vet. Res. 74, 185–192.

United Egg Producers, 2016. Animal Husbandry Guidelines for U.S. Egg Laying Flocks, 2016 ed. http://uepcertified.com/wp-content/uploads/2015/08/UEP-Animal-Welfare-Guidelines-20141.pdf.

United States Department of Agriculture, 2015. Food Safety Inspection Service, Serotypes Profile of *Salmonella* Isolates from Meat and Poultry Products: January 1998 through December 2013. Prepared November 17, 2015, http://www.fsis.usda.gov/wps/wcm/connect/c7b5903c-8e8b-4f85-9b5c-12eaf990d2dd/Salmonella-Serotype-Annual-2013.pdf?MOD=AJPERES.

Van Immerseel, F., De Buck, J., De Smet, I., Mast, J., Haesebrouck, F., Ducatelle, R., 2002. The effect of vaccination with *Salmonella* Enteritidis *aroA* mutant on early cellular responses in caecal lamina propria of newly-hatched chickens. Vaccine 20, 3034–3041.

Van Immerseel, F., Methner, U., Rychlik, I., Nagy, B., Velge, P., Martin, G., Foster, N., Ducatelle, R., Barrow, P.A., 2005. Vaccination and early protection against non-host-specific *Salmonella* serotypes in poultry: exploitation of innate immunity and microbial activity. Epidemiol. Infect. 133, 959–978.

Vazquez, M.E., 2014. Poultry emulsion vaccines. In: Avian Insight, Lohmann Animal Health News Brief, vol. 2, .

Williams, J.E., Benson, S.T., 1978. Survival of *Salmonella typhimurium* in poultry feed and litter at three temperatures. Avian Dis. 22, 742–747.

Xin, W., Wanda, S.-Y., Zhang, W., Santander, J., Scarpellini, G., Ellis, K., Alamuri, P., Curtiss III, R., 2012. The $Asd^{+}$-$DadB^{+}$ dual-plasmid system offers novel means to deliver multiple protective antigens by a recombinant attenuated *Salmonella* vaccine. Infect. Immun 80, 3621–3633.

Young, S.D., Olusanya, O., Jones, K.H., Liu, T., Liljebjelke, K.A., Hofacre, C.L., 2007. *Salmonella* incidence in broilers from breeders vaccinated with live and killed *Salmonella*. J. Appl. Poult. Res. 16, 521–528.

Zekarias, B., Mo, H., Curtiss III, R., 2008. Recombinant attenuated *Salmonella enterica* serovar Typhimurium expressing the carboxy-terminal domain from α-toxin from *Clostridium perfringens* induces protective responses against necrotic enteritis in chickens. Clin. Vaccine Immunol. 15, 805–816.

Zhang, X., Kelly, S.M., Bollen, W.S., Curtiss III, R., 1997. Characterization and immunogenicity of *Salmonella typhimurium* SL1344 and UK-1 Δ*crp* and Δ*cdt* deletion mutants. Infect. Immun. 65, 5381–5387.

CHAPTER

# 15 Use of Direct-Fed Microbials in Layer Hen Production—Performance Response and *Salmonella* Control*

**Todd R. Callaway[1], Tom S. Edrington[1], James A. Byrd[1], David J. Nisbet[1], Steven C. Ricke[2]**

*[1]USDA/ARS, College Station, TX, United States; [2]University of Arkansas, Fayetteville, AR, United States*

## 1. INTRODUCTION

Modern food production systems are aimed at producing as much food as efficiently and quickly as possible, while maintaining the safety and wholesomeness of the end product. The success of this effort can be seen in increasing concerns about caloric excess in the food supply, as well as in the high degree of safety of the food chain. However, improvements remain to be made in enhancing animal production efficiency, animal health, and food safety.

The US egg-laying flock is composed of more than 285 million hens that produce in excess of 81 billion eggs per year (Board, 2016). Eggs are produced by mature hens, and this presents the opportunity to improve production efficiency, poultry health, and food safety, not only during egg production, but during the growth phase of the hens until they reach maturity. A significant issue facing the poultry industry is that food-borne pathogenic bacteria can be harbored asymptomatically in or on poultry and their environments (Borland, 1975; Doyle and Erickson, 2006; Dunkley et al., 2009; Galiş et al., 2013). Each year, too many illnesses are associated with consumption of foods of animal origin (Scallan et al., 2011) and the total cost of food-borne illnesses in the United States is more than US$75 billion (Scharff, 2012). The indirect and direct cost each year of the five most common food-borne pathogenic bacteria in the United States totals more than US$8 billion (Hoffmann et al., 2015). Although enterohemorrhagic *Escherichia coli* (including *E. coli* O157:H7), *Salmonella*, *Campylobacter*, and

*Proprietary or brand names are necessary to report factually on available data; however, the USDA neither guarantees nor warrants the standard of the product, and the use of the name by the USDA implies neither approval of the product nor exclusion of others that may be suitable.

Producing Safe Eggs. http://dx.doi.org/10.1016/B978-0-12-802582-6.00015-X

*Listeria* have all been isolated from cattle, swine, and poultry (Borland, 1975; Callaway et al., 2006; Oliver et al., 2005), the biggest threats to the poultry and egg industry remain *Salmonella enterica* and *Campylobacter*. Although the presence of these zoonotic pathogens is largely asymptomatic, carriage of these bacteria represents a threat to the integrity as well as the efficiency and profitability of the food supply.

To improve growth production efficiency, many approaches have been tested to improve growth, health, and production efficiency in poultry from prophylactic antibiotic usage to a variety of housing systems (Weeks et al., 2016). One that has gained increasing recognition for long-term efficacy is the use of probiotics [in the animal industry called direct fed microbials (DFM)] (Nava et al., 2005; Patterson and Burkholder, 2003; Smith, 2014). The use of DFM in poultry seeks to harness the power of the gastrointestinal microbial ecosystem to both eliminate pathogens and improve growth and growth efficiency. In this chapter, we will discuss the theory behind probiotic approaches to improving egg production, as well as the benefits and challenges for future development and implementation.

## 2. MICROBIAL COMPETITIVE ENHANCEMENT WITHIN THE GUT

The gastrointestinal tract of mature poultry is a complex ecosystem that occupies all environmental niches and utilizes nearly all available fermentable nutrients. The symbiotic relationship between the host and its resident gastrointestinal microbial ecosystem is critical to animal health, food safety, and production efficiency (Fuller, 1989; Jayne-Williams and Fuller, 1971). The native gastrointestinal microbiota compete with each other fiercely for nutrients, and the winners of this competition directly affect animal performance and health by fermenting substrate in the gut to produce volatile fatty acids (VFAs) that are absorbed by the bird to provide energy for maintenance, growth, and production. In recent years, the role of the microbial ecosystem or "microbial organ" has been examined in regard to its role in health, well-being, skeletal structure, as well as in production parameters (Finegold, 2008; Ley et al., 2006; Lyte, 2010; McCabe et al., 2015; Murphy, 2004; Turnbaugh et al., 2006, 2009; Xu and Gordon, 2003). Increasingly, research has delved into the microbial organ (Bailey et al., 2011; Freestone and Lyte, 2010; Pullinger et al., 2010), for example, the two-way communication between the host and the resident microbiome (now called microbial endocrinology) offers a potential new mechanism to the animal industry (Lyte, 2010). Another new field that has lately evolved is that of "osteoimmunology," which demonstrates that the microbiome of the gut and microbial populations from DFM can alter skeletal fitness, bone density, and strength in humans and other animals (Charles et al., 2015; McCabe et al., 2015). With the possibility that microbial populations (native or added as a DFM) can be used as a type of drug delivery mechanism, tailored modification of animal performance and food safety via alterations in the microbial population is at last a real possibility (Price et al., 2010).

Unfortunately, the composition of the microbial ecosystem of an animal is not always ideal for peak production efficiency (from the point of view of the host or producer) or for excluding bacteria that are pathogenic to the bird or human consumers. When chicks are hatched, the naïve gastrointestinal tract contains a bacterial population that is very low in both numbers and diversity (Falk et al., 1998). Over time, environmental exposure beginning with the eggshell passage out of the hen populates the intestinal tract with a succession of microbial species and an increasing diversity, resulting in the establishment of a mature, complex, microbial ecosystem that occupies all ecological niches and is more resistant to penetration by pathogens and other opportunistic infections (Lu et al., 2003). Thus, the diversity and richness of the microbial population provides a beneficial inertia, preventing drastic changes while remaining remarkably plastic. Disturbances in the environment (host animal) such as dietary changes, environmental or social stress, or transport, can disrupt the microbial ecology in the intestinal ecosystem, allowing pathogen penetration or dramatically impacting the efficiency of production.

Historically, studies in food animals examining probiotic/DFM approaches have been hampered by a lack of understanding of the microbial ecology of the gastrointestinal tract, as well as a lack of understanding of the ecological role of the selected probiotic organisms (Barroga et al., 2007; LeJeune et al., 2006). Some interstudy challenge and variation can be attributed to the fact that mature animals contain a stable, relatively individualistic intestinal microbial population with which the probiotic must come to equilibrium. However, when probiotics are applied to newly hatched chicks that have a sparse or poorly established intestinal population, results are more consistent. All of these factors have contributed to difficulties in reproducing effects of some probiotics beyond the newly hatched phase (Patterson and Burkholder, 2003).

Molecular methodologies such as pyrosequencing and next-generation sequencing have now allowed researchers to begin to define "normal" and "most efficient" microbial ecosystem compositions (Benson et al., 2010; Callaway et al., 2009), which has allowed precise monitoring of specific changes caused by DFM feeding or environmental/production stresses. These scientific advances can lead to the development of highly tailored probiotic products for use in specific production situations (e.g., heat/cold stress, broiler production versus egg production).

Because antibiotics are cheap and effective but are often antagonistic to the action of DFM, the use of probiotic products in the poultry industry has been sporadic and relatively limited (Steer et al., 2000). However, as concerns over the dissemination of antimicrobial resistance have grown (Witte, 1999), it appears that prophylactic antimicrobial usage in food animals will become more closely regulated and expensive, causing probiotic strategies to become more economically viable (Taylor, 2001). Furthermore, future trade regulations within the European Union (EU 1003/2005) are expected to increase the use of nonantibiotic methods (such as DFM) to reduce *Salmonella* on eggs and in chicks shipped into the European Union.

Probiotic strategies that enhance the competition within the microbial population are diverse, and include (Fuller, 1989; Collins and Gibson, 1999): (1) addition of a microbial supplement to the ration (DFM) that enters an established gastrointestinal tract and improves gastrointestinal health and the diversity of the microbial population, (2) introduction of a "normal" adult microbial population to a naïve gastrointestinal tract (competitive exclusion, or CE), (3) adding a limiting, non–host digestible nutrient (prebiotic) that provides an existing (or introduced) commensal microbial population with a competitive advantage in the gastrointestinal tract, and (4) use of prebiotics along with a DFM or CE culture, known as "synbiotics." Each of these approaches seeks to harness the activities of the native microbial ecosystem by capitalizing on the natural microbial competition and offering a natural "green" method to improve animal production and food safety.

## 2.1 DIRECT-FED MICROBIALS

In this chapter, we will utilize the original definition of "Direct-Fed Microbials" (DFM) as "microbial feed supplements which beneficially affects the host animal by improving intestinal microbial balance" to enhance performance, or to reduce zoonotic pathogens (Fuller, 1989). A more recently proposed definition for probiotics is "a preparation or a product containing viable, defined microorganisms… which alter the micro-flora…and exerts beneficial health effects in this host" (Schrezenmeir and De Vrese, 2001). The microorganisms used in DFM are typically classified as generally recognized as safe, and do not have to be isolated from the host they will be used in. Probiotic cultures (DFM or CE) may be composed of a single organism or a mixture of organisms and these cultures may also be defined or undefined, meaning their composition is known or unknown. Long-term efficacy of undefined cultures has sometimes been problematic, thus their rate of use in the industry has decreased (Van Immerseel et al., 2005).

Cultures used in DFM can be: (1) live cultures of yeast or bacteria, (2) heat-treated (or otherwise inactivated) cultures of yeast or bacteria, or (3) fermentation end products from growth of yeast or bacteria. The ultimate goal of DFM feeding is to improve animal growth, performance, and health by filling all ecological niches in the gut to maximize nutrient degradation and fermentation by the bacteria and exclude or displace opportunistic pathogens (affecting animal or human health) from the animal (Endt et al., 2010). Stimulation of intestinal populations of *Bifidobacterium* in broilers has been shown by some DFM (Mountzouris et al., 2015), and higher populations of this bacterium are linked with healthier gut populations (Vandenplas et al., 2015).

## 2.2 COMPETITIVE EXCLUSION CULTURES

CE is a specific probiotic approach that is simply the addition of a nonpathogenic bacterial culture of a single or multiple strains derived from an adult of the same animal species to the naïve (or nearly so) intestinal tract to prevent pathogen colonization or improve growth (Fuller, 1989; Nisbet et al., 1993a; Nurmi et al., 1992). This harnesses the symbiotic relationship between the host animal and its native microbial ecosystem that developed evolutionarily and can cause the early establishment of a "normal" or "ideal" microbial population, prevent the establishment of a pathogenic bacterial population, and improve growth efficiency and/or rate (Nurmi et al., 1992; Steer et al., 2000). This is especially critical in broiler and egg production because eggs and newly hatched chicks are naïve microbiologically and can be quickly colonized at hatch by pathogens such as *Salmonella* and *Campylobacter* (Cox et al., 1990). However, when CE is used in older birds, it must compete against the previously discussed established native population, and therefore results are inconsistent. Best results have been obtained by utilizing CE shortly after hatch and then working with this established population as the bird matures. Therefore, the best mixture of bacteria (or yeast) chosen for use as a CE/DFM treatment regime will differ based on strain/species characteristics (Bozkurt et al., 2011), production stage, and scenario in which it will be utilized.

Although the mode of action of CE (and many DFM for that matter) remains unknown, there are several hypotheses for how these compounds improve animal performance and reduce pathogens, including: (1) direct and indirect nutrient competition, (2) physical attachment site competition, (3) antimicrobial compound production (including VFA, medium chain fatty acids, and lactic acid), (4) host immune system activity stimulation, and (5) a synergistic interaction of two or more of these (Chichlowski et al., 2007). Nutrients are utilized by the added species that produce VFA, which can be utilized by the host, preventing "inefficient" (from the host perspective) species or pathogens from flourishing. By denying physical binding sites to opportunistic pathogens or bacteria that are less efficient than the chosen culture, transient bacteria that may depend on epithelial adherence would be washed out of the gut (Collins and Gibson, 1999; Lloyd et al., 1977). VFAs and lactic acid produced by the normal microbial population (or from DFM) can also inhibit some opportunistic pathogens (such as *Salmonella* or *Campylobacter*) and may reduce the competitive fitness of pathogens, and VFAs serve as an additional source of energy to the bird (De Keersmaecker et al., 2006; Neal-McKinney et al., 2012; Prohaszka and Baron, 1983; Wolin, 1969). Some bacteria used in CE or DFM produce antimicrobial protein compounds, such as bacteriocins (including colicins), which can inhibit or eliminate species competing within the same niche (Al-Qumber and Tagg, 2006; Jack et al., 1995; Schamberger et al., 2004; Walsh et al., 2008). Collectively, the natural antipathogen and pro-"normal flora" activity of CE/DFM has been called "bacterial antagonism" or "bacterial interference" (Lloyd et al., 1974; Nurmi et al., 1992).

## 3. PROBIOTIC IMPACTS IN CHICKENS

*Growing birds*: Because laying hens must mature before they can produce eggs, they experience an extended growing period and share many of the same challenges and opportunities for improvement as broiler production. If the gastrointestinal tract is disturbed during development, subsequent impacts may last throughout the production cycle or even the life of the animal (Yeoman and White, 2014), potentially including the long-term colonization of a laying hen by *Salmonella* or *Campylobacter*. Therefore much of the probiotic research in broiler production is applicable not only to ensuring a pathogen-free laying flock, but also to animal health, welfare, and, ultimately, productivity of the laying hen.

Chicks are hatched with an essentially sterile intestinal tract, which is sequentially colonized by bacteria as the bird ages (Lu et al., 2003; Uyeno et al., 2010); however, the sterile intestinal tract is an "open field" for colonization by *Campylobacter* and *Salmonella*. In addition, if a pathogen-free microbial population can be inserted into the intestinal tract rapidly, then feedstuffs can be fermented to produce VFAs, which can improve the energy status of the chick, allowing a rapid start to growth. This has prompted CE cultures to be developed and used in many countries on day of hatch chicks (Bielke et al., 2003; Stavric, 1992; Stavric and D'Aoust, 1993; Stavric and Kornegay, 1995). In the United States, a mixed, defined (e.g., the exact species included are identified) commercial CE product that comprises 28 species of bacteria was developed and used to reduce *Salmonella* colonization in day of hatch chicks (Nisbet et al., 1994, 1996). Subsequently, alternative undefined CE products have also been relatively widely adopted in the poultry industry (Schneitz, 2005; Zhang et al., 2007). Further studies have found that *Campylobacter* colonization of poultry can be inhibited by the use of specific CE cultures in newly hatched chicks as well (Zhang et al., 2007). The use of these CE cultures has been dramatically impacted by the efficacy and cost of prophylactic antibiotic treatment.

Lactobacilli have been widely used as DFM components because they often improve animal performance and can inhibit pathogen populations (Neal-McKinney et al., 2012), and some yeast-based DFM have been shown to directly increase native *Lactobacillus* populations in the gut (Han et al., 1999). This is critical not only because Lactobacilli [and other lactic acid bacteria (LAB)] produce a strong acid (lactate) that can reduce the pH rapidly and impact the microbial ecosystem as a whole (Neal-McKinney et al., 2012), but also because these bacteria can produce antimicrobial peptides (bacteriocins) that are toxic to *Salmonella* and *Campylobacter* (Joerger, 2003; Lima et al., 2007; Van Coillie et al., 2007).

A culture of *Lactobacillus salivarius* fed to 1-day-old leghorn chicks by dosing in feed and water prevented *S.* Enteritidis long-term colonization; but more than one dose was needed as the DFM disappeared from the intestinal population after 21 days (Pascual et al., 1999). Other studies found that treatment with a combination of *Lactobacillus acidophilus* and *Streptococcus faecium* DFM culture, coupled with specific antibodies for key *Salmonella* serotypes, reduced the colonization of *S.* Enteritidis in market-aged broilers (Tellez et al., 2001). Similarly, a *Lactobacillus*-based DFM decreased the incidence of *S.* Enteritidis–positive crops and ceca in broiler chicks (Wolfenden et al., 2007). A four-species *Lactobacillus* DFM (*L. acidophilus*, *Lactobacillus fermentum*, *Lactobacillus reuteri*, and *L. salivarius*) fed to 1-week-old broilers resulted in a decrease in *S.* Enteritidis populations in the cecum of more than 10-fold (Penha Filho et al., 2015). Other LAB DFM significantly reduced *Salmonella* Typhimurium populations and fecal shedding in growing chicks (Hsu et al., 2016). *Campylobacter jejuni* colonization of day-old chicks was also reduced by nearly 50% through day 21 by PrimaLac (a *Lactobacillus*-based DFM) feeding (Willis and Reid, 2008). Lactobacilli isolated from the cloaca and vagina of hens reduced the growth of *S.* Enteritidis cultures, indicating a potential role of these organisms either as a dietary/environmental DFM additive, or as a method to reduce *S.* Enteritidis contamination of eggshells directly (Miyamoto et al., 2000). Other poultry cloacal and vaginal isolates of *L. salivarius*, *L. acidophilus*, and *L. reuteri* exhibited significant in vitro activity against *S.* Enteritidis (linked to lactate production) and these strains were further demonstrated as a DFM that reduced *S.* Enteritidis colonization in chicks (Van Coillie et al., 2007). Oddly, there have been sex differences observed in the effectiveness of some *Lactobacillus*-based DFM, and it has been shown that the feeding regimen can play a critical role in the efficacy of DFM treatment as it impacts growth (Willis and Reid, 2008). Horizontal transmission of *S.* Enteritidis among chicks was also reduced by LAB DFM treatment (Jarquin et al., 2007).

Cultures of *Bacillus cereus* have also been used as DFM in food animals to improve performance and food safety (Abudabos et al., 2014; Vilà et al., 2009). Although early research with *B. cereus* var. *toyoi* focused on piglets (Lodemann et al., 2008; Scharek et al., 2007; Schierack et al., 2007), other research with poultry has demonstrated that the use of these spores can inhibit *S.* Enteritidis colonization in poultry (Vilà et al., 2009). The use of *Bacillus subtilis* as a commercial DFM reduced cecal *Salmonella* populations in broilers by 1000-fold and slightly improved the feed conversion rate (Knap et al., 2011). *Bacillus* culture addition in growing chicks also demonstrated a reduction in *S.* Enteritidis colonization (Abudabos et al., 2014). Replacement of antibiotic treatment with *B. cereus* DFM treatment demonstrated an equal impact on production parameters as did antibiotic treatment (Abudabos et al., 2014; Duarte et al., 2014), Interestingly, the use of a specific aflatoxin-degrading *B. subtilis* strain reversed some of the deleterious effects of feeding aflatoxin B1 to laying hens, and improved liver and kidney function (Ma et al., 2012); these data underscore that in addition to production parameters, DFM feeding can also impact animal health and well-being and counteract known toxins in the ration.

*Saccharomyces* yeasts (including *Saccharomyces boulardii* and *Saccharomyces cerevisiae*) have been used as DFM products in many animal species, including poultry (El-Deek et al., 2009; Isik et al., 2004; Katoch et al., 2013). An *S. boulardii* DFM reduced *Salmonella* colonization but not *Campylobacter* populations in broiler chicks by more than 50% (Line et al., 1998). Interestingly, one of the modes of action suggested for another DFM from *S. boulardii* included the reduction in motility of a challenge strain of *S.* Enteritidis, thereby decreasing its invasiveness (Pontier-Bres et al., 2012).

A mixed DFM product comprising *Enterococcus faecium*, *Pediococcus acidilactici*, *L. salivarius*, and *L. reuteri* was dosed daily in the drinking water of 1-day-old chicks (Ghareeb et al., 2012). Following oral administration of *Campylobacter jejuni*, cecal colonization was reduced at both 8 and 15 days after challenge (Ghareeb et al., 2012). Day-of-hatch chicks were treated with the commercial probiotic compound Floramax an hour after challenge with *Salmonella* Heidelberg, and the probiotic reduced the incidence of *S.* Heidelberg in the ceca of chicks after 24 and 72 h (Menconi et al., 2011). One-day-old Ross broiler chicks treated with a probiotic mixture of *L. reuteri*, *B. subtilis*, and *S. cerevisiae* had a better feed conversion rate through day seven and had greater body weight gain from 0 to 21 days compared with controls, and was similar to levels from birds fed virginiamycin (Salim et al., 2013). Neither virginiamycin nor the DFM treatment affected *Salmonella* incidence in these birds (Salim et al., 2013). A mixed DFM (comprising lactobacilli, streptococci, and an unspecified yeast) demonstrated greater growth performance among broiler chicks and higher microbial populations in the gastrointestinal tract than controls fed a normal or a Ca/P-deficient diet (Katoch et al., 2013).

Lactose, as well as more traditional prebiotic oligosaccharides (fructo- and mannooligosaccharides; FOS and MOS), are indigestible by poultry, and have been used to significantly decrease *Salmonella* in broiler chicks (Corrier et al., 1990, 1991, 1997). Interestingly, the simultaneous use of lactose along with CE or a DFM-type culture (a synbiotic-type approach) yielded mixed results. Feeding of lactose along with used broiler litter reduced *Salmonella* colonization in broiler chicks, but not in mature hens (Corrier et al., 1993), indicating a role of the maturity of the gastrointestinal population. In another study investigating the use of CE along with lactose, a synergistic impact of both treatments was observed in colonization of broiler chicks by *S.* Typhimurium, and this was correlated to a reduced cecal pH and increased propionate concentration in the gut (Nisbet et al., 1993b).

Addition of a probiotic, and synbiotic mixtures at a variety of levels throughout the 42-day growing period improved feed conversion ratio (Midilli et al., 2008). A study comparing antibiotic treatment to probiotic, prebiotic, and synbiotic treatments in male broiler chicks found that the feed conversion was higher in probiotic-treated birds compared with controls or antibiotic-treated groups, although this was not statistically significant (Mokhtari et al., 2010). A synbiotic treatment including a lactate-producing probiotic and a proprietary prebiotic found that the probiotic alone had the lowest feed-to-gain ratio, and that the synbiotic treatment had the greatest weight gain (Falaki et al., 2010). A probiotic mixture of *L. acidophilus*, *S. faecium*, and *Bifidobacterium bifidum* coupled with inulin (a prebiotic) treatment demonstrated that both the probiotic and synbiotic treatment resulted in greater weight gain for the first 21 days of growth compared with flavomycin treatment (Da Silva et al., 2011). However, flavomycin treatment did reduce feed intake for the first 21 days compared with probiotic and synbiotic treatments (Da Silva et al., 2011).

*Mature laying hens*: Although immunization against *Salmonella* species have been effective in reducing horizontal and vertical transmission, probiotic approaches still have an important role in improving food safety in laying hens (Trampel et al., 2014). When 40-week-old laying hens were treated with either *B. subtilis*- or *Bacillus methylotrophicus*-based DFM and subsequently challenged with *Salmonella* Gallinarum, egg production was improved in the hens following treatment with the *Bacillus* probiotic and *Salmonella* populations were reduced (Upadhaya et al., 2016). Avian intestinal spirochetosis is an increasing problem in the European Union that is caused by *Brachyspira pilosicoli*; feeding laying hens an *L. reuteri*-based

DFM reduced *B. pilosicoli* populations and pathological changes and increased body weights and egg weights of hens fed this DFM (Mappley et al., 2013). Feeding of a *B. subtilis* dried culture increased egg production, feed consumption, as well as feed conversion while also reducing cholesterol levels in the yolk and improving yolk color (Xu et al., 2006). Synbiotic treatments improved egg weight, egg mass, and egg production as well as feed conversion ratio in hens from 24 to 36 weeks of age (Abdel-Wareth, 2016). In 25- to 45-week-old laying hens, supplementation with a *B. subtilis* probiotic increased egg production and egg mass significantly, and improved egg mass/kg feed ($P<.08$) (Ribeiro et al., 2014). Layers from 20 to 59 weeks that were fed a *Lactobacillus*-based DFM had greater feed consumption and egg size than did controls, and had improved N and Ca retention (Nahashon et al., 1996). Supplementation of an *E. faecium* DFM to mature laying hens produced an increase in egg production, egg weight, and eggshell thickness, and also increased fecal *Lactobacillus* populations (Zhang and Kim, 2013). Supplementation with a *Bacillus licheniformis* DFM in 28-week-old laying hens increased egg production and egg mass, but the feed conversion ratio remained unchanged (Lei et al., 2013).

Heat stress in laying hens takes a toll on egg production and impacts growth and production efficiency. A synbiotic approach using MOS and a probiotic mixture was shown to alleviate some of the growth-depressing effects of chronic heat stress on growing chicks (Sohail et al., 2012). A specific *Lactobacillus* DFM strain that produced γ-aminobutyric acid was fed to heat-stressed Hy line brown hens, and the DFM treatment improved egg production, egg weight, daily feed intake, feed conversion ratio, and misshaped egg percentage compared with control heat-stressed hens (Zhu et al., 2015). The deleterious effects of heat stress on egg production and feed intake were also mitigated largely by feeding a *B. licheniformis* DFM to 60-week-old hens, and DFM treatment restored the villus height impairment in these heat-stressed hens (Deng et al., 2012).

The use of lactic acid bacteria as DFM has a role in mature laying hens as well as in growing birds. A LAB-containing probiotic was fed to hens from 24 to 72 weeks of age, and during this period, probiotic treatment significantly increased the egg production, shelf weight, shell thickness, and serum calcium, and reduced the concentrations of cholesterol in the serum and yolk, but did not impact feed conversion rate (Panda et al., 2003). A marginal amino acid deficiency in 33- to 44-week-old laying hens was partially alleviated by DFM treatment, but the *Lactobacillus*-based DFM did improve intake-to-egg mass ratios (Applegate et al., 2009). Feeding a DFM of *Rhodobacter capsulatus* to 30-week-old hens increased polyunsaturated fatty acid and decreased the cholesterol content of egg yolks (Salma et al., 2012). In other studies, a *Lactobacillus*-based DFM increased egg weight, egg mass, egg size, and body weight, but did not affect feed consumption, feed conversion, or egg quality (Nahashon et al., 1994). Feeding a different type of *Lactobacillus*-based DFM resulted in an increase in egg size and lowered feed costs in caged hens (Davis and Anderson, 2002), and improved the cholesterol content of eggs without impacting egg quality (Tang et al., 2015). A study examining a *Bacillus*-based spore product found an increase in egg specific gravity, but no impact on feed consumption, egg production, egg weight, or body weight of hens (Sohail et al., 2002). Another *B. subtilis* DFM improved yolk color, albumen quality, and shell thickness, but growth parameters were not affected by DFM (Sobczak and Kozłowski, 2015).

*In molting hens*: As laying hens age, egg production decreases; to initiate a new egg-laying cycle characterized by increased egg production and size, aging hens are induced to molt (Bell, 2003; Berry, 2003). Traditionally, feed withdrawal (for up to 10 days) has been used to initiate molting, but feed withdrawal inhibits the immune system of layers (Berry, 2003; Klasing, 2007), especially reducing heterophil efficiency (Kogut et al., 1999), resulting in molting layers being more susceptible to colonization by *Salmonella*, especially *S.* Enteritidis (Holt, 1993, 2003; Porter and Holt, 1993; Ricke, 2003). Furthermore, starvation causes change in the intestinal microbial population of laying hens (Callaway et al., 2009; Durant et al., 1999; Ricke, 2003), likely due to the reduced amount of substrate present in the gut for bacteria to ferment. Starvation creates a selective pressure in the gut for bacteria that can survive long nutrient deprivation or can attach to epithelial tissues (Nettelbladt et al., 1997), and molting has been correlated with an increase in intestinal lesions, potentially providing a route of entry to the bird for *Salmonella* (Porter and Holt, 1993).

As an alternative approach to feed withdrawal, providing a fermentable substrate that is unavailable to the host (similar in concept to a prebiotic-type of approach) to layers during feed withdrawal has been shown to reduce *Salmonella* colonization (Corrier et al., 1997; Park et al., 2013; Ricke et al., 2013), as has feeding of alfalfa crumbles (Donalson et al., 2005; Dunkley et al., 2007b,c; Landers et al., 2005; Woodward et al., 2005). Fermentation of these substrates produces VFAs (Callaway et al., 2009), which can directly inhibit the growth of pathogenic bacteria in the crop and gut, such as *Salmonella* and *Campylobacter* (Durant et al., 1999), and it also provides a level of energy to the bird that may enhance immune function by reducing stress and subsequent inflammation (Dunkley et al., 2007a). Inclusion of alfalfa in growing chicks at levels as high as 25% and 50% of the diet reduced *Salmonella* populations but did not negatively impact average daily gain during a short (7-day) feeding period (Escarcha et al., 2012). The inclusion of a true prebiotic (FOS) along with alfalfa supplementation to molting layers during feed withdrawal reduced cecal *S.* Enteritidis populations and penetration of the liver and ovary (Donalson et al., 2008). It should be noted that there were no differences in intestinal VFA concentrations in these studies, but there were higher intestinal lactate concentrations in FOS-supplemented birds compared with layers undergoing feed withdrawal (Donalson et al., 2008).

## 3.1 SKELETAL FITNESS

Studies in animal species and humans have led to the development of "osteoimmunology," a theory that bone development and strength can be linked with the microbial population of the gut (Charles et al., 2015; McCabe et al., 2015). Rapidly developing information from this blossoming area of investigation indicates that DFM can play a significant role in altering skeletal fitness (Charles et al., 2015; McCabe et al., 2015).

In poultry, the use of *B. licheniformis*- and *B. subtilis*-based DFM fed broilers for 6 weeks showed increased tibial strength and medullary canal diameter (Mutuş et al., 2006). Although data are lacking that show dramatic impacts on bone health in poultry, it is clear to see that the interaction between microbiome and host may play a role in skeletal fitness, which can impact egg production and hen longevity and morbidity (Charles et al., 2015; McCabe et al., 2015).

## 3.2 GUT ARCHITECTURE

Although it is clear that feeding DFM of all types can impact gut microbial populations, it has also become increasingly clear that these changes can also impact gut health and the ability to take up nutrients. An *L. salivarius*- and *L. reuteri*- based DFM fed to growing broilers increased villus height and crypt depth in parts of the small intestine (Awad et al., 2010), as did a DFM comprising *L. johnsonii* and *L. crispatus* that resulted in increased bird body weight (Taheri et al., 2010). Furthermore, treatment with Lactobacilli caused an apparent increase in glucose transport in the small and large intestine (Awad et al., 2008) and enhanced the maintenance and function of the gut epithelium, and this also resulted in a slight improvement in weight gain (Awad et al., 2010). A DFM composed of *Clostridium butyricum* was linked to higher villus height and lower crypt depth in broiler chicks (Zhang et al., 2016). *Bacillus* cultures also improved ileal epithelial morphology (Abudabos et al., 2014). Other types of *Bacillus* cultures were fed to broiler chicks and caused increased villus height and crypt depth (Lee et al., 2010). Crypt depths were reduced and villus heights were increased in 28-week-old laying hens supplemented with a *B. licheniformis* DFM (Lei et al., 2013). Further studies with a *B. cereus/Aspergillus oryzae* DFM indicated that DFM treatment improved the intestinal integrity of broiler chickens (Murugesan et al., 2014). Intestinal length and weight were reduced by DFM treatment in hens from 7 to 59 weeks of age (Nahashon et al., 1996).

## 3.3 IMPACTS ON POULTRY IMMUNITY

Chicks have an immature immune system upon hatch, and the humoral response can be enhanced by feeding of probiotics, but this takes time. The innate immune system can respond more rapidly to pathogen threats, and this has been shown to be stimulated by probiotics. Dosing of chicks with probiotic cultures of *B. subtilis*, *Lactococcus lactis* sbsp. *lactis*, and *L. acidophilus* cultures at $10^8$ colony forming units per chick resulted in increased heterophil degranulation and oxidative burst size (Farnell et al., 2006). Probiotic treatment of both a *Lactobacillus* culture and chicory (a prebiotic) showed a decrease in the heterophil to lymphocyte ratios in adult birds subjected to transport stress (Ghareeb et al., 2008). This suggested that probiotic and prebiotic supplementation could attenuate the physiological impacts of stress (Ghareeb et al., 2008), such as the effects of a *Bacillus* DFM that reduced levels of the inflammatory marker α-1-acid glycoprotein (Lee et al., 2010). Other results have also indicated a role for DFM in mediating the effects of heat stress (Roul et al., 2015; Zulkifli et al., 2000).

The humoral immune system can also be impacted by probiotic treatments. It has been suggested by studies that DFM feeding in broilers can lead to a repartitioning of energy to the immune system, increasing (Qiu et al., 2012) or inducing (Haghighi et al., 2006) antibody production. Some probiotics have been shown to increase CD8 production, as well as IgG and IgM concentrations in the serum and the gut of swine (Duncker et al., 2006; Walsh et al., 2008; Zhang et al., 2008). High doses of Lactobacilli cultures given to young broiler chicks (up to 3 weeks) did modulate the immune system, but in older layers a lower dose administered sporadically improved immune response (Koenen et al., 2004). A proprietary culture of lactic acid bacteria decreased the expression levels of the inflammatory markers interferon-γ and interleukin (IL)-1β in growing chicks (Hsu et al., 2016). Chicks treated with a mixture of *L. reuteri*, *B. subtilis*, and *S. cerevisiae* had higher white blood cell counts, monocyte levels, and plasma immunoglobulin levels (Salim et al., 2013). Results can vary dramatically based on the composition of DFM fed to growing chicks; a DFM containing *Propionibacterium* downregulated ileal expression of toll like receptor ( TLR)-2b, IL-2, IL-4, IL-6, IL-10, and IL-13, whereas a DFM containing three *Bacillus* strains downregulated ileal expression of the same interleukins except for IL-10 and IL-13, but downregulated IL-10 and upregulated IL-13 in the spleen (Waititu et al., 2014). A DFM composed of *C. butyricum* was fed to growing broilers and increased tumor necrosis factor-aα and caused greater IL-4 concentrations than did controls (Zhang et al., 2016). A four-species mix of *Lactobacillus* resulted in a reduction of proinflammatory cytokines (IL-1β, lipopolysaccharide-induced tumor necrosis factor-alpha factor) and stimulated cecal tonsil expression of TLR2 (Penha Filho et al., 2015). In other studies, however, the addition of a probiotic, a prebiotic, and a synbiotic all failed to impact serum IgG in growing broilers (Midilli et al., 2008). Collectively, these data indicate that the type and dose of culture or synbiotic, as well as age of birds, can play critical roles in probiotic effectiveness and impact on both branches of the immune system.

## 4. CONCLUSIONS

Collectively, the data from around the world suggest that probiotic approaches can enhance poultry production efficiency, health, and, ultimately, food safety. Probiotics can alter the microbiome of the gut, which, in turn, impacts gut architecture, bone structure, and the immune system. All of these factors in turn affect egg production and efficiency, as well as animal health and well-being, all of which have an important bearing on the bottom line of producers. Increasing evidence has shown that probiotic approaches can reduce food-borne pathogens in laying hens and therefore on eggs, which can be a powerful adjunct to improve public health that can be performed on the farm (Steinmuller et al., 2006). Unfortunately, probiotics are not "one size fits all," and the type of probiotic approach (i.e., CE, DFM, or synbiotic), composition (e.g., *Lactobacillus*, *Bacillus*, *Saccharomyces*), and timing of application will vary based on the age of the hen, breed, environmental conditions, production demands, pathogens endemic to the farm, and the nutritional plane of the hen. Advances in understanding how the populations of the gut change and how these specifically impact animal health are changing how we manage the gastrointestinal microbiota, and this will allow the development of probiotic products that are tailored to meet specific production needs and conditions.

## REFERENCES


Abdel-Wareth, A.A.A., 2016. Effect of dietary supplementation of thymol, synbiotic and their combination on performance, egg quality and serum metabolic profile of Hy-Line Brown hens. Brit. Poult. Sci. 57, 114–122.

Abudabos, A.M., Yehia, H.M., Alotybi, M.N., Garelnabi, A.R., Alyemni, A.H., 2014. Effects of direct-fed probiotics on broiler performance and susceptibility to oral *Salmonella* Enteritidis challenge. J. Food Agric. Environ. 12, 30–34.

Al-Qumber, M., Tagg, J.R., 2006. Commensal bacilli inhibitory to mastitis pathogens isolated from the udder microbiota of healthy cows. J. Appl. Microbiol. 101, 1152–1160.

Applegate, T.J., Onyango, E.M., Angel, R., Powers, W.J., 2009. Effect of amino acid formulation and dietary direct-fed microbial supplementation on egg production and egg characteristics in laying hens. J. Appl. Poult. Res. 18, 552–561.

Awad, W.A., Ghareeb, K., Böhm, J., 2010. Effect of addition of a probiotic micro-organism to broiler diet on intestinal mucosal architecture and electrophysiological parameters. J. Anim. Physiol. Anim. Nutr. 94, 486–494.

Awad, W.A., Ghareeb, K., Nitsch, S., Pasteiner, S., Abdel-Raheem, S., Böhm, J., 2008. Effects of dietary inclusion of prebiotic, probiotic and synbiotic on the intestinal glucose absorption of broiler chickens. Int. J. Poult. Sci. 7, 686–691.

Bailey, M.T., Dowd, S.E., Galley, J.D., Hufnagle, A.R., Allen, R.G., Lyte, M., 2011. Exposure to a social stressor alters the structure of the intestinal microbiota: implications for stressor-induced immunomodulation. Brain Behav. Immun. 25, 397–407.

Barroga, A.J., Salas, R.C.D., Martin, E.A., Besa, W.G., Santos, C.A.D.C., 2007. Efficacy of probiotics, enzymes and dried porcine solubles in swine. Philip. Agric. Sci. 90, 71–74.

Bell, D.D., 2003. Historical and current molting practices in the U.S. table egg industry. Poult. Sci. 82, 965–970.

Benson, A.K., Kelly, S.A., Legge, R., Ma, F., Low, S.J., Kim, J., Zhang, M., Oh, P.L., Nehrenberg, D., Hua, K., Kachman, S.D., Moriyama, E.N., Walter, J., Peterson, D.A., Pomp, D., 2010. Individuality in gut microbiota composition is a complex polygenic trait shaped by multiple environmental and host genetic factors. Proc. Natl. Acad. Sci. U.S.A. 107, 18933–18938.

Berry, W.D., 2003. The physiology of induced molting. Poult. Sci. 82, 971–980.

Bielke, L.R., Elwood, A.L., Donoghue, D.J., Donoghue, A.M., Newberry, L.A., Neighbor, N.K., Hargis, B.M., 2003. Approach for selection of individual enteric bacteria for competitive exclusion in turkey poults. Poult. Sci. 82, 1378–1382.

Board, A.E., 2016. About the U. S. Egg Industry.

Borland, E.D., 1975. *Salmonella* infection in poultry. Vet. Rec. 97, 406–408.

Bozkurt, M., Küçükyilmaz, K., Ayhan, V., Çabuk, M., Çatlil, A.U., 2011. Performance of layer or broiler breeder hens varies in response to different probiotic preparations. Ital. J. Anim. Sci. 10, 162–169.

Callaway, T.R., Dowd, S.E., Wolcott, R.D., Sun, Y., McReynolds, J.L., Edrington, T.S., Byrd, J.A., Anderson, R.C., Krueger, K., Nisbet, D.J., 2009. Evaluation of the bacterial diversity in cecal contents of laying hens fed various molting diets using bacterial tag-encoded FLX amplicon pyrosequencing (bTEFAP). Poult. Sci. 88, 298–302.

Callaway, T.R., Edrington, T.S., Brabban, A.D., Keen, J.E., Anderson, R.C., Rossman, M.L., Engler, M.J., Genovese, K.J., Gwartney, B.L., Reagan, J.O., Poole, T.L., Harvey, R.B., Kutter, E.M., Nisbet, D.J., 2006. Fecal prevalence of *Escherichia coli* O157, *Salmonella, Listeria*, and bacteriophage infecting *E. coli* O157:H7 in feedlot cattle in the southern plains region of the United States. Foodborne Path. Dis. 3, 234–244.

Charles, J.F., Ermann, J., Aliprantis, A.O., 2015. The intestinal microbiome and skeletal fitness: connecting bugs and bones. Clin. Immunol. 159, 163–169.

Chichlowski, M., Croom, J., McBride, B.W., Havenstein, G.B., Koci, M.D., 2007. Metabolic and physiological impact of probiotics or direct-fed-microbials on poultry: a brief review of current knowledge. Int. J. Poult. Sci. 6, 694–704.

Collins, D.M., Gibson, G.R., 1999. Probiotics, prebiotics, and synbiotics: approaches for modulating the microbial ecology of the gut. Amer. J. Clin. Nutr. 69, 1052S–1057S.

Corrier, D.E., Hargis, B.M., Hinton, A., DeLoach, J.R., 1993. Protective effect of used poultry litter and lactose in the feed ration on *Salmonella enteritidis* colonization of leghorn chicks and hens. Avian Dis. 37, 47–52.

Corrier, D.E., Hinton, A., Ziprin, R.L., DeLoach, J.R., 1990. Effect of dietary lactose on *Salmonella* colonization of market age broiler chickens. Avian Dis. 34, 668–676.

Corrier, D.E., Hinton Jr., A., Ziprin, R.L., Beier, R.C., DeLoach, J.L., 1991. Effect of dietary lactose and anaerobic cultures of cecal flora on *Salmonella* colonization of broiler chicks. In: Blankenship, L.C. (Ed.), Colonization Control of Human Bacterial Enteropathogens in Poultry. Academic Press, Inc., New York, pp. 299–316.

Corrier, D.E., Nisbet, D.J., Hargis, B.M., Holt, P.S., DeLoach, J.R., 1997. Provision of lactose to molting hens enhances resistance to *Salmonella enteriditis* colonization. J. Food Prot. 60, 10–15.

Cox, N.A., Bailey, J.S., Mauldin, J.M., Blankenship, L.C., 1990. Research note: presence and impact of *Salmonella* contamination in commercial broiler hatcheries. Poult. Sci. 69, 1606–1608.

Da Silva, W.T.M., Nunes, R.V., Pozza, P.C., Dos Santos Pozza, M.S., Appelt, M.D., Eyng, C., 2011. Evaluation of inulin and probiotic for broiler chickens. Acta Scient. Anim. Sci. 33, 19–24.

Davis, G.S., Anderson, K.E., 2002. The effects of feeding the direct-fed microbial, PrimaLac, on growth parameters and egg production in single comb white leghorn hens. Poult. Sci. 81, 755–759.

De Keersmaecker, S.C.J., Verhoeven, T.L.A., Desair, J., Marchal, K., Vanderleyden, J., Nagy, I., 2006. Strong antimicrobial activity of *Lactobacillus rhamnosus* GG against *Salmonella* Typhimurium is due to accumulation of lactic acid. FEMS Microbiol. Lett. 259, 89–96.

Deng, W., Dong, X.F., Tong, J.M., Zhang, Q., 2012. The probiotic *Bacillus licheniformis* ameliorates heat stress-induced impairment of egg production, gut morphology, and intestinal mucosal immunity in laying hens. Poult. Sci. 91, 575–582.

Donalson, L.M., Kim, W.K., Woodward, C.L., Herrera, P., Kubena, L.F., Nisbet, D.J., Ricke, S.C., 2005. Utilizing different ratios of alfalfa and layer ration for molt induction and performance in commercial laying hens. Poult. Sci. 84, 362–369.

Donalson, L.M., McReynolds, J.L., Kim, W.K., Chalova, V.I., Woodward, C.L., Kubena, L.F., Nisbet, D.J., Ricke, S.C., 2008. The influence of a fructooligosaccharide prebiotic combined with alfalfa molt diets on the gastrointestinal tract fermentation, *Salmonella* Enteritidis infection, and intestinal shedding in laying hens. Poult. Sci. 87, 1253–1262.

Doyle, M.P., Erickson, M.C., 2006. Reducing the carriage of foodborne pathogens in livestock and poultry. Poult. Sci. 85, 960–973.

Duarte, K.F., Ibuki, M., Fukui, K., Kato, M., Santos, E.T., Junqueira, O.M., 2014. Effect of hydrolyzed copra meal separately or in combination with *Bacillus cereus* var. *toyoi* on growth performance of broiler chickens. Acta Scient. Anim. Sci. 36, 373–377.

Duncker, S.C., Lorentz, A., Schroeder, B., Breves, G., Bischoff, S.C., 2006. Effect of orally administered probiotic *E. coli* strain Nissle 1917 on intestinal mucosal immune cells of healthy young pigs. Vet. Immunol. Immunopath 111, 239–250.

Dunkley, C.S., McReynolds, J.L., Dunkley, K.D., Njongmeta, L.N., Berghman, L.R., Kubena, L.F., Nisbet, D.J., Ricke, S.C., 2007a. Molting in Salmonella Enteritidis-challenged laying hens fed alfalfa crumbles. IV. Immune and stress protein response. Poult. Sci. 86, 2502–2508.

Dunkley, K.D., McReynolds, J.L., Hume, M.E., Dunkley, C.S., Callaway, T.R., Kubena, L.F., Nisbet, D.J., Ricke, S.C., 2007b. Molting in *Salmonella* enteritidis-challenged laying hens fed alfalfa crumbles. I. *Salmonella* enteritidis colonization and virulence gene *hil*A response. Poult. Sci. 86, 1633–1639.

Dunkley, K.D., McReynolds, J.L., Hume, M.E., Dunkley, C.S., Callaway, T.R., Kubena, L.F., Nisbet, D.J., Ricke, S.C., 2007c. Molting in *Salmonella* enteritidis-challenged laying hens fed alfalfa crumbles. II. Fermentation and microbial ecology response. Poult. Sci. 86, 2101–2109.

Dunkley, K.D., Callaway, T.R., Chalova, V.I., McReynolds, J.L., Hume, M.E., Dunkley, C.S., Kubena, L.F., Nisbet, D.J., Ricke, S.C., 2009. Foodborne *Salmonella* ecology in the avian gastrointestinal tract. Anaerobe 15, 26–35.

Durant, J.A., Corrier, D.E., Byrd, J.A., Stanker, L.H., Ricke, S.C., 1999. Feed deprivation affects crop environment and modulates *Salmonella* Enteritidis colonization and invasion of leghorn hens. Appl. Environ. Microbiol. 65, 1919–1923.

El-Deek, A.A., Ghonem, K.M., Hamdy, S.M., Aser, M.A., Aljassas, F.M., Osman, M.M., 2009. Producing single cell protein from poultry manure and evaluation for broiler chickens diets. Int. J. Poult. Sci. 8, 1062–1077.

Endt, K., Stecher, B., Chaffron, S., Slack, E., Tchitchek, N., Benecke, A., Maele, L.V., Sirard, J.-C., Mueller, A.J., Heikenwalder, M., Macpherson, A.J., Strugnell, R., Mering, C.V., Hardt, W.-D., 2010. The microbiota mediates pathogen clearance from the gut lumen after non-typhoidal *Salmonella* diarrhea. PLoS Pathog. 6, 1–18.

Escarcha, J.F., Callaway, T.R., Byrd, J.A., Miller, D.N., Edrington, T.S., Anderson, R.C., Nisbet, D.J., 2012. Effects of dietary alfalfa inclusion on *Salmonella* Typhimurium populations in growing layer chicks. Foodborne Path. Dis. 9, 945–951.

Falaki, M., Shargh, M.S., Dastar, B., Zrehdaran, S., 2010. Effects of different levels of probiotic and prebiotic on performance and carcass characteristics of broiler chickens. J. Anim. Vet. Adv. 9, 2390–2395.

Falk, P.G., Hooper, L.V., Midtvedt, T., Gordon, J.I., 1998. Creating and maintaining the gastrointestinal ecosystem: what we know and need to know from gnotobiology. Microbiol. Mol. Biol. Rev. 62, 1157–1170.

Farnell, M.B., Donoghue, A.M., Solis de los Santos, F., Blore, P.J., Hargis, B.M., Tellez, G., Donoghue, D.J., 2006. Upregulation of oxidative burst and degranulation in chicken heterophils stimulated with probiotic bacteria. Poult. Sci. 85, 1900–1906.

Finegold, S.M., 2008. Therapy and epidemiology of autism-clostridial spores as key elements. Med. Hypoth 70, 508–511.

Freestone, P., Lyte, M., 2010. Stress and microbial endocrinology: prospects for ruminant nutrition. Animal 4, 1248–1257.

Fuller, R., 1989. Probiotics in man and animals. J. Appl. Bacteriol. 66, 365–378.

Galiş, A.M., Marcq, C., Marlier, D., Portetelle, D., Van, I., Beckers, Y., Théwis, A., 2013. Control of *Salmonella* contamination of shell eggs-preharvest and postharvest methods: a review. Comp. Rev. Food Sci. Food Safe 12, 155–182.

Ghareeb, K., Awad, W.A., Mohnl, M., Porta, R., Biarnés, M., Böhm, J., Schatzmayr, G., 2012. Evaluating the efficacy of an avian-specific probiotic to reduce the colonization of *Campylobacter jejuni* in broiler chickens. Poult. Sci. 91, 1825–1832.

Ghareeb, K., Awad, W.A., Nitsch, S., Abdel-Raheem, S., Böhm, J., 2008. Effects of transportation on stress and fear responses of growing broilers supplemented with prebiotic or probiotic. Int. J. Poult. Sci. 7, 678–685.

Haghighi, H.R., Gong, J., Gyles, C.L., Hayes, M.A., Zhou, H., Sanei, B., Chambers, J.R., Sharif, S., 2006. Probiotics stimulate production of natural antibodies in chickens. Clin. Vacc. Immunol. 13, 975–980.

Han, S.W., Lee, K.W., Lee, B.D., Sung, C.G., 1999. Effect of feeding *Aspergillus Oryzae* culture on fecal microflora, egg qualities, and nutrient metabolizabilities in laying hens. Asian-Austral. J. Anim. Sci. 12, 417–421.

Hoffmann, S., Maculloch, B., Batz, M.B., 2015. Economic Burden of Major Foodborne Illnesses Acquired in the United States, Economic Information Bulletin. Economic Research Service/USDA, Washington, DC, p. 59.

Holt, P.S., 1993. Effect of induced molting on the susceptibility of white leghorn hens to a Salmonella enteritidis infection. Avian Dis. 37, 412–417.

Holt, P.S., 2003. Molting and *Salmonella enterica* serovar *Enteritidis* infection: the problem and some solutions. Poult. Sci. 82, 1008–1010.

Hsu, Y.M., Yu, B., Tseng, C.S., Chang, C.H., Chen, D.S., Su, C.H., Chen, Y.S., 2016. Preventive activities of *Scutellariae Radix, Gardeniae Fructus*, and probiotics in *Salmonella enterica* serovar Typhimurium infection in chickens. Anim. Feed Sci. Technol. 214, 121–129.

Isik, M., Ekimler, F., Ozen, N., Firat, M.Z., 2004. Effects of using probiotics on the growth performance and health of dairy calves. Turk. J. Vet. Anim. Sci. 28, 63–69.

Jack, R.W., Tagg, J.R., Ray, B., 1995. Bacteriocins of gram-positive bacteria. Microbiol. Rev. 59, 171–200.

Jarquin, R.L., Nava, G.M., Wolfenden, A.D., Donoghue, A.M., Hanning, I., Higgins, S.E., Hargis, B.M., 2007. The evaluation of organic acids and probiotic cultures to reduce *Salmonella* Enteriditis horizontal transmission and crop infection in broiler chickens. Int. J. Poult. Sci. 6, 182–186.

Jayne-Williams, D.J., Fuller, R., 1971. The influence of the intestinal microflora on nutrition. In: Bell, D.J., Freeman, B.M. (Eds.), Physiology and Biochemistry of the Domestic Food. Academic Press, London, UK, pp. 74–92.

Joerger, R.D., 2003. Alternatives to antibiotics: bacteriocins, antimicrobial peptides and bacteriophages. Poult. Sci. 82, 640–647.

Katoch, S., Sharma, K.A., Chahota, R., Sharma, K.S., Radmila, M., Šefer, D., 2013. Performance of broiler chickens fed varied nutrient density diets supplemented with direct fed microbial. Acta Veterin 63, 643–653.

Klasing, K.C., 2007. Nutrition and the immune system. Brit. Poult. Sci. 48, 525–537.

Knap, I., Kehlet, A.B., Bennedsen, M., Mathis, G.F., Hofacre, C.L., Lumpkins, B.S., Jensen, M.M., Raun, M., Lay, A., 2011. *Bacillus subtilis* (DSM17299) significantly reduces *Salmonella* in broilers. Poult. Sci. 90, 1690–1694.

Koenen, M.E., Kramer, J., Van Der Hulst, R., Heres, L., Jeurissen, S.H.M., Boersma, W.J.A., 2004. Immunomodulation by probiotic lactobacilli in layer- and meat-type chickens. Brit. Poult. Sci. 45, 355–366.

Kogut, M.H., Genovese, K.J., Stanker, L.H., 1999. Effect of induced molting on heterophil function in white leghorn hens. Avian Dis. 43, 538–548.

Landers, K.L., Howard, Z.R., Woodward, C.L., Birkhold, S.G., Ricke, S.C., 2005. Potential of alfalfa as an alternative molt induction diet for laying hens: egg quality and consumer acceptability. Biores. Technol. 96, 907–911.

Lee, K.W., Lee, S.H., Lillehoj, H.S., Li, G.X., Jang, S.I., Babu, U.S., Park, M.S., Kim, D.K., Lillehoj, E.P., Neumann, A.P., Rehberger, T.G., Siragusa, G.R., 2010. Effects of direct-fed microbials on growth performance, gut morphometry, and immune characteristics in broiler chickens. Poult. Sci. 89, 203–216.

Lei, K., Li, Y.L., Yu, D.Y., Rajput, I.R., Li, W.F., 2013. Influence of dietary inclusion of *Bacillus licheniformis* on laying performance, egg quality, antioxidant enzyme activities, and intestinal barrier function of laying hens. Poult. Sci. 92, 2389–2395.

LeJeune, J.T., Kauffman, M.D., Amstutz, M.D., Ward, L.A., 2006. Limited effects of a commercial direct-fed microbial on weaning pig performance and gastrointestinal microbiology. J. Swine Health Prod. 14, 247–252.

Ley, R.E., Turnbaugh, P.J., Klein, S., Gordon, J.I., 2006. Human gut microbes associated with obesity. Nature 444, 1022–1023.

Lima, E.T., Andreatti Filho, R.L., Okamoto, A.S., Noujaim, J.C., Barros, M.R., Crocci, A.J., 2007. Evaluation in vitro of the antagonistic substances produced by *Lactobacillus* spp. isolated from chickens. Can. J. Vet. Res. 71, 103–107.

Line, J.E., Bailey, J.S., Cox, N.A., Stern, N.J., Tompkins, T., 1998. Effect of yeast-supplemented feed on *Salmonella* and *Campylobacter* populations in broilers. Poult. Sci. 77, 405–410.

Lloyd, A.B., Cumming, R.B., Kent, R.D., 1974. Competitive exclusion as exemplified by *Salmonella typhimurium*. Australasian Poult. Sci. Conv., World Poult. Sci. Assoc. Austral. Br. 155.

Lloyd, A.B., Cumming, R.B., Kent, R.D., 1977. Prevention of *Salmonella typhimurium* infection in poultry by pre-treatment of chickens and poults with intestinal extracts. Aust. Vet. J. 53, 82–87.

Lodemann, U., Lorenz, B.M., Weyrauch, K.D., Martens, H., 2008. Effects of *Bacillus cereus* var. *toyoi* as probiotic feed supplement on intestinal transport and barrier function in piglets. Arch. Anim. Nutr. 62, 87–106.

Lu, J., Idris, U., Hofacre, C., Maurer, J.J., Lee, M.D., Harmon, B., 2003. Diversity and succession of the intestinal bacterial community of the maturing broiler chicken. Appl. Environ. Microbiol. 69, 6816–6824.

Lyte, M., 2010. The microbial organ in the gut as a driver of homeostasis and disease. Med. Hypoth 74, 634–638.

Ma, Q.G., Gao, X., Zhou, T., Zhao, L.H., Fan, Y., Li, X.Y., Lei, Y.P., Ji, C., Zhang, J.Y., 2012. Protective effect of *Bacillus subtilis* ANSB060 on egg quality, biochemical and histopathological changes in layers exposed to aflatoxin B1. Poult. Sci. 91, 2852–2857.

Mappley, L.J., Tchórzewska, M.A., Nunez, A., Woodward, M.J., Bramley, P.M., la Ragione, R.M., 2013. Oral treatment of chickens with *Lactobacillus reuteri* LM1 reduces *Brachyspira pilosicoli*-induced pathology. J. Med. Microbiol. 62, 287–296.

McCabe, L., Britton, R.A., Parameswaran, N., 2015. Prebiotic and probiotic regulation of bone health: role of the intestine and its microbiome. Curr. Osteopor. Rep. 13, 363–371.

Menconi, A., Wolfenden, A.D., Shivaramaiah, S., Terraes, J.C., Urbano, T., Kuttel, J., Kremer, C., Hargis, B.M., Tellez, G., 2011. Effect of lactic acid bacteria probiotic culture for the treatment of *Salmonella enterica* serovar Heidelberg in neonatal broiler chickens and Turkey poults. Poult. Sci. 90, 561–565.

Midilli, M., Alp, M., Kocabagli, N., Muglali, O.H., Turan, N., Yilmaz, H., Cakir, S., 2008. Effects of dietary probiotic and prebiotic supplementation on growth performance and serum IgG concentration of broilers. S. Afr. J. Anim. Sci. 38, 21–27.

Miyamoto, T., Horie, T., Fujiwara, T., Fukata, T., Sasai, K., Baba, E., 2000. *Lactobacillus* flora in the cloaca and vagina of hens and its inhibitory activity against *Salmonella enteritidis* in vitro. Poult. Sci. 79, 7–11.

Mokhtari, R., Yazdani, A.R., Rezaei, M., Ghorbani, B., 2010. The effects of different growth promoters on performance and carcass characteristics of broiler chickens. J. Anim. Vet. Adv. 9, 2633–2639.

Mountzouris, K.C., Palamidi, I., Tsirtsikos, P., Mohnl, M., Schatzmayr, G., Fegeros, K., 2015. Effect of dietary inclusion level of a multi-species probiotic on broiler performance and two biomarkers of their caecal ecology. Anim. Prod. Sci. 55, 484–493.

Murphy, M., 2004. Bacteria could treat symptoms of autism. Chem. Indust. (London) 10, 6.

Murugesan, G.R., Gabler, N.K., Persia, M.E., 2014. Effects of direct-fed microbial supplementation on broiler performance, intestinal nutrient transport and integrity under experimental conditions with increased microbial challenge. Brit. Poult. Sci. 55, 89–97.

Mutuş, R., Kocabağli, N., Alp, M., Acar, N., Eren, M., Gezen, Ş.Ş., 2006. The effect of dietary probiotic supplementation on tibial bone characteristics and strength in broilers. Poult. Sci. 85, 1621–1625.

Nahashon, S.N., Nakaue, H.S., Mirosh, L.W., 1996. Performance of single comb white leghorn fed a diet supplemented with a live microbial during the growth and egg laying phases. Anim. Feed Sci. Technol. 57, 25–38.

Nahashon, S.N., Nakaue, H.S., Snyder, S.P., Mirosh, L.W., 1994. Performance of single comb white leghorn layers fed corn-soybean meal and barley-corn-soybean meal diets supplemented with a direct-fed microbial. Poult. Sci. 73, 1712–1723.

Nava, G.M., Bielke, L.R., Callaway, T.R., Castaneda, M.P., 2005. Probiotic alternatives to reduce gastrointestinal infections: the poultry experience. Anim. Health Res. Rev. 6, 105–118.

Neal-McKinney, J.M., Lu, X., Duong, T., Larson, C.L., Call, D.R., Shah, D.H., Konkel, M.E., 2012. Production of organic acids by probiotic lactobacilli can be used to reduce pathogen load in poultry. PLoS One 7.

Nettelbladt, C.G., Katouli, M., Volpe, A., Bark, T., Muratov, V., Svenberg, T., Mollby, R., Ljungqvist, O., 1997. Starvation increases the number of coliform bacteria in the caecum and induces bacterial adherence to caecal epithelium in rats. Eur. J. Surg. 163, 135–142.

Nisbet, D.J., Corrier, D.E., Ricke, S.C., Hume, M.E., Byrd, J.A., DeLoach, J.R., 1996. Maintenance of the biological efficacy in chicks of a cecal competitive-exclusion culture against *Salmonella* by continuous-flow fermentation. J. Food Prot. 59, 1279–1283.

Nisbet, D.J., Corrier, D.E., Scanlan, C.M., Hollister, A.G., Beier, R.C., DeLoach, J.R., 1993a. Effect of a defined continuous-flow derived bacterial culture and dietary lactose on *Salmonella typhimurium* colonization in broiler chickens. Avian Dis. 37, 1017–1025.

Nisbet, D.J., Corrier, D.E., Scanlan, C.M., Hollister, A.G., Beier, R.C., DeLoach, J.R., 1993b. Effect of a defined continuous flow derived bacterial culture and dietary lactose on Salmonella colonization in broiler chicks. Avian Dis. 37, 1017–1025.

Nisbet, D.J., Ricke, S.C., Scanlan, C.M., Corrier, D.E., Hollister, A.G., DeLoach, J.R., 1994. Inoculation of broiler chicks with a continuous-flow derived bacterial culture facilitates early cecal bacterial colonization and increases resistance to *Salmonella typhimurium*. J. Food Prot. 57, 12–15.

Nurmi, E., Nuotio, L., Schncitz, C., 1992. The competitive exclusion concept: development and future. Int. J. Food Microbiol. 15, 237–240.

Oliver, S.P., Jayarao, B.M., Almeida, R.A., 2005. Foodborne pathogens in milk and the dairy farm environment: food safety and public health implications. Foodborne Path. Dis. 2, 115–129.

Panda, A.K., Reddy, M.R., Rama Rao, S.V., Praharaj, N.K., 2003. Production performance, serum/yolk cholesterol and immune competence of white leghorn layers as influenced by dietary supplementation with probiotic. Trop. Anim. Health Prod. 35, 85–94.

Park, S.H., Hanning, I., Perrota, A., Bench, B.J., Alm, E., Ricke, S.C., 2013. Modifying the gastrointestinal ecology in alternatively raised poultry and the potential for molecular and metabolomic assessment. Poult. Sci. 92, 546–561.

Pascual, M., Hugas, M., Badiola, J.I., Monfort, J.M., Garriga, M., 1999. *Lactobacillus salivarius* CTC2197 prevents *Salmonella* Enteritidis colonization in chickens. Appl. Environ. Microbiol. 65, 4981–4986.

Patterson, J.A., Burkholder, K.M., 2003. Application of prebiotics and probiotics in poultry production. Poult. Sci. 82, 627–631.

Penha Filho, R.A.C., Díaz, S.J.A., Fernando, F.S., Chang, Y.F., Andreatti Filho, R.L., Berchieri Junior, A., 2015. Immunomodulatory activity and control of *Salmonella* Enteritidis colonization in the intestinal tract of chickens by *Lactobacillus* based probiotic. Vet. Immunol. Immunopathol. 167, 64–69.

Pontier-Bres, R., Prodon, F., Munro, P., Rampal, P., Lemichez, E., Peyron, J.F., Czerucka, D., 2012. Modification of *Salmonella* Typhimurium motility by the probiotic yeast strain *Saccharomyces boulardii*. PLoS One 7.

Porter, R.E., Holt, P.S., 1993. Effect of induced molting on the severity of intestinal lesions caused by Salmonella Enteritidis infection in chickens. Avian Dis. 37, 1009–1016.

Price, K.L., Totty, H.R., Lee, H.B., Utt, M.D., Fitzner, G.E., Yoon, I., Ponder, M.A., Escobar, J., 2010. Use of *Saccharomyces cerevisiae* fermentation product on growth performance and microbiota of weaned pigs during *Salmonella* infection. J. Anim. Sci. 88, 3896–3908.

Prohaszka, L., Baron, F., 1983. Antibacterial effect of volatile fatty acids on *Enterobacteriae* in the large intestine. Acta Vet. Hung 30, 9–16.

Pullinger, G.D., Van Diemen, P.M., Carnell, S.C., Davies, H., Lyte, M., Stevens, M.P., 2010. 6-hydroxydopamine-mediated release of norepinephrine increases faecal excretion of *Salmonella enterica* serovar Typhimurium in pigs. Vet. Res. 41.

Qiu, R., Croom, J., Ali, R.A., Ballou, A.L., Smith, C.D., Ashwell, C.M., Hassan, H.M., Chiang, C.C., Koci, M.D., 2012. Direct fed microbial supplementation repartitions host energy to the immune system. J. Anim. Sci. 90, 2639–2651.

Ribeiro, V., Albino, L.F.T., Rostagno, H.S., Barreto, S.L.T., Hannas, M.I., Harrington, D., de Araujo, F.A., Ferreira, H.C., Ferreira, M.A., 2014. Effects of the dietary supplementation of *Bacillus subtilis* levels on performance, egg quality and excreta moisture of layers. Anim. Feed Sci. Technol. 195, 142–146.

Ricke, S.C., 2003. The gastrointestinal tract ecology of *Salmonella* Enteritidis colonization in molting hens. Poult. Sci. 82, 1003–1007.

Ricke, S.C., Dunkley, C.S., Durant, J.A., 2013. A review on development of novel strategies for controlling *Salmonella* Enteritidis colonization in laying hens: fiber-based molt diets. Poult. Sci. 92, 502–525.

Roul, G., Panda, N., Mishra, P.K., Behera, P.C., Panigrahi, B., Pati, P.K., 2015. Effect of supplementation of probiotics and prebiotics on growth, lipid profile and slaughter traits of coloured broilers during heat stress. Ind. J. Anim. Sci. 85, 72–76.

Salim, H.M., Kang, H.K., Akter, N., Kim, D.W., Kim, J.H., Kim, M.J., Na, J.C., Jong, H.B., Choi, H.C., Suh, O.S., Kim, W.K., 2013. Supplementation of direct-fed microbials as an alternative to antibiotic on growth performance, immune response, cecal microbial population, and ileal morphology of broiler chickens. Poult. Sci. 92, 2084–2090.

Salma, U., Miah, A.G., Tsujii, H., Schellander, K., Südekum, K.H., 2012. Effect of dietary *Rhodobacter capsulatus* on lipid fractions and egg-yolk fatty acid composition in laying hens. J. Anim. Physiol. Anim. Nutr. 96, 1091–1100.

Scallan, E., Hoekstra, R.M., Angulo, F.J., Tauxe, R.V., Widdowson, M.-A., Roy, S.L., Jones, J.L., Griffin, P.L., 2011. Foodborne illness acquired in the United States—major pathogens. Emerg. Infect. Dis. 17, 7–15.

Schamberger, G.P., Phillips, R.L., Jacobs, J.L., Diez-Gonzalez, F., 2004. Reduction of *Escherichia coli* O157:H7 populations in cattle by addition of colicin E7-producing *E. coli* to feed. Appl. Environ. Microbiol. 70, 6053–6060.

Scharek, L., Guth, J., Filter, M., Schmidt, M.F.G., 2007. Impact of the probiotic bacteria *Enterococcus faecium* NCIMB 10415 (SF68) and *Bacillus cereus* var. *toyoi* NCIMB 40112 on the development of serum IgG and faecal IgA of sows and their piglets. Arch. Anim. Nutr. 61, 223–234.

Scharff, R.L., 2012. Economic burden from health losses due to foodborne illness in the United States. J. Food Prot. 75, 123–131.

Schierack, P., Wieler, L.H., Taras, D., Herwig, V., Tachu, B., Hlinak, A., Schmidt, M.F.G., Scharek, L., 2007. *Bacillus cereus* var. *toyoi* enhanced systemic immune response in piglets. Vet. Immunol. Immunopath. 118, 1–11.

Schneitz, C., 2005. Competitive exclusion in poultry–30 years of research. Food Cont. 16, 657–667.

Schrezenmeir, J., De Vrese, M., 2001. Probiotics, prebiotics, and synbiotics-approaching a definition. Am. J. Clin. Nutr. 73 (Suppl.), 354s–361s.

Smith, J.M., 2014. A review of avian probiotics. J. Avian Med. Surg. 28, 87–94.

Sobczak, A., Kozłowski, K., 2015. The effect of a probiotic preparation containing *Bacillus subtilis* ATCC PTA-6737 on egg production and physiological parameters of laying hens. Ann. Anim. Sci. 15, 711–723.

Sohail, M.U., Hume, M.E., Byrd, J.A., Nisbet, D.J., Ijaz, A., Sohail, A., Shabbir, M.Z., Rehman, H., 2012. Effect of supplementation of prebiotic mannan-oligosaccharides and probiotic mixture on growth performance of broilers subjected to chronic heat stress. Poult. Sci. 91, 2235–2240.

Sohail, S.S., Bryant, M.M., Voitle, R.A., Roland Sr, D.A., 2002. Influence of calsporin on commercial leghorns. J. Appl. Poult. Res. 11, 379–387.

Stavric, S., 1992. Defined cultures and prospects. Int. J. Food Microbiol. 55, 245–263.

Stavric, S., D'Aoust, J.-Y., 1993. Undefined and defined bacterial preparations for competitive exclusion of *Salmonella* in poultry. J. Food Prot. 56, 173–180.

Stavric, S., Kornegay, E.T., 1995. Microbial probiotics for pigs and poultry. In: Wallace, R.J., Chesson, A. (Eds.), Biotechnology in Animal Feeds and Animal Feeding. VCH, New York, NY, pp. 205–230.

Steer, T., Carpenter, H., Tuohy, K., Gibson, G.R., 2000. Perspectives on the role of the human gut microbiota and its modulation by pro and prebiotics. Nutr. Res. Rev. 13, 229–254.

Steinmuller, N., Demma, L., Bender, J.B., Eidson, M., Angulo, F.J., 2006. Outbreaks of enteric disease associated with animal contact: not just a foodborne problem anymore. Clin. Infect. Dis. 43, 1596–1602.

Taheri, H.R., Moravej, H., Tabandeh, F., Zaghari, M., Shivazad, M., 2010. Efficacy of combined or single use of *Lactobacillus crispatus* lt116 and *L. johnsonii* lt171 on broiler performance. Brit. Poult. Sci. 51, 580–585.

Tang, S.G.H., Sieo, C.C., Kalavathy, R., Saad, W.Z., Yong, S.T., Wong, H.K., Ho, Y.W., 2015. Chemical compositions of egg yolks and egg quality of laying hens fed prebiotic, probiotic, and synbiotic diets. J. Food Sci. 80, C1686–C1695.

Taylor, D.J., 2001. Effects of antimicrobials and their alternatives. Brit. Poult. Sci. 42 (Suppl.), S67–S68.

Tellez, G., Petrone, V.M., Escorcia, M., Morishita, T.Y., Cobb, C.W., Villasenõr, L., 2001. Evaluation of avian-specific probiotic and *Salmonella* Enteritidis-, *Salmonella* Typhimurium-, and *Salmonella* Heidelberg-specific antibodies on cecal colonization and organ invasion of *Salmonella* Enteritidis in broilers. J. Food Prot. 64, 287–291.

Trampel, D.W., Holder, T.G., Gast, R.K., 2014. Integrated farm management to prevent *Salmonella* Enteritidis contamination of eggs. J. Appl. Poult. Res. 23, 353–365.

Turnbaugh, P.J., Hamady, M., Yatsunenko, T., Cantarel, B.L., Duncan, A., Ley, R.E., Sogin, M.L., Jones, W.J., Roe, B.A., Affourtit, J.P., Egholm, M., Henrissat, B., Heath, A.C., Knight, R., Gordon, J.I., 2009. A core gut microbiome in obese and lean twins. Nature 457, 480–484.

Turnbaugh, P.J., Ley, R.E., Mahowald, M.A., Magrini, V., Mardis, E.R., Gordon, J.I., 2006. An obesity-associated gut microbiome with increased capacity for energy harvest. Nature 444, 1027–1131.

Upadhaya, S.D., Hossiendoust, A., Kim, I.H., 2016. Probiotics in *Salmonella*-challenged Hy-Line brown layers. Poult. Sci. 95, 1894–1897.

Uyeno, Y., Sekiguchi, Y., Kamagata, Y., 2010. rRNA-based analysis to monitor succession of faecal bacterial communities in Holstein calves. Lett. Appl. Microbiol. 51, 570–577.

Van Coillie, E., Goris, J., Cleenwerck, I., Grijspeerdt, K., Botteldoorn, N., Van Immerseel, F., De Buck, J., Vancanneyt, M., Swings, J., Herman, L., Heyndrickx, M., 2007. Identification of lactobacilli isolated from the cloaca and vagina of laying hens and characterization for potential use as probiotics to control *Salmonella* Enteritidis. J. Appl. Microbiol. 102, 1095–1106.

Van Immerseel, F., De Buck, J., Boyen, F., Pasmans, F., Bertrand, S., Collard, J.M., Saegerman, C., Hooyberghs, J., Haesebrouck, D., Ducatelle, R., 2005. *Salmonella* dans la viande de volaille et dans les oeufs: un danger pour le consommateur qui demande la mise en place d'un programme de lutte efficace. Ann. Med. Vet. 149, 34–48.

Vandenplas, Y., Huys, G., Daube, G., 2015. Probiotics: an update. J. Pediatria 91, 6–21.

Vilà, B., Fontgibell, A., Badiola, I., Esteve-Garcia, E., Jiménez, G., Castillo, M., Brufau, J., 2009. Reduction of *Salmonella enterica* var. Enteritidis colonization and invasion by *Bacillus cereus* var. *toyoi* inclusion in poultry feeds. Poult. Sci. 88, 975–979.

Waititu, S.M., Yitbarek, A., Matini, E., Echeverry, H., Kiarie, E., Rodriguez-Lecompte, J.C., Nyachoti, C.M., 2014. Effect of supplementing direct-fed microbials on broiler performance, nutrient digestibilities, and immune responses. Poult. Sci. 93, 625–635.

Walsh, M.C., Gardiner, G.E., Hart, O.M., Lawlor, P.G., Daly, M., Lynch, B., Richert, B.T., Radcliffe, S., Giblin, L., Hill, C., Fitzgerald, G.F., Stanton, C., Ross, P., 2008. Predominance of a bacteriocin-producing *Lactobacillus salivarius* component of a five-strain probiotic in the porcine ileum and effects on host immune phenotype. FEMS Microbiol. Ecol. 64, 317–327.

Weeks, C.A., Lambton, S.L., Williams, A.G., 2016. Implications for welfare, productivity and sustainability of the variation in reported levels of mortality for laying hen flocks kept in different housing systems: a meta-analysis of ten studies. PLoS One 11, e0146394. http://dx.doi.org/10.1371/journal.pone.0146394.

Willis, W.L., Reid, L., 2008. Investigating the effects of dietary probiotic feeding regimens on broiler chicken production and *Campylobacter jejuni* presence. Poult. Sci. 87, 606–611.

Witte, W., 1999. Antibiotic use in animal husbandry and resistance development in human infections. APUA 16, 1–7.

Wolfenden, A.D., Vicente, J.L., Higgins, J.P., Andreatti Filho, R.L., Higgins, S.E., Hargis, B.M., Tellez, G., 2007. Effect of organic acids and probiotics on *Salmonella* Enteritidis infection in broiler chickens. Int. J. Poult. Sci. 6, 403–405.

Wolin, M.J., 1969. Volatile fatty acids and the inhibition of *Escherichia coli* growth by rumen fluid. Appl. Microbiol. 17, 83–87.

Woodward, C.L., Kwon, Y.M., Kubena, L.F., Byrd, J.A., Moore, R.W., Nisbet, D.J., Ricke, S.C., 2005. Reduction of *Salmonella enterica* serovar Enteritidis colonization and invasion by an alfalfa diet during molt in leghorn hens. Poult. Sci. 84, 185–193.

Xu, C.L., Ji, C., Ma, Q., Hao, K., Jin, Z.Y., Li, K., 2006. Effects of a dried *Bacillus subtilis* culture on egg quality. Poult. Sci. 85, 364–368.

Xu, J., Gordon, J.I., 2003. Honor thy symbionts. Proc. Nat. Acad. Sci. (USA) 100, 10452–10459.

Yeoman, C.J., White, B.A., 2014. Gastrointestinal tract microbiota and probiotics in production animals. Ann. Rev. Anim. Biosci. 2, 469–486.

Zhang, G., Ma, L., Doyle, M.P., 2007. Potential competitive exclusion bacteria from poultry inhibitory to *Campylobacter jejuni* and *Salmonella*. J. Food Prot. 70, 867–873.

Zhang, L., Zhan, X., Zeng, X., Zhou, L., Cao, G., Chen, A., Yang, C., 2016. Effects of dietary supplementation of probiotic, *Clostridium butyricum*, on growth performance, immune response, intestinal barrier function, and digestive enzyme activity in broiler chickens challenged with *Escherichia coli* K88. J. Anim. Sci. Biotechnol. 7.

Zhang, W., Azevedo, M.S.P., Gonzalez, A.M., Saif, L.J., Van Nguyen, T., Wen, K., Yousef, A.E., Yuan, L., 2008. Influence of probiotic Lactobacilli colonization on neonatal B cell responses in a gnotobiotic pig model of human rotavirus infection and disease. Vet. Immunol. Immunopath 122, 175–181.

Zhang, Z.F., Kim, I.H., 2013. Effects of probiotic supplementation in different energy and nutrient density diets on performance, egg quality, excreta microflora, excreta noxious gas emission, and serum cholesterol concentrations in laying hens. J. Anim. Sci. 91, 4781–4787.

Zhu, Y.Z., Cheng, J.L., Ren, M., Yin, L., Piao, X.S., 2015. Effect of γ-aminobutyric acid-producing Lactobacillus strain on laying performance, egg quality and serum enzyme activity in Hy-line brown hens under heat stress. Asian-Austral. J. Anim. Sci. 28, 1006–1013.

Zulkifli, I., Abdullah, N., Mohd Azrin, N., Ho, Y.W., 2000. Growth performance and immune response of two commercial broiler strains fed diets containing *Lactobacillus* cultures and oxytetracycline under heat stress conditions. Brit. Poult. Sci. 41, 593–597.

CHAPTER

# Gastrointestinal Ecology of *Salmonella* Enteritidis in Laying Hens and Intervention by Prebiotic and Nondigestible Carbohydrate Dietary Supplementation

16

**Steven C. Ricke**

*University of Arkansas, Fayetteville, AR, United States*

## 1. INTRODUCTION

Food-borne *Salmonella* species and serovars and their association with food production continues to be an ongoing problem not just for poultry but domestic animals in general as well as fresh produce (Wray and Wray, 2000; Braden, 2006; Hanning et al., 2009; Finstad et al., 2012; Foley et al., 2008, 2011, 2013; Howard et al., 2012). Of the readily identified food-borne *Salmonella* serovars, *Salmonella enterica* subspecies *enterica* serovar Enteritidis (*S.* Enteritidis) has historically been associated with poultry and it has been suggested that part of this is due to its niche displacement of the closely related host-specific poultry *Salmonella* Gallinarum and Pullorum serovars that had previously been eradicated from commercial flocks (St. Louis et al., 1988; CDC, 2000; Bäumler et al., 2000; Rabsch et al., 2000, 2001; Porwollik et al., 2005; Martelli and Davies, 2012). Although *S.* Enteritidis can be found in most types of poultry production including broilers and layers, it is the infection of layers and subsequent contamination of eggs during production that has been the primary concern and where most of the research on potential control measures has been directed (Ricke, 2003a; Braden, 2006; Howard et al., 2012; Martelli and Davies, 2012; Galiş et al., 2013). Given the apparent tissue tropism toward the layer hen reproductive tract, much of the research effort has been focused on genetic mechanisms that *S.* Enteritidis possesses that can be directly linked to colonization, subsequent transovarian internal contamination of eggs during formation, and survival in the egg contents (Thiagarajan et al., 1994; Keller et al., 1995; Guard-Petter, 2001; Lu et al., 2003b; Gantois et al., 2009; Gast et al., 2009; Van Immerseel, 2010; Raspoet et al., 2011; Shah et al., 2012; De Vylder et al., 2013).

Producing Safe Eggs. http://dx.doi.org/10.1016/B978-0-12-802582-6.00016-1

Colonization of the layer hen gastrointestinal tract (GIT) has also been identified as a source of *S.* Enteritidis occurrence in flocks and eventual systemic infection leading to contaminated eggs (Dunkley et al., 2009; Howard et al., 2012). Of course young layer chicks, like their broiler chick counterparts, are highly susceptible to *Salmonella* colonization due to the underdeveloped GIT microbiota and lack of a GIT barrier (Ricke et al., 2004; Ricke, 2014). Consequently, considerable efforts have been undertaken to introduce probiotics and other dietary amendments to limit *Salmonella* colonization either directly or indirectly via early enhancement and subsequent development of the protective GIT microbiota. Likewise administering biologicals such as bacteriophage to eliminate the already established GIT *Salmonella* or the use of vaccines to trigger the host's immune responses to limit colonization have been suggested and/or put into practice (Galiş et al., 2013). Given the number of potential reservoirs including rodents, insects, feed, just to name a few, along with differences in layer house management (Davies and Breslin, 2003a,b; Garber et al., 2003; Maciorowski et al., 2004; Mollenhorst et al., 2005; Park et al., 2008; Carrique-Mas et al., 2009; De Vylder et al., 2009, 2011; Silva et al., 2012), it would appear to be critical to employ several interventions that mechanistically provide more of a multiple hurdle approach to live bird protection.

Although a mature layer hen possesses a diverse GIT microbial population that would normally be considered a viable barrier to invasion and colonization, circumstances do occur where extensive *Salmonella* colonization happens. Such circumstances would probably involve some sort of major perturbation or disruption of the GIT microbiota in the laying hen. The most noteworthy example of this was the historical use of complete feed withdrawal over an extended period of time to induce molt as an initiation for an additional egg laying cycle (Holt, 2003; Ricke, 2003a; Park, 2004; Golden et al., 2008; Mazzuco et al., 2011; Ricke et al., 2013; Ricke, 2014). Based on extensive research a better understanding of the role of the layer hen GIT was developed and a wide range of alternative nonfeed withdrawal strategies were developed over time to limit *S.* Enteritidis colonization and infection as well retain egg performance parameters (Ricke, 2003a; Park et al., 2004; Ricke et al., 2013). This in turn led to the concept that dietary modifications could impact the layer hen GIT and dietary amendments such as fermentable nondigestible carbohydrates and prebiotics were thus examined for possible *Salmonella* mitigation properties (Ricke et al., 2013). In this chapter the development of the layer hen GIT will be examined, followed by a discussion on the relationship with food-borne *Salmonella* colonization and the impact of fermentable nondigestible carbohydrates and prebiotic supplementation on the GIT microbiome, *Salmonella* ecology, and layer hen performance.

## 2. THE CHICKEN GIT MICROBIOTA

Much of the work on understanding the microbial ecology of the avian microbiota has focused on broilers. In particular, interest has centered on the development of the microbiome as the bird matures and how that process impacts the host health, performance, and nutritional responses (Gabriel et al., 2006). In the early stages most of the characterizations were based on cultural approaches employing anaerobic cultivation techniques originally developed for rumen microorganisms (Salanitro et al., 1974; Fan et al., 1995; Ricke and Pillai, 1999; Krause et al., 2013; Ricke, 2015a). Unfortunately such approaches only detected a fraction of the total microorganisms present in the GIT with the large majority uncultivated and/or unidentified (Ricke and Pillai, 1999; Gabriel et al., 2006; Stanley et al., 2014). With the emergence of molecular approaches that allowed for noncultivation characterization, more was learned about the GIT microbial profiles in chickens. One of the more commonly applied approaches, denatured gradient gel electrophoresis (DGGE), involved polymerase chain reaction (PCR) amplification of 16S rRNA to generate a range of amplified products from unique sequences that could be separated via a gradient gel and potentially be representative of each microorganism present (Hanning and Ricke, 2011; Ricke et al., 2015). Analyses of leghorn chick and adult laying hen GIT contents using DGGE was reported to be capable of detecting differences in intestinal, cecal, colonic, and fecal populations (Hume et al., 2003; Dunkley et al., 2007e).

Since the early 2000's there has been a virtual explosion in the development of various molecular and analytical tools for a more complete "omics" approach to delineating microbial ecosystems that has led to a better understanding of the GIT microbiota and host health (Marchesi et al., 2016). Much of this has derived from the sequencing technology originally developed for the human genome project (Venter et al., 2001; Choudhuri, 2003, 2006; Ricke et al., 2015). This led to the whole genome sequencing of a wide range of food-borne pathogens including those associated with poultry for applications such as tracking in food production systems, evolution of pathogen strain emergence and development of improved molecular-based detection systems (Kwon and Ricke, 2011; O'Flaherty and Klaenhammer, 2011; Douglas et al., 2013; Gerner-Smidt et al., 2013; Park et al., 2014; Ricke et al., 2015). On a separate track, 16S rRNA sequencing offered the opportunity for the routine characterization of a more complete set of the organisms within the GIT microbial community (Kwon and Ricke, 2011). This led to initial studies over several years on the human GIT microbiome followed by animal GIT microbiome studies that elucidated impact factors such as aging, nutrition, and diet (Ley et al., 2006; Gordon, 2012; Nicholson et al., 2012; O'Toole and Jeffery, 2012; Walker and Parkhill, 2013; Lahti et al., 2014; Firkins and Yu, 2015; Hanning and Diaz-Sanchez, 2015; Marchesi et al., 2016). As more of these interactions became known, functionality of individual commensal members of the GIT community were identified and characterized as well as the role that human and nonhuman immune systems play in the interchange with the GIT microbiome (Hooper et al., 2012; Nicholson et al., 2012; Smith et al., 2013; Chaucheyras-Durand and Ossa, 2014; Hanning and Diaz-Sanchez, 2015; Hevia et al., 2015; Sivan et al., 2015).

In the last few years, the avian GIT microbiota has become more of a focal point of next-generation sequencing. This interest has been driven in part by the impact of diet and nutrition on broiler performance and the potential influence of the GIT microbiome as it develops with maturity of the growing bird. It was known before sequencing that the avian ceca was the primary fermentation site for carbohydrates, particularly those considered nondigestible, and was the GIT site where most of the short chain fatty acids (SCFAs) were produced as well as methane from resident methanogens (McNab, 1973; Ricke, 2003b; Józefiak et al., 2004; Ricke et al., 2004; Saengkerdsub et al., 2007a,b; Saengkerdsub and Ricke, 2014), but much less was known about the detailed composition of the cecal microbiome as well as the rest of the avian GIT. Early work on molecular profiling summarized by Apajalahti et al. (2004) indicated that in addition to the large majority of bacteria being unculturable, environment and diet influenced the GIT microbiome composition and that the GIT ileal and cecal populations become fairly dense early in the bird's life, not long after hatching. Based on studies by Van der Wielen et al. (2002) and Lu et al. (2003a), Stanley et al. (2014) also pointed out that, although there are distinctive differences in the avian GIT compartments (crop, duodenum, ileum, and ceca), they are still highly connected and influence each other with compartmentalization continuing as the bird matures. As next-generation sequencing methods have been applied, the composition of these respective GIT microbial populations has become even clearer. Oakley et al. (2014) in an overview of published research on the chicken GIT microbiome concluded that the phyla Firmicutes and Bacteroides dominate the cecal microbial consortia. However, they also noted that as more in-depth taxonomic distinctions were made, even with more diversity being evident, similar metabolic functions such as carbohydrate metabolism still prevailed based on metagenomic comparisons. However, discernible phylogenetic differences are more prevalent among different avian species. For example, Wei et al. (2013) reported that chickens and turkeys possessed distinct intestinal populations sharing only 16% similarity at the species-equivalent level.

## 3. LAYING HEN GIT MICROBIOTA

Most of the chicken GIT microbiome work has been done on broiler chickens, which as Stanley et al. (2014) has noted are sent to slaughter generally at no more than 7 weeks and thus they are still relatively juvenile and immature. Even though the growing cycle can be slightly longer for nonconventional pasture flock broiler birds (Park et al., 2013), this is still a much shorter life span than the considerably older laying hens, which can undergo a complete laying cycle and sometimes multiple cycles resulting in a life span of more than a year (Bell, 2003). It is not really known how much this difference in maturity may be reflected in the corresponding birds as considerably less microbiome work has been conducted on mature laying hens.

It does appear that disturbances occurring at an early age of the layer chick can influence the host as well as the GIT microbiome later in life. Simon et al. (2016) demonstrated that administering antibiotics at an early age caused early life alterations in the GIT microbiome and impacted adaptive immunity in older layer hens as evidenced by the lower T-cell-dependent antibody titers. Establishment of pathogens in the GIT and the timing at which this occurs may also be a factor. For example, when birds of two different genetic lines were infected with *S.* Enteritidis on day one, the overall diversity of the layer chick GIT cecal microbiome was not only reduced but also was accompanied by an increase in the *Enterobacteriaceae* family (Mon et al., 2015). Although the genetic line of bird did not appear to be a factor, they did note an inverse relationship between the families *Enterobacteriaceae* and *Lachnospiraceae*. It would be interesting to conduct follow-up studies to see if other *Salmonella* serovars also had similar impacts on the layer chick GIT microbiome and what the long-term consequences of such shifts might be. It is conceivable that if virulence properties differ among *Salmonella* serovars, this could in turn impact the GIT microbiota such that it would be somewhat distinct for each particular serovar. Likewise, it would be important to establish the mechanism(s) of how this shift is occurring.

Juricova et al. (2013) noted that the greatest impact on the cecal microbiota was in chicks inoculated on day one with *S.* Enteritidis versus days 4 and 16 with an increase in *Enterobacteriales* and concomitant decreases in *Clostridiales*, *Bifidobacteriales*, and *Lactobacillales* on days 3 and 10 postinfection. However, they also concluded that these changes would be considered relatively minor and speculated that this may be because they were examining lumen GIT bacteria and not bacteria on the epithelial surfaces where *Salmonella*-induced inflammation could occur and be accompanied by more dramatic changes in microbiota composition. There is precedence for this as *Salmonella*-induced inflammation to alter the surrounding epithelial microbiota ecology in mammalian species has been reported and described by others (Barman et al., 2008; Ahmer and Gunn, 2011; Winter and Bäumler, 2011). Resolving the microbiome GIT site of the greatest impact and mechanism(s) by *Salmonella* could have practical implications for *Salmonella* vaccine applications and offer opportunities to further optimize their effectiveness.

In mature laying hens the most notable impact on the GIT microbiome and *S.* Enteritidis was documented to occur when a drastic alteration in nutritional management via complete feed withdrawal was used to induce molt. Historically, in the United States complete removal of feed over a series of days had been the means to consistently shift the reproductive tract into a resting state and thereby halt egg production until the layers were returned to full feed and a second egg laying cycle was initiated (Brake, 1993; Bell, 2003; Berry, 2003; Holt, 1995, 2003; Ricke, 2003a). Although there were economic incentives for molt induction, the increased incidence of *S.* Enteritidis in eggs appeared to coincide with the implementation of these types of layer management systems (Bell, 2003; Holt, 1999, 2003; Ricke, 2003a; Golden et al., 2008; Dunkley et al., 2009; Ricke et al., 2013). This connection was further established by the increased horizontal distribution of *S.* Enteritidis infection to molted hens in nearby cages and more eggs contaminated with the bacterium (Holt, 1995; Holt et al., 1995, 1998). The relationship between feed withdrawal, followed by GIT emptying, and systemic *S.* Enteritidis invasion into tissues such as the liver, spleen, and ovaries was more clearly established in a series of experimental infection studies (Holt, 1995; Durant et al., 1999; Kubena et al., 2005; Moore and Holt, 2006; Dunkley et al., 2007d). Although somewhat difficult to completely distinguish between molt physiology versus feed withdrawal responses, general layer hen physiology as a function of immune response, stress protein levels, and blood metabolite changes appeared to parallel overall host susceptibility and hen behavior patterns also followed suit (Holt and Gast, 2002; Dunkley et al., 2007a,b, 2008a,b).

As the connection between feed withdrawal and increases in systemic invasion of *S.* Enteritidis became more established the key question remained as to what factors were serving as signals in the laying hen GIT microenvironment to trigger invasion. Using DGGE, Hume et al. (2003) detected shifts in the cecal microbiome in hens undergoing feed withdrawal that were distinguishable from nonmolted birds. Dunkley et al. (2007e) also reported similar shifts in DGGE fecal and cecal profiles of hens undergoing feed withdrawal on samples collected over the entire molting time period. Studies that examined SCFA production in the ceca indicated a fairly uniform reduction in SCFA production in birds undergoing feed withdrawal (Corrier et al., 1997; Woodward et al., 2005; Dunkley et al., 2007e; Donalson et al., 2008a). Taken together these studies suggested that there was some sort of alteration occurring in the cecal microbiota and/or their metabolic activities that limited cecal microbial population's ability to not only serve as a barrier to prevent *S.* Enteritidis colonization but also actually enhance invasiveness of *S.* Enteritidis in some fashion. Durant et al. (1999) conducted an *S.* Enteritidis infection study in birds either full fed or molted via a 9-day feed withdrawal. Birds were challenged with an *S.* Enteritidis marker strain on day 4 of the molt period and cultured for *Salmonella* on day 9. *Salmonella* Enteritidis crop and cecal colonization along with spleen and liver invasion all increased in the molted hens when compared with the nonmolted controls. When the crop contents were cultured on days 4 and 9 of the molt period, lactobacilli populations and the concentrations of lactate, acetate, propionate, butyrate, and total SCFA decreased but pH increased in the molted hens compared with the controls.

Based on the findings of Durant et al. (1999) it was concluded that removal of feed did indeed, after a few days, not only decrease fermentation but actually reduce the crop lactobacilli populations. Consequently, high levels of fermentation acids and a low pH were both diminished as barriers to *S.* Enteritidis. So do these changes alter the crop microenvironment in such a way to directly influence *S.* Enteritidis pathogenesis? There was some precedent for suspecting that this might be the case. It was already well known that virulence phenotypes in pathogens can be induced by numerous environmental conditions (Mekalanos, 1992) and from what is now known this is also in line with the hypothesis that nutrient limitation may be one of the primary drivers for initiating pathogenesis (Rohmer et al., 2011). To test this, Durant et al. (1999) incubated a *hilA* fusion strain of *S.* Enteritidis in the crop contents from the molted and unmolted birds that had been pooled separately, centrifuged, and filter sterilized. Although the overall regulation of *Salmonella* pathogenesis is complex, the transcriptional activator HilA is a good indicator of virulence initiation as it regulates the invasion genes resulting in expression of the invasion phenotype and by using a *lacZY* transcriptional fusion to the *hilA* gene (structural gene of β-galactosidase combined with the *hilA* promoter), *hilA* expression level can be quantitated as a simple enzyme assay (Bajaj et al., 1995, 1996). When the sterile crop contents were incubated with the *S.* Enteritidis *hilA* fusion strain, *hilA* increased 1.6- to 2.1-fold in crop contents from feed withdrawal birds, which was consistent with observed increases in organ invasion in these birds. Nearly 10 years later, Dunkley et al. (2007d) used real-time PCR to quantitate *hilA* of a nonfusion strain of *S.* Enteritidis in an infection experiment from day 6 and 11 fecal samples and day 12 cecal samples. The advantage of the PCR assay compared with the earlier fusion studies was that it could be directly applied to *S.* Enteritidis recovered from the in vivo infection study. In general, *hilA* expression was higher in feed withdrawal layers, paralleling the corresponding levels of organ invasion and confirming the earlier results by Durant et al. (1999).

Although it appeared that the conditions in the laying hen GIT experiencing complete removal of feed for an extended period of time was conducive for *S.* Enteritidis colonization and subsequent invasion, it was less clear what signals were serving as the trigger for these activities. To determine nutrient and crop microenvironment impact, Durant et al. (2000a) used Luria broth to simulate crop contents and compared undiluted (LB) versus diluted (DLB, simulated nutrient poor conditions) and examined the effects of pH, carbohydrate sources, amino acids, and lactate on *S.* Enteritidis *hilA* expression. Expression levels of *hilA* were nearly threefold higher in DLB broth compared with LB broth but addition of 0.2% glucose, fructose, or mannose to LB and DLB reduced *hilA* expression 1.5- to 2-fold. Although 0.2% casaminoacids, arabinose, fucose, or lactose had little effect, lactate (25 and 50 mmol/L) reduced *hilA* expression at pH 6, 5, and 4 compared with the same lactate levels at pH 7, with the lowest expression occurring at pH 4, 50 mmol lactate per liter. Durant et al. (2000b) also demonstrated that SCFAs (acetate, propionate, and butyrate) could influence *hilA* and *invF–lacZY* transcriptional fusion expression in *S.* Typhimurium and Kwon and Ricke (1998) demonstrated that the SCFA could alter acid tolerance in *S.* Typhimurium.

In summary, it appears that even an adult laying hen can experience sufficient disruption in the GIT microbiome to become susceptible to colonization and infection by *S.* Enteritidis. Although there are indications that both the GIT microbiota and the associated fermentation activities are impacted, it is not certain whether these are mutually exclusive. Put another way, it is not clear how much of the observed reduction in fermentation is due to a shift/actual reduction in the GIT microbial population versus a limitation of fermentable substrates. Given the potential functional redundancies such as carbohydrate metabolism that exist in the avian GIT microbiota (Oakley et al., 2014) some resiliencies in the overall GIT population response is possible. However, it is also possible that some GIT compartments such as the crop are actually undergoing a microbial population reduction, whereas the cecal microbiota are simply fermenting less. Part of this could be the high variability in passage rate observed in the crop because depending on the diet, the crop can be empty within 4 h or ingested feed can remain much longer (Heuser, 1945; Bayer et al., 1978; May et al., 1988, 1990). In contrast, ceca are believed to empty once every 24–48 h with one cecum emptying at a time and completely empty ceca are rarely observed (McNab, 1973). Consequently, the ceca may be more resilient to dietary changes than the upper parts of the layer hen GIT. Particle size of the feed may also be a factor in controlling GIT passage rate

(Hetland et al., 2004). Although some passage rate work has been attempted with laying hens using rare earth markers (Dunkley et al., 2008c), more studies need to be done to delineate both compartmental as well as entire GIT passage rates for different layer hen diets as well as when feed intake is reduced. It would be particularly interesting to compare these values with microbiome composition in the different GIT compartments and if overall changes in passage rate are reflected in detectable microbiome differences. Individual passage rate changes among different GIT compartments might also play a role in this as well. Finally, how much each compartment influences the others could dictate the level of impact of feed removal over time.

## 4. DIETARY NONDIGESTIBLE CARBOHYDRATE MITIGATION STRATEGIES—MOLTING LAYER HENS

Once it was realized that the laying hen GIT microbiota and associated fermentation activities could serve as a barrier to *S.* Enteritidis colonization and subsequent infection, strategies that involved some means to retain these microorganisms and still achieve the physiological state associated with an induced molt were explored. Fundamentally, the concept was to provide the laying hen GIT microorganisms with fermentable dietary substrates that were nutritionally unavailable to the host (Ricke, 2003a). From the late 1990s onward a plethora of dietary schemes, either revisited from earlier alternative molt diets or newly developed dietary approaches, were examined for their capabilities of limiting *S.* Enteritidis under molt induction conditions. The criteria were fairly straightforward, namely, prevent *S.* Enteritidis establishment, shift the laying hen into a complete shutdown of egg production fairly soon after receiving the diet and allow for complete restoration of optimal egg production once the molt period had ended and the birds were once again being fed normal layer rations. Molt induction dietary strategies ranged from nutritional imbalances, incorporation of ingredients that decreased appetite, to feeding of low-energy bulking agent fillers (Bell, 2003; Berry, 2003; Park et al., 2004; Yousaf and Chaudhry, 2008; Mazzuco et al., 2011; Ricke et al., 2010, 2013). Although many of these approaches were effective in either limiting *S.* Enteritidis or inducing a successful molt, fewer were successful at accomplishing both and/or were considered commercially impractical by the egg industry (Ricke et al., 2013).

At this point in time the concept of introducing molt diets that consisted primarily of dietary fiber or nondigestible carbohydrates came into vogue. The idea was that these types of diets would be low-energy high-bulk "filler"-type ingredients that would be relatively unavailable nutritionally to the laying hen but still fermentable by the cecal GIT. Strategies varied from using molt diets that consisted entirely of these sources to partial supplementation in combination with conventional layer diets to ensure rapid adaptation and continued intake (Ricke et al., 2010, 2013). A myriad of nondigestible carbohydrate sources were examined including alfalfa, barley, cotton seed, grape pomace, rice hulls, tomato extract, wheat gluten, and middlings and the pros and cons of these sources have been extensively discussed in previous reviews (Ricke et al., 2010, 2013).

Among these sources, alfalfa and wheat middlings were shown to consistently reduce *S.* Enteritidis in molting layer hens while still inducing molt (Seo et al., 2001; Woodward et al., 2005; Ricke et al., 2010, 2013). In a series of in vitro studies it was demonstrated that alfalfa was fermentable by cecal microorganisms and could support the production of SCFA as well as methane as fermentation end products that would be expected to be present in live bird cecal contents (Saengkerdsub et al., 2006; Dunkley et al., 2007c; Donalson et al., 2008b). When alfalfa was used to induce molt in layer hen trials, an examination of cecal contents revealed a general increase in SCFAs along with reduction of *S.* Enteritidis and a corresponding shift in fecal and cecal microbial populations when profiled by DGGE (Dunkley et al., 2007d,e). Utilizing bacterial tag-encoded FLX amplicon pyrosequencing in a follow-up experiment, Callaway et al. (2009) reported that the greatest number of cecal bacterial genera were present in the alfalfa-fed layer hen group and the least number in the feed withdrawal group. In addition, they noted that microbial diversity was the least in the birds undergoing feed withdrawal and *Lactobacillus* populations could not be detected. In an effort to further optimize administration of dietary alfalfa, different physical forms were examined such as a crumbles product as well as different ratios of alfalfa to layer diet and combined with antimicrobials in an effort to improve consistency of intake, retain molt response, and enhance *S.* Enteritidis colonization limitation (Donalson et al., 2005; McReynolds et al., 2005, 2006; Woodward et al., 2005; Dunkley et al., 2007a,b,d,e, 2008a,b).

## 5. PREBIOTICS—DEFINITION AND GENERAL CONCEPTS

Despite the relative success of the nondigestible carbohydrate diets such as the alfalfa-based molt-induction diets, inconsistencies remained in terms of reducing *S.* Enteritidis under all conditions. Although combinations such as adding the antimicrobial chlorate to the drinking water were tested (McReynolds et al., 2005) as a possible means to further reduce *S.* Enteritidis during molt, there was interest in developing approaches that could incorporate dietary ingredients directly into the molt diet that would complement the benefits of these types of diets and perhaps further enhance antagonism to the establishment of *S.* Enteritidis. Intuitively, ingredients that would not only support general GIT fermentation as most nondigestible carbohydrate sources already do, but also selectively support the metabolism of resident GIT bacteria considered to particularly beneficial to the host could potentially offer a better, more consistent, and focused protection against pathogen colonization.

A prebiotic compound falls into this category according to the original definition by Gibson and Roberfroid (1995) as being "a non-digestible food ingredient that beneficially affects the host by selectively stimulating the growth and/or activity of one or a limited number of gastrointestinal (GI) microflora." Although this definition has continued to be further refined as more has become known about the microbiome composition and their corresponding responses to prebiotic compounds (Rastall and Maitin, 2002; Rastall, 2004; Patterson and Burkholder, 2003; Wang, 2009; Hutkins et al., 2016), the essence of this definition still holds true. They are essentially nondigestible carbohydrates that are resistant to the animal's digestive processes and thus pass onto the lower parts of the GIT such as the colon and the cecum where they can be fermented by the resident microbial population to produce SCFAs. In general, prebiotics elicit benefits to the host including, among others, altering the GIT microbiome, stimulating the immune system, and reducing growth and host invasion by pathogens such as *Salmonella* (Cummings and Macfarlane, 2002; Bengmark, 2012; Roto et al., 2015).

Bacteria in the GIT microbiome that are believed to be specifically stimulated by prebiotic supplements are lactobacilli and bifidobacteria but as the microbiome becomes more defined with increased availability of microbial community sequencing information there may be other unidentified bacteria that are also selectively supported by certain prebiotics (Kaplan and Hutkins, 2000; Ricke, 2015b; Hutkins et al., 2016). Likewise, the list of nondigestible carbohydrates that may be considered to possess prebiotic-like properties has continued to expand and includes the classically defined prebiotics such as fructooligosaccharide products (oligofructose, inulin-type fructans), as well as compounds such as *trans*-galactooligosaccharides, lactulose, maltooligosaccharides, xylooligosaccharides, stachyose, raffinose, and yeast cell walls (mannan oligosaccharides, MOS) to name a few, along with a multitude of other fermentable carbohydrates that at most are only minimally digested in the upper GIT (Monsan and Paul, 1995; Flickinger et al., 2003; Collins and Gibson, 1999; Bird et al., 2010; Ricke, 2015b; Roto et al., 2015; Hutkins et al., 2016). It is anticipated that this list will continue to expand as more becomes known about the interaction between the GIT microbiome, potential candidate compounds, and other nondigestible carbohydrates (Bird et al., 2010; Hutkins et al., 2016).

## 6. PREBIOTICS—LAYER HENS AND EGG PRODUCTION

For poultry production in general, considerably more work on prebiotics has been done with broiler production and correspondingly less so with layer hens and egg production. Application of prebiotics for layer hens has primarily focused on specific administration for particular management situations such as molting or more generally for potentially beneficial influences on overall egg production parameters. For example, Tang et al. (2012) looked at the influence of the prebiotic isomaltooligosaccharide (IMO) alone and in combination with a commercial probiotic on the chemical composition of egg yolks and quality of eggs from Hisex Brown pullets at 28, 32, and 36 weeks. They reported that traditional egg quality parameters such as Haugh units, yolk color, specific gravity, and shell thickness, among others, were not impacted. However, egg composition did change as both IMO and the combination with a probiotic decreased egg yolk cholesterol and total saturated fatty acids while increasing total unsaturated fatty acids, total omega 6 and polyunsaturated fatty acids including linoleic and alpha-linoleic acids. Tang et al. (2012) concluded that feeding hens IMO with and without the probiotic improved the nutritional quality of the eggs produced by these hens without compromising the corresponding egg production quality.

As a prebiotic, MOS is believed to behave somewhat differently than some of the other prebiotics as it appears to interfere with pathogen attachment to GIT epithelial cells by binding to the bacterial fimbriae, thus preventing them from binding to carbohydrate sites of the intestinal lining (Hooge, 2004). When Fernandez et al. (2002) supplemented 30-week-old hen diets with MOS, they observed cecal population decreases in Gram-negative Enterobacteria and increases in Gram-positive *Enterococcus* spp. Bozkurt et al. (2012) compared MOS and an essential oil mixture fed to 36-week-old layers under moderate and hot environmental conditions and did not observe improvements in the efficiency of egg production or humoral immune response to counteract the heat stress adversity experienced by these hens but they did see amelioration by both MOS and essential oils on eggshell characteristics. Jahanian and Ashnagar (2015) examined the impact of MOS in *Escherichia coli* challenged laying 55-week-old hens over 77 days (7 days for adaptation and 70 days as the experimental period). Although they did not observe differences in ileal *E. coli* and total GIT bacteria, they did note decreases in *Salmonella* and increases in *Lactobacillus*. In addition, they reported increases in egg production percentage and egg mass and a decrease in feed conversion ratio for birds fed MOS over the first 35 days.

Fructooligosaccharides (FOS) are considered nondigestible and nonabsorbable and are composed of short-chain polymers of linked fructose units that can be produced commercially either by hydrolysis of inulin or by enzymatic synthesis from sucrose or lactose (Patterson and Burkholder, 2003; Ricke, 2015b). Various forms of FOS have been examined for potential benefits in laying hens and early studies established precedent for possible impacts on the layer hen cecal microbiome. For example, Rada et al. (2001) demonstrated that feeding inulin at 5% of the diet increased inulin-fermenting bifidobacteria twofold in cecal contents when compared with control birds not fed inulin. To demonstrate that FOS could selectively alter the ceca microbiota to directly inhibit *Salmonella*, Donalson et al. (2007) used an in vitro cecal incubation system to assess the effect of combining FOS with poultry feeds on the growth of *S.* Typhimurium. In this study, cecal contents were pooled from three laying hens, diluted in an anaerobic dilution solution, and subsequently added to sterile test tubes containing either alfalfa or layer ration with and without FOS. These in turn were inoculated with an *S.* Typhimurium marker strain and plated at 0 and 24 h after inoculation. Another set was incubated for 24 h to adapt the cecal inocula to FOS and feed combinations in the absence of *S.* Typhimurium then inoculated with the pathogen at hour 24 followed by another 24 h incubation. The samples immediately inoculated with *S.* Typhimurium without prior cecal fermentation did not lower *S.* Typhimurium 24 h later but those preincubated for 24 h with cecal microorganisms and feed combinations before *S.* Typhimurium inoculation did result in a 2 log reduction of *S.* Typhimurium with the most dramatic decreases seen in incubations from alfalfa or layer ration combined with FOS. Donalson et al. (2007) concluded that addition of FOS to feed substrate diets in combination with cecal contents synergistically acted to decrease *S.* Typhimurium growth only after adaptation of the cecal microbiota to the feed and FOS. In a follow-up study, Donalson et al. (2008b) using the same in vitro cecal incubation system observed some inconsistencies between trials but still noted that the addition of FOS to both alfalfa and the layer rations appeared to further enhance fermentation as demonstrated by generating higher propionate, butyrate, total SCFA, and lactic acid concentrations.

Donalson et al. (2008a) used an *S.* Enteritidis marker strain challenge study in a series of four independent trials to determine whether FOS when added to a combination of 90% alfalfa and 10% layer ration at two levels (0.750% and 0.375%) would influence GIT fermentation, *S.* Enteritidis GIT colonization, and organ invasion. The impact of FOS was apparent in some trials but not others. For example, lactic acid was increased in the crop in birds fed the higher level of FOS in the alfalfa diet at concentrations comparable with full fed birds but this was not seen in other trials. In the ceca, total SCFA were generally increased for both alfalfa alone and alfalfa-FOS combination fed birds compared with feed withdrawal treated birds, whereas lactic acid was higher in birds fed all three molt diets versus full fed birds and birds undergoing feed withdrawal. Although birds fed molt diets in most trials exhibited less *S.* Enteritidis colonization in the crop and ceca, organ invasion, and intestinal shedding than birds undergoing feed withdrawal, there were minimal differences among alfalfa fed birds versus those fed alfalfa and FOS. Donalson et al. (2008a) concluded that, although FOS addition did reduce *S.* Enteritidis organ invasion in half of the trials, lack of uniform response to FOS addition may have been due to inconsistent intake of the alfalfa-based diets and reduced consumption of FOS. They also speculated that there may have simply been fewer microorganisms capable of specifically responding directly to FOS as a substrate and/or alfalfa providing sufficient levels of similar compounds to mask any direct response to the FOS. Certainly, comparing the overall fermentation profiles among treatments would support this, although in individual trials some increases in crop lactate were noted for FOS-fed birds.

The in vitro work with cecal inoculations by Donalson et al. (2007, 2008b) would suggest that there are cecal microorganisms capable of fermenting FOS but perhaps part of the issue in the bird feeding trials is that most of the FOS is fermented before reaching the ceca in GIT sites such as the crop that are known to harbor lactobacilli (Fuller and Brooker, 1974; Fuller, 1977). In summary, it is difficult to draw conclusions linking the GIT microbiome to fermentation without further characterizing the composition of the microbiome and identifying the impact on individual groups such as the lactobacilli and bifidobacteria that are known to respond to specific prebiotics. The ability to now accomplish this with next-generation sequencing and metabolomic assessments should provide the tools to evaluate and potentially optimize administration of prebiotics to achieve a consistent response from adult birds (Park et al., 2013).

## 7. CONCLUSIONS AND FUTURE DIRECTIONS

Gaining a better understanding of the laying hen GIT microbiota was one of the factors that led to understanding of the *S.* Enteritidis risks associated with management practices such as feed withdrawal molting. Another outcome of these research efforts was the appreciation of the importance of the adult laying hen GIT microbiome along with its interaction and responses to dietary changes. However, it also became apparent how complex and unpredictable this microbial ecosystem could be and that outcomes were not always the expected or projected scenario. Certainly the advent of molecular tools such as next-generation sequencing and other omics approaches along with the corresponding bioinformatic programs offer the possibility to conduct more in-depth analysis to unravel and interpret some of these complexities.

Prebiotics and nondigestible carbohydrates represent dietary ingredients that can be used as potential GIT microbiome modifiers either by direct selection/enrichment of certain GIT bacterial species known to be beneficial to the host or by indirectly serving as substrates that favor particular fermentation pathways of GIT microorganisms. Historically, it has always been viewed that a prebiotic such as FOS is being selectively fermented exclusively by certain specific bacteria and thus leading to enrichment of these organisms in the GIT. As more becomes known about the intricacies of the GIT it is now being realized that GIT ecology is probably much more complex with potentially many more GIT microorganisms involved in hydrolysis and utilization of these prebiotic compounds. This is not surprising since this sort of metabolic communication and cross-feeding of substrates and end products is commonly observed in other GIT microbial consortia involved in fiber and nondigestible carbohydrate breakdown such as the rumen in ruminant animals (Ricke et al., 1996; Mackie, 2002; Weimer et al., 2009). As this becomes better understood, this also changes the perception of what defines a "prebiotic" compound and a refined definition may be more of a function of the host and microbiome response rather than being exclusively classified only as traditional polymers such as FOS (Hutkins et al., 2016). Consequently, nondigestible carbohydrates such as resistant starch may qualify as having prebiotic properties along with a myriad of complex fibers with indeterminate structure and chemistry (Bird et al., 2010; Bengmark, 2012). The very nature of their being more chemically complex may actually be advantageous by providing a multitude of potentially different substrates to a much broader spectrum of GIT microorganisms.

The other aspect to consider is when to target the layer hen with the addition of the dietary prebiotic. Certainly in the adult layer during times of GIT microbiome upheaval such as during molting would represent a reasonable application, but may not have the greatest impact. It is well known that *Salmonella* can colonize the GIT early in the chick's life and often quite easily (Ricke et al., 2004; Ricke, 2014). Combine this with the possibility that *Salmonella* may have some control over the development of the GIT microbiota and an appreciation for mitigation at early stages of bird development becomes apparent. Consequently, rather than emphasizing addition of prebiotics in more mature birds where, based on some of the studies reported here, the results can be variable, it may make more sense to start layer chicks out on prebiotic-based diets long before they reach maturity. For example, in the same trial that Fernandez et al. (2002) examined mature hens fed MOS they also looked at MOS administered to young broiler chicks challenged with an *S.* Enteritidis marker strain and observed decreases in this pathogen over several weeks. Likewise, Santos et al. (2012) were able to demonstrate a synergistic protective effect of MOS and threonine administered to 1-day-old broiler chicks against *S.* Enteritidis infection based on numbers of organisms enumerated from the ceca.

Although young broiler chick studies may suggest similar GIT responses for layer chicks when fed prebiotics, this may in fact not be the case. When Walugembe et al. (2015) compared broiler and layer chicks fed corn-soybean meal diets containing wheat bran and dried distiller grains with solubles, they noted distinctive differences between the two in both cecal SCFA concentrations and corresponding microbiota composition. Consequently, it will be important to test prebiotics directly in young layer chicks to draw definitive conclusions on their GIT responses and the potential of candidate prebiotics to limit *Salmonella* colonization. In addition, long-term studies need to be conducted on early age layer chicks with immediate introduction of dietary prebiotics that continues throughout the grow out period and onto their entry into layer flocks to determine if long-term exposure not only is protective against *Salmonella* establishment throughout this time frame but also would consistently improve egg production performance qualities. In addition, it would be of interest to monitor the GIT microbiome to determine whether a particular profile is selected for after early exposure to prebiotics and if this profile remains stable over time or requires continued feeding of the same prebiotic. With the emergence of molecular tools to conduct in-depth analysis of the layer hen GIT microbiome, some of these questions can now be addressed.

## REFERENCES


Ahmer, B.M.M., Gunn, J.S., 2011. Interaction of *Salmonella* spp. with the intestinal microbiota. Front. Microbiol. 2, 1–9:Article 101.

Apajalahti, J., Kettunen, A., Graham, H., 2004. Characteristics of the gastrointestinal microbial communities, with special reference to the chicken. Worlds Poult. Sci. J. 60, 223–232.

Bajaj, V., Hwang, C., Lee, C.A., 1995. *hilA* is a novel *ompR/toxR* family member that activates the expression of *Salmonella typhimurium* invasion genes. Mol. Microbiol. 18, 715–727.

Bajaj, V., Lucas, R.L., Hwang, C., Lee, C.A., 1996. Co-ordinate regulation of *Salmonella typhimurium* invasion genes by environmental and regulatory factors is mediated by control of *hilA* expression. Mol. Microbiol. 22, 703–714.

Barman, M., Unold, D., Shifley, K., Amir, E., Hung, K., Bos, N., Salzman, N., 2008. Enteric salmonellosis disrupts the microbial ecology of the murine gastrointestinal tract. Infect. Immun. 76, 907–915.

Bäumler, A.J., Hargis, B.M., Tsolis, R.M., 2000. Tracing the origins of *Salmonella* outbreaks. Science 287, 50–52.

Bayer, R.C., Hoover, W.H., Muir, F.V., 1978. Dietary fiber and meal feeding influence on broiler growth and crop fermentation. Poult. Sci. 57, 1456–1459.

Bell, D.D., 2003. Historical and current molting practices in the U.S. table egg industry. Poult. Sci. 82, 965–970.

Bengmark, S., 2012. Integrative medicine and human health – the role of pre-, pro- and synbiotics. Clin. Transl. Med. 1 (6), 1–13. http://www.clintransmed.com/content/1/1/6.

Berry, W.D., 2003. The physiology of induced molting. Poult. Sci. 82, 971–980.

Bird, A.R., Conlon, M.A., Christopher, C.T., Topping, D.L., 2010. Resistant starch, large bowel fermentation and a broader perspective of prebiotics and probiotics. Benef. Microbes 1, 423–431.

Bozkurt, M., Küçükyilmaz, K., Çatli, A.U., Çinar, M., Bintaş, E., Çöven, F., 2012. Performance, egg quality, and immune response of laying hens fed diets supplemented with mannan-oligosaccharide or an essential oil mixture under moderate and hot environmental conditions. Poult. Sci. 91, 1379–1386.

Braden, C.R., 2006. *Salmonella enterica* serotype Enteritidis and eggs: a national epidemic in the United States. Clin. Infect. Dis. 43, 512–517.

Brake, J., 1993. Recent advances in induced molting. Poult. Sci. 72, 929–931.

Callaway, T.R., Dowd, S.E., Wolcott, R.D., Sun, Y., McReynolds, J.L., Edrington, T.S., Byrd, J.A., Anderson, R.C., Krueger, N., Nisbet, D.J., 2009. Evaluation of the bacterial diversity in cecal contents of laying hens fed various molting diets by using bacterial tag-encoded FLX amplicon pyrosequencing. Poult. Sci. 88, 298–302.

Carrique-Mas, J.J., Breslin, M., Snow, L., McLaren, I., Sayers, A.R., Davies, R.H., 2009. Persistence and clearance of different *Salmonella* serovars in buildings housing laying hens. Epidemiol. Infect. 137, 837–846.

Centers for Disease Control and Prevention (CDC), 2000. Outbreaks of *Salmonella* serotype Enteritidis infection associated with eating raw or undercooked shell eggs–United States, 1996–1998. MMWR. Morb. Mortal. Wkly. Rep. 49 (4), 73.

Chaucheyras-Durand, F., Ossa, F., 2014. Review: the rumen microbiome: composition, abundance, diversity, and new investigative tools. Prof. Anim. Sci. 30, 1–12.

Choudhuri, S., 2003. The path from nuclein to human genome: a brief history of DNA with a note on human genome sequencing and its impact on future research in biology. Bull. Sci. Technol. Soc. 23, 360–367.

Choudhuri, S., 2006. Some major landmarks in the path from nuclein to human genome 1. Toxicol. Mech. Methods 16, 137–159.

Collins, M.D., Gibson, G.R., 1999. Probiotics, prebiotics, and synbiotics: approaches for modulating the microbial ecology of the gut. Am. J. Clin. Nutr. 69 (Suppl.), 1052S–1057S.

Corrier, D.E., Nisbet, D.J., Hargis, B.M., Holt, P.S., DeLoach, J.R., 1997. Provision of lactose to molting hens enhances resistance to *Salmonella enteritidis* colonization. J. Food Prot. 60, 10–15.

Cummings, J.H., Macfarlane, G.T., 2002. Gastrointestinal effects of prebiotics. Br. J. Nutr. 87, S145–S151.

Davies, R., Breslin, M., 2003a. Observations on *Salmonella* contamination of commercial laying farms before and after cleaning and disinfection. Vet. Rec. 152, 283–287.

Davies, R.H., Breslin, M., 2003b. Persistence of *Salmonella* enteritidis phage type 4 in the environment and arthropod vectors on an empty free-range chicken farm. Environ. Microbiol. 5, 79–84.

De Vylder, J., Van Hoorebeke, S., Ducatelle, R., Pasmans, F., Haesebrouck, F., Dewulf, J., Van Immerseel, F., 2009. Effect of the housing system on shedding and colonization of gut and internal organs of laying hens with *Salmonella* Enteritidis. Poult. Sci. 88, 2491–2495.

De Vylder, J., Dewulf, J., Van Hoorebeke, S., Pasmans, F., Haesebrouck, F., Ducatelle, R., Van Immerseel, F., 2011. Horizontal transmission of *Salmonella* enteritidis in groups of experimentally infected laying hens housed in different housing systems. Poult. Sci. 90, 1391–1396.

De Vylder, J., Raspoet, R., Dewulf, J., Haesebrouck, F., Ducatelle, R., Van Immerseel, F., 2013. *Salmonella* Enteritidis is superior in egg white survival compared with other *Salmonella* serotypes. Poult. Sci. 92, 842–845.

Donalson, L.M., Kim, W.K., Woodward, C.L., Hererra, P., Kubena, L.F., Nisbet, D.J., Ricke, S.C., 2005. Utilizing different ratios of alfalfa and layer ration for molt induction and performance in commercial laying hens. Poult. Sci. 84, 362–369.

Donalson, L.M., Kim, W.K., Chalova, V.I., Herrera, P., Woodward, C.L., McReynolds, J.L., Kubena, L.F., Nisbet, D.J., Ricke, S.C., 2007. In vitro anaerobic incubation of *Salmonella enterica* serotype Typhimurum and laying hen cecal bacteria in poultry feed substrates and a fructooligosacharide prebiotic. Anaerobe 13, 208–214.

Donalson, L.M., McReynolds, J.L., Kim, W.K., Chalova, V.I., Woodward, C.L., Kubena, L.F., Nisbet, D.J., Ricke, S.C., 2008a. The influence of a fructooligosaccharide prebiotic combined with alfalfa molt diets on the gastrointestinal tract fermentation, *Salmonella* Enteritidis infection and intestinal shedding in laying hens. Poult. Sci. 87, 1253–1262.

Donalson, L.M., Kim, W.K., Chalova, V.I., Herrera, P., McReynolds, J.L., Gotcheva, V.G., Vidanović, D., Woodward, C.L., Kubena, L.F., Nisbet, D.J., Ricke, S.C., 2008b. In vitro fermentation response of laying hen cecal bacteria to combinations of fructooligosaccharide (FOS) prebiotic with alfalfa or a layer ration. Poult. Sci. 87, 1263–1275.

Douglas, G.L., Pfeiler, E., Duong, T., Klaenhammer, T.R., 2013. Chapter 39. Genomics and proteomics of foodborne microorganisms. In: Doyle, M.P., Buchanan, R.L. (Eds.), Food Microbiology – Fundamentals and Frontiers, fourth ed. American Society for Microbiology Press, Washington, D.C., pp. 975–995.

Dunkley, C.S., McReynolds, J.L., Dunkley, K.D., Kubena, L.F., Nisbet, D.J., Ricke, S.C., 2007a. Molting in *Salmonella* Enteritidis-challenged laying hens fed alfalfa crumbles. III. Blood plasma metabolite response. Poult. Sci. 86, 2492–2501.

Dunkley, C.S., McReynolds, J.L., Dunkley, K.D., Njongmeta, L.N., Berghman, L.R., Kubena, L.F., Nisbet, D.J., Ricke, S.C., 2007b. Molting in *Salmonella* Enteritidis-challenged laying hens fed alfalfa crumbles. IV. Immune and stress protein response. Poult. Sci. 86, 2502–2508.

Dunkley, K.D., Dunkley, C.S., Njongmeta, N.L., Callaway, T.R., Hume, M.E., Kubena, L.F., Nisbet, D.J., Ricke, S.C., 2007c. Comparison of in vitro fermentation and molecular microbial profiles of high-fiber feed substrates (HFFS) incubated with chicken cecal inocula. Poult. Sci. 86, 801–810.

Dunkley, K.D., McReynolds, J.L., Hume, M.E., Dunkley, C.S., Callaway, T.R., Kubena, L.F., Nisbet, D.J., Ricke, S.C., 2007d. Molting in *Salmonella* enteritidis-challenged laying hens fed alfalfa crumbles. I. *Salmonella* Enteritidis colonization and virulence gene *hilA* response. Poult. Sci. 86, 1633–1639.

Dunkley, K.D., McReynolds, J.L., Hume, M.E., Dunkley, C.S., Callaway, T.R., Kubena, L.F., Nisbet, D.J., Ricke, S.C., 2007e. Molting in *Salmonella* Enteritidis challenged laying hens fed alfalfa crumbles II. Fermentation and microbial ecology response. Poult. Sci. 86, 2101–2109.

Dunkley, C.S., Friend, T., McReynolds, J.L., Kim, W.K., Dunkley, K.D., Kubena, L.F., Nisbet, D.J., Ricke, S.C., 2008a. Behavior of laying hens on alfalfa crumble molt diets. Poult. Sci. 87, 815–822.

Dunkley, C.S., Friend, T.H., McReynolds, J.L., Woodward, C.L., Kim, W.K., Dunkley, K.D., Kubena, L.F., Nisbet, D.J., Ricke, S.C., 2008b. Behavioral responses of laying hens to different alfalfa-layer ration combinations during molting. Poult. Sci. 87, 1005–1011.

Dunkley, C.S., Kim, W.-K., James, W.D., Ellis, W.C., McReynolds, J.L., Kubena, L.F., Nisbet, D.J., Ricke, S.C., 2008c. Passage rates in poultry digestion using stable isotope tracer markers. J. Radioanal. Nucl. Chem. 276, 35–39.

Dunkley, K.D., Callaway, T.R., Chalova, V.I., McReynolds, J.L., Hume, M.E., Dunkley, C.S., Kubena, L.F., Nisbet, D.J., Ricke, S.C., 2009. Foodborne *Salmonella* ecology in the avian gastrointestinal tract. Anaerobe 15, 26–35.

Durant, J.A., Corrier, D.E., Byrd, J.A., Stanker, L.H., Ricke, S.C., 1999. Feed deprivation affects crop environment and modulates *Salmonella* Enteritidis colonization and invasion of leghorn hens. Appl. Environ. Microbiol. 65, 1919–1923.

Durant, J.A., Corrier, D.E., Stanker, L.H., Ricke, S.C., 2000a. Expression of the *hilA Salmonella typhimurium* gene in a poultry *Salm. enteritidis* isolate in response to lactate and nutrients. J. Appl. Microbiol. 89, 63–69.

Durant, J.A., Corrier, D.E., Ricke, S.C., 2000b. Short-chain volatile fatty acids modulate the expression of the *hilA* and *invF* genes of *Salmonella* Typhimurium. J. Food Prot. 63, 573–578.

Fan, Y.-Y., Ricke, S.C., Scanlan, C.M., Nisbet, D.J., Vargas-Moskola, A.A., Corrier, D.E., DeLoach, J.R., 1995. Use of a differential rumen fluid-based carbohydrate agar media for culturing lactose selected cecal bacteria from chickens. J. Food Prot. 58, 361–367.

Fernandez, F., Hinton, M., Van Gils, B., 2002. Dietary mannan-oligosaccharides and their effect on chicken caecal microflora in relation to *Salmonella* Enteritidis colonization. Avian Pathol. 31 (1), 49–58.

Finstad, S., O'Bryan, C.A., Marcy, J.A., Crandall, P.G., Ricke, S.C., 2012. *Salmonella* and broiler production in the United States: relationship to foodborne salmonellosis. Food Res. Int. 45, 789–794.

Firkins, J.L., Yu, Z., 2015. How to use data on the rumen microbiome to improve our understanding of ruminant nutrition. J. Anim. Sci. 93, 1450–1470.

Flickinger, E.A., Van Loo, J., Fahey Jr., G.C., 2003. Nutritional responses to the presence of inulin and oligofructose in the diets of domesticated animals: a review. Crit. Rev. Food Sci. Nutr. 43, 19–60.

Foley, S.L., Lynne, A.M., Nayak, R., 2008. *Salmonella* challenges: prevalence in swine and poultry and potential pathogenicity of such isolates. J. Anim. Sci. 86, E149–E162.

Foley, S.L., Nayak, R., Hanning, I.B., Johnson, T.J., Han, J., Ricke, S.C., 2011. Population dynamics of *Salmonella enterica* serotypes in commercial egg and poultry production. Appl. Environ. Microbiol. 77, 4273–4279.

Foley, S.L., Johnson, T.J., Ricke, S.C., Nayak, R., Danzeisen, J., 2013. *Salmonella* pathogenicity and host adaptation in chicken-associated serovars. Microbiol. Mol. Biol. Rev. 77, 582–607.

Fuller, R., Brooker, B.E., 1974. Lactobacilli which attach to the crop epithelium of the fowl. Am. J. Clin. Nutr. 27, 1305–1312.

Fuller, R., 1977. The importance of Lactobacilli in maintaining normal microbial balance in the crop. Br. Poult. Sci. 18, 85–94.

Gabriel, I., Lessire, M., Mallet, S., Guillot, J.F., 2006. Microflora of the digestive tract: critical factors and consequences for poultry. Worlds Poult. Sci. J. 62, 499–511.

Galiş, A.M., Marcq, C., Marlier, D., Portetelle, D., Van, I., Beckers, Y., Théwis, A., 2013. Control of *Salmonella* contamination of shell eggs—preharvest and postharvest methods: a review. Comp. Rev. Food Sci. Food Saf. 12, 155–182.

Gantois, I., Ducatelle, R., Pasmans, F., Haesebrouck, F., Gast, R., Humphrey, T.J., Van Immerseel, F., 2009. Mechanisms of egg contamination by *Salmonella* Enteritidis. FEMS Microbiol. Rev. 33, 718–738.

Garber, L., Smeltzer, M., Fedorka-Cray, P., Ladely, S., Ferris, K., 2003. *Salmonella enterica* serotype enteritidis in table egg layer house environments and in mice in U.S. layer houses and associated risk factors. Avian Dis. 47 (1), 134–142.

Gast, R.K., Guard-Bouldin, J., Guraya, R., Holt, P.S., 2009. Effect of prior passage through laying hens on invasion of reproductive organs by *Salmonella* Enteritidis. Int. J. Poult. Sci. 8, 116–212.

Gerner-Smidt, P., Hyytia-Trees, E., Barrett, T.J., 2013. Chapter 43 – Molecular source tracking and molecular subtyping. In: Doyle, M.P., Buchanan, R.L. (Eds.), Food Microbiology – Fundamentals and Frontiers, fourth ed. American Society for Microbiology Press, Washington, D.C., pp. 1059–1077.

Gibson, G.R., Roberfroid, M.B., 1995. Dietary modulation of the human colonic microflora: introducing the concept of prebiotics. J. Nutr. 125, 1401–1412.

Golden, N.J., Marks, H.H., Coleman, M.E., Schroeder, C.M., Bauer Jr., N.E., Schlosser, W.D., 2008. Review of induced molting by feed removal and contamination of eggs with *Salmonella enterica* serovar Enteritidis. Vet. Microbiol. 131, 215–228.

Gordon, J.I., 2012. Honor thy gut symbionts redux. Science 336, 1251–1253.

Guard-Petter, J., 2001. The chicken, the egg and *Salmonella enteritidis*. Environ. Microbiol. 3, 421–430.

Hanning, I., Diaz-Sanchez, S., 2015. The functionality of the gastrointestinal microbiome in non-human animals. Microbiome 3, 51. http://dx.doi.org/10.1186/s40168-015-0113-6.

Hanning, I., Ricke, S.C., 2011. Prescreening methods of microbial populations for the assessment of sequencing potential. In: Kwon, Y.M., Ricke, S.C. (Eds.), Methods in Molecular Microbiology 733-high-throughput Next Generation Sequencing: Methods and Applications. Springer Protocols, Humana Press, New York, NY, pp. 159–170.

Hanning, I., Nutt, J.D., Ricke, S.C., 2009. Salmonellosis outbreaks due to fresh produce: sources and potential intervention measures. Foodborne Pathog. Dis. 6, 635–648.

Hetland, H., Choct, M., Svihus, B., 2004. Role of insoluble non-starch polysaccharides in poultry nutrition. Worlds Poult. Sci. J. 60, 415–422.

Heuser, G.F., 1945. The rate of passage of feed from the crop of the hen. Poult. Sci. 24, 20–24.

Hevia, A., Delgado, S., Sánchez, B., Margolles, A., 2015. Molecular players involved in the interaction between beneficial bacteria and the immune system. Front. Microbiol. 18. http://dx.doi.org/10.3389/fmicb.2015.01285.

Holt, P.S., Gast, R.K., 2002. Comparison of the effects of infection with *Salmonella enteritidis* in combination with an induced molt, on serum levels of the acute phase protein, $\alpha_1$-acid glycoprotein, in hens. Poult. Sci. 81, 1295–1300.

Holt, P.S., Macri, N.P., Porter Jr., R.E., 1995. Microbiological analysis of the early Salmonella enteritidis infection in molted and unmolted hens. Avian Dis. 39, 55–63.

Holt, P.S., Mitchell, B.W., Gast, R.K., 1998. Airborne horizontal transmission of Salmonella enteritidis in molted laying chickens. Avian Dis. 42, 45–52.

Holt, P.S., 1995. Horizontal transmission of *Salmonella enteritidis* in molted and unmolted laying chickens. Avian Dis. 39, 239–249.

Holt, P.S., 1999. Impact of induced molting on immunity and *Salmonella enterica* serovar Enteritidis infection in laying hens. In: Saeed, A.M., Gast, R.K., Potter, M.E., Wall, P.G. (Eds.), *Salmonella enterica* Serovar Enteritidis in Humans and Animals – Epidemiology, Pathogenesis, and Control. Iowa State Press, Ames, IA, pp. 367–375.

Holt, P.S., 2003. Molting and *Salmonella enterica* serovar enteritidis infection: the problem and some solutions. Poult. Sci. 82, 1008–1010.

Hooge, D.M., 2004. Meta-analysis of broiler chicken pen trials evaluating dietary mannan oligosaccharide, 1993–2003. In. J. Poult. Sci. 3, 163–174.

Hooper, L.V., Littman, D.R., Macpherson, A.J., 2012. Interactions between the microbiota and the immune system. Science 336, 1268–1273.

Howard, Z.R., O'Bryan, C.A., Crandall, P.G., Ricke, S.C., 2012. *Salmonella* Enteritidis in shell eggs: current issues and prospects for control. Food Res. Int. 45 (2), 755–764.

Hume, M.E., Kubena, L.F., Edrington, T.S., Donskey, C.J., Moore, R.W., Ricke, S.C., Nisbet, D.J., 2003. Poultry digestive microflora diversity as indicated by denaturing gradient gel electrophoresis. Poult. Sci. 82, 1100–1107.

Hutkins, R.W., Krumbeck, J.A., Bindels, L.B., Cani, P.D., Fahey Jr., G.C., Goh, Y.J., Hamaker, B., Martens, E.C., Mills, D.A., Rastal, R.A., Vaughan, R.A., Sanders, M.E., 2016. Prebiotics: why definitions matter. Curr. Opin. Biotechnol. 37, 1–7.

Jahanian, R., Ashnagar, M., 2015. Effect of dietary supplementation of mannan-oligosaccharides on performance, blood metabolites, ileal digestibility, and gut microflora in *Escherichia coli*-challenged laying hens. Poult. Sci. 94, 2165–2172.

Józefiak, D., Rutkowski, A., Martin, S.A., 2004. Carbohydrate fermentation in the avian ceca: a review. Anim. Feed Sci. Technol. 113, 1–15.

Juricova, H., Videnska, P., Lukac, M., Faldynova, M., Babak, V., Havlickova, H., Sisak, F., Rychlik, I., 2013. Influence of *Salmonella enterica* serovar Enteritidis infection on the development of the cecum microbiota in newly hatched chicks. Appl. Environ. Microbiol. 79, 745–747.

Kaplan, H., Hutkins, R.W., 2000. Fermentation of fructooligosaccharides by lactic acid bacteria and bifidobacteria. Appl. Environ. Microbiol. 66, 2682–2684.

Keller, L.H., Benson, C.E., Krotec, K., Eckroade, R.J., 1995. *Salmonella enteritidis* colonization of the reproductive tract and forming and freshly laid eggs of chickens. Infect. Immun. 63, 2443–2449.

Krause, D.O., Nagaraja, T.G., Wright, A.D.G., Callaway, T.R., 2013. Rumen microbiology: leading the way in microbial ecology. J. Anim. Sci. 91, 331–341.

Kubena, L.F., Byrd, J.A., Moore, R.W., Ricke, S.C., Nisbet, D.J., 2005. Effects of drinking water treatment on susceptibility of laying hens to *Salmonella* enteritidis during forced molt. Poult. Sci. 84, 204–211.

Kwon, Y.M., Ricke, S.C., 1998. Induction of acid resistance of *Salmonella typhimurium* by exposure to short-chain fatty acids. Appl. Environ. Microbiol. 64, 3458–3463.

Kwon, Y.M., Ricke (Eds.), 2011. High-Throughput Next Generation Sequencing: Methods and Applications. Methods in Molecular Microbiology 733-Springer Protocols. Humana Press, New York, NY. 308 pp.

Lahti, L., Salojärvi1, J., Anne Salonen, A., Scheffer, M., de Vos, W.M., 2014. Tipping elements in the human intestinal ecosystem. Nat. Commun. 5, 4344. http://dx.doi.org/10.1038/ncomms5344l. www.nature.com/naturecommunications.

Ley, R.E., Turnbaugh, P.J., Klein, S., Gordon, J.I., 2006. Human gut microbes associated with obesity. Nature 444, 1022–1023.

Lu, J., Idris, U., Harmon, B., Hofacre, C., Maurer, J.J., Lee, M.D., 2003a. Diversity and succession of the intestinal bacterial community of the maturing broiler chicken. Appl. Environ. Microbiol. 69, 6816–6824.

Lu, S., Killoran, P.B., Riley, L.W., 2003b. Association of *Salmonella enterica* serovar Enteritidis YafD with resistance to chicken egg albumen. Infect. Immun. 71, 6734–6741.

Maciorowski, K.G., Jones, F.T., Pillai, S.D., Ricke, S.C., 2004. Incidence, sources, and control of foodborne *Salmonella* spp. in poultry feeds. Worlds Poult. Sci. J. 60, 446–457.

Mackie, R.I., 2002. Mutualistic fermentative digestion in the gastrointestinal tract: diversity and evolution. Integr. Comp. Biol. 42, 319–326.

Marchesi, J.R., Adams, D.H., Fava, F., Hermes, G.D.A., Hirschfield, G.M., Hold, G., Quraishi, M.N., Kinross, J., Smidt, H., Tuohy, L.V., Zoetendal, E.G., Hart, A., 2016. The gut microbiota and host health: a new clinical frontier. Gut 65, 330–339.

Martelli, F., Davies, R.H., 2012. *Salmonella* serovars from table eggs: an overview. Food Res. Int. 45, 745–754.

May, J.D., Branton, J.L., Deaton, J.W., Simmons, J.D., 1988. Effect of environmental temperature and feeding regimen on quantity of digestive tract contents in broilers. Poult. Sci. 67, 64–71.

May, J.D., Lott, B.D., Deaton, J.W., 1990. The effect of light and environmental temperature on broiler digestive tract contents after feed withdrawal. Poult. Sci. 69, 1681–1684.

Mazzuco, H., Avila, V.S., Coldebella, A., Mores, R., Jaenisch, F.R.F., Lopes, L.S., 2011. Comparison of the effect of different methods of molt: production and welfare evaluation. Poult. Sci. 90, 2913–2920.

McNab, J., 1973. The avian caeca. A review. Worlds Poult. Sci. J. 29, 251–263.

McReynolds, J., Kubena, L., Byrd, J., Anderson, R., Ricke, S., Nisbet, D., 2005. Evaluation of *Salmonella* Enteritidis (SE) in molting hens after administration of an experimental chlorate product (for nine days) in the drinking water and feeding an alfalfa molt diet. Poult. Sci. 84, 1186–1190.

McReynolds, J.L., Moore, R.W., Kubena, L.F., Byrd, J.A., Woodward, C.L., Nisbet, D.J., Ricke, S.C., 2006. Effect of various combinations of alfalfa and standard layer diet on susceptibility of laying hens to *Salmonella* Enteritidis during forced molt. Poult. Sci. 85, 1123–1128.

Mekalanos, J.J., 1992. Environmental signals controlling expression of virulence determinants in bacteria. J. Bacteriol. 174, 1–7.

Mollenhorst, H., van Woudenbergh, C.J., Bokkers, E.G.M., de Boer, I.J.M., 2005. Risk factors for *Salmonella* Enteritidis infections in laying hens. Poult. Sci. 84, 1308–1313.

Mon, K.K.Z., Saelao, P., Halstead, M.M., Chanthavixay, G., Chang, H.-C., Garas, L., Maga, E.A., Zhou, H., 2015. *Salmonella enterica* serovar Enteritidis infection alters the indigenous microbiota diversity in young layer chicks. Front. Vet. Sci. 2, 61. http://dx.doi.org/10.3389/fvets.2015.00061.

Monsan, P., Paul, F., 1995. Enzymatic synthesis of oligosaccharides. FEMS Microbiol. Rev. 16, 187–192.

Moore, R.W., Holt, P.S., 2006. The effect of feed deprivation on tissue invasion by *Salmonella* Enteritidis. Poult. Sci. 85, 1333–1337.

Nicholson, J.K., Holmes, E., Kinross, J., Burcelin, R., Gibson, G., Jia, W., Pettersson, S., 2012. Host-gut microbiota metabolic interactions. Science 336, 1262–1267.

O'Flaherty, S., Klaenhammer, T.R., 2011. The impact of omic technologies on the study of food microbes. Annu. Rev. Food Sci. Technol. 2, 16.1–16.19.

O'Toole, P.W., Jeffery, I.B., 2012. Gut microbiota and aging. Science 350, 1214–1215.

Oakley, B.B., Lillehoj, H.S., Kogut, M.H., Kim, W.K., Maurer, J.J., Pedroso, A., Lee, M.D., Collett, S.R., Johnson, T.J., Cox, N.A., 2014. The chicken gastrointestinal microbiome. FEMS Microbiol. Lett. 360, 100–112.

Park, S.H., Hanning, I., Perrota, A., Bench, B.J., Alm, E., Ricke, S.C., 2013. Modifying the gastrointestinal ecology in alternatively raised poultry and the potential for molecular and metabolomic assessment. Poult. Sci. 92, 546–561.

Park, S.H., Aydin, M., Khatiwara, A., Dolan, M.C., Gilmore, D.F., Bouldin, J.L., Ahn, S., Ricke, S.C., 2014. Current and emerging technologies for rapid detection and characterization of *Salmonella* in poultry and poultry products. Food Microbiol. 38, 250–262.

Park, S.Y., Kim, W.K., Birkhold, S.G., Kubena, L.F., Nisbet, D.J., Ricke, S.C., 2004. Induced moulting issues and alternative dietary strategies for the egg industry in the United States. Worlds Poult. Sci. J. 60, 196–209.

Park, S.Y., Woodward, C.L., Kubena, L.F., Nisbet, D.J., Birkhold, S.G., Ricke, S.C., 2008. Environmental dissemination of foodborne *Salmonella* in preharvest poultry production: reservoirs, critical factors and research strategies. Crit. Rev. Environ. Sci. Technol. 38, 73–111.

Patterson, J.A., Burkholder, K., 2003. Application of prebiotics and probiotics in poultry production. Poult. Sci. 82, 627–631.

Porwollik, S., Santiviago, C.A., Cheng, P., Florea, L., McClelland, M., 2005. Differences in gene content between *Salmonella enterica* serovar Enteritidis isolates and comparison to closely related serovars Gallinarum and Dublin. J. Bacteriol. 187, 6545–6555.

Rabsch, W., Hargis, B.M., Tsolis, R.M., Kingsley, R.A., Hinz, K.H., Tschäpe, H., Bäumler, A.J., 2000. Competitive exclusion of *Salmonella* Enteritidis by *Salmonella* Gallinarum in poultry. Emerg. Infect. Dis. 6, 443–448.

Rabsch, W., Tschäpe, H., Bäumler, A.J., 2001. Non-typhoidal salmonellosis: emerging problems. Microbes Infect. 3, 237–247.

Rada, V., Duškova, D., Marounek, M., Petr, J., 2001. Enrichment of bifidobacteria in the hen caeca by dietary inulin. Folia Microbiol. 46, 73–75.

Raspoet, R., Gantois, I., Devloo, R., Martel, A., Haesebrouck, F., Pasmans, F., Ducatelle, R., Van Immerseel, F., 2011. *Salmonella* Enteritidis universal stress protein (usp) gene expression is stimulated by egg white and supports oviduct colonization and egg contamination in laying hens. Vet. Microbiol. 153, 186–190.

Rastall, R.A., Maitin, V., 2002. Prebiotics and synbiotics: towards the next generation. Curr. Opin. Biotechnol. 13, 490–496.

Rastall, R., 2004. Bacteria in the gut: friends and foes and how to alter the balance. J. Nutr. 134, 2022S–2026S.

Ricke, S.C., Pillai, S.D., 1999. Conventional and molecular methods for understanding probiotic bacteria functionality in gastrointestinal tracts. Crit. Rev. Microbiol. 25, 19–38.

Ricke, S.C., Martin, S.A., Nisbet, D.J., 1996. Ecology, metabolism, and genetics of ruminal selenomonads. Crit. Rev. Microbiol. 22, 27–65.

Ricke, S.C., Woodward, C.L., Kwon, Y.M., Kubena, L.F., Nisbet, D.J., 2004. Chapter 11-Limiting avian gastrointestinal tract *Salmonella* colonization by cecal anaerobic bacteria and a potential role for methanogens. In: Beier, R.C., Pillai, S.D., Phillips, T.D., Ziprin, R.L. (Eds.), Pre-harvest and Post-harvest Food Safety: Contemporary Issues and Future Directions. Blackwell Publishing Professional, Ames, IA, pp. 141–150.

Ricke, S.C., Dunkley, C.S., McReynolds, J.L., Dunkley, K.D., Nisbet, D.J., 2010. Molting in laying hens and *Salmonella* infection. In: van der Aar, P.J., Doppenberg, J. (Eds.), Dynamics in Animal Nutrition. Wageningen Academic Publishers, Wageningen, The Netherlands, pp. 135–146.

Ricke, S.C., Dunkley, C.S., Durant, J.A., 2013. A review on development of novel strategies for controlling *Salmonella* Enteritidis colonization in laying hens: fiber-based molt diets. Poult. Sci. 92 (2), 502–525.

Ricke, S.C., Dawoud, T.M., Kwon, Y.M., 2015. Chapter 4. Application of molecular methods for traceability of foodborne pathogens in food safety systems. In: Ricke, S.C., Donaldson, J.R., Phillips, C.A. (Eds.), Food Safety: Emerging Issues, Technologies and Systems. Elsevier, Oxford, UK, pp. 37–63.

Ricke, S.C., 2003a. The gastrointestinal tract ecology of *Salmonella* Enteritidis colonization in molting hens. Poult. Sci. 82, 1003–1007.

Ricke, S.C., 2003b. Perspectives on the use of organic acids and short chain fatty acids as antimicrobials. Poult. Sci. 82, 632–639.

Ricke, S.C., 2014. Application of molecular approaches for understanding foodborne *Salmonella* Establishment in poultry production. Adv. Biol. 2014. http://dx.doi.org/10.1155/2014/813275.

Ricke, S.C., 2015a. Chapter 19. Application of anaerobic microbiology laboratory training and writing comprehension to food safety education. In: Ricke, S.C., Donaldson, J.R., Phillips, C.A. (Eds.), Food Safety: Emerging Issues, Technologies and Systems. Elsevier, Oxford, UK, pp. 395–419.

Ricke, S.C., 2015b. Potential of fructooligosaccharide prebiotics in alternative and nonconventional poultry production systems. Poult. Sci. 94, 1411–1418.

Rohmer, L., Hocquet, D., Miller, S.I., 2011. Are pathogenic bacteria just looking for food? Metabolism and microbial pathogenesis. Trends Microbiol. 19, 341–348.

Roto, S.M., Rubinelli, P.M., Ricke, S.C., 2015. An introduction to the avian gut microbiota and the effects of yeast-based prebiotic compounds as potential feed additives. Front. Vet. Sci. 2, 1–18.:Article 28. http://dx.doi.org/10.3389/fvets.2015.00028.

Saengkerdsub, S., Ricke, S.C., 2014. Ecology and characteristics of methanogenic archaea in animals and humans. Crit. Rev. Microbiol. 40, 97–116.

Saengkerdsub, S., Kim, W.-K., Anderson, R.C., Woodward, C.L., Nisbet, D.J., Ricke, S.C., 2006. Effects of nitrocompounds and feedstuffs on in vitro methane production in chicken cecal contents and rumen fluid. Anaerobe 12, 85–92.

Saengkerdsub, S., Anderson, R.C., Wilkinson, H.H., Kim, W.-K., Nisbet, D.J., Ricke, S.C., 2007a. Identification and quantification of methanogenic archaea in adult chicken ceca. Appl. Environ. Microbiol. 73, 353–356.

Saengkerdsub, S., Herrera, P., Woodward, C.L., Anderson, R.C., Nisbet, D.J., Ricke, S.C., 2007b. Detection of methane and quantification of methanogenic archaea in faeces from young broiler chickens using real-time PCR. Lett. Appl. Microbiol. 45, 629–634.

Salanitro, J.P., Fairchilds, I.G., Zgornicki, Y.D., 1974. Isolation, culture characteristics, and identification of anaerobic bacteria from the chicken cecum. Appl. Microbiol. 27, 678–687.

Santos, E.G., Costa, F.G.P., Silva, J.H.V., Martins, T.D.D., Figueiredo-Lima, D.F., Macari, M., Oliveira, C.J.B., Givisiez, P.E.N., 2012. Protective effect of mannan oligosaccharides against early colonization by *Salmonella* Enteritidis in chicks is improved by higher threonine levels. J. Appl. Microbiol. 114, 1158–1165.

Seo, K.H., Holt, P.S., Gast, R.K., 2001. Comparison of *Salmonella* Enteritidis infection in hens molted via long-term feed withdrawal versus full-fed wheat middling. J. Food Prot. 64, 1917–1921.

Shah, D.H., Zhou, X., Kim, H.Y., Call, D.R., Guard, J., 2012. Transposon mutagenesis of *Salmonella enterica* serovar Enteritidis identifies genes that contribute to invasiveness in human and chicken cells and survival in egg albumen. Infect. Immun. 80, 4203–4215.

Silva, C.A., Blondel, C.J., Quezada, C.P., Porwollik, S., Andrews-Polymenis, H.L., Toro, C.S., Zaldívar, M., Contreras, I., McClelland, M., Santiviago, C.A., 2012. Infection of mice by *Salmonella enterica* serovar Enteritidis involves additional genes that are absent in the genome of serovar Typhimurium. Infect. Immun. 80, 839–849.

Simon, K., Verwoolde, M.B., Zhang, J., Smidt, H., de Vries Reilingh, G., Kemp, B., Lammers, A.A., 2016. Long-term effects of early life microbiota disturbance on adaptive immunity in laying hens. Poult. Sci. 95, 1543–1554.

Sivan, A., Corrales, L., Hubert, N., Williams, J.B., Aquino-Michaels, K., Earley, Z.M., Benyamin, F.W., Lei, Y.M., Jabri, B., Alegre, M.-L., Chang, E.B., Gajewski, T.F., 2015. Commensal *Bifidobacterium* promotes antitumor immunity and facilitates anti–PD-L1 efficacy. Science 350, 1084–1089.

Smith, P.M., Howitt, M.R., Panikov, N., Michaud, M., Gallini, C.A., Bohlooly-Y, M., Glickman, J.N., Garrett, W.S., 2013. The microbial metabolites, short-chain fatty acids, regulate colonic Treg cell homeostasis. Science 341, 569–573.

St Louis, M.E., Morse, D.L., Potter, M.E., DeMelfi, T.M., Guzewich, J.J., Tauxe, R.V., Blake, P.A., The *Salmonella enteritidis* working group, 1988. The emergence of grade A eggs as a major source of *Salmonella enteritidis* infections: new implications for the control of salmonellosis. JAMA 259, 2103–2107.

Stanley, D., Hughes, R.J., Moore, R.J., 2014. Microbiota of the chicken gastrointestinal tract: influence of health, productivity and disease. Appl. Microbiol. Biotechnol. 98, 4301–4310.

Tang, S.G.H., Sieo, C.C., Kalavathy, R., Saad, W.Z., Yong, S.T., Wong, H.K., 2012. Chemical compositions of egg yolks and egg quality of laying hens fed prebiotic, probiotic, and symbiotic diets. J. Food Sci. 80, C1686–C1695.

Thiagarajan, D., Saeed, A.M., Asem, E.K., 1994. Mechanism of transovarian transmission of *Salmonella* enteritidis in laying hens. Poult. Sci. 73, 89–98.

Van der Wielen, P.W., Keuzenkamp, D.A., Lipman, L.J., van Knapen, F., Biesterveld, S., 2002. Spatial and temporal variation of the intestinal bacterial community in commercially raised broiler chickens during growth. Microb. Ecol. 44, 286–293.

Van Immerseel, F., 2010. Stress-induced survival strategies enable *Salmonella* Enteritidis to persistently colonize the chicken oviduct tissue and cope with antimicrobial factors in egg white: a hypothesis to explain a pandemic. Gut Pathog. 2, 23. http://www.gutpathogens.com/content/2/1/23.

Venter, J.C., Adams, M.D., Myers, E.W., Li, P.W., Mural, R.J., Sutton, G.G., et al., 2001. The sequence of the human genome. Science 291, 1304–1351.

Walker, A.W., Parkhill, J., 2013. Fighting obesity with bacteria. Science 341, 1069–1070.

Walugembe, M., Hsieh, J.C.F., Koszewski, N.J., Lamont, S.J., Persia, M.E., Rothschild, M.F., 2015. Effects of dietary fiber on cecal short-chain fatty acid and cecal microbiota of broiler and laying-hen chicks. Poult. Sci. 94, 2351–2359.

Wang, Y., 2009. Prebiotics: present and future science and technology. Food Res. Int. 42, 8–12.

Wei, S., Morrison, M., Yu, Z., 2013. Bacterial census of poultry intestinal microbiome. Poult. Sci. 92, 671–683.

Weimer, P.J., Russell, J.B., Muck, R.E., 2009. Lessons from the cow: what the ruminant animal can teach us about consolidated bioprocessing of cellulosic biomass. Bioresour. Technol. 100, 5323–5331.

Winter, S.E., Bäumler, A.J., 2011. *Salmonella* exploits suicidal behavior of epithelial cells. Front. Microbiol. 2, 1–2:Article 48.

Woodward, C.L., Kwon, Y.M., Kubena, L.F., Byrd, J.A., Moore, R.W., Nisbet, D.J., Ricke, S.C., 2005. Reduction of *Salmonella enterica* serovar Enteritidis colonization and invasion by an alfalfa diet during molt in Leghorn hens. Poult. Sci. 84, 185–193.

Wray, C., Wray, A. (Eds.), 2000. *Salmonella* in Domestic Animals. CABI Publishing, CAB International, Wallingford, Oxon OX10 8DE, UK. 463 pp.

Yousaf, M., Chaudhry, A.S., 2008. History, changing scenarios and future strategies to induce moulting in laying hens. Worlds Poult. Sci. J. 64, 65–75.

CHAPTER

# Preharvest Food Safety—Potential Use of Plant-Derived Compounds in Layer Chickens

# 17

**Anup Kollanoor Johny[1], Kumar Venkitanarayanan[2]**

*[1]University of Minnesota, Saint Paul, MN, United States; [2]University of Connecticut, Storrs, CT, United States*

## 1. INTRODUCTION

The use of antibiotic growth promoters (AGPs) in food animal agriculture has tremendously improved productivity by improving intestinal health and reducing disease incidence in animals (Hao et al., 2014). However, the emergence of antibiotic-resistant bacteria and their potential food-borne transmission to humans have raised concerns leading to restrictions or possible substitution of AGPs with alternatives. Both situations should be considered with caution since the broad food safety gap that the antibiotics leave has to be filled with options with similar antimicrobial efficacy and growth-supporting potential. Given the situation, researchers worldwide are exploring viable, safe alternatives to antibiotics that could work well with the antibiotics during the phasing out period, and eventually act independently against current and emerging pathogens in food animals without affecting production or economic viability of the industry. However, currently, no alternatives have been tested extensively for their efficacy, safety, and health outcomes in poultry. This situation warrants research efforts for developing viable alternative/adjuncts to antibiotics in animal agriculture, including poultry production. This chapter focuses on plant-derived compounds (PDCs) that have the potential to improve the preharvest microbiological safety of eggs, with proposed mechanisms of action, experimental challenges, and hurdles for adoption by the industry.

## 2. *SALMONELLA*—A CONTINUING THREAT TO LAYER SAFETY

Among the infectious agents causing food-borne diseases, *Salmonella* ranks high along with *Campylobacter* and Shigatoxigenic *Escherichia coli* O157. Since designated as a notifiable disease in 1943, the incidence of nontyphoidal salmonellosis

Producing Safe Eggs. http://dx.doi.org/10.1016/B978-0-12-802582-6.00017-3

(NTS) has steadily increased over the years (Angulo and Swerdlow, 1999; Tauxe, 1999). Poultry and eggs contribute to 22% of the outbreak-associated *Salmonella* infections in humans (Painter et al., 2013), indicating the significant role played by poultry as the source of salmonellosis.

*Salmonella* can colonize the intestines of domesticated poultry due to several potential contributing factors supporting pathogen survival on farms, including intensive production, contaminated feed and water, personnel, rodents, flies, dust, and others (Marin et al., 2011; Galiş et al., 2013). Live poultry serves as a natural reservoir host of *Salmonella* where the bacterium colonizes the gastrointestinal tract (GIT) of poultry, with the highest persistence noticed in the ceca. Once colonized, the pathogen could be excreted in the feces with birds showing no obvious signs of infection. Pathogen shedding via feces will eventually lead to the horizontal transmission of the infection to other healthy birds and flocks. Moreover, colonization of the pathogen results in the contamination of whole carcasses during the evisceration process. Furthermore, fresh eggs may get contaminated with feces containing the pathogen in the cloaca/vent, the common exit point for reproductive, digestive, and urinary systems (De Reu et al., 2006; Gantois et al., 2008), which in turn may result in the entry of the pathogen into the egg at the time of lay (oviposition).

The potential of *Salmonella* to contaminate eggs via systemic and extrasystemic routes has been discussed in detail (Borland, 1975; Baskerville, 1992; Miyamoto et al., 1997; Leach et al., 1999; Vazquez-Torres et al., 1999; Okamura et al., 2001; Timoney et al., 1989; Shivaprasad et al., 1990; Gantois et al., 2008). After entering poultry, *Salmonella* can invade the intestinal cells, cecal tonsils, and Peyer patches, which may ultimately result in their uptake by and multiplication in macrophages. The pathogen may end up in internal organs such as the crop, gizzard, lungs, liver, spleen, and reproductive organs such as ovaries and oviduct (Miyamoto et al., 1997; Okamura et al., 2001; Gast et al., 2007; Gantois et al., 2008; Kollanoor Johny et al., 2009, 2012a; Foley et al., 2011; Upadhyaya, 2015). *Salmonella* can reach the internal contents of the eggs by direct transmission from infected ovaries and/or several areas of oviduct such as infundibulum, magnum, isthmus, and/or vagina (Baskerville, 1992; Leach et al., 1999; Vazquez-Torres et al., 1999). In addition to the systemic spread, *Salmonella* colonization of hen's reproductive tract can result from an ascending infection from the cloaca (Reiber et al., 1995; Miyamoto et al., 1997). Cloaca may contain feces infected with *Salmonella* that may invade the oviduct cells resulting in a progressive upward movement of the pathogen to colonize the reproductive tract. Moreover, a potential for descending infection from the ovary to oviduct tissues, called a retrocontamination, has been reported (Keller et al., 1995). Even after laying, egg shells can come into contact with the pathogen surviving in the fecal material on the coop/farm floors, potentially resulting in the contamination of egg contents. Whichever route *Salmonella* enter eggs, consumption of contaminated eggs and egg products is epidemiologically linked to NTS incidence (Hennessy et al., 2004; Braden, 2006; CDC, 2010; reviewed by Foley et al., 2011).

It is known that *Salmonella* is versatile in employing different virulence mechanisms such as flagellar motility, multiplication in the GIT, pathogenicity island-mediated invasion, and secretion systems that induce pathogen uptake to the intestinal cells. Macrophages will carry the pathogen to cause systemic spread to internal organs, including the reproductive organs and eggs (Van Immerseel et al., 2004). Therefore strategies that would reduce *Salmonella* pathogenesis and systemic spread in layer chickens are of paramount importance. This situation is chiefly due to the multiple sources of *Salmonella* at the farm, and the multipronged strategies of the pathogen to survive and grow under adverse conditions (Davies and Breslin, 2004; Upadhyaya, 2015).

In the poultry industry, antibiotics are used to treat specific bacterial infections, and a few are used as growth promoters (Rabsch et al., 2001). However, the situation has been complicated by the emergence of multidrug-resistant strains of pathogenic bacteria with increased virulence potential (Shea, 2003; Bywater, 2005). Numerous investigators have reported *Salmonella* strains that are resistant to a variety of antibiotics (Chadfield and Hinton, 2003; Daly et al., 2005; Dias de Oliveira et al., 2005; Erdem et al., 2005; Kilmartin et al., 2005). Moreover, with restricted use of antibiotics, alternatives that have similar antibacterial potential as antibiotics and with sufficient growth-supporting capacity need to be explored.

## 3. PLANT-DERIVED COMPOUNDS

Historically, plants have served to support human and animal life by providing oxygen, food, wood, shade, and heat. In addition, plants have contributed significantly to the improvement of overall human health and well-being in the form of traditional medicines, health supplements, food preservatives, and flavor enhancers (Upadhyay et al., 2014; Savoia, 2012). A plethora of PDCs responsible for multiple benefits have been investigated extensively in vitro, but only selectively in vivo (Osbourn, 1996; Burt, 2004; Newman, 2008; Antony and Singh, 2011; Dixon, 2001; Upadhyay et al., 2014). The PDCs synthesized in response to the interaction of plants with invading microorganisms, stress conditions, and predators, are produced during the specific developmental period of plant growth (Hashemi and Davoodi, 2012; Han et al., 1981). Due to the volatility of PDCs and difficulty in mass manufacture, they are produced as essential oils (EOs) containing a few analyzable PDCs or as crude plant extracts/powders containing several different compounds in them.

Plant extracts and EOs containing these critical PDCs have been investigated for their antimicrobial, antiinflammatory, antioxidative, and antiparasitic activities in poultry (Vondruskova et al., 2010; Hashemi and Davoodi, 2011). A few EOs have been investigated against coccidia, *Clostridium perfringens*, and *Histomonas meleagridis* in turkeys (Oviedo-Rondon et al., 2006). However, several inconsistencies with the use of EOs across investigations have been identified that could be attributed to several factors. These factors include the differences in the concentration of active components of the tested EOs; their potential synergistic, additive, and counteractive effects with other components and feed ingredients; hygienic conditions in which the birds are kept; and the overall health condition of the flock (Zeng et al., 2015). Individual PDCs or EOs with known PDCs are being explored to avoid such discrepancies across experiments (Tiihonen et al., 2010). Due to the presence of multiple active sites in the chemical structure of PDCs, and their multiple mechanisms of action on bacteria, it is unlikely that bacteria develop resistance against them. It is interesting that among the new antimicrobial drugs approved from 1981 to 2006, 69% were naturally derivatized (Newman, 2008), underscoring the relevance of PDCs with high bioactivity for potential inclusion in the list of approved drugs. Deciphering and understanding the most active ingredient in the EOs are required for bringing up consistency in results. Some of the potential PDCs that may have significant implications for long-term use in chickens, especially layers, are discussed in the following sections.

## 3.1 CHEMICAL COMPOSITION

Windisch and Kroismayr (2007) classified phytobiotics, the plant-derived products added to feed to improve livestock health, into (1) herbs (flowering, non-woody, and nonpersistent plants), (2) botanicals (entire or processed parts of a plant such as roots, leaves, and bark), (3) EOs (hydrodistilled extracts of plant volatile compounds), and (4) oleoresins (extracts based on nonaqueous solvents) (Yang et al., 2009). The PDCs are derived from very complex natural mixtures of active secondary metabolites from plants, and may amount to 20–60 components at different concentrations. For example, 70 different kinds of PDCs along with some nonidentified, bioactive compounds are reported to be present in hawthorn fruit (Wang and Slavik, 1998). PDCs are responsible for giving specific odors, color (pigments), flavor, and therapeutic activity to the plants or their derivatives. For example, some terpenes and terpenoids contribute to odor, quinones and tannins give color, whereas other terpenoids are responsible for flavor (e.g., capsaicin from chili peppers) (Cowan, 1999). Although many in numbers, these compounds can be further classified into major and trace components. The major components may be present at fairly high concentrations (20–70%), whereas the trace components occur at low to very low (trace) concentrations. Some of the major components that are commonly present in plant extracts are cinnamaldehyde, eugenol, carvacrol (CR), thymol (THY), linalool, α and β thujones, camphor, 1,8-cineole, α phellandrene, limonene, carvone, menthol, and menthone. These compounds fall into two different groups based on the biosynthetic origin, the main group composed of terpenes and terpenoids, and the other group consisting of aromatic and aliphatic constituents (Bakkali et al., 2008). Some of the major plant sources and active PDCs that have potential in poultry production are listed in Table 17.1.

**Table 17.1** Potential PDCs for Use in Poultry

| Plant Source | Major PDC | Chemical Structure |
|---|---|---|
| Angelica root | α-Pinene | 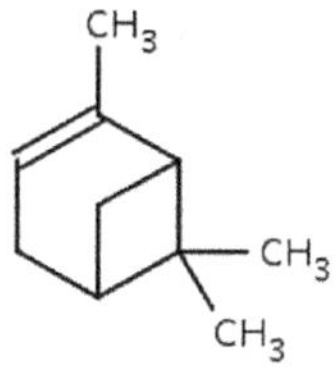 |
| Bergamot | β-Pinene | 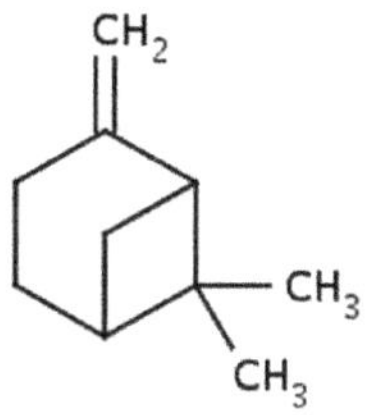 |
| Cinnamon bark | Cinnamaldehyde | 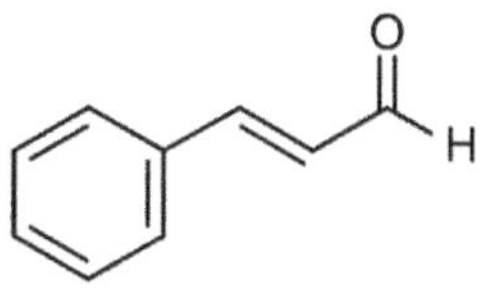 |
| Coriander | Linalool | 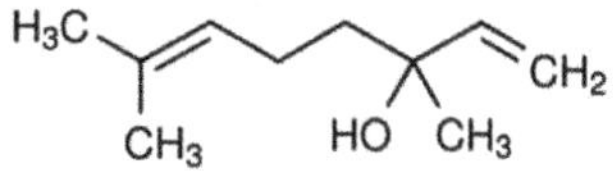 |
| Dill | Limonene | 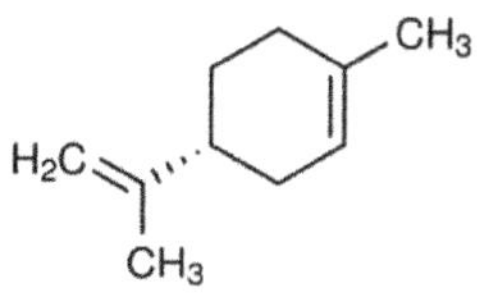 |
| Eucalyptus | Citronellal | 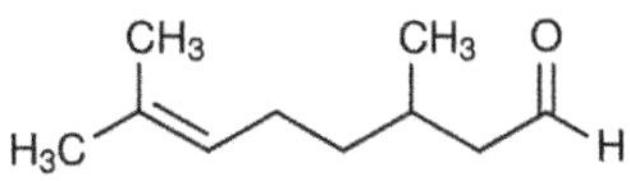 |
| Geranium | Citronellol | 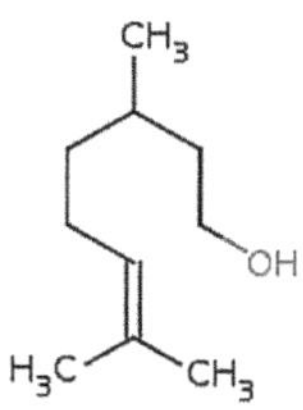 |
| Juniper berry | α-Pinene | 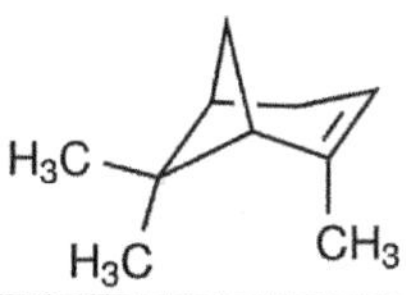 |

*Continued*

**Table 17.1** Potential EOs and Major PDCs for Use in Poultry—cont'd

| Plant Source | Major PDC | Chemical Structure |
| --- | --- | --- |
| Lime | Limonene | 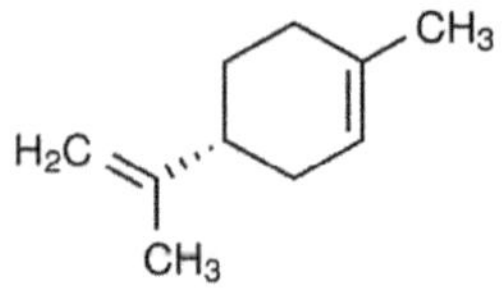 |
| Mandarine | Limonene | 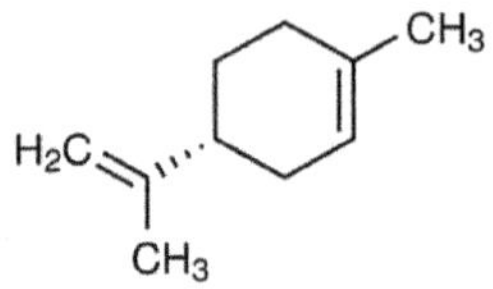 |
| Nutmeg | α-Pinene | 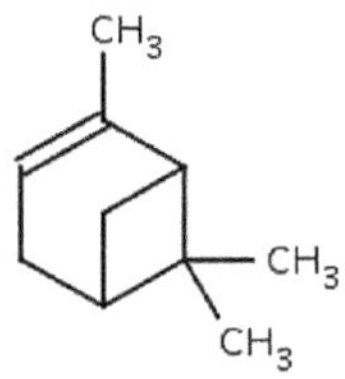 |
| Orange | Limonene | 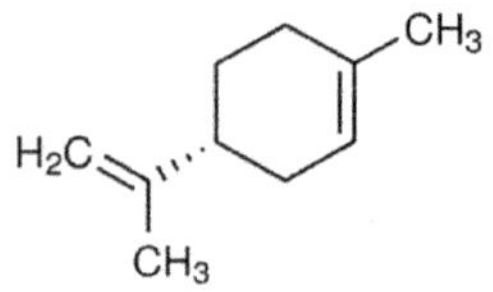 |
| Rosemary | 1,8-Cineole | 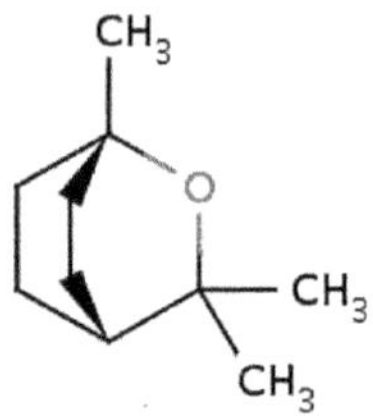 |
| Rosewood | Linalool | 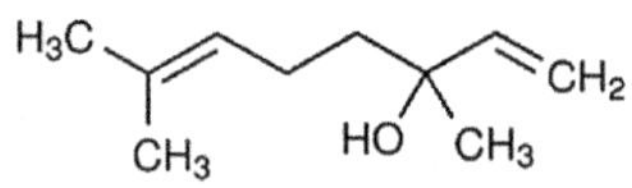 |
| Sage | α,β-Thujones | 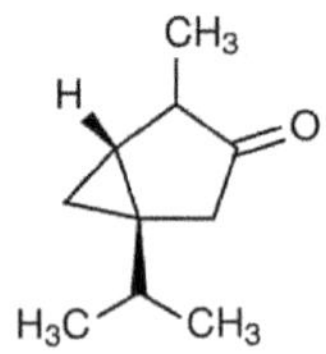 |
| Savory | Carvacrol | 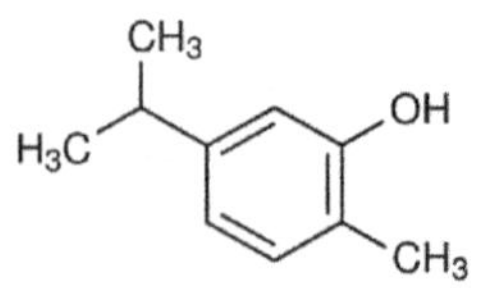 |

## 3.2 TERPENES, SESQUITERPENES, AND TERPENOIDS

Terpenes are made from combinations of several 5-carbon base units called isoprenes and are categorized as monoterpenes ($C_{10}$) and sesquiterpenes ($C_{15}$), which are the major classes. A monoterpene is formed by the coupling of two isoprene units, and constitute the most representative compounds in the group (90%) and allow a variety of structures (Bakkali et al., 2008). Monoterpenes are abundantly present in different plants such as *Pinus* sp., Coriander, Camphor, and Eucalyptus. On the other hand, sesquiterpenes are formed by the assembly of three isoprene units ($C_{15}$). A variety of structures can be obtained by the increase in the cyclization in each extension. They are commonly present in plants such as Angelica, Bergamot, Celery, Citronella, Coriander, Geranium, Lavender, Lemon, Orange, Pine, Rosemary, Sage and Thyme. Other minor classes include hemiterpenes ($C_5$), diterpenes ($C_{20}$), triterpenes ($C_{30}$), and tetraterpenes ($C_{40}$). A terpene containing oxygen is referred to as a terpenoid (Bakkali et al., 2008).

## 3.3 AROMATIC COMPOUNDS

Based on the biosynthetic origin, the second major group is aromatic compounds, which are derived from phenylpropane. They occur less frequently than terpenes. The aromatic group comprises aldehydes, alcohols, phenols, methoxy derivatives, and methylene dioxy compounds and occur abundantly in plant species such as Anise, Cinnamon, Clove, Fennel, Nutmeg, and other plant families such as *Apiaceae*, *Lamiaceae*, *Myrtaceae*, and *Rutaceae* (Bakkali et al., 2008). The majority of these aromatic compounds fall under phenolics and polyphenols, which includes simple phenols and phenolic acids, quinones, flavones, flavonoids, flavonols, tannins, and coumarins, and are being tested extensively against food-borne pathogens and improving shelf-life of foods (Upadhyay et al., 2014).

*Phenols and phenolic acids*: Phenols and phenolic acids are some of the natural phytochemicals consisting of a single substituted phenolic ring and could have high oxidation states that make them potential antimicrobial agents. It has been reported that the more highly oxidized the phenols are, the greater is their inhibitory activity against pathogenic bacteria. Similarly, hydroxylation increases their activity toward pathogenic organisms (Geissman, 1963). PDCs such as cinnamaldehyde, cinnamyl alcohol, catechol, and pyrogallol are phenolic derivatives that possess high oxidation potential and strong inhibitory effects on bacteria (reviewed in Cowan, 1999). Phenolic compounds that lack oxygen and thus remain at lower oxidation status are classified under EOs. Phenolic compounds that have $C_3$ side chains such as eugenol are representative EO compounds (Cowan, 1999).

*Quinones*: These are highly reactive compounds that have aromatic rings with two ketone substitutions. They are colored and give the materials their dyeing properties (reviewed by Upadhyay et al., 2014). These compounds are considered to be providers of stable free radicals, and complex irreversibly with nucleophilic amino acids in proteins, leading to inactivation of surface-exposed attachment proteins, cell wall polypeptides and membrane-bound enzymes, consequently leading to the inactivation of pathogens.

*Flavones, flavonoids, and flavonols*: Flavones are phenolic compounds containing one carbonyl group. The addition of 3-hydroxyl group yields flavonols and flavonoids, but the flavonoids are occurring as a $C_6$–$C_3$ unit linked to an aromatic ring. There are 14 classes of flavonoids, differentiated based on their chemical nature and the position of substituents on the different rings (Savoia, 2012). They have a similar mechanism of action as quinones. Catechins are the most reduced form of the $C_3$ unit in flavonoid compounds that are much more abundantly present in green teas. Common examples of this group include quercetin, naringin, hesperidin, chrysin, abyssinone, and totarol (Cowan, 1999).

*Tannins*: Found in every part of the plant, these polymeric phenolic compounds are divided into two groups—hydrolyzable and condensed tannins (Scalbert, 1991). Tannins are formed either by polymerization of quinone units or by condensation of flavan derivatives transported to wood tissues of plants. They are abundantly present in green teas and red wines (Serafini et al., 1994). Tannins such as ellagitannin bind to proteins such as adhesins, inhibit enzymes, deprive substrates, complex with cell wall leading to membrane disruption, and sometimes cause bacterial death by metal ion complexation (Schultz, 1988; Stern et al., 1996).

*Coumarins*: These are phenolic substances that contain benzene and α-pyrone rings (O'Kennedy and Thornes, 1997). Coumarins are well known for their antithrombotic, antiinflammatory, and vasodilatory activities (Thastrup et al., 1985; Piller, 1975; Namba et al., 1988). Coumarins such as scopoletin, chalcones, and phytoalexins have been shown to exert significant antimicrobial activity.

In addition to the above-mentioned classification, PDCs can be classified as those with and without nitrogen in their structure. The PDCs without nitrogen include monoterpenes, sesquiterpenes, diterpenes, triterpenes, saponins, steroids, tetrapenes, flavonoids, polyacetylenes, polyketides, and phenylpropanes. On the other hand, the PDCs with nitrogen include alkaloids, nonprotein amino acids, amines, cyanogenic glycosides, and glucosinolates (Acamovic and Brooker, 2005). Among all the biological activities attributed to PDCs, the antimicrobial activity of the compounds is of paramount importance and discussed in the following section.

## 4. ANTIMICROBIAL ACTIVITY OF PDCS IN BROILER CHICKENS

In our studies, we focused on two PDCs as preharvest in-feed supplements, namely *trans*-cinnamaldehyde (TC) and eugenol (EG) to control *S.* Enteritidis in broiler chickens. TC, the major PDC in cinnamon (*Cinnamomum zeylandicum*), and EG, the major PDC of clove oil (*Eugenia caryophillis*), possess antibacterial properties against gram-negative and gram-positive bacteria (Burt, 2004). Both of these compounds are generally recognized as safe chemicals for use in foods [generally recognized as safe (GRAS)] by the US Food and Drug Administration (FDA). After examining the anti-*Salmonella* potential of these PDCs in vitro (Kollanoor Johny et al., 2008, 2010a,b), studies were designed to assess their efficacy in reducing *S.* Enteritidis intestinal colonization in broiler chickens. The prophylactic efficacy of TC and EG was tested against *S.* Enteritidis populations in commercial day-old broiler chicks. Birds were supplemented with either 0.5% or 0.75% TC, and 0.75% or 1% EG through feed for 20 days. Results indicated that TC at 0.5% and 0.75% and EG at 1% reduced *S.* Enteritidis in the cecum [approximately 3 $\log_{10}$ colony forming units (CFU)/g]) and cloaca (approximately 2 $\log_{10}$ CFU/g), respectively, after 10 days of infection in challenged birds. Neither compound altered the cecal pH or the endogenous cecal microbiome populations. Feed intake and body weight were not significantly different for TC-supplemented groups. However, EG-treated groups had significantly lower body weight than the control birds (Kollanoor Johny et al., 2012d).

In follow-up studies, two experiments were conducted with market-age broiler chickens to determine the therapeutic efficacy of TC and EG for decreasing *S.* Enteritidis in birds. TC was added at 0.75% and EG at 1% as an antimicrobial additive in the feed given to market-age chickens for 5 days before slaughter. It was observed that TC and EG consistently reduced significant populations of *S.* Enteritidis in both experiments. These plant molecules reduced cecal colonization of *S.* Enteritidis by approximately 1.5 $\log_{10}$ CFU/g. In the cloacal contents, TC and EG decreased *S.* Enteritidis populations by approximately 1.5 and 2 $\log_{10}$ CFU/g, respectively (Kollanoor-Johny et al., 2012b).

Our studies with another PDC, CR, yielded inconsistent results. CR is the major PDC in oregano oil, obtained from *Origanum glandulosum.* The oil has been found effective against bacterial and fungal infections of the gastrointestinal and genitourinary tract (Blumenthal et al., 2000; Chun et al., 2005). CR at 0.9% reduced cecal and cloacal *S.* Enteritidis counts by approximately 4 $\log_{10}$ CFU/ml/g in trial 1, reduced 1.4 $\log_{10}$ CFU/ml/g of cloacal contents with no difference in cecal *S.* Enteritidis in trial 2 and reduced cecal *S.* Enteritidis by 0.9 $\log_{10}$ CFU/ml/g in trial 3. Compared with control birds, CR at 0.6% reduced cloacal counts in trial 1 by a log with no change in cecal *S.* Enteritidis, reduced approximately 2 $\log_{10}$ CFU/ml/g in ceca and cloaca in trial 2 and did not reduce *S.* Enteritidis counts in both the organ samples ($P>0.05$) in trial 3. Although CR did not alter the pH, endogenous bacterial counts in the cecum, feed intake, or body weights across trials, supplementation of the PDC through feed had an inconsistent effect on *S.* Enteritidis colonization and shedding (Kollanoor-Johny et al., 2011).

## 5. ANTIMICROBIAL ACTIVITY OF PDCS IN LAYERS

Following successful model studies in broiler chickens, we determined the efficacy of the PDCs, TC, and EG to reduce egg-borne transmission of *Salmonella* in layer chickens. In a related preliminary study, Upadhyaya et al. (2013) reported that the subinhibitory concentrations (concentration that does not inhibit bacterial growth) of four PDCs, namely, TC, CR, THY, and EG, reduced *S.* Enteritidis adhesion to and invasion of chicken oviduct epithelial cells, and survival in chicken macrophages. Found abundantly in the oil of thyme, THY, the isomer of CR, exerts significant antibacterial effects similar to other PDCs (Blumenthal et al., 2000; Chun et al., 2005). The effect of all four PDCs on *S.* Enteritidis genes critical for oviduct colonization and macrophage survival was also determined using quantitative real-time PCR (RT-qPCR). All four PDCs significantly reduced *S.* Enteritidis adhesion to and invasion of chicken oviduct epithelial cells. The PDCs, except THY, consistently decreased *S.* Enteritidis survival in macrophages. RT-qPCR results revealed that the PDCs downregulated the expression of *S.* Enteritidis genes involved in its colonization and macrophage survival (Upadhyaya et al., 2013).

Following the in vitro study, the efficacy of in-feed supplementation with TC in reducing *S.* Enteritidis cecal colonization and systemic spread was determined in layer chicken trials (Upadhyaya et al., 2015a). The consumer acceptability of eggs from TC-treated birds was also determined. Supplementation of TC in feed for 66 days at 1% or 1.5% (vol/wt) to 40- or 25-week-old layer chickens decreased *S.* Enteritidis on eggshell and in the yolk. In addition, *S.* Enteritidis persistence in the cecum, liver, and oviduct in TC-supplemented birds was decreased compared with that in controls. No significant differences in feed intake, body weight, or egg production in birds or consumer acceptability of eggs were observed. The results suggested that TC may potentially be used as a feed additive to reduce egg-borne transmission of *S.* Enteritidis in chickens (Upadhyaya et al., 2015a).

It was evident from our broiler and layer trials that TC could be a potentially effective PDC to control horizontal and vertical transmission of *S.* Enteritidis in chickens. However, TC at 0.3% was unable to reduce *Campylobacter jejuni* in broiler chickens in a seeder model (Hermans et al., 2011). In another study, Arsi et al. (2014) observed that THY, CR, or their combination resulted in significant reduction of *C. jejuni* in broiler chicks, although inconsistently.

# 6. POTENTIAL MECHANISMS OF ACTION OF PDCS

## 6.1 EFFECTS ON BACTERIAL CELL WALL

Although the literature proposes multiple ways of antimicrobial action of PDCs based on in vitro screening, they are not validated using bird trials. However, knowing the mechanisms of action of PDCs is essential to develop and refine PDC-based strategies for *Salmonella* control in poultry. It is possible that when used in appropriate concentrations, PDCs could exert significant bacteriostatic or bactericidal properties on the pathogens in layer chickens. The direct activity of PDCs on pathogens is reflected on the bacterial membranes. PDCs such as TC, CR, THY, and EG act on the bacterial cell membrane, resulting in the loss of membrane potential, impaired ATP production, inhibition of glucose uptake and utilization, and leakage of intracellular components. Other membrane-related bactericidal activities include chelation of metal ions, inhibition of ATPases, and altered membrane permeability leading to the death of the pathogen (Borneman et al., 1986; Hashimoto et al., 1999; Tsuchiya and Iinuma, 2000; Lambert et al., 2001; Gill and Holley, 2006a,b; Oussalah et al., 2006).

## 6.2 EFFECTS ON ATTENUATING VIRULENCE

Until recently, attenuation of bacterial virulence was not considered as a strategy to control food-borne pathogens. Pathogens use multiple strategies to infect a host and are mediated by several virulence factors (Upadhyay et al., 2014). It is thought that reducing the expression of these virulence factors would result in reduced infections by these pathogens. More recently, virulence factors have become the targets for designing therapeutic interventions (Wu et al., 2008). The major virulence factors in *Salmonella* include those responsible for quorum sensing, attachment and invasion of epithelial cells, motility, and multiplication and colonization in live hosts. In a follow-up investigation to broiler studies in which TC and EG exerted anti-*Salmonella* activity, Kollanoor-Johny et al. (2012d) reported that the subinhibitory concentrations of the PDCs could reduce *S.* Enteritidis motility and invasion of avian intestinal epithelial cells. Results also revealed that TC and EG reduced these virulence attributes of *S.* Enteritidis, and downregulated the expression of invasion genes *hilA*, *hilD*, and *invF*, and motility genes, *flhC* and *motA* (Kollanoor-Johny et al., 2012d). Later, we concluded using DNA microarray that these two PDCs exerted anti-*Salmonella* activity by downregulating several key genes including those involved in the regulation of *Salmonella* Pathogenicity Island-1, type three secretion systems, outer membrane proteins, metabolic pathways, and electron acceptors under anaerobiosis (Kollanoor-Johny et al., 2012c).

## 6.3 EFFECTS ON INCREASING PATHOGEN SENSITIVITY TO ANTIBIOTICS

PDCs have also been investigated for increasing the sensitivity of drug-resistant pathogens. This bioactivity is important since the knowledge on the combinatorial effects of low concentrations of antibiotics and PDCs could be helpful to devise strategies to control antibiotic-resistant bacterial pathogens in food animals, including poultry. Low concentrations of sesquiterpenes such as nerolidol, bisabolol, and apritone increased the sensitivity of *Staphylococcus aureus* and *E. coli* to clindamycin, tetracycline, vancomycin, and ciprofloxacin (Brehm-Stecher and Johnson, 2003). Similarly, geraniol was found to increase the efficacy of quinolones, chloramphenicol, and beta-lactams against multidrug-resistant pathogens, including *Acinetobacter baumannii* (Lorenzi et al., 2009). In another study, five PDCs, TC, CR, THY, EG, and beta resorcylic acid were found to increase the sensitivity of multidrug-resistant *Salmonella* Typhimurium DT104 to ampicillin, chloramphenicol, tetracycline, streptomycin, and sulfamethoxazole (Kollanoor Johny et al., 2010b). The aforementioned studies indicate the potential of PDCs to sensitize pathogens to lower concentrations of antibiotics to bring about synergistic outcomes.

## 6.4 EFFECTS ON METABOLIC ENZYMES

Stimulation of digestive enzymes is also considered to be one of the mechanisms of action of PDCs. Digestive enzymes such as lipases, amylases, and carbohydrases may be positively influenced by the PDCs, resulting in improved nutrient utilization. However, studies to validate the positive findings are scarce (Applegate et al., 2010). Although results suggest that PDCs exert improved effects on pancreatic trypsin, amylase, and intestinal maltase (Jang et al., 2007), other studies indicate no effect on alpha amylase, sucrase, or pancreatic amylase (Jang et al., 2004; Jamroz et al., 2005). Supplementation of garlic at 1% and 2%, and black tea at 2% resulted in reduction of total cholesterol, total triglyceride, and LDL- and HDL-cholesterol levels in eggs (Azeke and Ekpo, 2009). Other studies have also indicated the cholesterol-lowering effects of garlic in layers and eggs (Chowdhury et al., 2002; Yalcin et al., 2006). Garlic was found to be effective in reducing cholesterol content in broiler meat also. It was found that garlic specifically inhibited key enzymes in the cholesterol synthesis: 3-hydroxyl-3-methyl-glutaryl-coenzyme-A-reductase and cholesterol-7-hydroxylase (Stanacev et al., 2011; Puvaca et al., 2013).

## 6.5 EFFECTS ON INTESTINAL MICROBIOTA

Intestinal microbiota play a significant role in the defense against invading pathogenic microorganisms, in conjunction with the gut-associated immune system (Applegate et al., 2010). It is critical that chickens keep a balanced microbiota for inhibiting pathogenic microorganisms from attaching to the receptors in the intestinal epithelium and maintaining the integrity of the intestine. A perfect microbial balance will invoke a minimal immune response, reducing the burden on the health status of a bird given its intense nutritional requirement for growth (Applegate et al., 2010). Effective non-antibiotic alternative interventions such as PDCs should be able to stabilize the gut microbial system with minimal changes in the normal population and/or result in a significant shift toward beneficial organisms such as bifidobacteria and lactobacilli.

Potent PDCs may exert positive effects on the intestinal function and microbiota in chickens. Similar to antibiotics, the majority of benefits of PDCs are attributed to their pathogen suppressive activity, especially in the early life of chickens, where they are more prone to infections. Providing feed hygiene along with supplemental PDCs may improve overall gut health (Franz et al., 2010). Although PDCs show high antimicrobial activity in vitro at very low concentrations, their activity tends to be diminished or lowered in vivo, although alteration of intestinal microbiota composition is reported in some literature, especially toward selecting beneficial microorganisms (Gong et al., 2008; Manzanilla et al., 2004). Similar to these studies, Tiihonen et al. (2010) reported that inclusion of THY and cinnamaldehyde in feed at 15 and 5 g/ton, respectively, in two diet phases, namely, 0–21 days and 22–42 days, resulted in a 4–5% increase in the body weight gain in Ross broilers. An increase in the concentration of *Lactobacillus* at 41 days was also noticed. In another study, it was shown that supplementation of PDCs, including CR, TC, and capsaicin resulted in a positive effect on gut microbiota (Jamroz et al., 2005; Jamroz and Kamer, 2002). They also noticed a reduction in the pathogen attachment onto intestinal epithelial cells due to the increased production of mucus in response to PDC supplementation (Jamroz et al., 2006). In the Tiihonen study (2010), the researchers observed a notable difference in the microbiota concentration between the percentage G+C values of 46 and 52 in the control groups, which indicated potentially harmful pathogens such as *E. coli* and *Salmonella*. In PDC-treated groups, this concentration was very low, and there was an increased relative proportion of microorganisms at the high end of the percentage G+C range that included beneficial microbes. Similarly, Upadhyaya et al. (2015c) observed a relatively stable cecal microbiota population in layer chickens, with no significant changes to the Bacteroidetes, Firmicutes, and Proteobacteria populations after supplementation of 1.5% TC for 60 days.

Conversely, it has been reported that cinnamaldehyde exerted inhibition on *Bacteroides* and *Clostridium perfringens* populations (Lee and Ahn, 1998). Although EO blends were primarily tested against pathogens for their potential antimicrobial effect (Mitsch et al., 2004), one study reported that the lactobacilli population was also sensitive to EOs (Jamroz et al., 2005).

## 7. EFFECT OF PDCS ON PRODUCTION PARAMETERS

In poultry, a paradigm change would not occur in feed intake, but the inclusion of EOs and their constituent PDCs might result in improved weight gain and feed conversion ratio (FCR). It should be highlighted that not many experiments have used constituent PDCs to determine their effects on production parameters such as feed consumption, FCR, average daily gain, carcass weight, and egg production with very few exception. For example, adding 60 ppm of CR-rich thyme oil to the diet of quails resulted in improved FCR and decreased abdominal fat weight (Denli et al., 2004). Turkeys fed with 1.25–3.75 g/kg oregano resulted in a clearly improved FCR (Bampidis et al., 2005). However, this was not the case when THY was fed to broilers. THY at 200 mg/kg did not have any effect on feed intake, weight gain, or FCR. When fed with its isomer CR to broilers, it resulted in the lowering of values of all these parameters (Lee et al., 2003). Similarly, PDCs such as 50 mg/kg of CR, 25 mg/kg of terpinene and *p*-cymene, and 0.1–1% THY showed no significant effects on growth of broilers (Haselmeyer et al., 2015). Likewise, Upadhyaya et al. (2015a) observed feeding TC at 1% or 1.5% did not adversely affect feed intake, body weight gain, and egg production in layers.

In an earlier study, Aydin et al. (2008) determined the effects of various levels of dietary black cumin seed (1%, 2%, and 3%) on egg production, egg weight, FCR, egg shell quality, and egg yolk cholesterol. The researchers reported that the groups fed with 3% black cumin seed–supplemented diet had greater egg production and shell strength, compared with the controls. The two higher concentrations (2% and 3%) resulted in increased egg weight and shell thickness. Also, these two concentrations significantly decreased egg cholesterol per gram of yolk. All concentrations of cumin seed increased yolk weights of the eggs, compared with the control groups, without resulting in any adverse effects on live weight, feed consumption, FCR, organ weights, and abdominal adipose tissue. Similarly, when garlic and thyme were investigated for their effects on egg production and egg mass, Ghasemi et al. (2010)

reported no significant difference in these parameters, although the inclusion of the combination of herbs at 0.2% resulted in increased egg yolk color and blood lymphocyte counts. However, this concentration resulted in a decreased egg shell weight. In another study, Islam et al. (2011) found that *Nigella sativa* L. seed powder resulted in decreased serum triglycerides and egg cholesterol, without affecting feed intake, body weight, egg production, or physical properties of the eggs. A positive effect of *Nigella* seed powder on egg production, egg mass, and egg shell thickness in layer hens was reported previously (Akhtar et al., 2003).

Apart from improving production parameters, the effect of oregano supplementation on lipid oxidation in the shell eggs in Lohmann layers was examined by Florou-Paneri et al. (2005). Lipid oxidation was lowered in the oregano-supplemented group. Since oregano is composed of 78–82% of CR or THY (Adam et al., 1998; Yanishlieva et al., 1999), the delay in the oxidation could be attributed to the activity of these PDCs. In a previous study, dietary supplementation of extracts of thyme and rosemary had resulted in delaying lipid oxidation in eggs (Botsoglou et al., 1997; Galobart et al., 2001). Bozkurt et al. (2012) observed that a combination of an EO blend (CR, THY, 1:8-cineole, *p*-cymene, and limonene), and prebiotic mannan-oligosaccharide resulted in improved egg production and egg shell weight, without any influence on other performance indices and egg quality traits. The combination resulted in higher antioxidant activity in egg yolk and retained the oxidative stability in the liver that was indicated by higher liver antioxidant enzymes such as superoxide dismutase, and glutathione peroxidase. Similarly, in other studies, EOs of oregano and rosemary could increase the oxidative stability of egg yolk (Galobart et al., 2001; Botsoglou et al., 2005) when incorporated in layer diets.

# 8. CHALLENGES WITH USING PDCS

## 8.1 LACK OF CONSISTENCY OF CONSTITUENT PDC CONCENTRATIONS IN EOS

Although there exists abundant published research on the in vitro efficacy of PDCs on various food-borne pathogens of public health importance, studies validating their efficacy in animals are limited. Also, the exact quantity of the active ingredient present in the EOs or plant extracts are not similar across trials (Cross et al., 2007). This is because most research on EOs is, for the most part, product driven, but not function driven. There is a dearth of information on the absorption, distribution, metabolism, and excretion and mode of action of the EOs and PDCs as ingredients in the feed (Lahlou, 2004). There is a need for reliable and reproducible methods to study the bioactivity of PDCs for comparing the results obtained by different investigators on EOs and their ingredient PDCs. In addition, active PDCs in the EOs can vary in concentration depending on several factors, such as plant species; growing location; harvest conditions; different processes of manufacturing; storage conditions such as light, temperature, oxygen tension, and time; concentration of the compounds; and compatibility with other PDCs in EOs and interactions with other nutrients in the feed (Huyghebaert et al., 2011).

There is no sufficient evidence to suggest where EOs or their constituents are located in a plant when considered for PDC isolation. It is reported that the parts richer in EOs such as cones, fruits, leaves, and bark are to be included for preliminary chemical screening. Also, species belonging to aromatic classes need to be given due consideration (Lahlou, 2004). In addition, since the active compounds present in the EOs could be modified as a result of the extraction process, selection of the appropriate extraction process is important (Bruneton, 1995). Different extraction methods include hydrodistillation, steam distillation, hydrodiffusion, $CO_2$ extraction (Buchbauer, 2000), microwave irradiation-assisted process, and mechanical and thermochemical reactions (Bouzid and Chaaban, 1997). Once extracted, the chemical analysis of EOs is performed using gas chromatography (GC) and/or GC/MS (mass spectrometry) analysis. If the compounds are not easily separated by GC, $^{13}C$ nuclear magnetic resonance techniques could be used as well (reviewed by Lahlou, 2004).

## 8.2 POOR SOLUBILITY IN WATER

Another major problem associated with the use of PDCs is their poor solubility in water. Poor solubility is of particular importance when preliminary studies are carried out to determine their minimum inhibitory concentration against various test pathogens. Also, if the PDCs are soluble in water, they could be added as antimicrobials to the water given to poultry. Therefore various solvents such as acetone, alcohol, ethylene glycol, ethanol, methanol, dimethyl sulfoxide, Tween 20, or Tween 80 are used to dissolve them for bioactivity studies. However, ethanol, Tween 20, and Tween 80 would decrease the antimicrobial activity of tested EOs in a solid medium, in addition to the antagonist effect in a liquid medium. Moreover, the use of any of these diluents in large volumes in production animals, such as poultry is impractical (reviewed by Lahlou, 2004).

## 8.3 PROBLEM OF VOLATILITY AND OXIDATION OF PDCS

No research has been conducted to ascertain the stability of the PDCs in poultry feeds, their ability to withstand the pelleting process, and the storage/shelf life of the product (Darre et al., 2014). For accurate calculations on the amount of PDCs to be added to the diet, one has to determine the effective concentration of the compounds in the feed that would remain stable for the longer duration of time. The PDCs in the feed are relatively unstable due to their volatile nature. Also, PDCs may undergo reaction with air and lose their antimicrobial effect. Kollanoor Johny et al. (2012b,d) found that daily mixing of PDCs to poultry feed improved the antimicrobial effect compared with one-time addition in liquid fats while mixing feed in large amounts, underscoring the relative instability of PDCs in the mixed feed stored for a longer period of time. However, mitigation methods such as microencapsulation or coating can be can potentially overcome this problem to provide enhanced antimicrobial activity (Timbermont et al., 2010; Verlinden et al., 2013). Encapsulation may give sustained effect of the PDCs in a matrix to which they are applied. Since PDCs are lipophilic in nature, they can be mixed with liquid fats added to the diet (Darre et al., 2014). Another challenge with PDCs is that not many studies have focused on the palatability issues in poultry and the resultant effect on feed intake. It is advisable to give a period of adaptation to birds for circumventing the problem of reduced feed intake (Franz et al., 2010). In most of the studies involving broilers, stimulation of feed intake has been reported with PDC supplementation, which may not be just a consequence of improved palatability, but may be due to enhanced digestion as well (Giannenas et al., 2003). Another limitation is that much less is known about the interactions that the PDCs might have with other feed ingredients, especially those with high lipophilic properties. In addition, there is a high likelihood of potential interactions between PDCs and other essential nutrients in the feed. Adverse interactions between EOs and enzymes and/or proteins have been previously discussed (Sarica et al., 2005; Anadon et al., 2005).

## 8.4 DEARTH OF LONG-TERM TOXICOLOGY STUDIES

Most EOs have generally recognized as safe (GRAS) status approved by the Food and Drug Administration for use in foods as flavor ingredients. The level of inclusion of EOs at lower levels (less than 100 ppm) in food may not lead to adverse safety issues since the no-observed-adverse-effect levels are typically more than 100,000 times as their use as flavor ingredients (Adams et al., 1996; Smith et al., 2005; Lis-Balchin, 2006). In a study that explored the potential of EO ingredients, TC and EG against *S.* Enteritidis in broiler chicks, Kollanoor-Johny et al. (2012d) reported that the supplementation of EG for 21 days reduced feed consumption and body weight, significantly ($P<.05$). However, when EG was supplemented for 5 days before slaughter at 6 weeks, the EO ingredient did not cause any untoward effect. Although mild to severe congestion was noticed in the liver with 21-day prophylactic supplementation, no pathological lesions in the liver were noticed with 5-day therapeutic supplementation (Kollanoor-Johny et al., 2012b). However, studies that determined the long-term effects of PDCs on bird health are not available. Therefore increased knowledge about the pharmacokinetics, pharmacodynamics, and potential toxicity of PDCs administered by the oral route is required in poultry (Solórzano-Santos and Miranda-Novales, 2012). In addition, it is also difficult to identify and quantitate different activities of the compounds present in the extracts that determine growth, feed utilization, physiology, and health status of the animals (Cheng et al., 2014).

## 8.5 RAPID ABSORPTION IN INTESTINE

PDCs are easily absorbed in the stomach and may leave minimal amounts for action in the hind part of the intestine, especially if low concentrations are used (Michiels et al., 2008). Moreover, EOs are easily absorbed to feed particles (Chambers and Gong, 2011), which may necessitate protective delivery of the active components to the target site. As mentioned earlier, encapsulation of EOs and their ingredient PDCs will be beneficial to render them active. For example, in a human clinical study, microencapsulation of EO provided better performance in the treatment of irritable bowel syndrome compared with the uncoated formulation (Liu et al., 1997; Logan and Beaulne, 2002). However, it needs to be known if encapsulation of EOs or PDCs will result in improvements in intestinal health and performance by reducing the burden of enteric pathogens (Timbermont et al., 2010). However, since chicken digesta appears less liquid in nature, delivery of encapsulated EOs may present other challenges as a result of less beneficial interactions (Chambers and Gong, 2011). Although there are several potential methods for encapsulating EOs as reviewed by Bakry et al. (2016), evaluation of costs associated with each has not been ascertained by the industry for use in poultry feeds.

### 8.6 POTENTIAL CARRY-OVER PROBLEM IN PRODUCTS

Another factor to consider is the effect on the flavor of the animal products produced from animals raised with supplemental PDCs in the feed. The effects could be either beneficial or detrimental. Further research is needed to determine if PDCs could result in "off flavor" of poultry products. Upadhyaya et al. (2015a,d) and Upadhyaya (2015) concluded that a PDC, TC, and a medium chain fatty acid, caprylic acid, did not result in any off flavor in eggs produced by layers that were fed these compounds for 60–66 days during their active laying phase. Although this would indicate that these PDCs may leave negligible or no carry-over effect in the products, extended studies are warranted to determine the residue build up in poultry products. Moreover, a reliable analytical method for the identification and quantification of PDC traces in feeds, animal tissues, and carcasses, eggs, and milk is required (Brenes and Roura, 2010).

## 9. ECONOMIC CONSIDERATIONS

At the current market price, EOs or PDCs are not cheap; industry may be reluctant to introduce them in formulating feeds. For example, THY (21CFR 172.515) costs \$367.40 for conventional and \$484/kg for certified organic, CR (21CFR 172.515) costs \$48.40/kg, and TC (21CFR 182.60) costs \$4.40/kg. However, as the demand rises, it is likely that the prices on PDCs may become economical for inclusion in poultry rations. With less effective antimicrobial solutions against pathogens in the organic poultry industry, these farmers might be the early adopters of the use of PDCs because of their GRAS status with the FDA (Darre et al., 2014).

## 10. CONCLUSION AND FUTURE DIRECTIONS

A trend in the increased usage of herbal products as human dietary supplements has resulted in an emerging interest for their use in animal agriculture, specifically for improving feed and food safety (Franz et al., 2010). If PDCs are potential candidates to support health, they should be safe to feed animals and not expected to result in residual effects that may adversely affect humans. In this context, feed components such as PDCs with no direct nutritive value, yet having a high impact on animal and human health, has great significance. Current research indicates a tremendous opportunity for the use of PDCs in poultry production as feed and food safety enhancers. It is clear that PDCs have strong antibacterial and antioxidant properties, potentially improving the shelf life of feed or animal-derived foods. Similarly, research indicates that PDCs have a potential role in the overall stimulation of growth, especially due to their effects on FCR, attributed to the reduced gastric emptying, higher enzymatic activity, and, therefore, better absorption of nutrients. The PDCs also could cause increased mucus production thereby resulting in reduced attachment of the pathogens onto intestinal epithelial cells. It is also proved that they exert significant effects on pathogens that cause a significant economic impact on public health. However, further research on the stability of PDCs in feed, especially during storage, needs to be undertaken. In addition, although many of the PDCs have GRAS status, long-term toxicological studies in chickens, and research on the presence of potential residues in poultry products and their impact on product quality need to be performed.

## REFERENCES

Acamovic, T., Brooker, J.D., 2005. Biochemistry of plant secondary metabolites and their effects in animals. Proc. Nutr. Soc. 64, 403–412.

Adam, K., Sivropoulou, A., Kokkini, S., Lanaras, T., Arsenakis, M., 1998. Antifungal activities of *Origanum vulgare* subsp. *hirtum*, *Mentha spicata*, *Lavandula angustifolia*, and *Salviafruticosa* essential oils against human pathogenic fungi. J. Agric. Food Chem. 46, 1739–1745.

Adams, T.B., Hallagan, J.B., Putman, J.M., Gierke, T.L., Doull, J., Munro, I.C., Newberne, P.M., Portoghese, P.S., Smith, R.L., Wagner, B.M., Weil, C.S., Woods, L.A., Ford, R.A., 1996. The FEMA GRAS assessment of alicyclic substances used as flavor ingredients. Food Chem. Toxicol. 34, 763–828.

Akhtar, M.S., Nasir, Z., Abid, A.R., 2003. Effect of feeding powdered *Nigella sativa L.* seeds on poultry egg production and their suitability for human consumption. Vet. Arh. 73, 181–190.

Anadon, A., Abroix Arzo, M.G., Bories, P., Brantom, J., Brufau de Barbera, A., Chesson, P.S., Concconcelli, J., de Knecht, N., Dierick, G., Flachowsky, A., Franklin, J., Gropp, A.K., Haldorsen, I., Halle, A., Mantovani, K., Peltonen, G., Rychen, P., Sanders, A., Soares, P., Wester, W., Windisch, W., 2005. Opinion of the FEEDAP panel on the safety and efficacy of the product Farmatan for rabbits and piglets. EFSA J. 222, 1–20.

Angulo, F.J., Swerdlow, D.L., 1999. Epidemiology of human *Salmonella enterica* serovar Enteritidis infections in the United States. Salmonella enterica 33–41.

Antony, M.L., Singh, S.V., 2011. Molecular mechanisms and targets of cancer chemoprevention by garlic-derived bioactive compound diallyl trisulfide. Indian J. Exp. Biol. 49, 805.

Applegate, T.J., Klose, V., Steiner, T., Ganner, A., Schatzmayr, G., 2010. Probiotics and phytogenics for poultry: myth or reality? J. Appl. Poult. Res. 19, 194–210.

Arsi, K., Donoghue, A.M., Venkitanarayanan, K., Kollanoor-Johny, A., Fanatico, A.C., Blore, P.J., Donoghue, D.J., 2014. The efficacy of the natural plant extracts, thymol and carvacrol against *Campylobacter* colonization in broiler chickens. J. Food Saf. 34, 321–325.

Aydin, R., Karaman, M., Cicek, T., Yardibi, H., 2008. Black cumin (*Nigella sativa L.*) supplementation into the diet of the laying hen positively influences egg yield parameters, shell quality, and decreases egg cholesterol. Poult. Sci. 87, 2590–2595.

Azeke, M.A., Ekpo, K.E., 2009. Egg yolk cholesterol lowering effects of garlic and tea. J. Med. Plants Res. 3, 1113–1117.

Bakkali, F., Averbeck, S., Averbeck, D., Idaomar, M., 2008. Biological effects of essential oils–a review. Food Chem. Toxicol. 46, 446–475.

Bakry, A.M., Abbas, S., Ali, B., Majeed, H., Abouelwafa, M.Y., Mousa, A., Liang, L., 2016. Microencapsulation of oils: a comprehensive review of benefits, techniques and applications. Compr. Rev. Food Sci. Food Saf. 15, 143–182.

Bampidis, V.A., Christodoulou, V., Florou-Paneri, P., Christaki, E., Chatzopoulou, P.S., Tsiligianni, T., Spais, A.B., 2005. Effect of dietary dried oregano leaves on growth performance, carcass characteristics and serum cholesterol of female early maturing turkeys. Br. Poult. Sci. 46, 595–601.

Baskerville, A., Humphrey, T.J., Fitzgeorge, R.B., Cook, R.W., Chart, H., Rowe, B., Whitehead, A., 1992. Airborne infection of laying hens with *Salmonella* Enteritidis phage type 4. Vet. Rec. 130, 395–398.

Blumenthal, M., Goldberg, A., Brinckmann, J., 2000. Herbal Medicine: Expanded Commission E Monographs. Integrative Medicine Communications, Newton, MA. http://trove.nla.gov.au/work/28272199?selectedversion=NBD44500697.

Borland, E.D., 1975. *Salmonella* infection in poultry. Vet. Rec. 97, 406–408.

Borneman, W.S., Akin, D.E., VanEseltine, W.P., 1986. Effect of phenolic monomers on ruminal bacteria. Appl. Environ. Microbiol. 52, 1331–1339.

Botsoglou, N.A., Yannakopoulos, A.L., Fletouris, D.J., Tserveni-Goussi, A.S., portomaris, P.D., 1997. Effect of dietary thyme on the oxidative stability of egg yolk. J. Agric. Food Chem. 45, 3711–3716.

Botsoglou, N.A., Florou-Paneri, P., Nikolakakis, I., Giannenas, I., Dotas, V., Botsoglou, E.N., Aggelopoulos, S., 2005. Effect of dietary saffron *(Crocus sativus L.)* on the oxidative stability of egg yolk. Br. Poult. Sci. 46, 701–707.

Bouzid, A., Chaaban, A., 1997. An accurate method of evaluating relaxation in bolted flanged connections. J. Press. Vessel Technol. 119, 10–17.

Bozkurt, M., Küçükyilmaz, K., Pamukçu, M., Çabuk, M., Alçiçek, A., Çatli, A.U., 2012. Long-term effects of dietary supplementation with an essential oil mixture on the growth and laying performance of two layer strains. Ital. J. Anim. Sci. 11, 5.

Braden, C.R., 2006. *Salmonella enterica* serotype Enteritidis and eggs: a national epidemic in the United States. Clin. Infect. Dis. 43, 512–517.

Brehm-Stecher, B.F., Johnson, E.A., 2003. Sensitization of *Staphylococcus aureus* and *Escherichia coli* to antibiotics by the sesquiterpenoids nerolidol, farnesol, bisabolol, and apritone. Antimicrob. Agents Chemother. 47, 3357–3360.

Brenes, A., Roura, E., 2010. Essential oils in poultry nutrition: main effects and modes of action. Anim. Feed Sci. Technol. 158, 1–14.

Bruneton, J., 1995. Pharmacognosy, Phytochemistry, Medicinal Plants. Lavoisier Publishing, p. 1136.

Buchbauer, G., 2000. The detailed analysis of essential oils leads to the understanding of their properties. Perfumer & Flavorist. 25, 64–67.

Burt, S., 2004. Essential oils: their antibacterial properties and potential applications in foods—a review. Int. J. Food Microbiol. 94, 223–253.

Bywater, R.J., 2005. Identification and surveillance of antimicrobial resistance dissemination in animal production. Poult. Sci. 84, 644–648.

CDC, 2010. CDC Report on *Salmonella* Serotype Enteritidis Infection. http://www.cdc.gov/nczved/divisions/dfbmd/diseases/salmonella_enteritidis.

Chadfield, M.S., Hinton, M.H., 2003. Evaluation of treatment and prophylaxis with nitrofurans and comparison with alternative antimicrobial agents in experimental *Salmonella enterica* serovar Enteritidis infection in chicks. Vet. Res. Commun. 27, 257–273.

Chambers, J.R., Gong, J., 2011. The intestinal microbiota and its modulation for *Salmonella* control in chickens. Food Res. Int. 44, 3149–3159.

Cheng, G., Hao, H., Xie, S., Wang, X., Dai, M., Huang, L., Yuan, Z., 2014. Antibiotic alternatives: the substitution of antibiotics in animal husbandry? Front. Microbiol. 5, 217.

Chowdhury, S.R., Chowdhury, S.D., Smith, T.K., 2002. Effects of dietary garlic on cholesterol metabolism in laying hens. Poult. Sci. 81, 1856–1862.

Chun, S.-S., Vattem, D.A., Lin, Y.-T., Shetty, K., 2005. Phenolic antioxidants from clonal oregano (*Origanum vulgare*) with antimicrobial activity against *Helicobacter pylori*. Process. Biochem. 40, 809–816.

Cowan, M.M., 1999. Plant products as antimicrobial agents. Clin. Microbiol. Rev. 12, 564–582.

Cross, D.E., McDevitt, R.M., Hillman, K., Acamovic, T., 2007. The effect of herbs and their associated essential oils on performance, dietary digestibility and gut microflora in chickens from 7 to 28 days of age. Br. Poult. Sci. 48, 496–506.

Daly, M., Villa, L., Pezzella, C., Fanning, S., Carattoli, A., 2005. Comparison of multidrug resistance gene regions between two geographically unrelated *Salmonella* serotypes. J. Antimicrob. Chemother. 55, 558–561.

Darre, M.J., Kollanoor-Johny, A., Venkitanarayanan, K., Upadhyaya, I., 2014. Practical implications of plant-derived antimicrobials in poultry diets for the control of *Salmonella* Enteritidis. J. Appl. Poult. Res. 23, 340–344.

Davies, R., Breslin, M., 2004. Observations on *Salmonella* contamination of eggs from infected commercial laying flocks where vaccination for *Salmonella enterica* serovar Enteritidis had been used. Avian Pathol. 33, 133–144.

De Reu, K., Grijspeerdt, K., Messens, W., Heyndrickx, M., Uyttendaele, M., Debevere, J., Herman, L., 2006. Eggshell factors influencing eggshell penetration and whole egg contamination by different bacteria, including *Salmonella* Enteritidis. Intl. J. Food Microbiol. 112, 253–260.

Denli, M., Okan, F., Uluocak, A.N., 2004. Effect of dietary supplementation of herb essential oils on the growth performance, carcass and intestinal characteristics of quail *(Coturnix coturnix japonica)*. S. Afr. J. Anim. Sci. 34, 174–179.

Dias de Oliveira, S., Siqueira Flores, F., dos Santos, L.R., Brandelli, A., 2005. Antimicrobial resistance in *Salmonella* Enteritidis strains isolated from broiler carcasses, food, human and poultry-related samples. Intl. J. Food Microbiol. 97, 297–305.

Dixon, R.A., 2001. Natural products and plant disease. Nature 411, 843–847.

Erdem, B., Ercis, S., Hascelik, G., Gur, D., Gedikoglu, S., Aysev, A.D., Sumerkan, B., Tatman-Otkun, M., Tuncer, I., 2005. Antimicrobial resistance patterns and serotype distribution among *Salmonella enterica* strains in Turkey, 2000–2002. Eur. J. Clin. Microbiol. Infect. Dis. 24, 220–225.

Florou-Paneri, P., Nikolakakis, I., Giannenas, I., Koidis, A., Botsoglou, E., Dotas, V., Mitsopoulos, I., 2005. Hen performance and egg quality as affected by dietary oregano essential oil and tocopheryl acetate supplementation. Int. J. Poult. Sci. 4, 449–454.

Foley, S.L., Nayak, R., Hanning, I.B., Johnson, T.J., Han, J., Ricke, S.C., 2011. Population dynamics of *Salmonella* enterica serotypes in commercial egg and poultry production. Appl. Environ. Microbiol. 77, 4273–4279.

Franz, C., Baser, K.H.C., Windisch, W., 2010. Essential oils and aromatic plants in animal feeding–a European perspective. A review. Flavour Fragr. J. 25, 327–340.

Galiş, A.M., Marcq, C., Marlier, D., Portetelle, D., Van, I., Beckers, Y., Théwis, A., 2013. Control of *Salmonella* contamination of shell eggs—preharvest and postharvest methods: a review. Compr. Rev. Food Sci. Food Saf. 12, 155–182. http://dx.doi.org/10.1111/1541-4337.12007.

Galobart, J., Barroeta, A.C., Baucells, M.D., Codony, R., Ternes, W., 2001. Effect of dietary supplementation with rosemary extract and α-tocopheryl acetate on lipid oxidation in eggs enriched with ω3-fatty acids. Poult. Sci. 80, 460–467.

Gantois, I., Ducatelle, R., Pasmans, F., Haesebrouck, F., Van Immerseel, F., 2008. *Salmonella enterica* serovar Enteritidis genes induced during oviduct colonization and egg contamination in laying hens. Appl. Environ. Microbiol. 74, 6616–6622.

Gast, R.K., Guraya, R., Guard-Bouldin, J., Holt, P.S., Moore, R.W., 2007. Colonization of specific regions of the reproductive tract and deposition at different locations inside eggs laid by hens infected with *Salmonella* Enteritidis or *Salmonella* Heidelberg. Avian Dis. 51, 40–44.

Geissman, T.A., 1963. Flavonoid compounds, tannins, lignins and related compounds. In: Stotz, E.H. (Ed.), Pyrrole Pigments, Isoprenoid Compounds and Phenolic Plant Constituents. Elsevier, New York, NY, p. 265.

Ghasemi, R., Zarei, M., Torki, M., 2010. Adding medicinal herbs including garlic (*Allium sativum*) and thyme (*Thymus vulgaris*) to diet of laying hens and evaluating productive performance and egg quality characteristics. Am. J. Anim. Vet. Sci. 5, 151–154.

Giannenas, I., Florou-Paneri, P., Papazahariadou, M., Christaki, E., Botsoglou, N.A., Spais, A.B., 2003. Effect of dietary supplementation with oregano essential oil on performance of broilers after experimental infection with *Eimeria tenella*. Arch. Anim. Nutr. 57, 99–106.

Gill, A.O., Holley, R.A., 2006a. Inhibition of membrane bound ATPases of *Escherichia coli* and *Listeria monocytogenes* by plant oil aromatics. Int. J. Food Microbiol. 111, 170–174.

Gill, A.O., Holley, R.A., 2006b. Disruption of *Escherichia coli, Listeria monocytogenes* and *Lactobacillus sakei* cellular membranes by plant oil aromatics. Int. J. Food Microbiol. 108, 1–9.

Gong, J., Yu, H., Liu, T., Li, M., Si, W., de Lange, C.F., Dewey, C., 2008. Characterization of ileal bacterial microbiota in newly-weaned pigs in response to feeding lincomycin, organic acids or herbal extract. Livest. Sci. 116, 318–322.

Han, G.Y., Xu, P., Wang, X.P., Liu, M.Z., Xu, X.Y., Meng, L.N., Chen, Z.L., Zhu, D.Y., 1981. Study on the active principle of polyalthia-nemoralis. 1. The isolation and identification of natural zinc compound. Acta Chim. Sinica 39, 433–437.

Hao, H., Cheng, G., Iqbal, Z., Ai, X., Hussain, H.I., Huang, L., Dai, M., Wang, Y., Liu, Z., Yuan, Z., 2014. Benefits and risks of antimicrobial use in food producing animals. Front. Microbiol. 5, 288.

Haselmeyer, A., Zentek, J., Chizzola, R., 2015. Effects of thyme as a feed additive in broiler chickens on thymol in gut contents, blood plasma, liver and muscle. J. Sci. Food Agric. 95, 504–508.

Hashemi, S.R., Davoodi, H., 2011. Herbal plants and their derivatives as growth and health promoters in animal nutrition. Vet. Res. Commun. 35, 169–180.

Hashemi, S.R., Davoodi, H., 2012. Herbal plants as new immuno-stimulator in poultry industry: a review. Asian J. Anim. Vet. Adv. 7, 105–116.

Hashimoto, T., Kumazawa, S., Nanjo, F., Hara, Y., Nakayama, T., 1999. Interaction of tea catechins with lipid bilayers investigated with liposome systems. Biosci. Biotechnol. Biochem. 63, 2252–2255.

Hennessy, T.W., Cheng, L.H., Kassenborg, H., Ahuja, S.D., Mohle-Boetani, J., Marcus, R., Shiferaw, B., Angulo, F.J., Emerging Infections Program FoodNet Working Group, 2004. Egg consumption is the principal risk factor for sporadic *Salmonella* serotype Heidelberg infections: a case-control study in Food Net sites. Clin. Infect. Dis. 38, 237–243.

Hermans, D., Van Deun, K., Martel, A., Van Immerseel, F., Messens, W., Heyndrickx, F., Haesebrouck, Pasmans, F., 2011. Colonization factors of *Campylobacter jejuni* in the chicken gut. Vet. Res. 42, 82.

Huyghebaert, G., Ducatelle, R., Van Immerseel, F., 2011. An update on alternatives to antimicrobial growth promoters for broilers. Vet. J. 187, 182–188.

Islam, M.T., Selim, A.S.M., Sayed, M.A., Khatun, M.A., Siddiqui, M.N.E.A., Alam, M.S., Hossain, M., 2011. *Nigella sativa L.* supplemented diet decreases egg cholesterol content and suppresses harmful intestinal bacteria in laying hens. J. Anim. Feed Sci. 20, 587–598.

Jamroz, D., Kamel, C., 2002. Plant extracts enhance broiler performance. J. Anim. Sci. 80, 41.

Jamroz, D., Wiliczkiewicz, A., Wertelecki, T., Orda, J., Skorupińska, J., 2005. Use of active substances of plant origin in chicken diets based on maize and locally grown cereals. Br. Poult. Sci. 46, 485–493.

Jamroz, D., Wertelecki, T., Houszka, M., Kamel, C., 2006. Influence of diet type on the inclusion of plant origin active substances on morphological and histochemical characteristics of the stomach and jejunum walls in chicken. J. Anim. Physiol. Anim. Nutr. 90, 255–268.

Jang, I.S., Ko, Y.H., Yang, H.Y., Ha, J.S., Kim, J.Y., Kim, J.Y., Lee, C.Y., 2004. Influence of essential oil components on growth performance and the functional activity of the pancreas and small intestine in broiler chickens. Asian Aust. J. Anim. Sci. 17, 394–400.

Jang, I.S., Ko, Y.H., Kang, S.Y., Lee, C.Y., 2007. Effect of a commercial essential oil on growth performance, digestive enzyme activity and intestinal microflora population in broiler chickens. Anim. Feed Sci. Technol. 134, 304–315.

Keller, L.H., Benson, C.E., Krotec, K., Eckroade, R.J., 1995. *Salmonella* Enteritidis colonization of the reproductive tract and forming and freshly laid eggs of chickens. Infect. Immun. 63, 2443–2449.

Kilmartin, D., Morris, D., O'Hare, C., Corbett-Feeney, G., Cormican, M., 2005. Clonal expansion may account for high levels of quinolone resistance in *Salmonella enterica* serovar Enteritidis. Appl. Environ. Microbiol. 71, 2587–2591.

Kollanoor Johny, A., Darre, M.J., Hoagland, T.A., Schreiber, D.T., Donoghue, A.M., Donoghue, D.J., Venkitanarayanan, K., 2008. Antibacterial effect of *trans*-cinnamaldehyde on *Salmonella* Enteritidis and *Campylobacter jejuni* in chicken drinking water. J. Appl. Poult. Res. 17, 490–497.

Kollanoor Johny, A., Baskaran, S.A., Charles, A.S., Amalaradjou, M.A.R., Darre, M.J., Khan, M.I., Hoagland, T.A., Schreiber, D.T., Donoghue, A.M., Donoghue, D.J., Venkitanarayanan, K., 2009. Prophylactic supplementation of caprylic acid in feed reduces *Salmonella* Enteritidis colonization in commercial broiler chicks. J. Food Prot. 72, 722–727.

Kollanoor Johny, A., Darre, M.J., Donoghue, A.M., Donoghue, D.J., Venkitanarayanan, K., 2010a. Antibacterial effect of trans-cinnamaldehyde, eugenol, carvacrol, and thymol on *Salmonella* Enteritidis and *Campylobacter jejuni* in chicken cecal contents *in vitro*. J. Appl. Poult. Res. 19, 237–244.

Kollanoor Johny, A., Hoagland, T., Venkitanarayanan, K., 2010b. Effect of subinhibitory concentrations of plant-derived molecules in increasing the sensitivity of multidrug resistant *Salmonella enterica* serovar Typhimurium DT104 to antibiotics. Foodborne Pathog. Dis. 7, 1165–1170.

Kollanoor-Johny, A., Mattson, T., Baskaran, S.A., Amalaradjou, M.A.R., Darre, M.J., Khan, M.I., Donoghue, A.M., Donoghue, D.J., Venkitanarayanan, K., 2011. Effect of food grade carvacrol on cecal *Salmonella* Enteritidis colonization and cloacal shedding in 19-day old commercial broiler chicks. Poult. Sci. 90 (E-Suppl. 1), 23–24.

Kollanoor-Johny, A., Mattson, T., Baskaran, S.A., Amalaradjou, M.A.R., Darre, M.J., Schreiber, D.T., Khan, M.I., Donoghue, A.M., Donoghue, D.J., Venkitanarayanan, K., 2012a. Caprylic acid reduces *Salmonella* Enteritidis populations in various segments of digestive tract and internal organs of 3-and 6-week-old broiler chickens, therapeutically. Poult. Sci. 91, 1686–1694.

Kollanoor-Johny, A., Upadhyay, A., Baskaran, S.A., Upadhyaya, I., Mooyottu, S., Mishra, N., Venkitanarayanan, K., 2012b. Effect of therapeutic supplementation of the plant compounds *trans*-cinnamaldehyde and eugenol on *Salmonella enterica* serovar Enteritidis colonization in market-age broiler chickens. J. Appl. Poult. Res. 21, 816–822.

Kollanoor-Johny, A., Frye, J.G., Porwollik, S., Darre, M.J., Donoghue, A.M., Donoghue, D.J., McClelland, M., Venkitanarayanan, K., 2012c. Microarray analysis of *Salmonella* Enteritidis phage type 8 treated with subinhibitory concentrations of *trans*-cinnamaldehyde or eugenol. Poult. Sci. (Suppl. 1), 76.

Kollanoor-Johny, A., Mattson, T., Baskaran, S.A., Amalaradjou, M.A., Babapoor, S., March, B., Khan, M.I., 2012d. Reduction of *Salmonella enterica* serovar Enteritidis colonization in 20-day-old broiler chickens by the plant-derived compounds *trans*-cinnamaldehyde and eugenol. Appl. Environ. Microbiol. 78, 2981–2987.

Lahlou, M., 2004. Methods to study the phytochemistry and bioactivity of essential oils. Phytother. Res. 18, 435–448.

Lambert, R.J.W., Skandamis, P.N., Coote, P.J., Nychas, G.-J.E., 2001. A study of the minimum inhibitory concentration and mode of action of oregano essential oil, thymol and carvacrol. J. Appl. Microbiol. 91, 453–462.

Leach, S.A., Williams, A., Davies, A.C., Wilson, J., Marsh, P.D., Humphrey, T.J., 1999. Aerosol route enhances the contamination of intact eggs and muscle of experimentally infected laying hens by *Salmonella* Typhimurium DTI04. FEMS Microbiol. Lett. 171, 203–207.

Lee, H.S., Ahn, Y.J., 1998. Growth-inhibiting effects of *Cinnamomum cassia* bark-derived materials on human intestinal bacteria. J. Agric. Food Chem. 46, 8–12.

Lee, K.W., Everts, H., Kappert, H.J., Yeom, K.H., Beynen, A.C., 2003. Dietary carvacrol lowers body weight gain but improves feed conversion in female broiler chickens. J. Appl. Poult. Res. 12, 394–399.

Lis-Balchin, M., 2006. Aromatherapy Science: A Guide for Healthcare Professionals. Pharmaceutical Press, London, UK, p. 467.

Liu, J.H., Chen, G.H., Yeh, H.Z., Huang, C.K., Poon, S.K., 1997. Enteric-coated peppermint-oil capsules in the treatment of irritable bowel syndrome: a prospective, randomized trial. J. Gastroenterol. 32, 765–768.

Logan, A.C., Beaulne, T.M., 2002. The treatment of small intestinal bacterial overgrowth with enteric-coated peppermint oil: a case report. (peppermint oil). Altern. Med. Rev. 7, 410–417.

Lorenzi, V., Muselli, A., Bernardini, A.F., Berti, L., Pagès, J.M., Amaral, L., Bolla, J.M., 2009. Geraniol restores antibiotic activities against multidrug-resistant isolates from gram-negative species. Antimicrob. Agents Chemother. 53, 2209–2211.

Manzanilla, E.G., Perez, J.F., Martin, M., Kamel, C., Baucells, F., Gasa, J., 2004. Effect of plant extracts and formic acid on the intestinal equilibrium of early-weaned pigs. J. Anim. Sci. 82, 3210–3218.

Marin, C., Balasch, S., Vega, S., Lainez, M., 2011. Sources of *Salmonella* contamination during broiler production in Eastern Spain. Prev. Vet. Med. 98, 39–45.

Michiels, J., Missotten, J., Dierick, N., Fremaut, D., Maene, P., De Smet, S., 2008. *In vitro* degradation and *in vivo* passage kinetics of carvacrol, thymol, eugenol and *trans*-cinnamaldehyde along the gastrointestinal tract of piglets. J. Sci. Food Agric. 88, 2371–2381.

Mitsch, P., Zitterl-Eglseer, K., Köhler, B., Gabler, C., Losa, R., Zimpernik, I., 2004. The effect of two different blends of essential oil components on the proliferation of *Clostridium perfringens* in the intestines of broiler chickens. Poult. Sci. 83, 669–675.

Miyamoto, T., Baba, E., Tanaka, T., Sasai, K., Fukata, T., Arakawa, A., 1997. Salmonella enteritidis contamination of eggs from hens inoculated by vaginal, cloacal, and intravenous routes. Avian Dis. 41, 296–303.

Namba, T., Morita, O., Huang, S.-L., Goshima, K., Hattori, M., Kakiuchi, N., 1988. Studies on cardio-active crude drugs. I. Effect of coumarins on cultured myocardial cells. Planta Med. 54, 277–282.

Newman, D.J., 2008. Natural products as leads to potential drugs: an old process or the new hope for drug discovery? J. Med. Chem. 51, 2589–2599.

O'Kennedy, R., Thornes, R.D., 1997. Coumarins: Biology, Applications and Mode of Action. John Wiley & Sons, Inc., New York, NY, p. 360.

Okamura, M., Kamijima, Y., Miyamoto, T., Tani, H., Sasai, K., Baba, E., 2001. Differences among six Salmonella serovars in abilities to colonize reproductive organs and to contaminate eggs in laying hens. Avian Dis. 45, 61–69.

Osbourn, A.E., 1996. Preformed antimicrobial compounds and plant defense against fungal attack. Plant Cell 8, 1821.

Oussalah, M., Caillet, S., Lacroix, M., 2006. Mechanism of action of Spanish oregano, Chinese cinnamon, and savory essential oils against cell membranes and walls of *Escherichia coli* O157: H7 and *Listeria monocytogenes*. J. Food Prot. 69, 1046–1055.

Oviedo-Rondón, E.O., Hume, M.E., Hernández, C., Clemente-Hernández, S., 2006. Intestinal microbial ecology of broilers vaccinated and challenged with mixed *Eimeria* species, and supplemented with essential oil blends. Poult. Sci. 85, 854–860.

Painter, J.A., Hoekstra, R.M., Ayers, T., Tauxe, R.V., Braden, C.R., Angulo, F.J., March 2013. Attribution of foodborne illnesses, hospitalizations, and deaths to food commodities by using outbreak data, United States, 1998–2008. Emerg. Infect. Dis. [Internet] 19, 407–415 (cited on April 18, 2016).

Piller, N.B., 1975. A comparison of the effectiveness of some anti-inflammatory drugs on thermal oedema. Br. J. Exp. Pathol. 56, 554–559.

Puvača, N., Stanaćev, V., Glamočić, D., Lević, J., Perić, L., Stanaćev, V., Milić, D., 2013. Beneficial effects of phyto additives in broiler nutrition. World Poult. Sci. J. 69, 27–34.

Rabsch, W., Tschäpe, H., Bäumler, A.J., 2001. Non-typhoidal salmonellosis: emerging problems. Microbe. Infect. 3, 237–247.

Reiber, M.A., Mclnroy, J.A., Conner, D.E., 1995. Enumeration and identification of bacteria in chicken semen. Poult. Sci. 74, 795–799.

Sarica, S., Ciftci, A., Demir, E., Kilinc, K., Yildirim, Y., 2005. Use of an antibiotic growth promoter and two herbal natural feed additives with and without exogenous enzymes in wheat based broiler diets. S. Afr. J. Anim. Sci. 35, 61–72.

Savoia, D., 2012. Plant-derived antimicrobial compounds: alternatives to antibiotics. Future Microbiol. 7, 979–990.

Scalbert, A., 1991. Antimicrobial properties of tannins. Phytochemistry 30, 3875–3883.

Schultz, J.C., 1988. Tannin-insect interactions. In: Hemingway, R.W., Karchesy, J.J. (Eds.), Chemistry and Significance of Condensed Tannins. Plenum Press, New York, NY, p. 553.

Serafini, M., Ghiselli, A., Ferro-Luzzi, A., 1994. Red wine, tea and anti-oxidants. Lancet 344, 626.

Shea, K.M., 2003. Antibiotic resistance: what is the impact of agricultural uses of antibiotics on children's health? Pediatrics 112, 253–258.

Shivaprasad, H.L., Timoney, J.F., Morales, S., Lucio, B., Baker, R.C., 1990. Pathogenesis of *Salmonella* Enteritidis infection in laying chickens. I. Studies on egg transmission, clinical signs, fecal shedding, and serologic responses. Avian Dis. 34, 548–557.

Smith, R.L., Cohen, S.M., Doull, J., Feron, V.J., Goodman, J.I., Marnett, L.J., Portoghese, P.S., Waddell, W.J., Wagner, B.M., Adams, T.B., 2005. GRAS flavoring substances 22. Food Technol. 59, 24–62.

Sol6rzano-Santos, F., Miranda-Novales, M.G., 2012. Essential oils from aromatic herbs as antimicrobial agents. Curr. Opin. Biotechnol. 23, 136–141.

Stanaćev, V., Glamočić, D., Milošević, N., Puvača, N., Stanaćev, V., Plavša, N., 2011. Effect of garlic *(Allium sativum L.)* in fattening chicks' nutrition. Afr. J. Agric. Res. 6, 943–948.

Stern, J.L., Hagerman, A.E., Steinberg, P.D., Mason, P.K., 1996. Phlorotannin-protein interactions. J. Chem. Ecol. 22, 1887–1899.

Tauxe, R.V., 1999. Food-related illness and death in the United States. Emerging Infect. Dis. 5, 607–625.

Thastrup, O., Knudsen, J.B., Lemmich, J., Winther, K., 1985. Inhibitions of human platelet aggregation by dihydropyrano- and dihydrofurano-coumarins, a new class of cAMP phosphodiesterase inhibitors. Biochem. Pharmacol. 34, 2137–2140.

Tiihonen, K., Kettunen, H., Bento, M.H.L., Saarinen, M., Lahtinen, S., Ouweh, A.C., Rautonen, N., 2010. The effect of feeding essential oils on broiler performance and gut microbiota. Br. Poult. Sci. 51, 381–392.

Timbermont, L., Lanckriet, A., Dewulf, J., Nollet, N., Schwarzer, K., Haesebrouck, F., Van Immerseel, F., 2010. Control of *Clostridium perfringens*-induced necrotic enteritis in broilers by target-released butyric acid, fatty acids and essential oils. Avian Pathol. 39, 117–121.

Timoney, J.F., Shivaprasad, H.L., Baker, R.C., Row, B., 1989. Egg transmission after infection of hens with *Salmonella* Enteritidis phage type 4. Vet. Rec. 125, 600–601.

Tsuchiya, H., Iinuma, M., 2000. Reduction of membrane fluidity by antibacterial sophoraflavanone G isolated from *Sophora exigua*. Phytomedicine 7, 161–165.

Upadhyay, A., Upadhyaya, I., Kollanoor-Johny, A., Venkitanarayanan, K., 2014. Combating pathogenic microorganisms using plant-derived antimicrobials: a minireview of the mechanistic basis. Biomed. Res. Int. http://dx.doi.org/10.1155/2014/761741.

Upadhyaya, I., Upadhyay, A., Kollanoor-Johny, A., Darre, M.J., Venkitanarayanan, K., 2013. Effect of plant derived antimicrobials on *Salmonella* Enteritidis adhesion to and invasion of primary chicken oviduct epithelial cells in vitro and virulence gene expression. Int. J. Mol. Sci. 14, 10608–10625. http://www.mdpi.com/1422-0067/14/5/10608.

Upadhyaya, I., 2015. Investigating the Efficacy of Natural Antimicrobial Molecules in Reducing Egg-Borne Transmission of *Salmonella enterica* Serovar Enteritidis in Layer Hens. p. 199 (Ph.D. thesis). Submitted to the University of Connecticut http://digitalcommons.uconn.edu/cgi/viewcontent.cgi?article=6869&context=dissertations.

Upadhyaya, I., Upadhyay, A., Kollanoor-Johny, A., Mooyottu, S., Baskaran, S.A., Yin, H., Schreiber, D.T., Khan, M.I., Darre, M.J., Curtis, P.A., Venkitanarayanan, K., 2015a. In-feed supplementation of *trans*-cinnamaldehyde reduces layer-chicken egg-borne transmission of *Salmonella* enterica serovar Enteritidis. Appl. Environ. Microbiol. 81, 2985–2994. http://aem.asm.org/content/81/9/2985.full.

Upadhyaya, I., Chen, A.C., Upadhyay, A., Yin, H., Nair, M.S., Khan, M.I., Darre, M.J., Donoghue, A.M., Donoghue, D.J., Venkitanarayanan, K., 2015c. Effect of in-feed supplementation of *trans*-cinnamaldehyde and caprylic acid on chicken cecal microbiome in response to *Salmonella* Enteritidis. Poult. Sci. (E-Suppl. 1), 60.

Upadhyaya, I., Upadhyay, A., Yin, H., Nair, M.S., Bhattaram, V.K., Karumathil, D., Kollanoor-Johny, A., Khan, M.I., Darre, M.J., Curtis, P.A., Venkitanarayanan, K., 2015d. Reducing colonization and eggborne transmission of *Salmonella* Enteritidis in layer chickens by in-feed supplementation of caprylic acid. Foodborne Pathog. Dis. 12, 591–597. http://www.ncbi.nlm.nih.gov/pubmed/26135893.

Van Immerseel, F., De Buck, J., Boyen, F., Bohez, L., Pasmans, F., Volf, J., Sevcik, M., Rychlik, I., Haesebrouck, F., Ducatelle, R., 2004. Medium-chain fatty acids decrease colonization and invasion shortly after infection with *Salmonella* Enteritidis in chickens through *hilA* suppression. Appl. Environ. Microbiol. 70, 3582–3587.

Vazquez-Torres, A., Jones-Carson, J., Bäumler, A.J., Falkow, S., Valdivia, R., Brown, W., Le, M., Berggren, R., Parks, W.T., Fang, F.C., 1999. Extra intestinal dissemination of *Salmonella* by CD18-expressing phagocytes. Nature 401, 804–808.

Verlinden, M., Pasmans, F., Mahu, M., Maele, L.V., De Pauw, N., Yang, Z., Martel, A., 2013. *In vitro* sensitivity of poultry *Brachyspira intermedia* isolates to essential oil components and *in vivo* reduction of *Brachyspira intermedia* in rearing pullets with cinnamaldehyde feed supplementation. Poult. Sci. 92, 1202–1207.

Vondruskova, H., Slamova, R., Trckova, M., Zraly, Z., Pavlik, I., 2010. Alternatives to antibiotic growth promoters in prevention of diarrhoea in weaned piglets: a review. Vet. Med. (Praha) 55, 199–224.

Wang, H., Slavik, M.F., 1998. Bacterial penetration into eggs washed with various chemicals and stored at different temperatures and times. J. Food Prot. 61, 276–279.

Windisch, W., Kroismayr, A., 2007. The Effects of Phytobiotics on Performance and Gut Function in Monogastrics. Biomin World Nutrition Forum. http://en.engormix.com/MA-feed-machinery/articles/the-effect-phytobiotics-performance-t285/p0.htm.

Wu, H.J., Wang, A.H.J., Jennings, M.P., 2008. Discovery of virulence factors of pathogenic bacteria. Curr. Opin. Chem. Biol. 12, 93–101.

Yalcin, S., Onbasilar, I., Reisli, Z., Yalcin, S., 2006. Effect of garlic powder on the performance, egg traits, and blood parameters in laying hens. J. Sci. Food Agri. 86, 1336–1339.

Yang, Y., Iji, P.A., Choct, M., 2009. Dietary modulation of gut microflora in broiler chickens: a review of the role of six kinds of alternatives to in-feed antibiotics. World Poult. Sci. J. 65, 97–114.

Yanishlieva, N.V., Marinova, E.M., Gordon, M.H., Raneva, V.G., 1999. Antioxidant activity and mechanism of action of thymol and carvacrol in two lipid systems. Food Chem. 64, 59–66.

Zeng, Z., Zhang, S., Wang, H., Piao, X., 2015. Essential oil and aromatic oils as feed additives in non-ruminant nutrition: a review. J. Anim. Sci. Biotechnol. 6, 7.

CHAPTER

# Chemical and Physical Sanitation and Pasteurization Methods for Intact Shell Eggs

18

**Corliss A. O'Bryan, Steven C. Ricke, Philip G. Crandall**

*University of Arkansas, Fayetteville, AR, United States*

## 1. INTRODUCTION

It is well known that the main microbiological hazard derived from egg consumption is the presence of *Salmonella*, with the serovars Enteritidis and Typhimurium being the most frequently implicated in food-borne outbreaks (EFSA, 2010; Howard et al., 2012; Pande et al., 2016; Wright et al., 2016). Other enteric pathogens have been isolated from eggs and egg products including *Campylobacter* spp. and *Listeria* spp. (Leasor and Foegeding, 1989; Farber et al., 1992; Shane et al., 1986). Bacteria can reach the inside of eggs by transovarian passage in the hen, but the main path of contamination is most likely through the shell pores or when the egg is broken for use in cooking or processing (Hutchison et al., 2003). There are numerous preharvest methods in use for reducing egg contamination such as house sanitation and biosecurity (Arnold et al., 2010; Holt et al., 2011), vaccination (Van Immerseel et al., 2005), bacteriophages (Waseh et al., 2010), and feed components and additives (Tellez et al., 2001, 2012; Van Immerseel et al., 2002; Doyle and Erickson, 2006). These preharvest methods are not completely successful in eliminating contamination of shell eggs and therefore postharvest measures have been investigated to reduce or eliminate contamination in shell eggs. The aim of this chapter is to review many of the postharvest methods for reducing contamination of shell eggs (Table 18.1).

## 2. WASHING

In the United States, as well as Canada, Australia, and Japan, shell eggs are washed before distribution in commerce, although washing is not permitted in the European Union (Bartlett et al., 1993; Hutchison et al., 2003, 2004). Shell eggs are washed, rinsed and sanitized during the grading process; this process removes the cuticle from the outside of the shell increasing moisture loss and transfer of $CO_2$ into the

**Producing Safe Eggs. http://dx.doi.org/10.1016/B978-0-12-802582-6.00018-5**

**Table 18.1** Methods of Decontamination of Shell Eggs to Reduce the Risk of Food-borne Illness From Egg and Egg Product Consumption

| **Chemical Methods** |
|---|
| Washing with sanitizers |
| Electrolyzed water |
| Ozone |
| **Physical Methods** |
| Pasteurization |
| Irradiation |
| Microwaves |
| Ultraviolet light |
| Pulsed light |
| Gas plasma |
| Ultrasound |
| Thermoultrasonificatoin |
| **Biological Methods** |
| Chitosan |
| Nisin |
| Lauric arginate |
| Organic acids |

eggs and therefore the eggs are coated with food grade mineral oil and subsequently refrigerated to retard these processes (USDA, 2000).

US Department of Agriculture (USDA) regulations dictate that egg wash water must be at least 32.2°C, or 11.1°C warmer than the warmest egg entering the processing line (USDA, 2000). These regulations were put into place based on research that indicated that when eggs were placed in a bacterial suspension cooler than the egg, the resulting temperature gradient would draw bacteria into the egg (Haines and Moran, 1940) and on research that indicated that eggs washed in cold water were more likely to spoil than eggs washed in warm water (Lorenz and Starr, 1952). However, these experiments were conducted when eggs were washed by immersion, which is now banned in favor of spray washing (USDA, 1975). These high wash water temperatures can raise the internal temperature of the egg by at least 6°C providing an optimal environment for growth of any contaminating bacteria (Caudill et al., 2010) and making it difficult to quickly cool the eggs to 7.2°C or below where *Salmonella* will not grow (Gast and Holt, 2000). Caudill et al. (2010) investigated the use of lower water temperatures to wash eggs and the effect on egg quality and microbial growth within the egg. They determined that using cool water at a pH of 10–12 lowered postprocessing egg temperatures and allowed for more rapid cooling. Washing in the cool water did not result in a decline in egg quality and bacteria within the egg did not increase during 5 weeks of storage.

Wang and Slavik (1998) investigated the impact of using three commercial egg-washing chemicals, including a quaternary ammonium compound (QAC, pH 7.5), sodium carbonate ($Na_2CO_3$, pH 12), and sodium hypochlorite (NaOCl, 100 ppm, pH 7.5). They washed separate groups of intact shell eggs at 43.3°C in one of the three chemicals whereas a control group was washed with tap water at the same temperature. Subsequent to washing, the eggs were contaminated with *Salmonella* Enteritidis on the shell and allowed to dry; the washed and inoculated eggs were stored at 4°C and checked at 0, 7, 14, and 21 days for penetration of *Salmonella* into the interior of the egg. Control eggs (washed with tap water) had a bacterial penetration rate of less than 6.7% on day 1 and 20% on day 21. Treatment with QAC and sodium hypochlorite produced penetration rates of less than 3.4% and 6.7%, respectively, on day 1 and 16.7% on day 21, whereas the sodium carbonate treatment increased bacterial penetration during storage from 30% on day 1 to 76.7% on day 21. Electron microscopy revealed that sodium carbonate altered the eggshell surface, which allowed bacterial penetration into the egg.

Although egg washing removes feces and debris from the outside of the eggs, thus reducing the microbial load on the eggs, this does not prevent eggs from being the source of food-borne illness, especially salmonellosis (Kim and Slavik, 1996). Washing also damages or removes the cuticle of the egg, which is the primary defense of the egg against bacterial contamination (EFSA, 2005; Board and Halls, 1973). When contaminated eggs are washed under optimum conditions (conveyor belt speed of 111 cm/min, prewash water at 44°C and 138 kPa, wash water containing 3 g/L chlorowash at 44°C, 262 kPa, rinse water containing 2.5 mL/L Quat 800 at 48°C, 262 kPa, and eggs air dried for 2 min at 42°C) a 5 log reduction of *Salmonella* colony forming units (CFU) was achieved on the shell surface and *Salmonella* was not found inside the egg (Hutchison et al., 2004). However, any deviations in time or temperature enabled *Salmonella* to penetrate the egg shell and contaminate the egg contents.

## 3. ELECTROLYZED WATER

Electrolyzed oxidizing water (EOW) is generated by electrolysis of a weak salt water solution producing an acidic EOW (pH 2.6) and an alkaline EOW (pH 11.4). Use of EOW has been demonstrated to be effective at inactivating bacteria in solutions (Kim et al., 2000), in foods (Park et al., 2002), and on solid surfaces (Venkitanarayanan et al., 1999). Bialka et al. (2004) proposed that EOW could fit into the washing process for table eggs since alkaline EOW has a high pH making it suitable as a substitute for high pH detergents, whereas the acidic EOW with low pH, high oxidation reduction potential, and available chlorine would be a good sanitizer. They inoculated the shell of eggs with *S.* Enteritidis and *Escherichia coli* K12 and soaked them in alkaline EOW followed by soaking in acidic EOW for different times at different temperatures and compared the results with the typical commercial treatment. Commercial treatments achieved a decrease of *S.* Enteritidis and *E. coli* K12 of between less than 1 log and 2.6 logs, whereas the EOW treatment led to greater than 2 logs reduction for both organisms. However, the dissolved chlorine gas volatilizes rapidly from the acidic EOW because of the low pH, which decreases the bactericidal activity over time as well as affects the health of workers, plus the low pH was corrosive to the processing equipment (Len et al., 2000).

Slightly acidic EOW (SAEOW) is produced by electrolysis of dilute hydrochloric acid and has a pH value of 5.0–6.5. The predominant form of chlorine in SAEOW is hypochlorous acid (HOCl), which has a very strong antibacterial activity and also is less corrosive and does not lead to health issues from chlorine gas off-gassing (Yoshifumi, 2003; Guentzel et al., 2008). Cao et al. (2009) tested the efficacy of SAEOW on shell eggs artificially contaminated on the surface with *S.* Enteritidis. A reduction of 6.5 log of *S.* Enteritidis was achieved by SAEOW after 3 min as compared with 0.9 to 1.2 log reduction for eggs in water alone. Ni et al. (2014) compared SAEOW with chlorine dioxide, acidic EOW, and NaClO on eggs artificially inoculated with *S.* Enteritidis, *E. coli* O157:H7, or *Staphylococcus aureus*. They found no significant bactericidal difference for any treatment for food-borne pathogens or indigenous flora on shell eggs. They did determine that SAEOW was more effective when used with acidic EOW, and higher reductions were obtained with immersion treatment, rather than spray. Effect of the SAEOW on the penetration of *Salmonella* into the interior of the eggs was not explored in any of these experiments.

## 4. PASTEURIZATION

Pasteurization is the process of using heat to destroy bacteria in primarily liquid foods. Heat is an effective means of eliminating *Salmonella* from numerous food products (Bermúdez-Aguirre and Corradini, 2012; Silva and Gibbs, 2012; Jarvis et al., 2016). Pasteurization of liquid whole eggs, egg whites, and yolks is widely practiced and has been extensively investigated but there are fewer studies of in-shell pasteurization of eggs using heat (Jarvis et al., 2016). Pasteurization is known to reduce the growth of *Salmonella* in shell eggs and has been approved for use in eggs sold to consumers in the United States by the Food and Drug Administration (FDA, 2011). The Food Safety and Inspection Service of the USDA conducted a risk assessment of *S.* Enteritidis in shell eggs and predicted that a reduction of just 3 logs would reduce illness caused by this organism by 70% (USDA-FSIS, 2005). USDA-FSIS (2011) requires a 5 log reduction of *Salmonella* in shell eggs for them to be labeled as pasteurized.

### 4.1 HOT AIR PASTEURIZATION

Hou et al. (1996) inoculated intact shell eggs with *S.* Enteritidis and investigated pasteurization methods including use of a water bath or a hot air oven. Eggs were immersed in a circulating water bath set at 57°C for 25 min, which resulted in a 3 log reduction of the inoculated *Salmonella*. However, use of a hot air oven set at 55°C for 180 min led to a 5 log reduction of *S.* Enteritidis. A combination of the two methods (water-bath heating at 57°C for 25 min followed by hot-air heating at 55°C for 60 min) produced 7 log reductions in *Salmonella*. Egg white functionality was found to be acceptable after all treatments. Although these methods were promising, they would be difficult to apply on line and the time for the hot air oven would be too long. Brackett et al. (2001) investigated the use of a humidity-controlled oven for pasteurizing shell eggs inoculated with a *Salmonella* cocktail of six strains of Enteritidis, Heidelberg, and Typhimurium. They determined that treating the eggs at 57.2°C at an unspecified humidity for 70 min produced at least a 5 log decrease in inoculated *Salmonella*.

Pasquali et al. (2010) developed a hot air gun prototype and tested it on eggshells contaminated with *S.* Enteritidis; they were able to achieve at least a 1 log reduction in *Salmonella* without affecting egg quality. Manfreda et al. (2010) tested this prototype on the eggshell surfaces of table eggs experimentally infected with *S.* Enteritidis, *E. coli*, or *Listeria monocytogenes*. They also determined the impact of the treatment on egg quality traits as compared with untreated samples. The hot air pasteurization was accomplished by treating eggs, which were rotating on rolling cylinders, with two blasts of 8 s each of hot air at 600°C from two hot air generators positioned above the rolling cylinders, and cold air (20–25°C) from a generator positioned under the rolling cylinders; in between blasts of hot air the eggs received only cold air for 32 s. Treated eggs were evaluated for levels of bacteria as well as quality parameters periodically during a storage period of 1 month at a temperature of 20–25°C. Levels of *Salmonella* and *Listeria* were significantly reduced on treated eggs as compared with untreated, but *E. coli* loads were not different for treated as compared with untreated. There were also no detrimental effects on quality recorded immediately after treatment or after storage. To minimize regrowth of any bacteria remaining after the process, this hot air pasteurization could be coupled with refrigeration and/or modified atmosphere packaging.

## 4.2 HOT WATER PASTEURIZATION

Schuman et al. (1997) followed up on the work done by Hou et al. (1996) by determining the effects of immersion in water at 57°C or 58°C on contaminating *S.* Enteritidis as well as effects on egg quality attributes. They inoculated the center of eggs with 8.5 logs of a six-strain cocktail of *S.* Enteritidis and immersed the inoculated eggs in a preheated circulating water bath at 57°C or 58°C and held them for varying time periods. They observed that no *Salmonella* was recovered by direct plating when eggs were heated for 57.5 min at 58°C or 75 min at 57°C (Schuman and others 1997). However, albumen functionality was negatively impacted by these treatments.

A method for pasteurizing shell eggs in a heat transfer medium at a temperature between 53.3°C and 61.1°C was patented by Davidson (1998) and there have been other patents obtained in this area (Cox et al., 1996; Polster, 1999; Vandepopuliere and Cotterill, 2001). A hot water immersion process has been commercialized and is being used to pasteurize 1 million shell eggs per day (Sinclair, 2012). Although these eggs compare favorably with nonheat pasteurized eggs where taste is concerned, the albumen is still negatively affected and does not form stiff peaks when whipped, even with the addition of cream of tartar (Day, 2010). Egg white is well known for having excellent foaming properties, which are challenging to match with other ingredients. Damage to the proteins in the white affects the volume and stability of egg white foams that are necessary for foods such as angel food cakes, sponge cakes, meringues, soufflés, and omelets (Cunningham, 1995). Herald and Smith (1989) also observed that pasteurization of whole eggs at higher temperatures decreased pie filling expansion.

### 4.3 MICROWAVE PASTEURIZATION

Conventional heating methods may result in the outside of the egg being overheated before the interior reaches the appropriate temperature (Dev et al., 2008). Dev et al. (2008) therefore investigated the use of microwaves, which heat the food from the inside out. Theoretically the albumen should heat faster in a microwave than yolk, and this was confirmed in studies with the individual components (Dev et al., 2008). However, in trials with intact shell eggs Dev et al. (2008) demonstrated that in actuality the yolk and albumen heated at a similar rate. Shenga et al. (2010) inoculated the yolk of intact shell eggs with *Salmonella* Typhimurium and compared the effects of moist heat, dry heat, or microwave to inactivate the pathogen. Moist heat was applied in a circulating water bath at 57°C for 15 min, dry heat with a hot air oven at 55°C for 2 h and microwave at power nine for 20 s (make and model of microwave were not specified). Both moist and dry heat resulted in a 2 log reduction of *Salmonella*, whereas the microwave treatment generated only a 1 log reduction.

Lakins et al. (2008) investigated the use of a new directional microwave technology for the inactivation of *S.* Enteritidis inoculated into shell eggs. The directional microwave has both horizontal and rotary movement as well as several sources of microwaves as compared with traditional microwaves that use only rotary movement and have only one source of microwaves (Lakins et al., 2008). The directional microwave technology resulted in a 2 log reduction of *S.* Enteritidis regardless of inoculation level. They detected no differences in water activity, albumen pH, or combined pH as compared with controls. However, there were significant changes in yolk pH, with treated eggs having a yolk pH of 6.53 as compared with 6.13 for controls, indicating a possibility of protein denaturation.

Dev et al. (2010) compared the effects of microwave pasteurization and hot water bath pasteurization of in-shell eggs on the functional properties of the egg white as compared with untreated eggs. They determined that, although both means of pasteurization affected viscosity and clarity of egg white as well as foam density and stability, the microwave pasteurization produced less change than did water bath pasteurization. They stated that the viscosity, foam density, and foam stability of the microwave heated egg white were not statistically significantly different from those of the untreated ones, as opposed to the water bath–heated egg whites.

## 5. THERMOULTRASONICATION

Burgos et al. (1972) discovered that when they used ultrasound to clean bacterial spores of debris the spores were subsequently more sensitive to heat treatment than were spores that did not receive the ultrasound treatment. Later they observed that a combination of ultrasound and heat were more effective against vegetative cells than either treatment alone (Ordonez et al., 1984). This new process, which they named thermoultrasonication, decreased the decimal reduction time for heat treatment of *S. aureus* by 63% in buffer and 43% in milk (Ordonez et al., 1987). Cabeza et al. (2004) attempted to develop this process for use in sanitizing shell eggs. They studied the effects of thermoultrasonication on *S.* Enteritidis in a liquid medium as well as on the shells of intact eggs. In their laboratory-scale equipment the combined process demonstrated a higher killing effect than heat treatment alone; they obtained a 99.5% reduction of the original bacterial load as compared with a 90% reduction for the heat treatment alone. When they applied the process to *Salmonella* Senftenberg, which is known to have greater resistance to heat than other *Salmonella* serovars, they determined that thermoultrasonication had more effect than heat alone but they did not achieve as great a reduction as with *S.* Enteritidis (Cabeza et al., 2005a). Cabeza et al. (2005b) extended their work to determine the effects of thermoultrasonication on functional properties of treated eggs including shelf-life, emulsifying and foaming capacities, emulsion and foam stabilities, texture properties of egg white gel, breakage resistance of shell, and sensory properties of cooked egg. No significant differences were observed in any of these properties between treated and untreated eggs. Although this methodology appears to be promising, it has yet to be commercialized, and it is effective only on bacteria on the surface of the egg and does not affect those inside the egg.

## 6. GAS PLASMA

Gas plasma is an ionized gas containing equal numbers of positively and negatively charged particles (Moisan et al., 2001). A gas plasma can be produced by exposing gas to an electric field producing charged particles and free radicals, which, separately or in synergistic combination, react with essential components of the bacteria to disrupt the metabolism of the bacteria (Laroussi and Leipold, 2004). The use of atmospheric pressure and low temperatures render the process inexpensive and practical to use and it has some applications in health care (Baier et al., 1992; Lloyd et al., 2009). Studies have been done on application of this technology to the food industry including research into packaging sterilization (Deilmann et al., 2008), disinfecting almonds contaminated with *E. coli* (Deng et al., 2007), inactivating pathogens in water from poultry chill tanks (Rowan et al., 2007), and decontamination of produce (Critzer et al., 2007).

Davies and Breslin (2003) investigated the use of a gas plasma air ionizer (HSC Associates, London) on eggs artificially contaminated with *S.* Enteritidis PT4. They determined that the ionized air did not reduce the number of *S.* Enteritidis-positive eggs as compared with controls, although treatment for 20 min as opposed to 5 min appeared to marginally reduce the number of positive eggs.

Intecon Systems, Inc. (Carlsbad, CA) developed a cleaning and disinfection process consisting of passing a mist through gas plasma, which they named Binary Ionization Technology (US Patent 6343425, 2002; US Patent 6343425, 2004). This system works at standard atmospheric conditions and does not require a vacuum and thus does not require containment, and by-products include only water and oxygen. Higgins et al. (2005) used this technology to treat shell eggs that had been inoculated on the shell with *S.* Enteritidis and achieved a greater than 7 log reduction as compared with water treatment.

Ragni et al. (2010) tested the effectiveness of a prototype of a resistive barrier discharge technique for disinfecting eggs artificially contaminated with *S.* Typhimurium or *S.* Enteritidis. They determined that 90 min of treatment yielded reductions of 2.5 log at low relative humidity and 4.5 log at high relative humidity. No detrimental effects of the gas plasma were seen on the egg quality.

## 7. PULSED LIGHT

Pulsed light processing uses broad-spectrum high-intensity light in short bursts to kill microorganisms and has applications in food packaging, processing equipment, and medical devices (Dunn, 1996). Dunn (1996) inoculated shell eggs with *S.* Enteritidis and subsequently treated the inoculated eggs with pulsed light; he determined that the treatment was able to eradicate the inoculum from the shells of the eggs. In a further study, Dunn (1996) used a temperature differential between eggs and inoculum to potentially draw *Salmonella* into the pores of the eggs. He determined that when the inoculum was colder than the egg, *Salmonella* was drawn through the pores of the shell into the interior of the egg, and although reduction of *Salmonella* was lowered in this case, he still observed reductions ranging from 2 to 4 logs as compared with untreated controls.

Hierro et al. (2009) used pulsed light on eggs inoculated with *S.* Enteritidis and compared the effects on washed and unwashed eggs. They noted that treated washed eggs had greater numbers of *Salmonella* surviving the treatment. They attributed this better survival on washed eggs to the damage incurred by the cuticle during the washing process, which in turn allowed the bacteria to penetrate within the pores of the eggs. These authors concluded that the pulsed light treatment should be applied soon after the eggs are laid and on unwashed eggs for maximum effectiveness.

Keklik et al. (2010) also evaluated effectiveness of pulsed light on eggs inoculated with *S.* Enteritidis and also determined temperature of eggshells during treatment and effect on quality parameters of egg components. A maximum log reduction of 5.3 logs/$cm^2$ of *Salmonella* was observed with a 20-s treatment and a dose of 0.1 J/$cm^2$, without any visual damage to the egg. Egg surface temperatures rose proportionate to the distance of the lamp from the egg, with a distance of 9.5 cm raising the temperature by 16.3°C.

No negative quality effects were observed. Lasagabaster et al. (2011) further investigated the use of pulsed light to inactivate *S.* Typhimurium intentionally inoculated onto the eggs. Pulsed light reduced *Salmonella* from 6 logs on untreated eggs to 2.5, 1.7, and 1.1 logs on eggs treated with doses of 0.35, 0.7, and 2.1 J $cm^2$, respectively. There were no adverse effects on albumen quality and a sensory panel was not able to detect any differences between treated and untreated eggs.

## 8. OZONE

Ozone ($O_3$) is a very strong antimicrobial agent active against microorganisms at relatively low concentrations (Khadre et al., 2001). Ozone has been researched for potential uses in the food industry (Kim et al., 1999, 2003) and it has been approved by the US FDA for use as an antimicrobial in foods (FDA, 2001). Ozone is unstable and decomposes quickly to oxygen and is also effective at low temperatures (Khadre et al., 2001). Rodriguez-Romo and Yousef (2005) noted that $O_3$ had been tested for disinfection of hatching eggs (Ito et al., 1999; Whistler and Sheldon, 1989) and subsequently tested the use of $O_3$ to disinfect shell eggs. Shell eggs were externally contaminated with *S.* Enteritidis and then treated with $O_3$ at various concentrations for times from 0 to 20 min. Three minutes of $O_3$ treatment caused a rapid decrease in *Salmonella* of approximately 5 logs as compared with controls up to 3 min of treatment but longer treatment times did not provide more inactivation. This group was also able to demonstrate that $O_3$ would penetrate the egg shell and inactivate *Salmonella* located in the yolk of the egg (Rodriguez-Romo et al., 2007).

Foods with a high fat content require a higher dose of $O_3$ and therefore may be subject to more rapid oxidation leading to off flavors and odors (Kim et al., 2003). Since the yolk of eggs contains a high amount of fat, Kamotani et al. (2010) used sensory panels to evaluate and compare eggs treated with either $O_3$ or heat with untreated eggs. On visual inspection the yolk and albumen of treated eggs were perceived to be cloudier than that of controls, although the $O_3$-treated eggs were considered closer to control eggs than thermally treated eggs. Hedonic scales and just-about-right scales were used to evaluate several attributes of cooked eggs. Overall liking, appearance, aroma, flavor, color, and texture were not significantly different but heat-treated and $O_3$-treated eggs were perceived as less moist than the control.

## 9. ULTRAVIOLET LIGHT

Ultraviolet (UV) light occupies a wide band of wavelengths in the nonionizing region of the electromagnetic spectrum between X-rays (200 nm) and visible light (400 nm), but only UV in the range of 250–260 nm (short-wave UV radiation, or UVC) may be lethal to most microorganisms (Sastry et al., 2000). Among its practical applications several should be noted including inhibition of microorganisms on surfaces, destruction of microorganisms in the air, and sterilization of liquids (Bintsis et al., 2000).

Chavez et al. (2002) treated eggs with UV light while they were being turned on a manually operated egg roller. They assessed total aerobic plate counts (APC) to determine reduction of natural microbial load on the eggs and determined that with an exposure of 30 s there was a significant reduction of 1–2 log CFU/egg as compared with untreated eggs. Eggs rotated for 60 s had significantly greater reductions of APC than the other time intervals of exposure (15 and 30 s) compared with controls (Chavez et al., 2002). De Reu et al. (2006) studied the effect of UV light for the decontamination of egg surfaces. Eggs were contaminated with *E. coli* or *S. aureus* and treated on a double-roller conveyor belt with standard UV light (254 nm) at an intensity of 10 mW/cm$^2$; UV treatment for 4.7 and 18.8 s resulted in reductions of 3–4 logs for *E. coli* and *S. aureus*, respectively. The natural bacterial load of uninoculated eggs that were visually clean was reduced from 4.47 to 3.57 log by standard UV treatment of 4.7 s, but those eggs that were visually dirty had no reduction in bacterial load. In addition, *E. coli* placed in the interior of eggshells was not inactivated. Although UV treatment can be a reliable, nonthermal alternative to traditional treatments for the exterior of shell eggs, the shells are impermeable to UV light and therefore no internal bacteria will be inactivated by this treatment. Combinations of UV with other nonthermal methods or with a mild heat treatment should be investigated.

## 10. IRRADIATION

In 2000 FDA approved the use of ionizing radiation up to 3 kGy for fresh shell eggs (FDA, 2015). Serrano et al. (1997) investigated the sensitivity to gamma irradiation of five *S.* Enteritidis isolates inoculated on or inside of whole shell eggs. They irradiated the inoculated eggs at doses of 0, 0.5, 1.0, or 1.5 kGy. The dose of 0.5 kGy eliminated all the isolates from the surface of eggs, but there was a great variation in sensitivity among the isolates on the inside of the egg. They concluded that an irradiation dose of 1.5 kGy should be sufficient to reduce *Salmonella* counts by approximately 4 logs in both whole shell and liquid eggs. They also determined that color and quality of egg proteins was not affected by a 1.5 kGy dose of gamma irradiation.

Wong and Kitts (2003) used electron beam irradiation (EBI) to treat shell eggs inoculated with *L. monocytogenes*, *E. coli*, or *S.* Typhimurium. Dosages of EBI at 3 and 4 kGy were effective at reducing all bacteria in eggs to an undetectable level. However, *S.* Typhimurium was determined to be slightly more resistant to EBI than were the other bacteria. The irradiated eggs exhibited loss of quality of the albumen, which translated to reduced foaming and gelling capacity. Yolk also lost color and the vitelline membrane was weakened by irradiation, which decreased the eggs' usefulness for further processing due to yolk breakage during shell breaking because of the weakened membrane.

# 11. STORAGE TO MINIMIZE GROWTH

## 11.1 REFRIGERATION

One recommendation to restrict the growth of *Salmonella* and other contaminants in shell eggs is to refrigerate the eggs as soon as possible at temperatures that do not allow the growth of these pathogens. Gast and Holt (2000) contaminated different components of eggs with small numbers of *S.* Enteritidis (15–150 CFU) and then stored them at different temperatures for up to 3 days. In contents stored at 25°C that had been contaminated with 150 cells the *Salmonella* grew rapidly, especially in yolk. Less growth was observed with the lower inocula, with storage for only 1 day, when storage was at lower temperatures (10–17.5°C) and when *S.* Enteritidis was only introduced into the albumen. In the Final Egg Rule (FDA, 2009), the FDA specifies a maximum ambient temperature of 7.2°C not only during storage, but also during transport, beginning 36 h after the time of lay. In Canada, shell eggs must be kept under refrigeration (20°C or less) for a maximum of 6 days or they may be stored at 20–30°C for a maximum of 2 days (Health Canada, 2011). However, eggs are not allowed to be refrigerated in the European Union (EC, 2008).

## 11.2 MODIFIED-ATMOSPHERE PACKAGING

Modified-atmosphere packaging using $CO_2$, $O_2$, or $N_2$ gases is frequently applied to extend the shelf-life of foods by inhibiting chemical, enzymatic, and microbial spoilage (Rajkovic et al., 2010). Packaging in a high-$CO_2$ atmosphere has been shown to maintain the quality aspects of fresh shell eggs, particularly on albumen factors that lend stability to foams (Rocculi et al., 2009, 2011). Pasquali et al. (2012) artificially inoculated table eggs with *S.* Enteritidis, *E. coli*, or *L. monocytogenes* and stored the eggs in air, 100% $CO_2$, or 100% $O_2$ filled packaging. Eggs were stored for 30 days at 4°C, 25°C, or 37°C and pathogen counts as well as total aerobic counts were determined periodically during storage. Using these parameters they determined that the temperature of the storage was more influential on bacterial survival than the type of gas used in the package. Spoilage bacteria and *E. coli* were controlled best at 4°C, but *L. monocytogenes* counts were lower on eggs stored at 37°C regardless of the gas used. The $CO_2$ packaging and storage at 4°C was the best treatment for reducing *Salmonella* loads, and $CO_2$ was better than either air or $O_2$ for eggs inoculated with *L. monocytogenes* and stored at 4°C as well as eggs containing only spoilage bacteria and stored at 25°C. These studies involved packing egg cartons individually into pouches, which were flushed and filled with the modified atmosphere. Individual pouches of this nature with 100% $CO_2$ would be a novel intervention for maintaining egg characteristics during transport and retail storage, but once these packages are in the domestic refrigerator, the modified atmosphere would be lost.

## 12. CONCLUSIONS

Although washing of shell eggs is mandated in many countries, it still carries a risk of allowing pathogens to enter the egg because the cuticle is lost. In addition, washing of eggs is prohibited in the European Union. The methods reviewed in this chapter, both thermal and nonthermal, are able to reduce the microbial numbers on the eggshell. However, few of these approaches were able to effectively inactivate bacteria within the egg itself, and those that did inactivate internal bacteria also had some negative effects on the functionality of the egg. However, irradiation and pasteurization could be satisfactory options for high-risk populations, such as the elderly, immunocompromised, children, and pregnant women. Other decontamination options need further investigation, especially into the feasibility and cost of scaling up for commercial use. In particular, combination methods should be explored as with thermoultrasonifcation where the use of ultrasound allows for less heat to be used thus mitigating the negative effects of heat on the functional properties of egg whites.

## REFERENCES


Arnold, M.E., Carrique-Mas, J.J., Davies, R.H., 2010. Sensitivity of environmental sampling methods for detecting *Salmonella* Enteritidis in commercial laying flocks relative to the within-flock prevalence. Epidemiol. Infect. 138, 330–339.

Baier, R.E., Carter, J.M., Sorensen, S.E., Meyer, A.E., McGowan, B.D., Kasprzak, S.A., 1992. Radiofrequency gas plasma (glow discharge) disinfection of dental operative instruments, including handpieces. J. Oral Implantol. 18, 236–242.

Bartlett, F.M., Laird, J.M., Addison, C.L., McKellar, R.C., 1993. The analysis of egg wash water for the rapid assessment of microbiological quality. Poult. Sci. 72, 1584–1591.

Bermudez-Aguirre, D., Corradini, M.G., 2012. Inactivation kinetics of *Salmonella* spp. under thermal and emerging treatments: a review. Food Res. Int 45, 700–712.

Bialka, K.L., Demirci, A., Knabel, S.J., Patterson, P.H., Puri, V.M., 2004. Efficacy of electrolyzed oxidizing water for the microbial safety and quality of eggs. Poult. Sci. 83, 2071–2078.

Bintsis, T., Litopoulou-Tzanetaki, E., Robinson, R.K., 2000. Existing and potential applications of ultraviolet light in the food industry – a critical review. J. Sci. Food Agric 80, 637–645.

Board, R.G., Halls, N.A., 1973. The cuticle: a barrier to liquid and particle penetration of the shell of the hen's egg. Br. Poult. Sci. 14, 69–97.

Burgos, J., Ordonez, J.A., Sala, F., 1972. Effect of ultrasonic waves on heat resistance of *Bacillus cereus* and *Bacillus licheniformis* spores. Appl. Microbiol. 24, 497–498.

Brackett, R.E., Schuman, J.D., Ball, H.R., Scouten, A.J., 2001. Thermal inactivation kinetics of *Salmonella* spp. within intact eggs heated using humidity-controlled air. J. Food Prot. 64, 934–938.

Cabeza, M.C., Garcia, M.L., De la Hoz, L., Cambero, I., Ordonez, J.A., 2005a. Destruction of *Salmonella* Senftenberg on the shells of intact eggs by thermoultrasonication. J. Food Prot. 68, 841–844.

Cabeza, M.C., Garcia, M.L., de la Hoz, L., Cambero, I., Ordonez, J.A., 2005b. Thermoultrasonication eliminates salmonellae from intact eggshells without changing the functional properties of their components. J. Food Sci. 70, M292–M295.

Cabeza, M.C., Ordonez, J.A., Cambero, I., De La Hoz, L., Garcia, M.L., 2004. Effect of thermoultrasonication on *Salmonella enterica* serovar Enteritidis in distilled water and intact shell eggs. J. Food Prot. 67, 1886–1891.

Cao, W., Zhu, Z.W., Shi, Z.X., Wang, C.Y., Li, B.M., 2009. Efficiency of slightly acidic electrolyzed water for inactivation of *Salmonella enteritidis* and its contaminated shell eggs. Int. J. Food Microbiol. 130, 88–93.

Caudill, A.B., Curtis, P.A., Anderson, K.E., Kerth, L.K., Oyarazabal, O., Jones, D.R., Musgrove, M.T., 2010. The effects of commercial cool water washing of shell eggs on Haugh unit, vitelline membrane strength, aerobic microorganisms, and fungi. Poult. Sci. 89, 160–168.

Chavez, C., Knape, K.D., Coufal, C.D., Carey, J.B., 2002. Reduction of eggshell aerobic plate counts by ultraviolet irradiation. Poult. Sci. 81, 1132–1135.

Cox, J.P., Cox, R.W.D., Cox, J.M., 1996. Method for Processing Poultry Shell Eggs (US Patent 5589211).

Critzer, F.J., Kelly-Winterberg, K., South, S.L., Golden, D.A., 2007. Atmospheric plasma inactivation of foodborne pathogens on fresh produce surfaces. J. Food Prot. 70, 2290–2296.

Cunningham, F.E., 1995. In: Stadelman, W.J., Cotterill, O.J. (Eds.), Egg Science and Technology, fourth ed. The Haworth Press Inc., Binghamton, NY, pp. 289–321.

Davidson, L.J., 1998. Method for Production of Pasteurized In-shell Chicken Eggs (US Patent 5843505).

Davies, R.H., Breslin, M., 2003. Investigations into possible alternative decontamination methods for *Salmonella enteritidis* on the surface of table eggs. J. Vet. Med. B. 50, 38–41.

Day, J., 2010. Pasteurized Eggs Put to the Test. Available at: http://articles.chicagotribune.com/2010-09-08/features/sc-food-0910-pasteurized-eggs-20100908_1_regular-eggs-whites-proteins.

Deilmann, M., Halfmann, H., Bibinov, N., Wunderlich, J., Awakowicz, P., 2008. Low-pressure microwave plasma sterilization of polyethylene terephtalate bottles. J. Food Prot. 71, 2119–2123.

Deng, S.B., Ruan, R., Mok, C.K., Huang, G.W., Lin, X.Y., Chen, P., 2007. Inactivation of *Escherichia coli* on almonds using nonthermal plasma. J. Food Sci. 72, M62–M66.

De Reu, K., Grijspeerdt, K., Herman, L., Heyndrickx, M., Uyttendaele, M., Debevere, J., Putirulan, F.F., Bolder, N.M., 2006. The effect of a commercial UV disinfection system on the bacterial load of shell eggs. Lett. Appl. Microbiol. 42, 144–148.

Dev, S.R.S., Orsat, V., Gariepy, Y., Raghavan, G.S.V., Ruiz-Feria, C., 2010. Selected post-heating properties of microwave or hot water heated egg white for in-shell pasteurization. Int. J. Food Prop. 13, 778–788.

Dev, S.R.S., Raghavan, G.S.V., Gariepy, Y., 2008. Dielectric properties of egg components and microwave heating for in-shell pasteurization of eggs. J. Food Eng. 86, 207–214.

Doyle, M.P., Erickson, M.C., 2006. Reducing the carriage of foodborne pathogens in livestock and poultry. Poult. Sci. 85, 960–973.

Dunn, J., 1996. Pulsed light and pulsed electric field for foods and eggs. Poult. Sci. 75, 1133–1136.

EC, 2008. Commission Regulation (EC) No 589/2008 of 23 June 2008 Laying Down Detailed Rules for Implementing Council Regulation (EC) No 1237/2007 as Regards Marketing Standards for Eggs. Official Journal of the European Union L163/6:6. Available from: http://eur-lex.europa.eu/LexUriServ/LexUriServ.do?uri=OJ:L:2008:163:0006:0023:EN:PDF.

EFSA, 2005. Opinion of the scientific panel on biological hazards on the request from the commission related to microbiological risks on washing of table eggs. EFSA J. 269, 1–39.

EFSA, 2010. Scientific opinion on a quantitative estimation of the public health impact of setting a new target for the reduction of *Salmonella* in laying hens. EFSA J. 8, 42–49.

Farber, J.M., Daley, E., Coates, F., 1992. Presence of *Listeria* spp. in whole eggs and wash water samples from Ontario and Quebec. Food Res. Int. 25, 143–145.

FDA, 2001. Secondary Direct Food Additives Permitted in Food for Human Consumption. Available at: http://www.accessdata.fda.gov/scripts/cdrh/cfdocs/cfcfr/CFRSearch.cfm?CFRPart=173.

FDA, 2009. Code of Federal Regulations, Title 21, Parts 16 and 118. Federal Register Final Rule: Guidance for Industry. Prevention of *Salmonella* Enteritidis in Shell Eggs during Production, Storage and Transportation. Available from: http://www.fda.gov/downloads/Food/GuidanceComplianceRegulatoryInformation/GuidanceDocuments/FoodSafety/UCM285137.pdf.

FDA, 2011. Prevention of *Salmonella Enteritidis* in Shell Eggs during Production, Storage, and Transportation. http://www.fda.gov/Food/GuidanceRegulation/GuidanceDocuments RegulatoryInformation/Eggs/ucm285101.htm.

FDA, 2015. Irradiation in the Production, Processing and Handling of Food. Available at: http://www.accessdata.fda.gov/scripts/cdrh/cfdocs/cfcfr/CFRSearch.cfm?CFRPart=179.

Gast, R.K., Holt, P.S., 2000. Influence of the level and location of contamination on the multiplication of *Salmonella* enteritidis at different storage temperatures in experimentally inoculated eggs. Poult. Sci. 79, 559–563.

Guentzel, J.L., Lam, K.L., Callan, M.A., Emmons, S.A., Dunham, V.L., 2008. Reduction of bacteria on spinach, lettuce, and surfaces in food service areas using neutral electrolyzed oxidizing water. Food Microbiol. 25, 36–41.

Haines, R.B., Moran, T., 1940. Porosity of, and bacterial invasion through, the shell of the hen's egg. J. Hyg. (Lond.) 40, 453–461.

Health Canada, 2011. Guidance for Industry on Reducing the Risk of *Salmonella* Enteritidis in Canadian Shell Eggs. Available from: http://www.hc-sc.gc.ca/fn-an/legislation/guide-ld/salmonella-enteritidis-eng.php.

Herald, T.J., Smith, D.M., 1989. Functional properties and composition of liquid whole egg proteins as influenced by pasteurization and frozen storage. Poult. Sci. 68, 1461–1469.

Hierro, E., Manzano, S., Ordonez, J.A., de la Hoz, L., Fernandez, M., 2009. Inactivation of *Salmonella enterica* serovar Enteritidis on shell eggs by pulsed light technology. Int. J. Food Microbiol. 135, 125–130.

Higgins, S.E., Wolfenden, A.D., Bielke, L.R., Pixley, C.M., Torres-Rodriguez, A., Vicente, J.L., Bosseau, D., Neighbor, N., Hargis, B.M., Tellez, G., 2005. Application of ionized reactive oxygen species for disinfection of carcasses, table eggs, and fertile eggs. J. Appl. Poult. Res. 14, 716–720.

Holt, P.S., Davies, R.H., Dewulf, J., Gast, R.K., Huwe, J.K., Jones, D.R., Waltman, D., Willian, K.R., 2011. The impact of different housing systems on egg safety and quality. Poult. Sci. 90, 251–262.

Hou, H., Singh, R.K., Muriana, P.M., Stadelman, W.J., 1996. Pasteurization of intact shell eggs. Food Microbiol. 13, 93–101.

Howard, Z.R., O'Bryan, C.A., Crandall, P.G., Ricke, S.C., 2012. *Salmonella* enteritidis in shell eggs: current issues and prospects for control. Food Res. Int. 45 (2), 755–764.

Hutchison, M.L., Gittins, J., Walker, A., Moore, A., Burton, C., Sparks, N., 2003. Washing table eggs: a review of the scientific and engineering issues. Worlds Poult. Sci. J. 59, 233–248.

Hutchison, M.L., Gittins, J., Walker, A., Sparks, N., Humphrey, T.J., Burton, C., Moore, A., 2004. An assessment of the microbiological risks involved with egg washing under commercial conditions. J. Food Prot. 67, 4–11.

Ito, H., Nakatani, H., Bamba, H., Hayashi, T., 1999. Influence of the hatchability of Japanese quail eggs (hatching eggs) and sterilization effect for *Salmonella* by ozone gas sterilization. Res. Bull. Aichi Agric. Res. 31, 305–310.

Jarvis, N.A., O'Bryan, C.A., Dawoud, T., Park, S.H., Kwon, Y.M., Crandall, P.G., Ricke, S.C., 2016. The thermal destruction of *Salmonella* in foods: a review of current perspectives. Food Control 68, 280–290.

Kamotani, S., Hooker, N., Smith, S., Lee, K., 2010. Consumer acceptance of ozone-treated whole shell eggs. J. Food Sci. 75, S103–S107.

Keklik, N.M., Demirci, A., Puri, V.M., 2010. Decontamination of unpackaged and vacuum-packaged boneless chicken breast with pulsed ultraviolet light. Poultry Sci 89, 570–581.

Khadre, M.A., Yousef, A.E., Kim, J., 2001. Microbiological aspects of ozone applications in food: a review. J. Food Sci. 6, 1242–1252.

Kim, C., Hong, Y.C., Brackett, R.E., 2000. Efficacy of electrolyzed oxidizing (EO) and chemically modified water on different types of foodborne pathogens. Int. J. Food Microbiol. 61, 199–207.

Kim, J.-G., Yousef, A.E., Dave, S., 1999. Application of ozone for enhancing the microbiological safety and quality of foods: a review. J. Food Prot. 62, 1071–1087.

Kim, J.-G., Yousef, A.E., Khadre, M.A., 2003. Ozone and its current and future application in the food industry. Adv. Food Nutr. Res. 45, 167–218.

Kim, J.W., Slavik, M.F., 1996. Cetylpyridinium chloride (CPC) treatment on poultry skin to reduce attached *Salmonella*. J Food Prot 59, 322–326.

Lakins, D.G., Alvarado, C.Z., Thompson, L.D., Brashears, M.T., Brooks, J.C., Brashears, M.M., 2008. Reduction of *Salmonella enteritidis* in shell eggs using directional microwave technology. Poult. Sci. 87, 985–991.

Laroussi, M., Leipold, F., 2004. Evaluation of the roles of reactive species, heat, and UV radiation in the inactivation of bacterial cells by air plasmas at atmospheric pressure. Int. J. Mass Spectrom. 233, 81–86.

Lasagabaster, A., Arboleya, J.C., de Maranon, I.M., 2011. Pulsed light technology for surface decontamination of eggs: impact on *Salmonella* inactivation and egg quality. Innov. Food Sci. Emerg. Technol. 12, 124–128.

Leasor, S.B., Foegeding, P.M., 1989. *Listeria* species in commercially broken raw liquid whole egg. J. Food Prot. 52, 777–780.

Len, S.V., Hung, Y.-C., Erickson, M.C., Kim, C., 2000. Ultraviolet spectrophotometric characterizations and bacterial properties of electrolyzed oxidizing water as influenced by amperage and pH. J. Food Prot. 63, 1534–1537.

Lloyd, G., Friedman, G., Jafri, S., Schultz, G., Fridman, A., Harding, K., 2009. Gas plasma: medical uses and developments in wound care. Plasma Process. Polym. 7, 194–211.

Lorenz, F.W., Starr, P.B., 1952. Spoilage of washed eggs: effect of sprayed versus static water under different washing temperatures. Poult. Sci. 31, 204–213.

Manfreda, G., Cevoli, C., Lucchi, A., Pasquali, F., Fabbri, A., Franchini, A., 2010. Hot air treatment for surface decontamination of table eggs experimentally infected with *Salmonella*, *Listeria*, and *Escherichia coli*. Vet. Res. Comm. 34, S179–S182.

Moisan, M., Barbeau, B., Moreau, S., Pelletier, J., Tabrizian, M., Yahia, L.H., 2001. Low-temperature sterilization using gas plasmas: a review of the experiments and an analysis of the inactivation mechanism. Int. J. Pharm. 226, 1–21.

Ni, L., Cao, W., Zheng, W.C., Chen, H., Li, B.M., 2014. Efficacy of slightly acidic electrolyzed water for reduction of foodborne pathogens and natural microflora on shell eggs. Food Sci. Technol. Res. 20, 93–100.

Ordonez, J.A., Aguilera, M.A., Garcia, M.L., Sanz, B., 1987. Effect of combined ultrasonic and heat treatment (thermoultrasonication) on the survival of a strain of *Staphylococcus aureus*. J. Dairy Res. 54, 61–67.

Ordonez, J.A., Sanz, B., Hernandez, P.E., Lopez-Lorenzo, P., 1984. A note on the effect of combined ultrasonic and heat treatments on the survival of thermoduric *Streptococci*. J. Appl. Bacteriol. 56, 175–177.

Pande, V.V., Devon, R.L., Sharma, P., McWhorter, A.R., Chousalkar, K.K., 2016. Study of *Salmonella* Typhimurium infection in laying hens. Front. Microbiol. 7, 203. http://dx.doi.org/10.3389/fmicb.2016.00203.

Park, H., Hung, Y.C., Brackett, R.E., 2002. Antimicrobial effect of electrolyzed oxidizing water for inactivating *Campylobacter jejuni* during poultry washing. Int. J. Food Microbiol. 72, 77–83.

Pasquali, F., Fabbri, A., Cevoli, C., Manfreda, G., Franchini, A., 2010. Hot air treatment for surface decontamination of table eggs. Food Control 21, 431–435.

Pasquali, F., Manfreda, G., Olivi, P., Rocculij, P., Sirri, F., Meluzzi, A., 2012. Modified-atmosphere packaging of hen table eggs: effects on pathogen and spoilage bacteria. Poult. Sci. 91, 3253–3259.

Polster, L.S., 1999. Method and Control System for Controlling Pasteurization of In-shell Eggs (US Patent 5993886).

Ragni, L., Berardinelli, A., Vannini, L., Montanari, C., Sirri, F., Guerzoni, M.E., Guarnieri, A., 2010. Non-thermal atmospheric gas plasma device for surface decontamination of shell eggs. J. Food Eng. 100, 125–132.

Rajkovic, A., Tomic, N., Smigic, N., Uyttendaele, M., Ragaert, P., Devlieghere, F., 2010. Survival of *Campylobacter jejuni* on raw chicken legs packed in high-oxygen or high carbon dioxide atmosphere after decontamination with lactic acid/sodium lactate buffer. Int. J. Food Microbiol. 140, 201–206.

Rocculi, P., Cocci, E., Sirri, F., Cevoli, C., Romani, S., Dalla Rosa, M., 2011. Modified atmosphere packaging of hen table eggs: effects on functional properties of albumen. Poult. Sci. 90, 1791–1798.

Rocculi, P., Tylewicz, U., Pekoslawska, A., Romani, S., Sirri, F., Siracusa, V., Dalla Rosa, M., 2009. MAP storage of shell hen eggs, Part 1: effect on physico-chemical characteristics of the fresh product. LWT-Food Sci. Technol. 42, 758–762.

Rodriguez-Romo, L.A., Vurma, M., Lee, K., Yousef, A.E., 2007. Penetration of ozone gas across the shell of hen eggs. Ozone Sci. Eng. 29, 147–150.

Rodriguez-Romo, L.A., Yousef, A.E., 2005. Inactivation of *Salmonella enterica* serovar Enteritidis on shell eggs by ozone and UV radiation. J. Food Prot. 68, 711–717.

Rowan, N.J., Espie, S., Harrower, J., Anderson, J.G., Marsili, L., MacGregor, S.J., 2007. Pulsed-plasma gas-discharge inactivation of microbial pathogens in chilled poultry wash water. J. Food Prot. 70, 2805–2810.

Sastry, S.K., Datta, K., Worobo, R.W., 2000. Ultraviolet light. J. Food Saf. 65, 90–92.

Schuman, J.D., Sheldon, B.W., Vandepopuliere, J.M., Ball Jr., H.R., 1997. Immersion heat treatments for inactivation of *Salmonella enteritidis* with intact eggs. J. Appl. Bacteriol 83, 438–444.

Serrano, L.E., Murano, E.A., Shenoy, K., Olson, D.G., 1997. D values of *Salmonella enteritidis* isolates and quality attributes of shell eggs and liquid whole eggs treated with irradiation. Poult. Sci. 76, 202–206.

Shane, S.M., Gifford, D.H., Yogasundram, K., 1986. *Campylobacter jejuni* contamination of eggs. Vet. Res. Commun. 10, 487–492.

Shenga, E., Singh, R.P., Yadav, A.S., 2010. Effect of pasteurization of shell egg on its quality characteristics under ambient storage. J. Food Sci. Technol. 47, 420–425.

Silva, F.M., Gibbs, P.A., 2012. Thermal pasteurization requirements for the inactivation of *Salmonella* in foods. Food Res. Int 45, 695–699.

Sinclair, A., 2012. Making Cookie Dough Safer. Available at: http://www.chicagobusiness.com/article/20120225/ISSUE01/302259975/making-cookie-dough-safer.

Tellez, G., Petrone, V.M., Escorcia, M., Morishita, T.Y., Cobb, C.W., Villasenor, L., Promsopone, B., 2001. Evaluation of avian-specific probiotic and *Salmonella* enteritidis, *Salmonella* typhimurium and *Salmonella* heidelberg-specific antibodies on cecal colonization and organ invasion of *Salmonella* enteritidis in broilers. J. Food Prot. 64 (3), 287–291.

Tellez, G., Pixley, C., Wolfenden, R.E., Layton, S.L., Hargis, B.M., 2012. Probiotics/direct fed microbials for *Salmonella* in poultry. Food Res. Int. 45, 628–633.

US Patent 6,343,425 February 5, 2002. Measurement and Cleaning of Elastomeric Articles Having Particulate Adhered Thereto (US Patent 6343425). Intecon Systems, Inc., Carlsbad, CA.

US Patent 6,343,425 March 16, 2004. Apparatus and Method for Cleaning Particulate Matter and Chemical Contaminants from a Hand (US Patent 6706243). Intecon Systems, Inc., Carlsbad, CA.

USDA, 1975. Regulations Governing the Inspection of Eggs (Egg Products Inspection Act). 7 CFR Part 57. USDA, Washington, DC.

USDA, 2000. Egg Grading Manual, Agricultural Handbook Number 75. Washington, DC.

USDA-FSIS, 2005. Risk Assessment for *Salmonella* Enteritidis in Shell Eggs and *Salmonella* spp. In Egg Products, October 2005. Available at: http://www.fsis.usda.gov/wps/wcm/connect/16abcb5f-21fc-4731-9346-fd8401582ba4/SE_Risk_Assess_Oct2005.pdf?MOD=AJPERES.

USDA-FSIS, 2011. Egg Products and Food Safety. Available at: http://www.fsis.usda.gov/shared/PDF/Egg_Products_and_Food_Safety.pdf.

Vandepopuliere, J.M., Cotterill, O.J., 2001. Method of Controlling *Salmonella* in Shell Eggs (US Patent 6303176).

Van Immerseel, F., Cauwerts, K., Devirese, L.A., Haesebrouck, F., Ducatelle, R., 2002. Feed additives to control *Salmonella* in poultry. Worlds Poult. Sci. J. 58, 501–513.

Van Immerseel, F., Methner, U., Rychlik, I., Nagy, B., Velge, P., Martin, G., Foster, N., Ducatelle, R., Barrow, P.A., 2005. Vaccination and early protection against non-host-specific *Salmonella* serotypes in poultry: exploitation of innate immunity and microbial activity. Epidemiol. Infect. 133, 959–978.

Venkitanarayanan, K.S., Ezeike, G.O., Hung, Y.C., Doyle, M.P., 1999. Inactivation of *Escherichia coli* O157:H7 and *Listeria monocytogenes* on plastic kitchen cutting boards by electrolyzed oxidizing water. J. Food Prot. 62, 857–860.

Wang, H., Slavik, M.F., 1998. Bacterial penetration into eggs washed with various chemicals and stored at different temperatures and times. J. Food Prot. 61, 276–279.

Waseh, S., Hanifi-Moghaddam, P., Coleman, R., Masotti, M., Ryan, S., Foss, M., MacKenzie, R., Henry, M., Szymanski, C.M., Tanha, J., 2010. Orally administered P22 phage tailspike protein reduces *Salmonella* colonization in chickens: prospects of a novel therapy against bacterial infections. PLoS One 5, e13904. http://dx.doi.org/10.1371/journal.pone.0013904.

Whistler, P.E., Sheldon, B.W., 1989. Bactericidal activity, eggshell conductance, and hatchability effects of ozone versus formaldehyde disinfection. Poult. Sci. 68, 1074–1077.

Wong, P.Y.Y., Kitts, D.D., 2003. Physicochemical and functional properties of shell eggs following electron beam irradiation. J. Sci. Food Agric. 83, 44–52.

Wright, A.P., Richardson, L., Mahon, B.E., Rothenberg, R., Cole, D.J., 2016. The rise and decline in *Salmonella enterica* serovar Enteritidis outbreaks attributed to egg-containing foods in the United States, 1973–2009. Epidemiol. Infect. 144, 810–819.

Yoshifumi, H., 2003. Improvement of the electrolysis equipment and application of slightly acidic electrolyzed water for dairy farming. J. Jpn. Soc. Agric. Machin. 65, 27–29.

CHAPTER

# Natural Approaches for Improving Postharvest Safety of Egg and Egg Products

# 19

**Indu Upadhyaya[1], Hsin-Bai Yin[2], Meera Surendran Nair[2], Kumar Venkitanarayanan[2]**

[1]*University of Arkansas, Fayetteville, AR, United States;* [2]*University of Connecticut, Storrs, CT, United States*

## 1. INTRODUCTION

Eggs represent a globally popular, inexpensive, and nutritionally rich food product. Eggs have a high nutritive value, particularly being a less expensive source of excellent animal protein than meat and milk (Friedman, 1996). Moreover, research has highlighted additional benefits associated with egg consumption, including immunomodulatory, anticarcinogenic, and antihypertensive activities that further contribute to human health and well-being (Kovacs-Nolan et al., 2005). This has led to an increase in world egg production from 51 million tons per year to 63.8 million tons between 2000 and 2010 (FAO, 2010). The US egg industry is the second largest producer of chicken eggs in the world after China (NASS, 2009), and on an average, Americans consume 250 eggs annually (Anonymous, 2015). Eggs are used as an economical food source in the form of shell eggs, liquid, frozen, and dried products (Ricke et al., 2001; Howard et al., 2012). According to United Egg Producers, more than 95% of egg production in the United States comes from only 240 egg producing companies that have flocks of more than 75,000 layers per flock. Despite this fact, the US shell egg production totaled 6.91 billion in December of last year (Anonymous, 2015; AEB, 2016) with table egg producing flock size reaching 277 million layers. Currently, there are approximately 63 egg producing companies with more than 1 million hens that contribute approximately 86% of the total production, and 17 companies with greater than 5 million hens (Anonymous, 2015). Although eggs are low priced, healthy, and form a critical part of the American diet, the microbiological safety of eggs is a concern, since they constitute the primary source of *Salmonella* spp., the most common bacterial agents causing disease in humans (Guard-Petter, 2001; Latimer et al., 2002; Bialka et al., 2004; Namata et al., 2008; Thomas et al., 2009; De Vylder et al., 2009; Li et al., 2013; Ricke et al., 2013).

**Producing Safe Eggs. http://dx.doi.org/10.1016/B978-0-12-802582-6.00019-7**

The egg industry primarily relies on Good Agricultural Practices and Hazard Analysis Critical Control Point programs (HACCP) for improving the microbiological safety of eggs. Cleaning and sanitation of shell eggs by washing is a common practice mandatory for plants operating under the Federal Grading Service (Anonymous, 2013). Currently, egg washers use a variety of detergents and sanitizers to reduce the microbial load on the eggshell surface (USDA, 2008; Zeidler, 2002). These practices have met with varying degrees of success against different pathogens, and the commonly employed chemicals have been shown to possess limited antimicrobial effect, thereby not rendering eggs pathogen-free (Moats, 1978; Wang and Slavik, 1998).

This chapter focuses on the major food-borne pathogens that affect the safety of eggs and egg products, with special emphasis on *Salmonella* spp. contamination. Moreover, traditional and natural postharvest intervention methods for reducing egg and egg product contamination are addressed.

## 2. MAJOR CONTAMINANTS IN EGGS

### 2.1 SALMONELLA

Epidemiological investigations around the world suggest that *Salmonella* Enteritidis is the most common serotype of *Salmonella* isolated from poultry products, including eggs (Machado and Bernardo, 1990; Plummer et al., 1995; Uyttendaele et al., 1999; Antunes et al., 2003). This is true for United States as well, where an estimated 2.2 million *S.* Enteritidis contaminated eggs are produced each year (Braden, 2006).

#### *2.1.1 Major Outbreaks*

Analysis of several epidemiological investigations during the last 20 years have highlighted a strong correlation between consumption of contaminated eggs and human salmonellosis (Anonymous, 1988, 1990; Angulo and Swerdlow, 1999; Braden, 2006; Guard-Petter, 2001). During the period from 1990 to 2001, the US Centers for Disease Control and Prevention (CDC) reported 677 outbreaks of egg-borne *S.* Enteritidis that caused 23,366 illnesses, 1988 hospitalizations, and 33 deaths in the United States (Anonymous, 2003). However, the largest egg-borne *Salmonella* outbreak reported in the United States was in Iowa in 2010 from May to November 2010, leading to a total of 3578 reported illnesses with significant economic losses (CDC, 2010). More recently, an ongoing outbreak due to *Salmonella* contamination of shelled eggs in the United States has led to multiple confirmed cases of salmonellosis in humans (Anonymous, 2016). Similarly, eggs and egg products have been documented as a major source of *S.* Enteritidis in the European Union (EFSA, 2007). This is further underscored by a report from the United Kingdom, which implicated eggs as the primary source of *S.* Enteritidis in humans (Anonymous, 2014). These aforementioned reports collectively indicate that contaminated eggs are a major source of human salmonellosis, emphasizing the need for effective control strategies.

In light of the mounting evidence linking human salmonellosis with shell eggs, the US Department of Agriculture (USDA)-Food Safety and Inspection Service (FSIS), and Food and Drug Administration (FDA), issued a "farm-to-table" risk assessment of *S.* Enteritidis in eggs in 1996, which served as the basis for the Federal and State Egg Safety Action Plan (Braden, 2006). During the same time, the US President's Council on Food Safety identified egg safety as a major public health problem warranting immediate federal and interagency action. The council published a report with the objective of reducing, and ultimately eliminating, egg-borne salmonellosis (Anonymous, 2000). In July 2009, to further control egg contamination, the FDA issued a final rule that requires egg producers to implement strict measures to prevent *S.* Enteritidis egg contamination on the farm and subsequent growth during storage and transportation. In addition, egg producers are required to maintain records concerning compliance with the rule and to register with the FDA (FDA, 2009).

### *2.1.2 Mechanism of Egg Contamination in* Salmonella

*S.* Enteritidis is the most frequently isolated *Salmonella* from layer flocks (Baird-Parker, 1990; Braden, 2006; Gast et al., 2005; EFSA, 2007). The primary colonization site of *S.* Enteritidis in chicken is the ceca (Allen-Vercoe and Woodward, 1999; Filho et al., 2000; Stern, 2008), with cecal carriage of *S.* Enteritidis leading to horizontal transmission of the infection, contamination of eggshell with feces, and probably retrocontamination of ovaries (Keller et al., 1995; Gantois et al., 2009). Egg contamination with *S.* Enteritidis occurs due to penetration of the eggshell from contaminated feces during or after oviposition (De Reu et al., 2006; Gast and Beard, 1990; Messens et al., 2005, 2006). Trans-shell route of egg contamination with *S.* Enteritidis can also occur from other sources such as farmers, pets, and rodents (Latimer et al., 2002). Following oviposition, *Salmonella* survival and/or growth on the outer shell surface of eggs is facilitated by the presence of chicken manure and other organic materials (Gantois et al., 2009). Researchers have suggested that *S.* Enteritidis can penetrate the eggshell (Miyamoto et al., 1998; Padron, 1990). In addition, egg becomes exposed to temperatures cooler than the chicken body temperature, potentially creating a negative pressure that allows bacteria to penetrate the eggshell and the membranes (Board, 1966; Howard et al., 2012). Moreover, it has been hypothesized that when the warm egg encounters a moist, cool environment, conditions are ideal for penetration of the shell by bacteria (Berrang et al., 1999). Apart from temperature, it is believed that the cuticle, which plays the first line of defense against bacterial penetration of the egg, can potentially dry and shrink with time resulting in bacterial penetration through the pores (Mayes and Takeballi, 1983). In addition, studies have indicated that in the absence of cuticle deposition, bacterial penetration can occur (De Reu et al., 2006; Messens et al., 2007). However, other investigations have found no significant correlation between the deposition of the cuticle and penetration of *Salmonella* through the shell (Messens et al., 2005; Nascimento et al., 1992). Furthermore, the quality of the eggshell as defined by shell specific gravity, shell weight, or shell thickness, has been hypothesized to have a role in bacterial penetration into the egg (Howard et al., 2012). It has been reported that selecting strains of birds for higher egg production and greater egg weight potentially results in poorer quality shells (Roberts and Brackpool, 1994; Howard et al., 2012), which are more susceptible to contamination (Jones et al., 2002). Likewise, age of the hen is a factor affecting shell quality and contamination of shells as air cells and contents have been found to increase in older hens thereby contributing to increased bacterial penetration (Jones et al., 2002).

## 2.2 *CAMPYLOBACTER* SPP.

*Campylobacter* spp. is another major food-borne pathogen that is the leading cause of diarrheal illness in the United States. According to the Foodborne Diseases Active Surveillance Network (FoodNet), approximately 14 cases are being diagnosed per 100,000 people in the population (CDC, 2014). Among the different species in the genera, *Campylobacter jejuni* is most commonly implicated in outbreaks associated with the consumption of contaminated poultry products. The domestic poultry population harbors the bacterium as a part of their endogenous flora, thereby acting as an important source of transmission (Izat et al., 1986; Lin, 2009; Keener et al., 2004). The source of several sporadic illnesses associated with poultry products were traced back to horizontal transmission of the pathogen and carcass contamination. However, the prevailing information on vertical transmission is weak, indicating that egg-borne transmission of the pathogen is rare as *Campylobacter* is not a strong invasive pathogen.

Table eggs and hatching eggs are not generally reported to be contaminated with *Campylobacter* (Jacobs-Reitsma et al., 2008). Studies conducted during the later 1900s reported frequent occurrence of a high population of *Campylobacter* in turkey feces, but apparently it was not transmitted through fertile eggs (Acuff et al., 1982; Izat et al., 1986). However, several other studies showed that following the experimental inoculation in eggs, *Campylobacter* was recovered from both the contents of unhatched eggs and from the newly hatched chicks (Shanker et al., 1986; Clark and Bueschkens, 1985). Furthermore, Doyle (1984) and Shanker et al. (1986) isolated the pathogen from the inner and outer shells and membranes of eggs laid by naturally infected commercial layers and broilers, but not from the internal contents. Studies have also shown the persistence of *Campylobacter* in the reproductive tract of healthy broiler and laying breeder hens and in the semen of healthy broiler breeder roosters (Camarda et al., 2000; Cox et al., 2002). Moreover, the presence of *Campylobacter* DNA in hatchery, embryos, and hatched chicks indicates the possibility of vertical transmission (Chuma et al., 1994, 1997; Hiett et al., 2002; Silva et al., 2011). However, Sahin et al. (2003) and Fonseca et al. (2014) showed that the ability of *C. jejuni* to penetrate eggs and survive within eggs was comparatively poor. Therefore despite the different potential sites of colonization in the host and contamination in eggs, the relevance of vertical transmission in campylobacteriosis is less compared with other sources of infection.

## 2.3 OTHER CONTAMINANTS: FUNGUS AND VIRUSES

Apart from bacterial contamination, mycotoxins produced by toxigenic molds and viruses such as avian influenza viruses (AIVs) and Newcastle disease viruses

(NDVs) can potentially contaminate eggs and egg products. Although the prevalence of mycotoxins and viruses in eggs or egg products is relatively lower than bacterial infections, these contaminants should be carefully controlled due to the potential public health concerns.

### 2.3.1 Mycotoxins

Mycotoxins are secondary metabolites produced by molds, which contaminate food crops or animal feed either before or after harvest and their concentrations increase during postharvest storage. Once chickens consume mycotoxin-contaminated feed, these mycotoxins can be transferred to eggs. Moreover, the normal industrial processing cannot remove mycotoxins from food products since they are heat resistant due to their stable chemical structure (Kabak, 2009; da Cruz Cabral et al., 2013). The major mycotoxins detected in eggs include aflatoxin, ochratoxin A, and zearalenone.

Aflatoxins are the most commonly occurring mycotoxin contaminants on foods. They are produced by *Aspergillus flavus* and *Aspergillus parasiticus*, and are considered as one of the most potent mycotoxins because of their toxicity and their ability to biotransform in animal and human systems. Moreover, aflatoxin B1 is listed as a group 1 human carcinogen by the International Agency for Research on Cancer (IARC, 2002), and it mainly targets liver (Yunus et al., 2011). Since *A. flavus* and *A. parasiticus* primarily contaminate peanut, corns, and cottonseeds, which are often used as feed for animals (Gong et al., 2004), aflatoxin residues can be transferred to chicken eggs (Hussain et al., 2010) when the birds consume aflatoxin-contaminated feed. A study conducted by Aly and Anwer (2009) demonstrated that aflatoxin transmission ratios from aflatoxin-contaminated feed to egg after 60 days' exposure at the levels of 25, 50, and 100 ppb aflatoxin were 625:1, 500:1, and 1428:1, respectively. Moreover, aflatoxin is heat resistant in eggs, where only 0.2–1% detoxification was found in eggs after boiling for 20 min (Aly and Anwer, 2009). Ochratoxin A is another common mycotoxin produced by *Aspergillus ochraceus* and *Penicillium verrucosum*, which contaminates predominantly foods and feedstuffs of cereal origins (Basilico and Basilico, 1999; Höhler, 1998) after harvest. Ochratoxin A can cause renal toxicity, nephropathy, and immunosuppression in several animal species, and it has been found to be carcinogenic in experimental animals (Stein et al., 1985; Richard et al., 1999). A prevalence study conducted in Pakistan reported that, of 80 egg samples collected in the region, 35% were found to be contaminated with ochratoxin A (Iqbal et al., 2014). Zearalenone is yet another mycotoxin primarily produced by *Fusarium graminearum*, which can contaminate wheat, barley, and rye. Zearalenone binds to estrogen receptors and results in hormonal changes (Mirocha et al., 2013). Although zearalenone has reduced toxicity to chickens compared with aflatoxin and ochratoxin A, studies have shown that zearalenone residues can be transferred from contaminated feed to chicken eggs (Dailey et al., 1980; Sypecka et al., 2004). The presence of zearalenone in eggs is critical to human health because of its estrogenic and genotoxic properties.

### 2.3.2 Viral Contamination of Eggs

Egg or egg product contamination by viruses is uncommon; however, AIV and NDV are potential viruses transmitted through eggs. According to the World Organization for Animal Health, AIV of high pathogenicity and NDV of the mesogenic and velogenic pathotypes are considered to cause diseases of high significance affecting poultry health and economy (Dortmans et al., 2011; Costa-Hurtado et al., 2015). AIV is a heat-labile single-stranded RNA virus of the family Orthomyxoviridae, genus *Influenza virus* A, whereas NDV is a single-stranded RNA virus of the family Paramyxoviridae, genus *Avulavirus*, serogroup avian paramyxovirus 1 (APMV-1) (Büchen-Osmond, 2002). These viruses can potentially be present in the internal egg contents or on the egg surface from virus-infected feces. These viruses can be partially protected from heat inactivation by the presence of organic materials (Alexander, 2003).

## 3. TRADITIONAL METHODS OF IMPROVING EGG SAFETY

Currently, there are no mandatory regulations that detail specific requirements for disinfection of shell eggs before sale as table eggs in the United States. However, the Agricultural Marketing Service has a voluntary shell egg-grading program, which includes specific egg washing and sanitizing requirements (Howard et al., 2012). Mostly, egg washers use a variety of detergents and sanitizers for reducing the microbial load on the eggshell surface (USDA, 2008; Zeidler, 2002). Generally, eggs are subjected to a wash cycle on a conveyor belt using recirculated water and brushes (Howard et al., 2012). Alkaline detergents are added to wash water to clean the egg surface and maintain a high pH for bacterial control. However, organic matter accumulates in the recirculated water and reduces the ability of the detergent to kill bacteria (Kinner and Moats, 1981). Immediately following the detergent wash, operations participating in the grading program may sanitize shell eggs with a potable water rinse containing a chlorine concentration of 100–200 ppm, or quaternary sanitizers compatible with the washing compound. In addition to the currently employed routine, government agencies and the egg industry are interested in alternative decontamination methods that could be more effective and economical, thus benefiting both the public and the industry. In this regard, numerous chemicals and methods have been investigated as post wash sanitation methods for reducing bacteria (Howard et al., 2012). However, since there are no effective postharvest interventions available to reduce mycotoxins, reduced mycotoxin exposure to chickens appears to be the most practical method to ensure egg safety for human consumption (Wood, 1988; Park et al., 2007). Current methods for reducing mycotoxin contaminations in eggs include mostly synthetic fungicides at the preharvest level, but these fungicides can accumulate in the environment such as soil, plants, and water, thereby potentially resulting in egg contamination because of their nonbiodegradable nature. In addition, the inclusion of aflatoxin-binding adsorbent in feed is employed to protect birds from the harmful effects of mycotoxins. However, several adsorbents have been shown to impair nutrient utilization (Chung et al., 1990; Kubena et al., 1993; Scheideler, 1993) and mineral absorption in chickens (Chestnut et al., 1992; Edrington et al., 1997) thereby limiting their use.

In selecting a suitable disinfectant to clean eggshells, factors such as the antimicrobial effectiveness of the agent to eliminate target bacterium from eggshell, safety, and cost should be considered (Scott and Sweetnam, 1993). The chemicals used to wash eggs are considered potential food additives and hence are regulated by the FDA. An ideal egg wash antimicrobial should be effective in reducing large populations of the target pathogen in a rapid time frame, even in the presence of organic matter. Furthermore, it should be safe to workers and the environment, cost-effective (Scott and Sweetnam, 1993) and should be easily incorporated in an HACCP plan. A variety of disinfectants in egg wash water, including hydrogen peroxide (Padron, 1995), chlorine and iodine-based sanitizers (Knape et al., 1999), ozone (Koidis et al., 2000), quaternary ammonium and sodium carbonate (Wang and Slavik, 1998), zinc sulfate and formaldehyde fumigation (Bierer and Barnett, 1962), and electrolyzed oxidizing (EO) water (Russel, 2003) have been investigated to reduce or eliminate pathogens on table eggs and hatching eggs. However, many of the aforementioned chemicals have been shown to possess limited antimicrobial effect, especially in the presence of organic matter, and many did not render eggs pathogen-free (Frank and Wright, 1956; Moats, 1978; Wang and Slavik, 1998).

**Table 19.1** Traditional and Natural Methods for Improving Postharvest Safety of Egg and Egg Products

| | Egg Product | Target Organism | References |
|---|---|---|---|
| **1. Traditional Methods** | | | |
| ***1.1 Wash Treatment With Sanitizers*** | | | |
| Peroxidase-catalyzed compound | Shell eggs | *S.* Enteritidis | Kuo et al. (1997a,b) |
| Distilled deionized water | Shell eggs | *Salmonella* spp. | Knape et al. (2001) |
| Iodine-based detergent | Shell eggs | *Salmonella* spp. | Knape et al. (2001) |
| Chlorine (200 ppm) | Shell eggs | *Salmonella* spp. | Knape et al. (2001) |
| Quaternary ammonium compound | Shell eggs | *S.* Enteritidis | Wang and Salvik (1998) |
| Sodium hypochlorite | Shell eggs | *S.* Enteritidis | Wang and Salvik (1998) |
| ***1.2 Ultraviolet (UV) Radiation*** | | | |
| UV radiation | Shell eggs | Aerobic bacteria, molds, *S.* Typhimurium | Gao et al. (1997) and Keklik et al. (2010) |

*Continued*

**Table 19.1** Traditional and Natural Methods for Improving Postharvest Safety of Egg and Egg Products—cont'd

| | Egg Product | Target Organism | References |
|---|---|---|---|
| ***1.3 Electrolyzed Water*** | | | |
| Electrolyzed oxidizing water | Shell eggs | *S.* Enteritidis | Bialka et al. (2004) and Howard et al. (2012) |
| ***1.4 Whole-Egg Pasteurization*** | | | |
| 57°C 25min in water bath | Shell eggs | *S.* Enteritidis | Barbour et al. (2001) |
| 100°C 2s steam | Shell eggs | *Salmonella* spp. | James et al. (2002) |
| Hot water alone, hot air alone, and the combination | Shell eggs | *Salmonella* spp. | Jeng et al. (1987) and Van Lith et al. (1995) |
| ***1.5 Ionizing Radiation*** | | | |
| Ionizing radiation | Shell eggs | *Salmonella* spp., *Campylobacter* spp. | Howard et al. (2012) |
| ***1.6 Ozone*** | | | |
| Gaseous ozone | Shell eggs | *S.* Enteritidis | Rodriguez-Romo et al. (2005, 2007) |
| Aqueous ozone | Shell eggs | *S.* Enteritidis | Davies and Breslin (2003) |
| ***1.7 Antimicrobial Coating*** | | | |
| Chitosan-based coating | Shell eggs | *S.* Enteritidis | Leleu et al. (2011) |
| Nisin, allyl isothiocynate, lauric arginate ester, and organic acids applied as a chitosan or polylactic acid coating | Shell eggs | *S.* Enteritidis | Jin et al. (2013) |
| ***1.8 Other Methods*** | | | |
| Microwave technology | Shell eggs | *S.* Enteritidis | Lakins et al. (2008) |
| Pulsed light technology | Shell eggs | *S.* Enteritidis | Dunn (1996) and Hierro et al. (2009) |
| Gas plasma technology | Shell eggs | *S.* Typhimurium, *S.* Enteritidis | Ragni et al. (2010) |
| Ultrasound | Shell eggs | *S.* Enteritidis | Cabeza et al. (2011) |
| **2. Natural Approaches** | | | |
| ***2.1 Phytochemicals*** | | | |
| *Trans*-cinnamaldehyde carvacrol, eugenol (washing treatment) | Shell eggs | *S.* Enteritidis | Upadhyaya et al. (2013) |

**Table 19.1** Traditional and Natural Methods for Improving Postharvest Safety of Egg and Egg Products—cont'd

| | Egg Product | Target Organism | References |
|---|---|---|---|
| Pomegranate rind (washing treatment) | Shell eggs | *S.* Enteritidis | Pohuang et al. (2011) |
| *Trans*-cinnamaldehyde and carvacrol (antimicrobial coating) | Liquid egg albumen | *S.* Enteritidis | Jin and Gurtler (2011) |
| *Trans*-cinnamaldehyde, eugenol (fumigation) | Embryonated eggs | *S.* Enteritidis | Upadhyaya et al. (2015) |
| Carvacrol, eugenol, beta-resorcylic acid (pectin- or gum arabic–based coating) | Shell eggs | *S.* Enteritidis | Upadhyaya et al. (2016) |
| ***2.2 Organic Compound*** | | | |
| DBS (1-bromo-3-chloro-2,2,5,5-tetramethylimidazolidin-4-one) and DC (1,3-dichloro-2,2,5,5,-tetramethylimidazolidin-4-one) | Shell eggs | *S.* Enteritidis | Worley et al. (1992) |
| Disodium ethylenediamine tetra acetic acid | Egg white | *S.* Typhimurium | Garibaldi et al. (1969) |
| β-Propiolactone, ethylene oxide, and butadiene dioxide | Liquid whole egg | *S.* Typhimurium | Lategan and Vaughn (1964) |
| Acetic acid, lactic acid, citric acid, or malic acid with heat | Whole egg, egg yolk, egg white, whole egg+10% salt, and egg yolk + 10% salt | *S.* Typhimurium DT104 and non-DT104 strains | Jung and Beuchat (2000) |
| ***2.3 Probiotics*** | | | |
| *Streptomyces*, *Amycolaptosis*, *Micromomospora*, *Plantactinospora*, and *Solwaraspora* | Endangered Sea turtle egg | *Fusarium falciforme* | Sarmiento-Ramírez et al. (2014) |
| ***2.4 Bacteriophages*** | | | |
| *Salmonella*-lytic bacteriophages | Fertile eggs | *S.* Enteritidis | Henriques et al. (2013) |
| | Egg yolk | *S.* Typhimurium | Guenther et al. (2012) |
| | Liquid egg | *S.* Typhimurium | Zinno et al. (2014) |

The different approaches for improving the microbiological safety of eggs are summarized in Table 19.1.

### 3.1 WASH TREATMENTS WITH SANITIZERS

Kuo et al. (1997a) tested the use of a peroxidase-catalyzed compound (PCC) for sanitizing shell eggs. They determined that dipping eggs in PCC reduced *S.* Enteritidis by almost 4 log CFU as compared with less than 1 log for water only, but PCC-mediated reduction was not significantly greater than that due to 200 ppm chlorine. Knape et al. (2001) found that washing in distilled deionized water, an iodine-based detergent, and chlorine (200 ppm) decreased *Salmonella* populations inoculated on eggs compared with dry egg controls, but the efficacy of egg sanitizers appeared to depend on the level of total dissolved solids in the egg wash water. In another study by Wang and Slavik (1998), three commercial egg-washing chemicals, including a quaternary ammonium compound, sodium carbonate, and sodium hypochlorite, were compared for reducing *S.* Entertidis. The results showed that both quaternary ammonium compound and sodium hypochlorite treatments reduced bacterial penetration; however, sodium carbonate treatment facilitated bacterial penetration during egg storage due to altered eggshell surface allowing further recontamination.

### 3.2 UV RADIATION

Radiation using UV wavelengths has been investigated as a means of sanitizing shell eggs especially since it does not damage the cuticle (Gritz et al., 1990). Previous research demonstrated that UV radiation could kill a variety of microorganisms on different surfaces such as contact lenses (Gritz et al., 1990), poultry carcasses (Wallner-Pendleton et al., 1994), fiber or plastic belts, metal, or eggshells (Gao et al., 1997). Kuo et al. (1997b) determined that UV radiation significantly reduced aerobic bacteria, molds, and *S.* Typhimurium inoculated on shell eggs. An extension of the aforementioned technology known as pulsed UV light was tested for inactivating pathogens on food surfaces, and initially approved by the FDA (Federal Register, 1999). The effectiveness of pulsed UV light for 1–30 s was evaluated for decontamination of eggs inoculated with *S.* Enteritidis. Briefly, the eggs were placed at a distance of 9.5 and 14.5 cm from the UV lamp. A 20-s treatment at 9.5 cm produced a log reduction of 5.3 CFU/cm$^2$ without any visual damage to the egg (Keklik et al., 2010). Longer exposure times resulted in an increase of the egg temperature. However, due to health hazards associated with UV, this methodology is not commonly used in the egg industry.

### 3.3 ELECTROLYZED WATER

EO water is generated by combining electrolysis and membrane separation to produce an acidic and an alkaline component from a weak salt-water solution (Venkitanarayanan et al., 1999). A few studies have demonstrated the effectiveness of EO water for inactivation of pathogens in suspension solutions (Kim et al., 2000; Venkitanarayanan et al., 1999), and in foods (Bari et al., 2003; Russell, 2003). Bialka et al. (2004) compared EO water treatment with a commercial detergent-sanitizer treatment, both in vitro and using a pilot-scale egg washer. These researchers reported that EO treatment decreased *S.* Enteritidis by 2.3 CFU/g compared with 2.0 CFU/g reduction brought about by commercial detergent-sanitizer treatments (Bialka et al., 2004; Howard et al., 2012). Neither treatment significantly affected albumen height or eggshell strength, but both had significant effects on the cuticle.

### 3.4 TEMPERATURE AND STORAGE

Howard et al. (2006) demonstrated that *S.* Typhimurium survives within the egg and even exhibits growth during 8 weeks of storage under refrigeration conditions. In further studies, Howard et al. (2007) inoculated egg components with *S.* Enteritidis and studied its survival and growth during refrigerated storage for 8 weeks. The results suggested that egg components recovered from refrigerated eggs only inconsistently supported *S.* Enteritidis growth, although they did not inhibit survivability. With regards to the survival of *Campylobacter* in eggs, Stern and Kazmi (1989) reported that heating at 60°C was adequate for completely inactivating the pathogen. *C. jejuni* was also reported highly susceptible to freezing conditions ≤−15°C. In addition, Doyle and Roman (1982) observed that *Campylobacter* is sensitive to drying and storage at room temperature, in an anhydrous environment, and in the presence of skim milk on a glass surface. But studies investigating mycotoxins in eggs have shown that once eggs are contaminated by mycotoxins, the thermal processing of eggs is not effective for detoxification of these toxins in eggs, thereby underscoring that preventing the contamination at the preharvest level is the most effective control measure (Wood, 1988; Park et al., 2007).

Prompt refrigeration of eggs has been recommended as a viable approach to reduce *S.* Enteritidis and further prevent transmission to humans (Gast and Holt, 2000; Galiş et al., 2013). Refrigeration reduces the metabolic activities of the pathogen, thereby decreasing the risk of horizontal transmission of *S.* Enteritidis (Hammack et al., 1993; Chen et al., 2005; Martelli and Davies, 2012). Hence, FDA (2009) in its final egg rule emphasized that commercially available eggs must be stored at refrigeration temperature in United States (less than 45°F; FDA, 2009; USDA, 2011).

### 3.5 WHOLE-EGG PASTEURIZATION

The USDA-FSIS risk assessment of *S.* Enteritidis in shell eggs has predicted that pasteurization of shell eggs resulting in a 3 log reduction of *S.* Enteritidis would reduce illness caused by this organism by 70% (USDA-FSIS, 2005). There have been several studies done to evaluate the effects of pasteurization and dry heat treatments of intact shell eggs on *S.* Enteritidis. In one study, eggs were inoculated internally with a five-strain *S.* Enteritidis cocktail, and treated for 25 min at 57°C in a water bath followed by 57 min at 55°C in a hot oven (Barbour et al., 2001). This treatment resulted in a 6 log reduction of *S.* Enteritidis with no effect on the overall functionality of the eggs

(Barbour et al., 2001). Similarly, James et al. (2002) demonstrated that shell eggs subjected to steam exposure for 2 s at 100°C yielded significant reductions in bacterial numbers on the shell without increasing the interior temperature of the egg contents. Other methods such as hot water, hot air, or their combinations have been used with some success for reducing bacteria on eggs (Jeng et al., 1987; Van Lith et al., 1995).

### 3.6 IONIZING RADIATION

The FDA has approved ionizing radiation up to 3 kGy for the reduction of pathogens in fresh eggs (FDA, 2000). However, when eggs were irradiated with doses in the range of 0.5–3.0 kGy, Meszaros et al. (2006) observed changes in different characteristics of eggs, including the flow behavior of egg white, brittleness of yolk membrane, whippability and foam stability of egg white, and sensory changes of raw and soft-boiled eggs. It was concluded that a minimal dose of 1.5 kGy would be required for radiation inactivation of *Salmonella*, without significantly affecting the quality of shell eggs. In the case of *Campylobacter* spp., since egg and egg products could possibly become cross-contaminated with the pathogen by infected rearing environments, irradiation has been used to eliminate the pathogen from eggs without adversely affecting the egg quality. It was shown that 0.2 kGy is required to reduce *Campylobacter* by one decimal log or 1 kGy to reduce it by 5 decimal logs (Tauxe, 2001). Verde et al. (2004) showed that gamma radiation at the dose rate of 1.0 kGy/h could eliminate *Campylobacter* in artificially contaminated eggs.

### 3.7 OZONE

Ozone is another potent sanitizer known to be active against all forms of microorganisms at relatively low concentrations (Khadre et al., 2001; Galiş et al., 2013). Ozone in aqueous phase has been demonstrated as a strong microbicidal agent to effectively inactivate *Salmonella* in shell eggs (Galiş et al., 2013). Rodriguez-Romo et al. (2007) showed that *S.* Enteritidis was reduced by ≥5 log on the surface of shell eggs by high ozone concentrations (12–14% wt/wt $O_3$ in $O_2$ mix). In addition, Rodriguez-Romo et al. (2005) demonstrated that application of pressurized gaseous ozone for up to 20 min resulted in significant decrease in *S.* Enteritidis population on shell eggs. However, when Davies and Breslin (2003) used dry and moist ozonated air, 95.8% of eggs remained contaminated after treatment compared with 91.7% controls for the dry form, and 33.3% treated eggs were contaminated compared with 75% control eggs subjected to the moist ozone treatment. Therefore the application of ozone in either type of environment was only partially effective. Furthermore, due to its low stability, ozone cannot be stored, and has to be produced on demand (Khadre et al., 2001; Galiş et al., 2013).

### 3.8 ANTIMICROBIAL COATING

Significant research has been done on the use of materials with film-forming or coating capacity together with antimicrobial properties for improving the microbiological safety and shelf life of foods (Dutta et al., 2009; Valencia-Chamorro et al., 2011). Antimicrobial agents have been successfully incorporated into edible composite coatings based on polysaccharides or proteins (Falguera et al., 2011; Ponce et al., 2008; Martínez-Camacho et al., 2010). In this regard, chitosan, a polysaccharide obtained from crustaceans, has been employed as an effective antimicrobial coating for reducing *S.* Enteritidis on eggs (Leleu et al., 2011). Moreover, coating eggshells with chitosan has been shown effective in preserving the internal quality and extending the shelf life of eggs (Bhale et al., 2003; Caner, 2005; Caner and Cansiz, 2007; Kim et al., 2006). Jin et al. (2013) determined the antimicrobial efficacy of nisin, allyl isothiocyanate, lauric arginate ester (LAE), and organic acids applied as a chitosan or polylactic acid coating in reducing *S.* Enteritidis on shell eggs. These researchers observed that chitosan coatings with 0.5% and 1% LAE decreased *S.* Enteritidis counts on eggs by more than 5 log CFU/cm$^2$. In another study, Leleu et al. (2011) reported that chitosan (2%) coating of eggs significantly reduced trans-shell penetration of *S.* Enteritidis, but failed to decrease the pathogen load on shell.

In addition to the aforementioned methods, other approaches such as microwaves (Lakins et al., 2008), pulsed light (Dunn, 1996; Hierro et al., 2009), gas plasma technology (Kayes et al., 2007; Ragni et al., 2010), and ultrasound treatment (Cabeza et al., 2011) have been explored with limited success to improve the microbiological safety of eggs.

## 4. NATURAL APPROACHES

Antibiotics have been traditionally used in poultry operations for maintaining health and production performance of layers as a preharvest measure. However, their continued use, especially at subtherapeutic levels has led to the emergence of antimicrobial resistant bacteria, including multidrug-resistant food-borne pathogens in layers. Due to the inability of the commonly employed postharvest interventions to reduce the risk of egg contamination and the current threat of antibiotic-resistant bacteria in the poultry industry, there has been an increased interest for an alternative strategy for improving the microbiological safety of egg. In this regard, the use of natural, environment-friendly, and biological approaches have gained importance in the recent past as these interventions pose fewer risks to the consumer compared with chemical methods. The major natural approaches employed for improving postharvest egg safety are discussed in the following sections and also summarized in Table 19.1.

### 4.1 PHYTOCHEMICALS

Plant-derived essential oils are a group of natural and environmental-friendly antimicrobials that have traditionally been used as food preservatives and flavor enhancers (Pitasawat et al., 2007; Upadhyay et al., 2014). Plants are capable of synthesizing a large number of molecules, many of which are phenolic compounds or their derivatives (Geissman, 1963). More recently, the use of phytochemicals has gained significant attention due to increasing concerns over the safety of synthetic chemicals and the emergence of antibiotic-resistant strains of microorganisms (Salamci et al., 2007). The antimicrobial activities of many phytochemicals against a wide range of Gram-positive and Gram-negative bacteria have been documented (Blumenthal et al., 2000; Adams et al., 2004; Burt, 2004; Chun et al., 2005; Gill and Holley, 2006), with many compounds applied to improve postharvest egg safety. Since limited studies have investigated the effect of phytochemicals on *Campylobacter* and viruses on eggs and egg products, this chapter primarily focuses on the effect of phytochemicals on *Salmonella*.

Upadhyaya et al. (2013) investigated the efficacy of *trans*-cinnamaldehyde, a major component of bark extract of cinnamon (*Cinnamomum zeylandicum*), carvacrol obtained from oregano oil (*Origanum glandulosum*), and eugenol derived from cloves (*Eugenia caryophillis*) as wash treatments for reducing *S.* Enteritidis on the eggshell surface (Upadhyaya et al., 2013). Eggs treated with the aforementioned phytochemical solutions at 42°C for 30 s significantly reduced *S.* Enteritidis on eggs, where *trans*-cinnamaldehyde (0.75%) was the most effective treatment that completely inactivated the pathogen on eggs. These phytochemicals were also found to be effective in the presence of organic matter. In another study, extracts from pomegranate rind (*Punica granatum* L.) used as antimicrobial wash treatments at 1.25 or 2.5% effectively inactivated *S.* Enteritidis on eggs (Pohuang et al., 2011). Jin and Gurtler (2011) investigated the efficacy of antimicrobial coating of phytochemicals in reducing *Salmonella* in liquid egg albumen. Four-ounce glass jars were coated with a mixture of polylactic acid polymer and phytochemicals (*trans*-cinnamaldehyde or carvacrol), and then liquid egg white inoculated with *S.* Enteritidis was stored at 10°C for 28 days. However, only 1 log CFU/mL reduction was observed in the presence of *trans*-cinnamaldehyde and carvacrol as compared with control during 28 days of storage. Similarly, the efficacy of *trans*-cinnamaldehyde and eugenol applied as a fumigation treatment to reduce *S.* Enteritidis on embryonated eggs was investigated (Upadhyaya et al., 2015). Day-old embryonated eggs were spot inoculated with *S.* Enteritidis (~6.5 log CFU) and subjected to fumigation with the aforementioned phytochemicals (0 or 1% concentration) for 20 min in a hatching incubator, and pathogen populations on shell and embryo were enumerated on days 1, 3, 6, 9, 13, 16, and 18 of incubation. On day 13, the eggs were reinoculated, followed by fumigation treatment for 20 min. It was found that both phytochemicals were more effective in reducing *S.* Enteritidis on the shell and embryo compared with controls (Upadhyaya et al., 2015). In a follow-up study, Upadhyaya et al. (2016) determined the efficacy of carvacrol, eugenol, and β-resorcylic acid (derived from cherries) as a pectin- or gum arabic–based coating for reducing *S.* Enteritidis on shell eggs. *S.* Enteritidis was spot inoculated on shelled eggs followed by coating with pectin or gum arabic solution containing each phytochemical (0, 0.25, 0.5, or 0.75%), and stored at 4°C for 7 days. Approximately 4.0 log CFU/egg of the pathogen was recovered from inoculated and pectin- or gum arabic–coated eggs on day 0. All coating treatments containing carvacrol, eugenol, and β-resorcylic acid at 0.75% reduced *Salmonella* to undetectable levels on day 3 itself, suggesting that the aforementioned phytochemicals could effectively be used as a coating to reduce *S.* Enteritidis on shell eggs.

Yin et al. (2015a) reported that carvacrol and *trans*-cinnamaldehyde were effective in reducing growth and aflatoxin production in *A. flavus* and *A. parasiticus* in vitro. In addition to reducing the fungal growth, these compounds decreased the aflatoxin production in poultry feed (Yin et al., 2015a). Follow-up in vivo studies showed that in-feed supplementation of carvacrol and *trans*-cinnamaldehyde was effective in reducing aflatoxicosis in chickens fed with aflatoxin-contaminated feed (Yin et al., 2015b). However, more studies to establish the effect of phytochemicals in reducing mycotoxins and virus contamination of eggs are warranted.

## 4.2 ORGANIC COMPOUNDS

Several organic compounds have been evaluated as potential replacement disinfectants for chlorine in egg processing. For instance, several *N*-halamine compounds were tested against *S.* Enteritidis on the surface of egg shells. Compounds DBS (1-bromo-3-chloro-2,2,5,5-tetramethylimidazolidin-4-one) and DC (1,3-dichloro-2,2,5,5,-tetramethylimidazolidin-4-one) significantly reduced *S.* Enteritidis on egg shells and were more effective than chlorine as a spray (Worley et al., 1992). Garibaldi et al. (1969) investigated the effect of disodium ethylenediamine tetra acetic acid (EDTA) on *S.* Typhimurium in egg white. Results revealed that EDTA supplementation at 5 and 50 mg/mL significantly reduced *S.* Typhimurium in egg white at 2°C and completely inactivated the pathogen after 28 days of storage. Moreover, EDTA was found more effective in killing *S.* Typhimurium in egg white at 28°C and decreased the bacterium by 6 log CFU after 60 h of storage. Another study investigated the efficacy of β-propiolactone, ethylene oxide, and butadiene dioxide in reducing *S.* Typhimurium in liquid whole egg and found that these compounds significantly reduced the bacterium in the whole egg at 30°C (Lategan and Vaughn, 1964).

Organic acids are potential antimicrobials for prevention of *Salmonella* outbreaks due to consumption of contaminated poultry meat and eggs (Jung and Beuchat, 2000; Mani-Lopez et al., 2012). Jung and Beuchat (2000) investigated the thermal inactivation of *S.* Typhimurium DT104 and non-DT104 strains when preexposed to a reduced pH media supplemented with acetic acid, lactic acid, citric acid, or malic acid (adjusted pH at 5.4, 4.4, 4.0, and 3.7) in five liquid egg products, including whole egg, egg yolk, egg white, whole egg+10% salt, and egg yolk+10% salt. Both DT104 and non-DT104 strains were more sensitive to heat when suspended in egg white compared with the other four liquid egg products. Addition of 10% salt protected bacteria against heat inactivation. However, when bacteria were preexposed to lactic acid supplemented media at pH 4.3, they became more sensitive to heat compared with the untreated control.

## 4.3 PROBIOTICS

The use of beneficial bacteria or probiotics is another potential strategy than can be used to improve postharvest safety of egg and egg products. Probiotics are defined as live microorganisms that, when administered in adequate amounts, confer a health benefit on the host (Morelli and Capurso, 2012). Numerous studies have investigated the efficacy of probiotics in improving the immune status (Toms and Powrie, 2001), intestinal health (Fuller, 1989; Coates and Fuller, 1977; Gilliland and Kim, 1984; Saarela et al., 2000; Prins, 1977), productivity (Kurtoglu et al., 2004; Van Immerseel et al., 2006; Panda et al., 2003, 2008; Youssef et al., 2013), and egg quality (Nahashon et al., 1994; Panda et al., 2008; Hassanein and Soliman, 2010) in chickens. In addition, the potential of probiotics for reducing pathogens in chickens preharvest has been documented (Kabir, 2009; Perumalla et al., 2012). However, their postharvest antimicrobial efficacy in improving egg safety has not been explored. Recent microbiome research on eggshell of various avian species such as homing pigeon (Grizard et al., 2014) and red-carp larks (Grizard et al., 2015) has revealed that the egg shell harbors a variety of bacterial and fungal communities, some of which might confer protection against pathogens. Sarmeinto-Ramírez et al. (2014) analyzed the microbiome on egg shells of the endangered sea turtle *Eretmochelys imbricate* infected with the fungal pathogen *Fusarium falciforme*. Metagenomic analysis revealed several genera (*Streptomyces*, *Amycolaptosis*, *Micromomospora*, *Plantactinospora*, and *Solwaraspora*) that inhibited the growth of the pathogen in vitro. Similar concepts need to be explored to improve the microbiological safety of chicken eggs.

## 4.4 BACTERIOPHAGES

Bacteriophages have emerged as potential tools for biocontrol of bacterial contamination of foods and balancing environmental microflora (Tiwari et al., 2011). Bacteriophages are viruses that infect and multiply in bacteria (Delbrock, 1946). For many bacteriophages, release into the environment after replication is accompanied by lysis of the host bacterium (Delbrock, 1946; Cohen, 1948). In light of the emergence of microbial antibiotic resistance in animals and humans and food safety concerns over residual effects of antibiotics, bacteriophages have been proposed as potential candidates to serve as an alternative to antibiotics in animal disease prevention and control, and can be particularly very effective in treating drug-resistant bacterial infections both in humans and in animals (Jassim and Limoges, 2014). Previous research has mainly focused on the application of bacteriophages for the treatment of enteric and respiratory infections in livestock and in poultry and food products as a preharvest intervention. For instance, *Salmonella*-lytic bacteriophages (Fiorentin et al., 2004) have been isolated previously and a few of these bacteriophages have been administered in vivo to *S.* Enteritidis PT4-infected broilers for reducing the pathogen up to 3.5 (log CFU/g) (Fiorentin et al., 2005). Other researchers have also successfully reported reductions in *Salmonella* counts by using bacteriophages in chicken internal organs and feces (Toro et al., 2005), skin (Goode et al., 2003), or poultry products

(Whichard et al., 2003). A number of studies have indicated the applicability of bacteriophages both as prophylactics and as therapeutic agents in fighting various bacterial pathogens in poultry, including *Escherichia coli*, *S.* Enteritidis, *S.* Typhimurium, *C. jejuni*, *Camplyobacter coli* and *Listeria monocytogenes*. Due to the wide range of antimicrobial activity possessed by phages, the feasibility of using bacteriophages in food products also has been studied. Henriques et al. (2013) reported the potential efficacy of bacteriophages for reducing *Salmonella* on fertile eggs, wherein a cocktail of two phages isolated from chicken litter, F1055S and F12013S, was applied by aerosol spray on fertile eggs challenged with *S.* Enteritidis. The phage treatment of *Salmonella*-challenged eggs was found to reduce the disease symptoms in the hatching chicks. Bacteriophages are promising agents that could complement and potentially replace current antibiotics, but their use as postharvest egg safety applications needs to be investigated in depth.

## 5. EGG PRODUCTS

Eggs provide a reliable source of nutrition and in conjunction with egg products, serve a variety of functions in other food products (Howard et al., 2012). The emulsifying properties of lecithin and cholesterol within the egg yolk make eggs valuable components of mayonnaise and other food systems requiring an emulsifier (Baker and Bruce, 1994). Moreover, albumen or egg white is used for its ability to form heat-stable foams in cakes, meringues, and other baked products. Moreover, a number of other foods containing egg products include noodles, candy, and ice creams (Ricke et al., 2001). Eggs used in products with other primary ingredients are referred to as hidden eggs. Egg products are popular in foodservice operations due to convenience of use, cost savings (for labor and storage), and for portion control (Messens et al., 2002). The various egg products include whole eggs, egg whites, and egg yolks in frozen, refrigerated liquid, and dried forms, as well as specialty egg products. Specialty egg products include prepeeled hard-cooked eggs, omelets, egg patties, quiches, quiche mixes, scrambled eggs, fried eggs, and others (Howard et al., 2012; Ricke et al., 2001). Due to the widespread use of eggs as a food source, the safety of various egg products is important.

In addition to table and hatching eggs, several frozen and fresh poultry products have been found to be contaminated with *Campylobacter* (Jacobs-Reitsma et al., 2008). However, Izat and Gardner (1988) collected eggs from two commercial egg-processing facilities to determine if *C. jejuni* could be isolated from the raw product or from further-processed egg products intended for human consumption. They could not detect the pathogen in raw eggs or in any processed eggs. This suggested that properly processed egg products were unlikely to be a source of *C. jejuni*. Clark and Bueschkens (1986) showed that *C. jejuni* could withstand lysozyme-mediated antimicrobial effect in egg white, whereas yolk was found to be toxic to the pathogen. Sato and Sashihara (2010) isolated *Campylobacter* isolates from unpasteurized liquid egg samples collected from egg-breaking facilities in Japan. However, these researchers found that pasteurized liquid egg samples were devoid of the bacterium due to its sensitivity to heat, suggesting that the current legal pasteurization conditions could eliminate the pathogen efficiently.

Traditionally, pasteurization of egg products has been employed to improve egg safety since the 1930s by inactivating *Salmonella* from eggshell or from the internal contents of the egg (Froning et al., 2002). In addition, a side benefit of pasteurization of egg products includes inactivation of other bacterial species as well as various viruses or fungi that could potentially affect human health. Currently, egg products are subjected to low temperature treatments (<70°C) to eliminate AIV and NDV contaminations (Zuber et al., 2013). In liquid egg products such as whole-egg blends, fresh liquid egg white, and fat-free egg products (FFEP), heat inactivation of H5N2 was shown to be more than 5 log with the application of time–temperature combinations according to industry pasteurization standards (Froning et al., 2002). King (1991) investigated the survivability of NDV and AIV in eggs when subjected to 57°C for 5, 10, 15, 30, 40, 50, and 60 min. Results revealed approximately 5.2 log reduction of AIV in the albumen samples when treated between 5 and 10 min, however, prolonged treatment was required to inactivate NDV, where approximately 7 log reduction of NDV was observed when treated for 40 and 50 min (King, 1991). For the yolk samples, NDV was not inactivated even after a treatment of 1 h. Another study showed that when H5N2 virus was heat treated at 60°C for 6.2 min in whole-egg blends, the viral load was predicted to decrease by 13.7 log reduction (Swayne and Beck, 2004). In addition, approximately 5.7 log reduction of H5N2 was observed when FFEP was heated at 57.7°C for 6.3 min (Chmielewski and Swayne, 2011). However, heat treatment (54.4°C for 7–10 days) was not effective for inactivating high pathogenic AIV in dried egg white (Swayne and Beck, 2004), indicating that the moisture content of different egg products is also one of the factors influencing the efficacy of heat-inactivation treatment.

In addition to pasteurization, the antimicrobial effect of bacteriophages against *Salmonella* in liquid eggs has been investigated (Guenther et al., 2012; Zinno et al., 2014). Guenther et al. described the application of the bacteriophage FO1-E2 to control *S.* Typhimurium in ready-to-eat foods, including egg yolk, and observed a reduction in bacterial counts up to 2 days after the application of bacteriophage. Similar results were observed by Zinno et al. (2014), wherein bacteriophage-22 reduced *S.* Typhimurium in liquid eggs within 48 h of application.

## 6. CONCLUDING REMARKS

Shelled egg has both physical barriers and several antimicrobial constituents in the egg contents to prevent microbial contamination. Under natural conditions, the presence of microorganisms inside the egg is rare; however, *S.* Enteritidis is potentially unique in its ability to migrate to the interior of the egg by the transovarian route and multiply without inducing noticeable changes. Although postharvest interventions exist and are relatively successful, there is a need for continued research on the potential application of alternative natural and biocontrol measures to ensure egg safety, especially under field and commercial settings.

## REFERENCES

Adams, T.B., Cohen, S.M., Doull, J., Feron, V.J., Goodman, J.I., Marnett, L.J., Munro, I.C., Portoghese, P.S., Smith, R.L., Waddell, W.J., Wagner, B.M., 2004. The FEMA GRAS assessment of cinnamyl derivatives used as flavor ingredients. Food Chem. Toxicol. 42, 157–185.

Acuff, G.R., Vanderzant, C., Gardner, F.A., Golan, F.A., 1982. Evaluation of an enrichment-plating procedure for the recovery of *Campylobacter jejuni* from turkey eggs and meat. J. Food Prot. 45, 1276–1278.

Alexander, D.J., 2003. Newcastle disease. In: Saif, Y.M., Barnes, H.J., Fadly, A.M., Glisson, J.R., McDougald, L.R., Swayne, D.E. (Eds.), Diseases of Poultry. Iowa State University Press Ames Iowa, pp. 64–99.

Allen-Vercoe, E., Woodward, M.J., 1999. Colonization of the chicken caecum by afimbriate and aflagellate derivatives of *Salmonella enterica* serotype Enteritidis. Vet. Microbiol. 69, 265–275.

Aly, S.A., Anwer, W., 2009. Effect of naturally contaminated feed with aflatoxins on performance of laying hens and the carryover of aflatoxin B1 residues in table eggs. Pak. J. Nutr. 8, 181–186.

AEB, 2016. American Egg Board, Industry Overview. http://www.aeb.org/farmers-and-marketers/industry-overview.

Angulo, F.J., Swerdlow, D.L., 1999. Epidemiology of human *Salmonella enterica* serovar Enteritidis infections in the United States. In: *Salmonella enterica* serovar Enteritidis in Human and Animals: Epidemiology, Pathogenesis, and Control, pp. 33–41.

Anonymous, 1988. Update: *Salmonella* Enteritidis infections and grade A shell eggs — United States. Morb. Mortal. Wkly. Rep. 37, 495–498.

Anonymous, 1990. Epidemiologic notes and reports update: *Salmonella* Enteritidis infections and grade A shell eggs — United States, 1989. Morb. Mortal. Wkly. Rep. 38, 877–880.

Anonymous, 2000. Federal Register, Food and Drug Administration. http://www.fda.gov/ohrms/dockets/dockets/06p0394/06p-0394-cp00001-37-Tab-34-FR-Rules-Regulations-vol1.pdf.

Anonymous, 2003. Outbreaks of *Salmonella* serotype Enteritidis infection associated with eating shell eggs – United States, 1999–2001. Morb. Mort. Wkly. Rep. 51, 1149–1152.

Anonymous, 2013. Shell Eggs From Farm to Table. http://www.fsis.usda.gov/wps/portal/fsis/topics/food-safety-education/get-answers/food-safety-fact-sheets/egg-products-preparation/shell-eggs-from-farm-to-table/CT_Index.

Anonymous, 2014. Update: UK *Salmonella* Outbreak Linked to Three Deaths Likely Came From a Single Source of Eggs. http://www.foodsafetynews.com/2014/08/update-uk-salmonella-outbreak-linked-to-three-deaths-traced-to-single-egg-source/#.VKoUwGTF_xg.

Anonymous, 2015. General US Stats. United Egg Product. http://www.unitedegg.org/GeneralStats/default.cfm.

Anonymous, 2016. Shell Eggs Recalled; Linked to *Salmonella* Outbreak. http://www.foodsafetynews.com/2016/03/shell-eggs-recalled-linked-to-salmonella-outbreak/#.VvioExIrLEY.

Antunes, P., Réu, C., Sousa, J.C., Peixe, L., Pestana, N., 2003. Incidence of *Salmonella* from poultry products and their susceptibility to antimicrobial agents. Int. J. Food Microbiol. 82, 97–103.

Baird-Parker, A.C., 1990. Foodborne salmonellosis. Lancet 336, 1231–1235.

Baker, R.C., Bruce, C., 1994. Effects of processing on the microbiology of eggs. In: Microbiology of the Avian Egg. Springer US, pp. 153–173.

Barbour, E.K., El Jurdi, L., Issa, C., Tannous, R., 2001. Preliminary attempts towards production of table eggs free from *Salmonella* Enteritidis. J. Clean. Prod. 9, 69–73.

Bari, M.L., Sabina, Y., Isobe, S., Uemura, T., Isshiki, K., 2003. Effectiveness of electrolyzed acidic water in killing *Escherichia coli* O157: H7, *Salmonella* Enteritidis, and *Listeria monocytogenes* on the surfaces of tomatoes. J. Food Prot. 66, 542–548.

Basilico, M.Z., Basilico, J.C., 1999. Inhibitory effects of some spice essential oils on *Aspergillus ochraceus* NRRL 3174 growth and ochratoxin A production. Lett. Appl. Microbiol. 29, 238–241.

Berrang, M.E., Cox, N.A., Frank, J.F., Buhr, R.J., 1999. Bacterial penetration of the eggshell and shell membranes of the chicken hatching egg: a review. J. Appl. Poult. Res. 8, 499–504.

Bhale, S., No, H.K., Prinyawiwatkul, W., Farr, A.J., Nadarajah, K., Meyers, S.P., 2003. Chitosan coating improves shelf life of eggs. J. Food Sci. 68, 2378–2383.

Bialka, K.L., Demirci, A., Knabel, S.J., Patterson, P.H., Puri, V.M., 2004. Efficacy of electrolyzed oxidizing water for the microbial safety and quality of eggs. Poult. Sci. 83, 2071–2078.

Bierer, B.W., Barnett, B.D., 1962. Nihydrazone and the *Salmonella* infections. Poult. Sci. 41, 1291–1294.

Blumenthal, M., Goldberg, A., Brinckmann, J., Foster, S., 2000. Herbal medicine: Expanded commission E monographs. American Botanical Council, Integr. Med. Commun.

Board, R.G., 1966. The course of microbial infection of the hen's egg. J. Appl. Bacteriol. 29, 319–341.

Braden, C.R., 2006. *Salmonella enterica* serotype Enteritidis and eggs: a national epidemic in the United States. Clin. Infect. Dis. 43, 512–517.

Büchen-Osmond, C., 2002. The Universal Virus Database of the International Committee on Taxonomy of Viruses. Columbia University. Web page: http://www.ncbi.nlm.nih.gov/ICTVdb/index.htm.

Burt, S., 2004. Essential oils: their antibacterial properties and potential applications in foods—a review. Int. J. Food Microbiol. 94, 223–253.

Cabeza, M.C., Cambero, M.I., De la Hoz, L., Garcia, M.L., Ordonez, J.A., 2011. Effect of thermoultrasonic treatment on the eggshell integrity and their impact on the microbial quality. Innov. Food Sci. Emerg. Technol. 12, 111–117.

Caner, C., 2005. Whey protein isolate coating and concentration effects on egg shelf life. J. Sci. Food Agric. 85, 2143–2148.

Caner, C., Cansiz, O., 2007. Effectiveness of chitosan-based coating in improving shelf-life of eggs. J. Sci. Food Agric. 87, 227–232.

Camarda, A., Newell, D.G., Nasti, R., Di Modugno, G., 2000. Genotyping *Campylobacter jejuni* strains isolated from the gut and oviduct of laying hens. Avian Dis. 907–912.

CDC, 2014. Campylobacter. http://www.cdc.gov/foodsafety/diseases/campylobacter/index.html.

Centers for Disease Control and Prevention (CDC), 2010. Investigation Update: Multistate Outbreak of Human *Salmonella* Enteritidis Infections Associated With Shell Eggs. CDC, Atlanta, GA.

Chen, J., Shallo-Thesmar, H., Kerr, W.L., 2005. Outgrowth of Salmonellae and the physical property of the albumen and vitelline membrane as influenced by egg storage conditions. J. Food Prot. 68, 2553–2558.

Chestnut, A.B., Anderson, P.D., Cochran, M.A., Fribourg, H.A., Gwinn, K.D., 1992. Effects of hydrated sodium calcium aluminosilicate on fescue toxicosis and mineral absorption. J. Anim. Sci. 70, 2838–2846.

Chmielewski, R., Swayne, D.E., 2011. Avian influenza: public health and food safety concerns. Annu. Rev. Food Sci. Tech. 2, 37–57.

Chun, S.S., Vattem, D.A., Lin, Y.T., Shetty, K., 2005. Phenolic antioxidants from clonal oregano (*Origanum vulgare*) with antimicrobial activity against *Helicobacter pylori*. Process Biochem. 40, 809–816.

Chung, T.K., Ekdman, J., Baker, D.H., 1990. Hydrated sodium calcium aluminosilicate: effects on zinc, manganese, vitamin A, and riboflavin utilization. Poult. Sci. 69, 1364–1370.

Chuma, T., Yamada, T., Yano, K., Okamoto, K., Yugi, H., 1994. A survey of *Campylobacter jejuni* in broilers from assignment to slaughter using DNA-DNA hybridization. J. Vet. Med. Sci. 56, 697–700.

Chuma, T., Yano, K., Omori, H., Okamoto, K., Yugi, H., 1997. Direct detection of *Campylobacter jejuni* in chicken cecal contents by PCR. J. Vet. Med. Sci. 59, 85–87.

Clark, A.G., Bueschkens, D.H., 1985. Laboratory infection of chicken eggs with *Campylobacter jejuni* by using temperature or pressure differentials. Appl. Environ. Microbiol. 49, 1467–1471.

Clark, A.G., Bueschkens, D.H., 1986. Survival and growth of *Campylobacter jejuni* in egg yolk and albumen. J. Food Prot. 49, 135–141.

Coates, M.E., Fuller, R., 1977. The Gnotobiotic Animal in the Study of Gut Microbiology. Microbial Ecology of the Gut. Academic Press, London, pp. 311–346.

Cohen, S.S., 1948. The synthesis of bacterial viruses I. The synthesis of nucleic acid and protein in *Escherichia coli* B infected with T2r+ bacteriophage. J. Biol. Chem. 174, 281–293.

Costa-Hurtado, M., Afonso, C.L., Miller, P.J., Shepherd, E., Cha, R.M., Smith, D., Spackman, E., Kapczynski, D.R., Suarez, D.L., Swayne, D.E., Pantin-Jackwood, M.J., 2015. Previous infection with virulent strains of Newcastle disease virus reduces highly pathogenic avian influenza virus replication, disease, and mortality in chickens. Vet. Res. 46, 1–15.

Cox, N.A., Stern, N.J., Hiett, K.L., Berrang, M.E., 2002. Identification of a new source of *Campylobacter* contamination in poultry: transmission from breeder hens to broiler chickens. Avian Dis. 46, 535–541.

da Cruz Cabral, L., Pinto, V.F., Patriarca, A., 2013. Application of plant derived compounds to control fungal spoilage and mycotoxin production in foods. Int. J. Food Microbiol. 166, 1–14.

Dailey, R.E., Reese, R.E., Brouwer, E.A., 1980. Metabolism of [14C] zearalenone in laying hens. J. Agri. Food Chem. 28, 286–291.

Davies, R.H., Breslin, M., 2003. Investigations into possible alternative decontamination methods for *Salmonella* Enteritidis on the surface of table eggs. J. Vet. Med. 50, 38–41.

De Reu, K., Grijspeerdt, K., Messens, W., Heyndrickx, M., Uyttendaele, M., Debevere, J., Herman, L., 2006. Eggshell factors influencing eggshell penetration and whole egg contamination by different bacteria, including *Salmonella* Enteritidis. Int. J. Food Microbiol. 112, 253–260.

De Vylder, J., Van Hoorebeke, S., Ducatelle, R., Pasmans, F., Haesebrouck, F., Dewulf, J., Van Immerseel, F., 2009. Effect of the housing system on shedding and colonization of gut and internal organs of laying hens with *Salmonella* Enteritidis. Poult. Sci. 88, 2491–2495.

Delbrock, M., 1946. Bacterial viruses or bacteriophages. Biol. Rev. 21, 30–40.

Dortmans, J.C., Koch, G., Rottier, P.J., Peeters, B.P., 2011. Virulence of Newcastle disease virus: what is known so far. Vet. Res. 42, 122.

Doyle, M.P., 1984. Association of *Campylobacter jejuni* with laying hens and eggs. Appl. Environ. Microbiol. 47, 533–536.

Doyle, M.P., Roman, D.J., 1982. Recovery of *Campylobacter jejuni* and *Campylobacter coli* from inoculated foods by selective enrichment. Appl. Environ. Microbiol. 43, 1343–1353.

Dunn, J., 1996. Pulsed light and pulsed electric field for foods and eggs. Poult. Sci. 75, 1133–1136.

Dutta, P.K., Tripathi, S., Mehrotra, G.K., Dutta, J., 2009. Perspectives for chitosan based antimicrobial films in food applications. Food Chem. 114, 1173–1182.

Edrington, T.S., Kubena, L.F., Harvey, R.B., Rottinghaus, G.E., 1997. Influence of a superactivated charcoal on the toxic effects of aflatoxin or T-2 toxin in growing broilers. Poult. Sci. 76, 1205–1211.

EFSA, 2007. The community summary report on trends and sources of zoonoses, zoonotic agents, antimicrobial resistance and foodborne outbreaks in the European Union in 2006. EFSA J. 130, 34–117.

Filho, A.R.L., Silva, E.N.D., Ribeiro, A.R., Kondo, N., Curi, P.R., 2000. Use of anaerobic cecal microflora, lactose and acetic acid for the protection of broiler chicks against experimental infection with *Salmonella* Typhimurium and *Salmonella* enteritidis. Bra. J. Microbiol. 31, 107–112.

FAO, 2010. From Food and Agricultural Organization, Rome, Italy; 2010. www.faostat.fao.org/site/339/default.aspx.

Falguera, V., Quintero, J.P., Jiménez, A., Muñoz, J.A., Ibarz, A., 2011. Edible films and coatings: structures, active functions and trends in their use. Trends Food Sci. Technol. 22, 292–303.

Federal Register, 1999. Pulsed light treatment of food. Fed. Regist. 66, 338829–338830.

Food and Drug Administration (FDA), 2009. Improves Egg Safety. Available online: http://www.fda.gov/ForConsumers/ConsumerUpdates/ucm170640.htm.

Fiorentin, L., Vieira, N.D., Barioni Júnior, W., Barros, S., 2004. In vitro characterization and in vivo properties of Salmonellae lytic bacteriophages isolated from free-range layers. Rev. Bras. Ciênc. Avíc. 6, 121–128.

Fiorentin, L., Vieira, N.D., Barioni Jr., W., 2005. Oral treatment with bacteriophages reduces the concentration of *Salmonella* Enteritidis PT4 in caecal contents of broilers. Avian Pathol. 34, 258–263.

Food and Drug Administration (FDA), 2000. Irradiation in the production, processing and handling of food. In: FDA Fed. Reg, 65, pp. 45280–45282.

Fonseca, B.B., Beletti, M.E., Melo, R.T.D., Mendonça, E.P., Coelho, L.R., Nalevaiko, P.C., Rossi, D.A., 2014. *Campylobacter jejuni* in commercial eggs. Braz. J. Microbiol. 45, 76–79.

Frank, J.F., Wright, G.W., 1956. The disinfection of eggs contaminated with *Salmonella typhimurium*. Can. J. Comp. Med. Vet. Sci. 20, 406.

Friedman, M., 1996. Nutritional value of proteins from different food sources. A review. J. Agric. Food Chem. 44, 6–29.

Froning, G.W., Peters, D., Muriana, P., Eskridge, K., Travnicek, D., Sumner, S.S., 2002. Presented in cooperation with the United Egg Association (Atlanta, GA) and the American Egg Board (Park Ridge, IL). Int. Egg Past. Manual. In: Santos-Garcia, J. (Ed.), Microbiologically Safe Foods. Wiley Publishing.

Fuller, R., 1989. Probiotics in man and animals. J. Appl. Bacteriol. 66, 365–378.

Garibaldi, J.A., Straka, R.P., Ijichi, K., 1969. Heat resistance of *Salmonella* in various egg products. Appl. Microbiol. 17, 491–496.

Galiş, A.M., Marcq, C., Marlier, D., Portetelle, D., Van, I., Beckers, Y., Théwis, A., 2013. Control of *Salmonella* contamination of shell eggs—preharvest and postharvest methods: a review. Compr. Rev. Food Sci. Food Saf. 12, 155–182.

Gantois, I., Ducatelle, R., Pasmans, F., Haesebrouck, F., Gast, R., Humphrey, T.J., Van Immerseel, F., 2009. Mechanisms of egg contamination by *Salmonella* Enteritidis. FEMS Microbiol. Rev. 33, 718–738.

Gao, G., Deng, Y., Kispert, L.D., 1997. Photoactivated ferric chloride oxidation of carotenoids by near-UV to visible light. J. Phys. Chem. 101, 7844–7849.

Gast, R.K., Beard, C.W., 1990. Isolation of *Salmonella* Enteritidis from internal organs of experimentally infected hens. Avian Dis. 991–993.

Gast, R.K., Holt, P.S., 2000. Influence of the level and location of contamination on the multiplication of *Salmonella* Enteritidis at different storage temperatures in experimentally inoculated eggs. Poult. Sci. 79, 559–563.

Gast, R.K., Guard-Bouldin, J., Holt, P.S., 2005. The relationship between the duration of fecal shedding and the production of contaminated eggs by laying hens infected with strains of *Salmonella* Enteritidis and *Salmonella* Heidelberg. Avian Dis. 49, 382–386.

Geissman, T.A., 1963. Flavonoid compounds, tannins, lignins and related compounds. In: Pyrrole Pigments, Isoprenoid Compounds and Phenolic Plant Constituents, 9, p. 265.

Gill, A.O., Holley, R.A., 2006. Disruption of *Escherichia coli*, *Listeria monocytogenes* and *Lactobacillus sakei* cellular membranes by plant oil aromatics. Int. J. Food Microbiol. 108, 1–9.

Gilliland, S.E., Kim, H.S., 1984. Effect of viable starter culture bacteria in yogurt on lactose utilization in humans. J. Dairy Sci. 67, 1–6.

Gong, Y., Hounsa, A., Egal, S., Turner, P.C., Sutcliffe, A.E., Hall, A.J., Cardwell, K., Wild, C.P., 2004. Postweaning exposure to aflatoxin results in impaired child growth: a longitudinal study in Benin, West Africa. Environ. Health Perspect. 1334–1338.

Goode, D., Allen, V.M., Barrow, P.A., 2003. Reduction of experimental *Salmonella* and *Campylobacter* contamination of chicken skin by application of lytic bacteriophages. App. Environ. Microbiol. 69, 5032–5036.

Gritz, D.C., Lee, T.Y., McDonnell, P.J., Shih, K., Baron, N., 1990. Ultraviolet radiation for the sterilization of contact lenses. Eye Contact Lens 16, 294–298.

Grizard, S., Dini-Andreote, F., Tieleman, B.I., Salles, J.F., 2014. Dynamics of bacterial and fungal communities associated with eggshells during incubation. Ecol. Evol. 4, 1140–1157.

Grizard, S., Versteegh, M.A., Ndithia, H.K., Salles, J.F., Tieleman, B.I., 2015. Shifts in bacterial communities of eggshells and antimicrobial activities in eggs during incubation in a ground-nesting passerine. PLoS One 10, 0121716.

Guard-Petter, J., 2001. The chicken, the egg and *Salmonella* Enteritidis. Environ. Microbiol. 3, 421–430.

Guenther, S., Herzig, O., Fieseler, L., Klumpp, J., Loessner, M.J., 2012. Biocontrol of *Salmonella* Typhimurium in RTE foods with the virulent bacteriophage FO1-E2. Int. J. Food Microbiol. 154, 66–72.

Hammack, T.S., Sherrod, P.S., Bruce, V.R., June, G.A., Satchell, F.B., Andrews, W.H., 1993. Research note: growth of *Salmonella* Enteritidis in grade A eggs during prolonged storage. Poult. Sci. 72, 373–377.

Hassanein, S.M., Soliman, N.K., 2010. Effect of probiotic (*Saccharomyces cerevisiae*) adding to diets on intestinal microflora and performance of Hy-Line layers hens. J. Am. Sci. 6, 159–169.

Henriques, A., Sereno, R., Almeida, A., 2013. Reducing *Salmonella* horizontal transmission during egg incubation by phage therapy. Foodborne Pathog. Dis. 10, 718–722.

Hierro, E., Manzano, S., Ordonez, J.A., De la Hoz, L., Fernandez, M., 2009. Inactivation of *Salmonella enterica* serovar Enteritidis on shell eggs by pulsed light technology. Int. J. Food Microbiol. 135, 125–130.

Hiett, K.L., Stern, N.J., Fedorka-Cray, P., Cox, N.A., Musgrove, M.T., Ladely, S., 2002. Molecular subtype analyses of *Campylobacter* spp. from Arkansas and California poultry operations. Appl. Environ. Microbiol. 68, 6220–6236.

Höhler, D., 1998. Ochratoxin A in food and feed: occurrence, legislation and mode of action. Z. Ernahrungswiss. 37, 2–12.

Howard, Z.R., Moore, R.W., Diaz, I.Z., Kim, W.K., Birkhold, S.G., Byrd, J.A., Kubena, L.F., Nisbet, D.J., Ricke, S.C., 2006. In vitro survival and growth of *Salmonella* Typhimurium inoculated on yolk membrane after long term refrigerated storage of shell eggs. J. Verbrauch. Lebensm. 1, 28–32.

Howard, Z.R., Moore, R.W., Diaz, I.Z., Kim, W.K., Birkhold, S.G., Byrd, J.A., Kubena, L.F., Nisbet, D.J., Ricke, S.C., 2007. Inoculation of a poultry isolate *Salmonella* Enteritidis on egg vitelline membrane: survival and growth in egg components after different refrigeration storage times. Am. J. Agri. Biol. Sci. 2, 123–129.

Howard, Z.R., O'Bryan, C.A., Crandall, P.G., Ricke, S.C., 2012. *Salmonella* Enteritidis in shell eggs: current issues and prospects for control. Food Res. Int. 45, 755–764.

Hussain, Z., Khan, M.Z., Khan, A., Javed, I., Saleemi, M.K., Mahmood, S., Asi, M.R., 2010. Residues of aflatoxin B1 in broiler meat: effect of age and dietary aflatoxin B1 levels. Food Chem. Toxicol. 48, 3304–3307.

International Agency for Research on Cancer (IARC), 2002. IARC Monographs on the Evaluation of Carcinogenic Risks to Humans: Traditional Herbal Medicines, Some Mycotoxins, Naphthalene and Styrene. IARC Scientific Publication, No. 82. IARC, Lyon.

Iqbal, S.Z., Nisar, S., Asi, M.R., Jinap, S., 2014. Natural incidence of aflatoxins, ochratoxin A and zearalenone in chicken meat and eggs. Food Control 43, 98–103.

Izat, A.L., Gardner, F.A., 1988. Incidence of *Campylobacter jejuni* in processed egg products. Poult. Sci. 67, 1431–1435.

Izat, A.L., Gardner, F.A., Mellor, D.B., 1986. The effects of age of bird and season of the year on egg quality. II. Haugh units and compositional attributes. Poult. Sci. 65, 726–728.

Jacobs-Reitsma, W., Lyhs, U., Wagenaar, J., Nachamkin, I., Szymanski, C.M., Blaser, M.J., 2008. *Campylobacter* in the food supply. Campylobacter 3, 627–644.

James, C., Lechevalier, V., Ketteringham, L., 2002. Surface pasteurisation of shell eggs. J. Food Eng. 53, 193–197.

Jassim, S.A., Limoges, R.G., 2014. Natural solution to antibiotic resistance: bacteriophages 'The Living Drugs'. World J. Microbiol. Biotechnol. 30, 2153–2170.

Jeng, D.K., Kaczmarek, K.A., Woodworth, A.G., Balasky, G., 1987. Mechanism of microwave sterilization in the dry state. Appl. Environ. Microbiol. 53, 2133–2137.

Jin, T., Gurtler, J.B., 2011. Inactivation of *Salmonella* in liquid egg albumen by antimicrobial bottle coatings infused with allyl isothiocyanate, nisin and zinc oxide nanoparticles. J. Appl. Microbiol. 110, 704–712.

Jin, T.Z., Gurtler, J.B., Li, S.Q., 2013. Development of antimicrobial coatings for improving the microbiological safety and quality of shell eggs. J. Food Prot. 76, 779–785.

Jones, D.R., Anderson, K.E., Curtis, P.A., Jones, F.T., 2002. Microbial contamination in inoculated shell eggs: I. Effects of layer strain and hen age. Poult. Sci. 81, 715–720.

Jung, Y.S., Beuchat, L.R., 2000. Sensitivity of multidrug-resistant *Salmonella* Typhimurium DT104 to organic acids and thermal inactivation in liquid egg products. Food Microbiol. 17, 63–71.

Kabak, B., 2009. The fate of mycotoxins during thermal food processing. J. Sci. Food Agric. 89, 549–554.

Kabir, S.M., 2009. The role of probiotics in the poultry industry. Int. J. Mol. Sci. 10, 3531–3546.

Kayes, M.M., Critzer, F.J., Kelly-Wintenberg, K., Reece Roth, J., Montie, T.C., Golden, D.A., 2007. Inactivation of foodborne pathogens using a one atmosphere uniform glow discharge plasma. Foodborne Pathog. Dis. 4, 50–59.

Keener, K.M., Bashor, M.P., Curtis, P.A., Sheldon, B.W., Kathariou, S., 2004. Comprehensive review of *Campylobacter* and poultry processing. Compr. Rev. Food Sci. Food Saf. 3, 105–116.

Keklik, N.M., Demirci, A., Patterson, P.H., Puri, V.M., 2010. Pulsed UV light inactivation of *Salmonella* Enteritidis on eggshells and its effects on egg quality. J. Food Prot. 73, 1408–1415.

Keller, L.H., Benson, C.E., Krotec, K., Eckroade, R.J., 1995. *Salmonella* Enteritidis colonization of the reproductive tract and forming and freshly laid eggs of chickens. Infect. Immun. 63, 2443–2449.

Khadre, M.A., Yousef, A.E., Kim, J.G., 2001. Microbiological aspects of ozone applications in food: a review. J. Food Sci. 66, 1242–1252.

King, D.J., 1991. Evaluation of different methods of inactivation of Newcastle disease virus and avian influenza virus in egg fluids and serum. Avian Dis. 505–514.

Kinner, J.A., Moats, W.A., 1981. Effect of temperature, pH, and detergent on survival of bacteria associated with shell eggs. Poult. Sci. 60, 761–767.

Kim, C., Hung, Y.C., Brackett, R.E., 2000. Efficacy of electrolyzed oxidizing (EO) and chemically modified water on different types of foodborne pathogens. J. Food Microbiol. 61, 199–207.

Kim, Y., Cook, S., Tuladhar, S.M., Choulis, S.A., Nelson, J., Durrant, J.R., Bradley, D.D., Giles, M., McCulloch, I., Ha, C.S., Ree, M., 2006. A strong regioregularity effect in self-organizing conjugated polymer films and high-efficiency polythiophene: fullerene solar cells. Nat. Mater. 5, 197–203.

Knape, K.D., Carey, J.B., Burgess, R.P., Kwon, Y.M., Ricke, S.C., 1999. Comparison of chlorine with an iodine-based compound on eggshell surface microbial populations in a commercial egg washer. J. Food Saf. 19, 185–194.

Knape, K.D., Carey, J.B., Ricke, S.C., 2001. Comparison of chlorine with an iodine based compound on eggshell surface *Salmonella* Typhimurium and *S.* Enteritidis populations in a commercial egg washer. J. Environ. Sci. Health 36, 219–227.

Koidis, P., Bori, M., Vareltzis, K., 2000. Efficacy of ozone treatment to eliminate *Salmonella* Enteritidis from eggshell surface. Archiv Lebensm. 51, 4–6.

Kovacs-Nolan, J., Phillips, M., Mine, Y., 2005. Advances in the value of eggs and egg components for human health. J. Agric. Food Chem. 53, 8421–8431.

Kubena, L.F., Harvey, R.B., Huff, W.E., Elissalde, M.H., Yersin, A.G., Phillips, T.D., Rottinghaus, G.E., 1993. Efficacy of a hydrated sodium calcium aluminosilicate to reduce the toxicity of aflatoxin and diacetoxyscirpenol. Poult. Sci. 72, 51–59.

Kuo, F.L., Kwon, Y.M., Carey, J.B., Hargis, B.M., Krieg, D.P., Ricke, S.C., 1997a. Reduction of *Salmonella* contamination on chicken egg shells by a peroxidase-catalyzed sanitizer. J. Food Sci. 62, 873–884.

Kuo, F.L., Carey, J.B., Ricke, S.C., 1997b. UV irradiation of shell eggs: effect on populations of aerobes, molds, and inoculated *Salmonella* Typhimurium. J. Food Prot. 60, 639–643.

Kurtoglu, V., Kurtoglu, F., Seker, E., Coskun, B., Balevi, T., Polat, E.S., 2004. Effect of probiotic supplementation on laying hen diets on yield performance and serum and egg yolk cholesterol. Food Addit. Contam. 21, 817–823.

Lakins, D.G., Alvarado, C.Z., Thompson, L.D., Brashears, M.T., Brooks, J.C., Brashears, M.M., 2008. Reduction of *Salmonella* Enteritidis in shell eggs using directional microwave technology. Poult. Sci. 87, 985–991.

Lategan, P.M., Vaughn, R.H., 1964. The influence of chemical additives on the heat resistance of *Salmonella* Typhimurium in liquid whole egg. J. Food Sci. 29, 339–344.

Latimer, H.K., Jaykus, L.A., Morales, R.A., Cowen, P., Crawford-Brown, D., 2002. Sensitivity analysis of *Salmonella* Enteritidis levels in contaminated shell eggs using a biphasic growth model. Int. J. Food Microbiol. 75, 71–87.

Leleu, S., Herman, L., Heyndrickx, M., De Reu, K., Michiels, C.W., De Baerdemaeker, J., Messens, W., 2011. Effects on *Salmonella* shell contamination and trans-shell penetration of coating hens' eggs with chitosan. Int. J. Food Microbiol. 145, 43–48.

Li, H., Wang, H., D'Aoust, J.Y., Maurer, J., 2013. *Salmonella* species. In: Doyle, M.P., Buchanan, R.L. (Eds.), Food Microbiology – Fundamentals and Frontiers, fourth ed. American Society for Microbiology Press, Washington DC, pp. 222–261.

Lin, J., 2009. Novel approaches for *Campylobacter* control in poultry. Foodborne Pathog. Dis. 6, 755–765.

Machado, J., Bernardo, F., 1990. Prevalence of *Salmonella* in chicken carcasses in Portugal. J. App. Bacteriol. 69, 477–480.

Mani-Lopez, E., García, H.S., López-Malo, A., 2012. Organic acids as antimicrobials to control *Salmonella* in meat and poultry products. Food Res. Int. 45, 713–721.

Martelli, F., Davies, R.H., 2012. *Salmonella* serovars isolated from table eggs: an overview. Food Res. Int. 45, 745–754.

Martínez-Camacho, A.P., Cortez-Rocha, M.O., Ezquerra-Brauer, J.M., Graciano-Verdugo, A.Z., Rodriguez-Félix, F., Castillo-Ortega, M.M., Yépiz-Gómez, M.S., Plascencia-Jatomea, M., 2010. Chitosan composite films: thermal, structural, mechanical and antifungal properties. Carbohydr. Polym. 82, 305–315.

Mayes, F.J., Takeballi, M.A., 1983. Microbial contamination of the hen's egg: a review. J. Food Prot. 46, 1091–1098.

Messens, W., Grijspeerdt, K., Herman, L., Billiet, L., 2002. A survey on institutional users of shell eggs and egg products in Flanders. J. Food Saf. 22, 273–290.

Messens, W., Grijspeerdt, K., Herman, L., 2005. Eggshell penetration by *Salmonella*: a review. World's Poult. Sci. 61, 71–86.

Messens, W., Grijspeerdt, K., Herman, L., 2006. Eggshell penetration of hen's eggs by *Salmonella enterica* serovar Enteritidis upon various storage conditions. Br. Poult. Sci. 47, 554–560.

Messens, W., Grijspeerdt, K., De Reu, K., De Ketelaere, B., Mertens, K., Bamelis, F., Kemps, B., De Baerdemaeker, J., Decuypere, E., Herman, L., 2007. Eggshell penetration of various types of hens' eggs by *Salmonella enterica* serovar Enteritidis. J. Food Protect. 70, 623–628.

Meszaros, L., Horti, K., Farkas, J., 2006. Changes of hen eggs and their components caused by non-thermal pasteurizing treatments I. Gamma irradiation of shell eggs. Acta Aliment. 35, 229–236.

Mirocha, C.J., Christensen, C.M., Nelson, G.H., 2013. F-2 (zearalenone) estrogenic mycotoxin from Fusarium. Microb. Toxin 7, 107–138.

Miyamoto, T., Horie, T., Baba, E., Sasai, K., Fukata, T., Arakawa, A., 1998. *Salmonella* penetration through eggshell associated with freshness of laid eggs and refrigeration. J. Food Prot. 61, 350–353.

Moats, W.A., 1978. Egg washing-a review. J. Food Prot. 41, 919–925.

Morelli, L., Capurso, L., 2012. FAO/WHO guidelines on probiotics: 10 years later. J. Clin. Gastroenterol. 46, S1–S2.

Nahashon, S.N., Nakaue, H.S., Mirosh, L.W., 1994. Production variables and nutrient retention in Single Comb White Leghorn laying pullets fed diets supplemented with direct-fed microbials. Poult. Sci. 73, 1699–1711.

Namata, H., Méroc, E., Aerts, M., Faes, C., Abrahantes, J.C., Imberechts, H., Mintiens, K., 2008. *Salmonella* in Belgian laying hens: an identification of risk factors. Prev. Vet. Med. 83, 323–336.

NASS, 2009. Chickens and Eggs. http://www.nass.usda.gov/Surveys/Guide_to_NASS_Surveys/Chickens_and_Eggs/index.asp.

Nascimento, V.P., Cranstoun, S., Solomon, S.E., 1992. Relationship between shell structure and movement of *Salmonella* Enteritidis across the eggshell wall. Br. Poult. Sci. 33, 37–48.

Padron, M., 1995. Egg dipping in hydrogen peroxide solution to eliminate *Salmonella* Typhimurium from eggshell membranes. Avian Dis. 627–630.

Panda, A.K., Reddy, M.R., Rao, S.R., Praharaj, N.K., 2003. Production performance, serum/yolk cholesterol and immune competence of white leghorn layers as influenced by dietary supplementation with probiotic. Trop. Anim. Health Prod. 35, 85–94.

Panda, A.K., Rama Rao, S.S., Raju, M.V., Sharma, S.S., 2008. Effect of probiotic (*Lactobacillus sporogenes*) feeding on egg production and quality, yolk cholesterol and humoral immune response of White Leghorn layer breeders. J. Sci. Food Agric. 88, 43–47.

Padron, M., 1990. *Salmonella* Typhimurium penetration through the eggshell of hatching eggs. Avian Dis. 463–465.

Pitasawat, B., Champakaew, D., Choochote, W., Jitpakdi, A., Chaithong, U., Kanjanapothi, D., Rattanachanpichai, E., Tippawangkosol, P., Riyong, D., Tuetun, B., Chaiyasit, D., 2007. Aromatic plant-derived essential oil: an alternative larvicide for mosquito control. Fitoterapia 78, 205–210.

Park, B.J., Takatori, K., Sugita-Konishi, Y., Kim, I.H., Lee, M.H., Han, D.W., Chung, K.H., Hyun, S.O., Park, J.C., 2007. Degradation of mycotoxins using microwave-induced argon plasma at atmospheric pressure. Surf. Coat. Technol. 201, 5733–5737.

Perumalla, A.V.S., Hettiarachchy, N.S., Ricke, S.C., 2012. Current perspectives on probiotics in poultry preharvest food safety. In: Callaway, T.R., Ricke, S.C. (Eds.), Direct-Fed Microbials and Prebiotics for Animals. Springer, New York, pp. 89–120.

Plummer, R.A., Blissett, S.J., Dodd, C.E., 1995. *Salmonella* contamination of retail chicken products sold in the UK. J. Food Prot. 58, 843–846.

Pohuang, T., Chuachan, K., Sarachu, K., 2011. Effect of *Punica granatum* L. rind extract against *Salmonella* Enteritidis on eggshells and eggshell membranes. KKU Vet. J. 19, 38–47.

Ponce, A.G., Roura, S.I., del Valle, C.E., Moreira, M.R., 2008. Antimicrobial and antioxidant activities of edible coatings enriched with natural plant extracts: in vitro and in vivo studies. Postharvest Biol. Technol. 49, 294–300.

Prins, R.A., 1977. Biochemical activities of gut microorganisms. In: Clarke, R.T.J., Bouchop, T. (Eds.), Microbial Ecology of the Gut. Academic Press, London, pp. 73–183.

Ragni, L., Berardinelly, A., Vannini, L., Montanari, C., Sirri, F., Guerzoni, M.E., Guarneri, A., 2010. Non-thermal atmospheric gas plasma device for surface decontamination of shell eggs. J. Food Eng. 100, 125–132.

Richard, J.L., Plattner, R.D., May, J., Liska, S.L., 1999. The occurrence of ochratoxin A in dust collected from a problem household. Mycopathologia 146, 99–103.

Ricke, S.C., Birkhold, S.G., Gast, R.K., 2001. Eggs and egg products. In: Downes, F.P., Ito, K. (Eds.), Compendium of Methods for the Microbiological Examination of Foods, fourth ed. Am. Public Health Assoc., Washington, DC, pp. 473–479.

Ricke, S.C., Koo, O.K., Foley, S., Nayak, R., 2013. Salmonella. In: Labbe, R., Garcia, S. (Eds.), Guide to Foodborne Pathogens, second ed. Wiley-Blackwell, Oxford, UK, pp. 112–137 (Chapter 7).

Roberts, J.R., Brackpool, C.E., 1994. The ultrastructure of avian egg shells. Poult. Sci. Rev. 5, 45–272.

Rodriguez-Romo, L.A., Vurma, M., Lee, K., Yousef, A.E., 2007. Research note: penetration of ozone gas across the shell of hen eggs. Ozone Sci. Eng. 29, 147–150.

Rodriguez-Romo, L., Yousef, A., Griffiths, M., 2005. Cross-Protective Effects of Bacterial Stress. In: Griffiths, M. (Ed.), Understanding Pathogen Behavior. Woodhead Publishing Ltd., Boca Raton, Boston, New York, Washington DC, pp. 128–129.

Russell, S.M., 2003. The effect of electrolyzed oxidative water applied using electrostatic spraying on pathogenic and indicator bacteria on the surface of eggs. Poult. Sci. 82, 158–162.

Saarela, M., Mogensen, G., Fondén, R., Mättö, J., Mattila-Sandholm, T., 2000. Probiotic bacteria: safety, functional and technological properties. J. Biotechnol. 84, 197–215.

Sahin, O., Kobalka, P., Zhang, Q., 2003. Detection and survival of *Campylobacter* in chicken eggs. J. Appl. Microbiol. 95, 1070–1079.

Salamci, E., Kordali, S., Kotan, R., Cakir, A., Kaya, Y., 2007. Chemical compositions, antimicrobial and herbicidal effects of essential oils isolated from Turkish *Tanacetum aucheranum* and *Tanacetum chiliophyllum* var. chiliophyllum. Biochem. Syst. Ecol. 35, 569–581.

Sarmiento-Ramírez, J.M., van der Voort, M., Raaijmakers, J.M., Diéguez-Uribeondo, J., 2014. Unravelling the microbiome of eggs of the endangered sea turtle *Eretmochelys imbricata* identifies bacteria with activity against the emerging pathogen *Fusarium falciforme*. PLoS One 9(4), e95206.

Sato, M., Sashihara, N., 2010. Occurrence of *Campylobacter* in commercially broken liquid egg in Japan. J. Food Prot. 73, 412–417.

Scheideler, S.E., 1993. Effects of various types of aluminosilicates and aflatoxin B1 on aflatoxin toxicity, chick performance, and mineral status. Poult. Sci. 72, 282–288.

Scott, T.A., Sweetnam, C., 1993. Screening sanitizing agents and methods of application for hatching eggs I. Environmental and user friendliness. J. Appl. Poult. Res. 2, 1–6.

Shanker, S., Lee, A., Sorrell, T.C., 1986. *Campylobacter jejuni* in broilers: the role of vertical transmission. J. Hyg. 96, 153–159.

Silva, J., Leite, D., Fernandes, M., Mena, C., Gibbs, P.A., Teixeira, P., 2011. *Campylobacter* spp. as a Foodborne Pathogen: A Review. Front. Microbiol. 2, 200. http://dx.doi.org/10.3389/fmicb.2011.00200.

Stein, A.F., Phillips, T.D., Kubena, L.F., Harvey, R.B., 1985. Renal tubular secretion and reabsorption as factors in ochratoxicosis: effects of probenecid on nephrotoxicity. J. Toxicol. Environ. Health 16, 593–605.

Stern, N.J., 2008. *Salmonella* species and *Campylobacter jejuni* cecal colonization model in broilers. Poult. Sci. 87, 2399–2403.

Stern, N.J., Kazmi, S.U., 1989. Campylobacter jejuni. Foodborne Bact. Pathog. 3, 71–110.

Sypecka, Z., Kelly, M., Brereton, P., 2004. Deoxynivalenol and zearalenone residues in eggs of laying hens fed with a naturally contaminated diet: effects on egg production and estimation of transmission rates from feed to eggs. J. Agric. Food Chem. 52, 5463–5471.

Swayne, D.E., Beck, J.R., 2004. Heat inactivation of avian influenza and Newcastle disease viruses in egg products. Avian Pathol. 33, 512–518.

Tauxe, R.V., 2001. Food safety and irradiation: protecting the public from foodborne infections. Emerg. Inf. Dis. 7, 516–521.

Tiwari, R., Dhama, K., Wani, M.Y., Verma, V., Vaid, R.K., Chauhan, R.S., 2011. Bacteriophage therapy: a novel tool for combating bacterial diseases of poultry-a review. J. Immunol. Immunopathol. 13, 55–66.

Thomas, M.E., Klinkenberg, D., Ejeta, G., Van Knapen, F., Bergwerff, A.A., Stegeman, J.A., Bouma, A., 2009. Quantification of horizontal transmission of *Salmonella enterica* serovar Enteritidis bacteria in pair-housed groups of laying hens. Appl. Environ. Microbiol. 75, 6361–6366.

Toms, C., Powrie, F., 2001. Control of intestinal inflammation by regulatory T cells. Microb. Infect. 3, 929–935.

Toro, H., Price, S.B., McKee, S., Hoerr, F.J., Krehling, J., Perdue, M., Bauermeister, L., 2005. Use of bacteriophages in combination with competitive exclusion to reduce *Salmonella* from infected chickens. Avian Dis. 49, 118–124.

Upadhyaya, I., Upadhyay, A., Kollanoor-Johny, A., Baskaran, S.A., Mooyottu, S., Darre, M.J., Venkitanarayanan, K., 2013. Rapid inactivation of *Salmonella* Enteritidis on shell eggs by plant-derived antimicrobials. Poult. Sci. 92, 3228–3235.

Upadhyay, A., Upadhyaya, I., Kollanoor-Johny, A., Venkitanarayanan, K., 2014. Combating pathogenic microorganisms using plant-derived antimicrobials: a minireview of the mechanistic basis. BioMed Res. Int. 2014.

Upadhyaya, I., Yin, H.B., Nair, M.S., Chen, C.H., Upadhyay, A., Darre, M.J., Venkitanarayanan, K., 2015. Efficacy of fumigation with Trans-cinnamaldehyde and eugenol in reducing *Salmonella enterica* serovar Enteritidis on embryonated egg shells. Poult. Sci. 94(7), 1685–1690.

Upadhyaya, I., Yin, H., Nair, M.S., Chen, C., Lang, R., Darre, M.J., Venkitanarayanan, K., 2016. Inactivation of *Salmonella* Enteritidis on shell eggs by coating with phytochemicals. Poult. Sci. http://dx.doi.org/10.3382/ps/pew152.

USDA-FSIS, October 2005. Risk assessment for *Salmonella* Enteritidis in shell eggs and *Salmonella* spp. In: Egg Products Washington, DC http://www.fsis.usda.gov/Science/Risk_Assessments/index.asp#eggs.

USDA, 2011. Shell Eggs From Farm to Table. http://www.fsis.usda.gov/wps/wcm/connect/5235aa20-fee1-4e5b-86f5-8d6e09f351b6/Shell_Eggs_from_Farm_to_Table.pdf?MOD=AJPERES.

USDA, 2008. Guidance for Shell Egg Cleaners and Sanitizers. www.fsis.usda.gov/Regulations_&_Policies/Shell_Egg_Cleaners_&_Sanitizers_Guidance/index.asp.

Uyttendaele, M., De Troy, P., Debevere, J., 1999. Incidence of *Salmonella*, *Campylobacter jejuni*, *Campylobacter coli*, and *Listeria monocytogenes* in poultry carcasses and different types of poultry products for sale on the Belgian retail market. J. Food Prot. 62, 735–740.

Van Immerseel, F., Russell, J.B., Flythe, M.D., Gantois, I., Timbermont, L., Pasmans, F., Haesebrouck, F., Ducatelle, R., 2006. The use of organic acids to combat *Salmonella* in poultry: a mechanistic explanation of the efficacy. Avian Pathol. 35, 182–188.

Van Lith, L.A.J.T., Putirulan, F.F., Mulder, R.W.A.W., 1995. Pasteurization of table eggs to eliminate salmonellae. Archiv Geflügelkd. 59, 157–160.

Valencia-Chamorro, S.A., Palou, L., del Río, M.A., Pérez-Gago, M.B., 2011. Antimicrobial edible films and coatings for fresh and minimally processed fruits and vegetables: a review. Crit. Rev. Food Sci. Nutr. 51, 872–900.

Venkitanarayanan, K.S., Ezeike, G.O., Hung, Y.C., Doyle, M.P., 1999. Efficacy of electrolyzed oxidizing water for inactivating *Escherichia coli* O157:H7, *Salmonella* Enteritidis, and Listeria monocytogenes. Appl. Environ. Microbiol. 65, 4276–4279.

Verde, S.C., Tenreiro, R., Botelho, M.L., 2004. Sanitation of chicken eggs by ionizing radiation: HACCP and inactivation studies. Radiat. Phys. Chem. 71, 29–33.

Wallner-Pendleton, E.A., Sumner, S.S., Froning, G.W., Stetson, L.E., 1994. The use of ultraviolet radiation to reduce *Salmonella* and psychrotrophic bacterial contamination on poultry carcasses. Poult. Sci. 73, 1327–1333.

Wang, H., Slavik, M.F., 1998. Bacterial penetration into eggs washed with various chemicals and stored at different temperatures and times. J. Food Prot. 61, 276–279.

Whichard, J.M., Sriranganathan, N., Pierson, F.W., 2003. Suppression of *Salmonella* growth by wild-type and large-plaque variants of bacteriophage Felix O1 in liquid culture and on chicken frankfurters. J. Food Prot. 66, 220–225.

Wood, G.E., 1988. Aflatoxins in domestic and imported foods and feeds. J. Assoc. Off. Anal. Chem. 72, 543–548.

Worley, B.S., Wheatley, W.B., Lauten, S.D., Williams, D.E., Mora, E.C., Worley, S.D., 1992. Inactivation of *Salmonella* Enteritidis on shell eggs by novel N-halamine biocidal compounds. J. Ind. Microb. 11, 37–42.

Yin, H.B., Chen, C.H., Kollanoor-Johny, A., Darre, M.J., Venkitanarayanan, K., 2015a. Controlling *Aspergillus flavus* and *Aspergillus parasiticus* growth and aflatoxin production in poultry feed using carvacrol and trans-cinnamaldehyde. Poult. Sci. 94, 9. (Abstract No. 23).

Yin, H., Chen, C., Upadhyaya, I., Upadhyay, A., Fancher, S., Li, J., Nair, M.S., Mooyottu, S., Khan, M.I., Darre, M.J., Venkitanarayanan, K., 2015b. Efficacy of in-feed supplementation of plant-derived antimicrobials in reducing aflatoxicosis in chickens. In: Meeting Abstract, Vol. 94, No. 9.

Youssef, A.W., Hassan, H.M.A., Ali, H.M., Mohamed, M.A., 2013. Effect of probiotics, prebiotics and organic acids on layer performance and egg quality. Asian J. Poult. Sci. 7, 65–74.

Yunus, A.W., Razzazi-Fazeli, E., Bohm, J., 2011. Aflatoxin B1 in affecting broiler's performance, immunity, and gastrointestinal tract: a review of history and contemporary issues. Toxins 3, 566–590.

Zeidler, G., 2002. Shell eggs and their nutritional value. In: Bell, D.D., Weaver, W.D. (Eds.), Commercial Chicken Meat and Egg Production. Springer US, pp. 1109–1128.

Zinno, P., Devirgiliis, C., Ercolini, D., Ongeng, D., Mauriello, G., 2014. Bacteriophage P22 to challenge *Salmonella* in foods. Int. J. Food Microbiol. 191, 69–74.

Zuber, S., Butot, S., Baert, L., 2013. Effects of treatments used in food processing on viruses. In: Frans, J.M., Smulders, B.N., Budka, H. (Eds.), Foodborne Viruses and Prions and Their Significance for Public Health Food Safety Assurance and Veterinary Public Health, Vol. 6. Wageningen Academic Publishers, Wageningen, pp. 113–136.

# Index

'*Note:* Page numbers followed by "f" indicate figures, "t" indicate tables.'

## B



## C

## D



## E

## F

## Q



## R

## T